NEUROMETHODS ☐ 20

Intracellular Messengers

NEUROMETHODS

Program Editors: Alan A. Boulton and Glen B. Baker

NEUROMETHODS □ 20

Intracellular Messengers

Edited by

Alan A. Boulton
University of Saskatchewan, Saskatoon, Canada

Glen B. Baker
University of Alberta, Edmonton, Canada

and

Colin W. Taylor
University of Cambridge, Cambridge, UK

Humana Press • Totowa, New Jersey

© 1992 The Humana Press Inc.
999 Riverview Drive, Suite 208
Totowa, New Jersey 07512

Printed in the United States of America.

Library of Congress Cataloging-in-Publication Data

Main entry under title:

Intracellular messengers / edited by Alan A. Boulton, Glen B. Baker, and Colin W. Taylor.
   p. cm. -- (Neuromethods : 20)
   Includes bibliographical references and index.
   ISBN 0-89603-207-8
   1. Second messengers (Biochemistry)--Research--Methodology.
2. Neurochemistry--Research--Methodology. I. Boulton, A. A. (Alan A.) II. Baker, Glen B., 1947– . III. Taylor, Colin W.
IV. Series.
   [DNLM: 1. Second Messenger Systems. W1 NE337G v. 20 / QU 120 I645]
QP517.C45I59 1991
612.8'042--dc20
DNLM/DLC
for Library of Congress   91-20801
                 CIP

Preface to the Series

When the President of Humana Press first suggested that a series on methods in the neurosciences might be useful, one of us (AAB) was quite skeptical; only after discussions with GBB and some searching both of memory and library shelves did it seem that perhaps the publisher was right. Although some excellent methods books have recently appeared, notably in neuroanatomy, it is a fact that there is a dearth in this particular field, a fact attested to by the alacrity and enthusiasm with which most of the contributors to this series accepted our invitations and suggested additional topics and areas. After a somewhat hesitant start, essentially in the neurochemistry section, the series has grown and will encompass neurochemistry, neuropsychiatry, neurology, neuropathology, neurogenetics, neuroethology, molecular neurobiology, animal models of nervous disease, and no doubt many more "neuros." Although we have tried to include adequate methodological detail and in many cases detailed protocols, we have also tried to include wherever possible a short introductory review of the methods and/or related substances, comparisons with other methods, and the relationship of the substances being analyzed to neurological and psychiatric disorders. Recognizing our own limitations, we have invited a guest editor to join with us on most volumes in order to ensure complete coverage of the field. These editors will add their specialized knowledge and competencies. We anticipate that this series will fill a gap; we can only hope that it will be filled appropriately and with the right amount of expertise with respect to each method, substance or group of substances, and area treated.

Alan A. Boulton
Glen B. Baker

Preface

Despite the success of earlier *Neuromethods* volumes, I was initially reluctant to edit a further volume because my own research is concerned with nonneural tissues. I changed my mind for two simple reasons. First, though the sheer diversity of extracellular signal molecules is staggering, still more impressive is the remarkably small number of transmembrane signaling processes they recruit—their receptors either have integral ion channels or enzyme activities, or else they catalytically activate G proteins. Likewise, when we look to the final intracellular targets of these signaling pathways, they are stucturally diverse, but again there are common themes: the response may either be the gating of an ion channel, or else the phosphorylation of a target protein. Such conservation of signaling mechanisms is both impressive and convenient, and provides my justification for asking authors with interests in diverse tissues to contribute their methodological expertise to this volume.

Second, I think it would be difficult to overestimate the extent to which our understanding of intracellular signaling has been transformed by new and improved methodology. Thus, simple methods for measuring inositol phosphates have revealed the profound and widespread importance of the phosphoinositide pathways, the techniques of contemporary molecular biology have provided unrivaled opportunities to relate structure and function, and the complex spatial and temporal characteristics of intracellular signaling pathways were barely imaginable before the introduction of fluorescent indicators and single cell-imaging technology.

These are exciting times for students of intracellular messengers and much of that excitement comes from the opportunities to apply new methods. I hope this volume will provide an introduction to many of these methods and encourage neurobiologists to apply them.

Colin W. Taylor

Dedication

As this book neared completion, two people who contributed to it in their different ways died. Philip Godfrey, in his brief career, published a number of original papers on phosphoinositide pathways, notably on their roles in the brain; his chapter is testimony to his expertise in this area. In a very different way Elsie Philipson made her contribution for she made my career in science possible.

This volume is dedicated to them.

Colin W. Taylor

Contents

Identification and Analysis of Function of Heterotrimeric Guanine Nucleotide-Binding Proteins Expressed in Neural Tissue
Graeme Milligan

Methods for the Analysis of Phosphoinositides and Inositol Phosphates
Philip P. Godfrey

Contents

Inositol Trisphosphate Receptors and Intracellular Calcium:
Experimental Approaches
Colin W. Taylor, Jennifer M. Bond, David L. Nunn, and Katherine A. Oldershaw

Measurement of Intracellular Calcium with Fluorescent Calcium Indicators
Martin Poenie

Single-Cell Imaging Technology
Roger B. Moreton

Analysis of Protein Kinase C Function
Gillian M. Burgess

Synthetic Analogs of Intracellular Messengers
Barry V. L. Potter

Methods in Cyclic Nucleotide Research
Kenneth J. Murray

Caged Intracellular Messengers and the Inositol Phosphate Signaling Pathway
Ian Parker

Protein Phosphorylation
David B. Morton

Investigation of the Intracellular Regulators and Components of the Exocytotic Pathway
Robert D. Burgoyne

Intracellular Messengers in Vertebrate Photoreceptors:
Electrophysiological Techniques
Hugh R. Matthews and Trevor D. Lamb

**Intracellular Messengers in Invertebrate Photoreceptors
Studied in Mutant Flies**
Baruch Minke and Zvi Selinger

Contributors

JENNIFER M. BOND • *Department of Pharmacology, University of Cambridge, Cambridge, UK*

GILLIAN M. BURGESS • *Sandoz Institute for Medical Research, London, UK*

ROBERT D. BURGOYNE • *Department of Physiology, University of Liverpool, Liverpool, UK*

PHILIP P. GODFREY (DECEASED) • *Glaxo Institute for Molecular Biology, Geneva, Switzerland*

T. D. LAMB • *Physiological Laboratory, University of Cambridge, Cambridge, UK*

H. R. MATTHEWS • *Physiological Laboratory, University of Cambridge, Cambridge, UK*

GRAEME MILLIGAN • *Department of Biochemistry, University of Glasgow, Glasgow, Scotland*

BARUCH MINKE • *Department of Physiology, The Hebrew University, Jerusalem, Israel*

ROGER B. MORETON • *Department of Zoology, AFRC Laboratory of Molecular Signalling, Cambridge, UK*

DAVID B. MORTON • *Department of Zoology, University of Washington, Seattle, Washington*

KENNETH J. MURRAY • *SmithKline Beecham Pharmaceuticals, Herts, UK*

DAVID L. NUNN • *Department of Pharmacology, University of Cambridge, Cambridge, UK*

KATHERINE A. OLDERSHAW • *Department of Pharmacology, University of Cambridge, Cambridge, UK*

IAN PARKER • *Department of Psychobiology, University of California, Irvine, Irvine, California*

MARTIN POENIE • *Department of Zoology, University of Texas, Austin, Texas*

BARRY V. L. POTTER • *School of Pharmacy and Pharmacology, University of Bath, Bath, UK*

ZVI SELINGER • *Department of Biological Chemistry, The Hebrew University, Jerusalem, Israel*

COLIN W. TAYLOR • *Department of Pharmacology, University of Cambridge, Cambridge, UK*

Identification and Analysis of Function of Heterotrimeric Guanine Nucleotide-Binding Proteins Expressed in Neural Tissue

Graeme Milligan

1. Introduction

Most neurotransmitters, hormones, and growth factors that cause alterations in the rate of synthesis of intracellular messengers interact with cell surface receptors that are members of a family of single polypeptide proteins. Secondary and tertiary structure predictions for such receptors indicate that they are glycoproteins in which highly hydrophobic blocks of amino acids produce seven *trans*-plasma membrane-spanning elements. In all known cases, such receptors are required to interact with members of a family of heterotrimeric guanine nucleotide-binding proteins (G proteins) before agonist-induced alterations in the enzymatic activity of intracellular messenger-generating systems can be detected.

G proteins, which have been implicated in cellular signaling processes, are found widely throughout evolution. Highly conserved G proteins have been identified either via cDNA cloning or immunological means in mammals, birds, amphibia, invertebrates, yeast and slime molds, and green plants. It should be remembered, however, that not all proteins that bind and hydrolyze GTP are likely to be involved in the regulation of intracellular messenger production. For example, fac-

From: *Neuromethods, Vol. 20: Intracellular Messengers*
Eds: A. Boulton, G. Baker, and C. Taylor © 1992 The Humana Press Inc.

tors involved in protein synthesis initiation and elongation require GTP, as do the α and β subunits of the microtubule-forming protein, tubulin. Further, a series of low M_r (21–28 kDa), GTP-utilizing polypeptides has been identified, including members of the *ras*, *ral*, *rho*, ARF, and *smg* families of proteins. A number of these proteins have been implicated in regulation of intracellular messenger generation, particularly in relation to the control of phosphoinositidase C activity and mitogenesis, but further roles, especially in the control of secretory processes, appear likely, particularly when genetic information from yeast systems is taken into consideration.

Each of the classical heterotrimeric G proteins, of which some 10 have been identified to date, consists of distinct α, β, and γ polypeptides. The α subunits define the identity of the G protein and vary in size from 39–46 kDa, as calculated from primary sequence information derived from corresponding cDNA clones. Although the α subunits are distinct, they are highly homologous, with sequence identity varying from 94% between $G_i1α$ and $G_i3α$ to 41% between $G_sα$ and $G_zα$. Further, there is remarkable sequence conservation of the individual G protein α subunits between species; for example, in $G_i1α$ from human and bovine tissues, there is complete conservation of sequence (354 amino acids), and there is but a single amino acid substitution in $G_sα$ (394 amino acids) between rat and human. Even in *Drosophila melanogaster*, a form of $G_sα$ is 71% identical with that from mammalian species, whereas the $G_oα$ homolog is 81% identical with mammalian $G_oα$.

With the exceptions of receptor control of phospholipase A_2 activity and, to a degree, inhibition of adenylyl cyclase activity, it is believed that the α subunits of G proteins define the nature of the interactions between G protein and both receptor and effector moieties, although the interaction of α subunits with receptors requires the presence of βγ subunits. The individual G protein α subunits must therefore possess domains able to interact selectively with the other components of the signal transduction cascade. In addition to these functions, the α subunit of a G protein must also interact with the βγ complex, and is the site of GTP binding and hydrolysis.

It is this GTPase activity that defines the state of activation of the G protein and, hence, that of the effector system. None of the classical signal-transducing G proteins have yet been crystallized, but crystallization of the single polypeptide GTP-binding proteins, $p21^{N-ras}$, and the bacterial elongation factor, EFT has provided information that allows a tentative model for the structure of the α subunit of an average G protein to be constructed. This can then be used to test hypotheses relating to the location of functional domains of these proteins (Masters et al., 1986). Substantial evidence exists to map the site of interaction with the βγ subunit complex to the extreme N-terminal region of the α subunits. The extreme C-terminal region of all G protein α subunits appears to represent the site of contact with receptors, and evidence from the construction of chimeric G protein α subunits suggests that the site of interaction with effector systems is also likely to be toward the C-terminus of the α subunit (Masters et al., 1988).

Genetic diversity has also been recorded at the level of the β subunit (Gao et al., 1987). Three distinct cDNAs have been isolated that correspond to highly homologous polypeptides of 35–36 kDa. However, there is no strong evidence to date to indicate that these three polypeptides serve distinct functions, or that they interact either exclusively or even selectively with distinct subpopulations of α subunits. Although genetic diversity is also known to occur at the level of the γ subunit, relatively little attention has been directed to date toward the function of this (these) polypeptide(s). Part of the reason for this omission is the small size of the γ subunits (<10 kDa). Moreover, since the β and γ subunits do not dissociate from one another under physiological conditions, the two subunits are often regarded as forming a single composite entity. Given this fact and the number of distinct β and γ subunits, there is considerable potential for the formation of a large number of unique βγ complexes. The recent observation that a G protein γ subunit has homology with $p21^{N-ras}$ (Gautam et al., 1989) is likely to promote a more detailed examination of their role. For a comprehensive review of the mechanism of action of G proteins, *see* Gilman (1987).

Brain is the tissue that contains the highest levels of a number of the heterotrimeric G proteins. As such, it has frequently been utilized as a source of material for purification studies. Further, the expression of a considerable number of distinct G proteins in neuronal tissues (Table 1) means that many studies on the function and potential specificity of interaction of these proteins with both receptors and effector systems have been performed either with brain membranes or with permanent cell lines established from cells of the central nervous system. This chapter will discuss approaches that have been shown to be useful in the identification of individual G proteins and will attempt to assess the specific roles of each G protein. Further, the closing section will indicate directions of research that are likely to be fruitful for such studies in the near future.

2. Purification Studies

Initial purification of "G_i" from rabbit liver had identified only a single protein that was a heterotrimer of 41, 35, and 8 kDa subunits. It was the purification of "G_i" from bovine brain using either the binding of [^{35}S]GTPγS or pertussis toxin-catalyzed ADP-ribosylation as an assay for "G_i" that provided the first evidence that there were multiple pertussis toxin substrates. Purification of the stimulatory G protein of the adenylyl cyclase cascade (G_s) was facilitated by the availability of membranes of the cyc⁻ variant of the murine S49 lymphoma cell line, which was deficient in the α subunit of G_s and, hence, could act as an acceptor system for a reconstitution assay. By contrast, the purifications of "G_i" were performed without a functional assay for the protein. Indeed, identification of selective functional reconstitution assays for the pertussis toxin-sensitive G proteins remains a problem and has hindered the assignment of function to individual G proteins that are substrates for ADP-ribosylation catalyzed by this toxin.

Purification of all heterotrimeric G proteins tends to follow a very similar approach (*see* Spiegel, 1990 for review). The G proteins are solubilized from plasma membranes using the ionic detergent sodium cholate. Purification generally involves sequential chromatography by ion exchange (usually DEAE-Sephacel),

Table 1
Heterotrimeric G Proteins Expressed in Neural Tissue

G Protein	Sensitivity to ADP-ribosylation by bacterial toxins	Function	Distribution
G_s	Cholera toxin	Stimulation of adenylyl cyclase, activation of dihydropyridine-sensitive Ca^{2+} channels	Universal
G_{olf}	Cholera toxin	Stimulation of adenylyl cyclase	Only olfactory sensory neurons
G_i1	Pertussis toxin	Undefined	High levels in brain
G_i2	Pertussis toxin	Inhibition of adenylyl cyclase, stimulation of phosphoinositase C?	Universal
G_i3	Pertussis toxin	Regulation of K^+ channels	Universal
G_o	Pertussis toxin	Regulation of Ca^{2+} channels	Limited
$G_o{}^*$	Pertussis toxin	Regulation of Ca^{2+} channels?	Limited
G_z	No	Undefined	Undefined, but restricted

gel filtration (Ultrogel AcA34), and hydrophobic chromatography (either heptylamine or octylamine Sepharose). A further series of ion-exchange steps, generally based on HPLC or FPLC columns of resins, such as Mono-Q or DEAE Toyo-Pearl, are then routinely used to separate different G proteins. To resolve α and $\beta\gamma$ subunits, it is appropriate to include G protein activators, such as $AlCl_3$, NaF, and $MgCl_2$, with the extracts and in column buffers, since these would be anticipated to produce dissociation of the G protein subunits. In the case of $G_{s'}$ the protein appears to be relatively unstable in the presence of sodium cholate, and during purification, this detergent is usually exchanged for Lubrol during the late stages of purification.

3. Bacterial ADP-Ribosylation

3.1. Basic Aspects

The α subunits of many of the heterotrimeric G proteins are substrates for an NAD^+-dependent ADP-ribosylation catalyzed by certain bacterial toxins (*see* Milligan, 1988 for review). The studies of Ui and collaborators (Katada and Ui, 1982) originally indicated that pertussis toxin might modify the inhibitory G protein of the adenylyl cyclase cascade. Ui and coworkers were able to demonstrate that pertussis toxin (initially called islet-activating protein) was able to attenuate α-adrenergic control of insulin release from pancreatic islets, and that there was an increase in adenylyl cyclase activity in membranes of rat glioma C6 cells that had been pretreated with the toxin. Time courses of the effects of the toxin correlated highly with the [^{32}P] NAD^+-dependent ADP-ribosylation of a 41 kDa polypeptide in membranes of these cells. Pertussis toxin was demonstrated to attenuate hormone- and neurotransmitter-mediated inhibition of adenylyl cyclase in all tissues examined, and in all cases, an apparently equivalent 40–41 kDa polypeptide was a substrate for pertussis toxin-catalyzed ADP-ribosylation. This polypeptide was thus identified as the inhibitory G protein of the adenylyl cyclase cascade (G_i). The apparent universal identification, at that time, of a single polypeptide substrate for pertussis toxin-catalyzed ADP-ribosylation implied that "G_i" was the sole substrate for this reaction. Further, by extension, it was implied that any effect produced by pertussis toxin must be the result of covalent modification of "G_i." However, the observation, originally in systems derived from hemopoietic stem cells, that the effects of a number of Ca^{2+}-mobilizing receptors were also attenuated by treatment of the cells with pertussis toxin (Ohta et al., 1985) implied that "G$_i$" was either able to control the activity of a number of distinct effector systems or that the pertussis toxin substrate must display some degree of heterogeneity. It should be noted that a number of effects of pertussis toxin do not appear to be the consequence of the activity of the ADP-ribosyltransferase activity of the A subunit of the toxin. Rather, these effects are attributable to the B-oligomer (Banga et al., 1987), which is responsible for the initial binding of the holotoxin to a cell prior to internalization of the A subunit. Such results demand that the effect of the holotoxin on intact cells be combined with a demonstration of ADP-ribosylation of polypeptides of about 40 kDa before concluding that a pertussis toxin-sensitive G protein is implicated in controlling the response.

The acceptor site for pertussis toxin-catalyzed ADP-ribosylation is a cysteine residue that is located four amino acids from the C-terminus of the α subunit of the relevant G proteins. Six of the G proteins for which cDNA clones are available (rod transducin, cone transducin, G$_i$1, G$_i$2, G$_i$3, and G$_o$) have this characteristic signature, and all of these have been demonstrated directly to be substrates for pertussis toxin-catalyzed ADP-ribosylation under appropriate conditions. In each case, the state of association of the G protein affects the kinetics of the reaction. Analogs of GTP, such as GMP-PNP, reduce the rate of incorporation of ADP-ribose, whereas analogs of GDP enhance incorporation. This is generally taken to imply that the holomeric form of the G protein is either a better, or indeed the only substrate for the toxin-catalyzed reaction. It should be noted, however, that the activity of the toxin itself is modified by nucleotides. The isolated α subunit of G$_o$ is a poor substrate for pertussis toxin, and back titration of $\beta\gamma$ subunits increases ADP-ribosylation of the α subunit. These observations have led to the unfounded perception that the ADP-ribosylated, holomeric G protein is no longer able to dissociate into α + $\beta\gamma$ subunits and that this explains the inability of receptors that interact with these G proteins to function following pertussis toxin treatment. However, it has been clearly demonstrated, using sucrose density gradient experiments, that the addition of poorly hydrolyzed analogs of GTP will promote the dissociation of G protein subunits following pertussis toxin-catalyzed ADP-ribosylation (Huff and Neer, 1986).

The functional consequences of pertussis toxin-catalyzed ADP-ribosylation are produced by preventing productive contact between receptor and G protein. This is because the extreme C-terminal region of the α subunit of all G proteins is of vital importance in receptor–G protein interactions (*see* Section 7.4.).

Pertussis toxin-catalyzed ADP-ribosylation cannot be used to discriminate which of the various pertussis toxin-sensitive G proteins interact functionally with a particular receptor, even though the rates of ADP-ribosylation of the various α subunits are somewhat different. These variations in rate of ADP-ribosylation may be a reflection of the different affinities of the individual α subunits for βγ subunits.

Improved resolution of SDS-PAGE gels in the 40 kDa region, which can be achieved by lowering the crosslinking of the gel matrix (Mitchell et al., 1989), in combination with prior alkylation of the membranes (Sternweis and Robishaw, 1984), can be used to demonstrate the coexpression of a number of pertussis toxin-sensitive G proteins in a single cell type or tissue. The rank order of apparent M_r for the pertussis toxin-sensitive G proteins under such conditions is $G_i1 > G_i3 > G_i2 > G_o$.

In comparison to cholera toxin-catalyzed ADP-ribosylation in which radioactivity is often incorporated into a variety of polypeptides other than G protein α subunits (*see below*), pertussis toxin-catalyzed ADP-ribosylation is a relatively specific enzymatic reaction. However, in a number of systems, particularly brain, a number of polypeptides may incorporate radioactivity in a toxin-independent manner (Milligan et al., 1987a). This is usually the consequence of NAD^+-glycohydrolyase activity. Stability of $[^{32}P]NAD^+$ in the assay should be assessed routinely and can be performed easily by thin-layer chromatography on polyethylene imine plates using 0.4M LiCl as solvent (Milligan et al., 1987b).

If stability of $[^{32}P]NAD^+$ in the presence of the membranes is a problem, this can usually be overcome by solubilizing the G proteins from the membrane with sodium cholate (usually 1% [w/v] for 1 h) prior to assay. The glycohydrolyase either is not solubilized or is inactivated during this process. Sodium cholate, however, is a potent inhibitor of pertussis toxin-catalyzed ADP-ribosylation, and must be diluted out or otherwise removed prior to assay. Good results can be obtained by diluting 1% sodium cholate extracts of brain membranes at least 20-fold into buffer containing 0.05% Lubrol PX, prior to performing the ADP-ribosylation assay (Milligan et al., 1987a). Indeed, addition of low concentrations of Lubrol PX (0.05%) can significantly enhance pertussis toxin-catalyzed ADP-ribosylation of toxin substrates when added directly to membranes. This approach is not appropriate for studies of cholera toxin-catalyzed ADP-ribosylation, which requires the presence of the GTP binding protein, ADP-ribosylation factor (ARF), and cannot be performed in the absence of membrane structure.

Improved resolution of the various pertussis toxin-sensitive G proteins in SDS-PAGE can be achieved either by alkylation of the membranes with freshly prepared *N*-ethyl maleimide prior to addition of Laemmli buffer, by reducing the degree of crosslinking of the gel matrix, or by the inclusion of deionized urea in the resolving gel. However, prior treatment of membranes with *N*-ethyl maleimide will prevent pertussis toxin-catalyzed ADP-ribosylation as the acceptor site for such alkylation is the cysteine residue, which acts as the ADP-ribose acceptor.

Pertussis toxin-catalyzed ADP-ribosylation is a relatively inefficient process when performed in vitro. As such, attempts to quantitate levels of the pertussis toxin substrates by the excision of radiolabeled polypeptides from gels and liquid scintillation counting are likely to provide minimum estimates.

3.2. Recent Advances in the Use of Bacterial Toxins to Study Pertussis Toxin-Sensitive G Proteins

Although both rod and cone transducins can be substrates, under appropriate conditions, for both pertussis and cholera toxins, it is generally considered that the other pertussis toxin-sensitive G proteins are not substrates for ADP-ribosylation catalyzed by cholera toxin. However, if cholera toxin-catalyzed ADP-ribosylation is performed on well-washed membranes derived from a range of tissues, in the absence of added guanine nucleotides, then it is possible to detect toxin-dependent incorporation of radioactivity into a 40 kDa polypeptide, as well as into the polypeptides corresponding to forms of $G_s\alpha$. In membranes of neuroblastoma × glioma hybrid, NG108-15 cells, the addition of receptor-saturating concentrations of the synthetic enkephalin [D-Ala^2 D-Leu^5] enkephalin (DADLE) produced a marked increase in cholera toxin-catalyzed ADP-ribosylation of this 40 kDa protein, but had no effect on the cholera toxin-catalyzed ADP-ribosylation of G_s (Milligan and McKenzie, 1988). Further, DADLE had no effect on pertussis toxin-catalyzed ADP-ribosylation of G proteins in these membranes. The most likely explanation of these results is that the 40 kDa polypeptide represents the G protein with which the opioid receptor on these cells interacts. A possible explanation of these results is that receptor activation promotes the release of GDP from the nucleotide binding site. As under the conditions of these experiments, if GTP was not available to replace the GDP, then a conformational change occurred such that the G protein became a (weak) substrate for cholera toxin (Fig. 1). It is pertinent to this argument that all G proteins identified to date have an invariant arginine residue in a position in the primary sequence equivalent to that which is the site of cholera toxin-catalyzed ADP-ribosylation in $G_s\alpha$ (Table 2). Further, this amino acid is close to a section of the primary sequence that forms part of the nucleotide-binding site, and the maintenance of this arginine residue is of extreme importance for G protein GTPase activity (Landis et al., 1989).

The opioid receptor in NG108-15 cells is of the δ subtype and is known to interact with a pertussis toxin-sensitive G protein to cause inhibition of adenylyl cyclase activity. Since pertussis toxin prevents effective coupling among receptors and relevant G proteins, we argued that pertussis toxin treatment of NG108-15 cells prior to cell harvest and preparation of the membranes would prevent DADLE stimulation of cholera toxin-catalyzed ADP-ribosylation of the 40 kDa protein. This indeed was the case. Further, pretreatment of the cells with cholera toxin did not prevent opioid peptide stimulation of cholera toxin-catalyzed ADP-ribosylation of the 40 kDa polypeptide. Since cholera toxin pretreatment had presumably produced ADP-ribosylation of G_sa

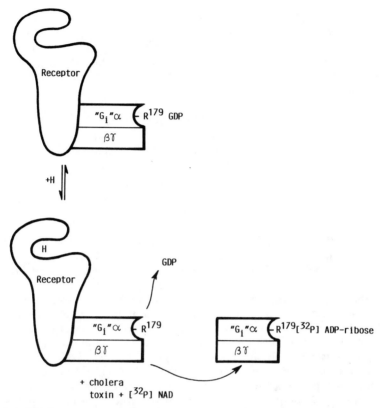

Fig. 1. Cholera toxin-catalyzed ADP-ribosylation of "G_i," a diagrammatic representation. Hormone-activation of a receptor will promote the release of GDP from the nucleotide-binding pocket of the G protein(s) with which the receptor interacts. In the absence of guanine nucleotides, the nucleotide binding site will remain empty. Under such conditions, it can be noted that the G protein(s) α subunit can become ADP-ribosylated by cholera toxin in the presence of [32P] NAD+ (*see* Fig. 2). It is likely, although untested, that the site of ADP-ribosylation will be the arginine residue in the G protein α subunit sequence (Arg 179 in $G_i2α$, for example) that is equivalent to the site for cholera toxin-catalyzed ADP-ribosylation (Arg 201) in $G_sα$. Such experiments allow the direct visualization of receptor–G protein contact.

using endogenous nonradioactive NAD+ as substrate, then the 40 kDa polypeptide was essentially the only band in membranes from these cells now able to incorporate radioactivity from [32P]NAD+ in a cholera toxin-dependent manner. This is a reflection that cholera toxin does not catalyze ADP-ribosylation of

Table 2

The α Subunits of All Pertussis Toxin-Sensitive G-Proteins
Have a Conserved Arginine Residue at the Position Equivalent to the One
That is the Site for Cholera Toxin-Catalyzed ADP-Ribosylation in G_s[a]

	C.T.↓ ↓	
	195	207
G_s	QDLLRCRVLTSGI	
	172	184
G_i1	QDVLRTRVKTTGI	
	173	185
G_i2	QDVLRTRVKTTGI	
	172	184
G_i3	QDVLRTRVKTTGI	
	173	185
G_o	QDILRTRVKTTGI	

[a]The site of cholera toxin-catalyzed ADP-ribosylation in $G_s\alpha$ is Arg 201. All of the G protein α subunits that are substrates for pertussis toxin-catalyzed ADP-ribosylation have this arginine residue conserved in the equivalent position, since it plays a key role in the GTPase activity of the G protein. It might thus be anticipated that all of these G proteins will also be substrates for cholera toxin-catalyzed ADP-ribosylation. However, they are not unless conditions are arranged such that the guanine nucleotide binding pocket is empty. This can be achieved by agonist-stimulated receptor activation of the G protein in the absence of added nucleotides. Herein, agonist-promoted release of GDP will leave the nucleotide binding site empty (*see* Fig. 1, 2). All pertussis toxin-sensitive G proteins can be ADP-ribosylated by cholera toxin in an agonist-dependent manner (although with varying efficiency) when reconstituted into membranes of pertussis toxin-treated HL-60 cells (Iiri et al., 1989). Coupled with immunoprecipitation studies, such an approach should allow definition of the selectivity/specificity of receptor–G protein interactions in a variety of tissues (*see text*).

the "G_i-like" proteins in whole cells. Indeed, cholera toxin-cata-lyzed ADP-ribosylation of "G_i" cannot be observed when the nucleotide-binding pocket of the polypeptide contains either GDP or GDP + AlF_4^-, which provides a mimic for GTP (Fig. 2), but only under these artificial conditions. Similar observations have also been noted in rat glioma C6 cells, where a small, heat-stable factor in fetal calf serum promotes the cholera toxin-de-pendent ADP-ribosylation of a 40-kDa polypeptide that appears to be the pertussis toxin-sensitive G protein G_i2 (Milligan, 1989). Recent experiments on the human leukemia cell line HL60

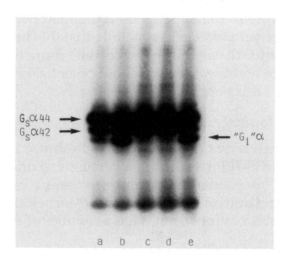

Fig. 2. Receptor stimulation of the cholera toxin-mediated ADP-ribosylation of forms of "G_i" requires that the nucleotide-binding pocket of G protein α subunit is empty. Rat 11 fibroblasts were transfected with the gene for the human α_2 A-adrenergic receptor and selected on the basis of resistance to G418 sulfate. One clone of cells (α_2A:1C) was shown to express the receptor at some 3 pmol/mg membrane protein as assessed by ligand binding studies with [^{3}H] yohimbine. Membranes of these cells (60 μg) were treated with thiol-activated cholera toxin and [^{32}P] NAD$^+$ in the absence of guanine nucleotides, but in the presence (b–e) or absence (a) of the selective α_2 receptor agonist U.K. 14304. The samples also contained (a,b) no other additions; (c) GDP (100 μM); (d) GDP (100 μM) + Al^{3+} (20 μM), F$^-$ (10 mM); (e) Al^{3+} (20 μM), F$^-$ (10mM). Autoradiography of the samples following SDS-PAGE demonstrated incorporation in an agonist-dependent manner into a 40 kDa polypeptide only when neither GDP nor GDP + AlF$_4^-$ was present. As such, the presence of either GDP (c) or a mimetic of GTP (d) prevented cholera toxin-catalyzed ADP-ribosylation of this 40 kDa protein.

have provided convincing evidence that a single agonist, the bacterial chemotactic factor, *N*-formyl-methionyl-leucyl-phenylalanine (FMLP), can promote cholera toxin-catalyzed ADP-ribosylation of two distinct pertussis toxin-sensitive G proteins in a single cell (Gierschik et al., 1989). Based on the immunological profile of pertussis toxin-sensitive G proteins in these cells (Murphy et al., 1987), these are likely to be G_i2 and G_i3. Although agonist-promoted, cholera toxin-catalyzed ADP-ribosylation of these two G proteins demonstrated very similar dose–response curves, it is not yet possible to conclude that this implies that a single receptor for the agonist interacts with more than a single G protein, since nothing is known about potential heterogeneity of receptors for this peptide.

3.3. Methodological Considerations

The approach described above has been used to identify coupling of the FMLP receptor to a pertussis toxin-sensitive G protein(s) in both human HL60 leukemic cells (Gierschik and Jakobs, 1987) and human U937 monocytes (Mitchell and Milligan, unpublished observations), but it has a number of limitations.

1. It is not universally applicable. In NG108-15 cells, although opioid peptides, α_2-adrenergic agonists, and somatostatin all function to inhibit adenylyl cyclase, we have been able to observe receptor-mediated, cholera toxin-catalyzed ADP-ribosylation of the 40-kDa polypeptide only when using opiate ligands (Milligan, unpublished observations). This may be a reflection that the opioid receptor is either present at higher levels than the others or is more "tightly" coupled to the G protein signaling system.

2. There has been no rigorous proof to demonstrate that the polypeptide labeled is indeed the relevant G protein. It should be noted, however, that Gierschik and Jakobs (1987) have demonstrated that cholera toxin treatment of membranes of HL60 cells will produce an attenuation of the FMLP-receptor stimulation of high affinity GTPase activity, which is strictly dependent upon the presence of NAD^+.

3. Although it has yet to be demonstrated directly, it is assumed that ADP-ribose is incorporated into the equivalent arginine residue in the pertussis toxin-sensitive G protein (e.g., amino acid 179 in $G_i2\alpha$) as is the acceptor amino acid (187 or 201 in the short and long forms, respectively, of G_s) of $G_s\alpha$. As such, it might be anticipated that the functional consequences might be similar. Cholera toxin-

catalyzed ADP-ribosylation of $G_s\alpha$ produces a persistent activation of the G protein and, therefore, of adenylyl cyclase by inhibiting the intrinsic GTPase activity (*see* Gilman, 1987 for details) that normally functions to switch off the activated state of the G protein. It is intriguing to note that site-directed mutagenesis of the arginine, which is the ADP-ribose acceptor in G_s, to any other possible amino acid also leads to a constitutive activation of the protein (Landis et al., 1989; Freissmuth and Gilman, 1989; Graziano and Gilman, 1989; Masters et al., 1989). It remains to be addressed whether pertussis toxin-sensitive G proteins will be persistently activated following cholera toxin-catalyzed ADP-ribosylation and whether site-directed mutagenesis of the arginine residue will generate a G protein with substantially reduced GTPase activity, which is thus persistently activated.

4. Assays for Receptor–G Protein Interactions

4.1. Alterations in Receptor–Agonist Binding Interactions in the Presence of Analogs of GTP

Early experiments on the binding of ligands to receptors that produce stimulation of adenylyl cyclase noted that GTP interfered with the binding of glucagon to its receptor (Rodbell et al., 1971). It was subsequently noted that guanine nucleotide effects on β-adrenergic receptor binding were limited to agonists and not antagonists (Maguire et al., 1976). Detailed analyses of ligand binding in a range of systems indicated that agonist displacement of [3H] antagonist binding curves had pseudo-Hill coefficients significantly less than 1.0. In contrast, antagonist displacement of [3H] antagonist binding curves was characterized by pseudo-Hill coefficients close to 1.0. Full agonists characteristically had lower pseudo-Hill coefficients than did partial agonists. The conclusions drawn from these studies were that agonists were able to recognize two states or conformations of the receptor with different affinities, whereas antagonist affinity for these two forms must either be identical, or at least extremely similar (*see* Birdsall et al., 1980, for example). Furthermore, addition of poorly hydrolyzed analogs of GTP, such as GMP-PNP or GTPγS, to the binding incubations reduced the ability of agonists, but not antagonists, to compete for [³H] antagonist binding sites, and under these conditions, the agonist displacement curves had a pseudo-Hill coefficient close to 1. These data were consistent with a model whereby agonists interacted with receptors that were in intimate association with a GDP-liganded, and hence unstimulated, G protein with higher affinity than they did with receptors that were not in such contact with the G protein. Similar experiments prompted the conclusion that receptors that mediated inhibition of adenylyl cyclase must also interact with a G protein (*see* Rodbell, 1980), and arguments of this nature were later extended to provide evidence for the interaction of the Ca^{2+}-mobilizing receptors with G proteins. Thus, historically, guanine nucleotide-sensitivity of agonist binding affinity has often provided the initial suggestion that a particular receptor interacts with a G protein.

The limitation of this type of approach is that it can offer no information as to the molecular nature of the G protein involved. Binding studies performed on membranes derived from tissues pretreated with a toxin isolated from supernatant cultures of *Bordetella pertussis* have been used to further subdivide the nature of receptor-linked G proteins (Kurose et al., 1983; Hsia et al., 1984). In these experiments, agonist affinity for the displacement of [³H] antagonist binding was reduced in membranes of pertussis toxin-treated cells in comparison to that in membranes from untreated cells. Furthermore, addition of poorly hydrolyzed analogs of GTP was not able to further reduce agonist affinity for the receptor, indicating that pertussis toxin pretreatment had modified the relevant G protein in such a manner that it now appeared to be functionally uncoupled from the receptor. As discussed earlier (Section 3.1.), however, these studies are of restricted usefulness because of the limited specificity of this toxin. They were, however, of use in demonstrating that receptors that mediate inhibition of adenylyl cyclase do so by interacting with a pertussis toxin-sensitive G protein(s), and that Ca^{2+}-mobilizing receptors in a number (Nakamura and Ui, 1985; Ohta et al., 1985), but not all, tissues (Helper and Harden, 1986; Martin et al., 1986) also interact with a pertussis toxin-sensitive G protein. Unfortunately, the prospectively erroneous, if most simple conclusion, i.e., that the G protein involved in inhibition of adenylyl cyclase and stimulation of inositol phospholipid turnover was one and the same, was derived from this approach.

In studies that analyze saturation binding isotherms of [³H] agonists, it is usual to perform the assays in the presence of a high concentration (5–20 mM) of Mg^{2+}, since this promotes the association of receptor and G protein in the basal state and limits the potential biphasic nature of Scatchard analysis of the binding, which would result from the basal state consisting of a complex mixture of receptors either coupled to or uncoupled from the G protein signaling system.

Methodologies have recently been developed that utilize either antibodies against specific G proteins or synthetic peptides that are believed to represent sequences of the primary sequence of G protein α subunits important for contact between receptors and G proteins to interfere with receptor–G protein interactions (*see* Section 7.4.). These allow a more direct assessment of the molecular nature of the G protein(s) that interact with a particular receptor.

4.2. GTPase Studies

The interaction of agonist with receptor promotes the release of GDP from a relevant G protein and its exchange by GTP. This is followed by the hydrolysis of the nucleotide by the GTPase activity of the G protein. Measurement of this enhanced rate of GTPase activity of a membrane in response to agonist thus provides a simple and convenient assessment of the interaction of a receptor with a G protein(s) (Fig. 3). This approach was first employed by Cassel and Selinger (1976) in an avian erythrocyte system to demonstrate interaction of the β-adrenergic receptor with a G protein. Similar experiments have been performed on a wide range of systems with similar conclusions. Despite the simplicity of the experimental protocol, in many systems, however, it has not been possible to demonstrate receptor stimulation of GTPase activity, even though other evidence indicates that a particular receptor interacts with the G protein

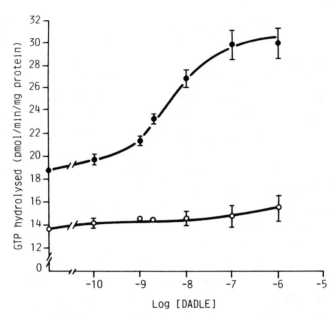

Fig. 3. Opioid agonists stimulate high-affinity GTPase activity in membranes of NG 108-15 cells. NG 108-15 cells were treated either with pertussis toxin (O-O) or with vehicle (●-●) and high-affinity GTPase activity measured in membranes of these cells in response to differing concentrations of the opioid peptide DADLE. Data represent the means ± SD of quadruplicate assays. These data, which are taken from McKenzie et al. (1988), demonstrate that the opioid receptor of these cells interacts with a G protein and that this G protein is a substrate for pertussis toxin-catalyzed ADP-ribosylation.

signaling system. It is thus worth commenting upon why this may be so. The basal "GTPase" activity of any membrane system results from the hydrolysis of GTP by all of the G proteins within that membrane, as well as by other enzymatic reactions. G protein-related GTP hydrolysis is such that the K_m of these enzymes for GTP is low. Thus, it is usual to subtract from the total GTPase activity a blank in which hydrolysis in the presence of a high concentration of GTP (50–100 μM) is assessed. Also, the contribution of any particular G protein to the basal low K_m GTPase activity will be dependent on (1) the relative proportions of the various G proteins present in the membrane and (2) their relative rates of GTP hydrolysis. Thus, the observations that receptor–G_s interactions cannot be measured using this type of assay in a number of systems and have only been reproducibly noted in a few systems, for example, in avian erythrocytes (Cassel and Selinger, 1976) and in platelets (Houslay et al., 1986), indicate that levels of G_s tend to be lower than that of many other G proteins and that the rate of GTP hydrolysis by purified G_s is extremely low. As such, receptor-mediated stimulation of this rate may still allow the situation in which the signal of increased hydrolysis of GTP by activated G_s remains lost within the "noise" because of the basal rates of hydrolysis by the other G proteins present.

In general, GTPase studies of receptors linked to pertussis toxin-sensitive G proteins have been more successful (Koski and Klee, 1981). This is presumably owing to the higher abundance of these proteins and their greater enzymatic capacity. As with binding studies, however, it has not generally been possible to further subdivide receptor interactions with particular pertussis toxin-sensitive G proteins (except with the use of specifc antisera, *see* Section 7.4.) because of the lack of specificity of this toxin.

Assessments of the specificity of receptor–G protein interactions within the native membrane have, however, been addressed by performing GTPase additivity experiments. The rationale for these is that activation of the entire population of receptors in a membrane preparation with a saturating concentration of a full agonist will prospectively lead to the activation of the full complement of G protein(s) with which that receptor is able to interact. Thus, additivity of receptor-stimulated GTPase activity following addition of two agonists that interact with independent receptors would indicate the activation of separate pools of G protein, and, by extension, different G proteins. This approach has been elegantly employed by, for example, Houslay et al. (1986) to examine a range of receptor–G protein interactions in human platelets. A further example of the usefulness of this technique has been the demonstration (McKenzie et al., 1988) that, in membranes of neuroblastoma × glioma hybrid cells, individual receptors for opioid peptides and for a growth factor interact with separate, distinct pertussis toxin-sensitive G proteins. Of course, if agonist interaction with a particular receptor is able to activate only a small proportion of the population of a particular G protein, then GTPase additivity would be observed to a second receptor able to activate the same G protein. This is likely to be the case in nervous tissue, where the pertussis toxin-sensitive G proteins may represent some 1–2% of the total membrane protein (Sternweis and Robishaw, 1984; Neer et al., 1984; Milligan and Klee, 1985). Furthermore, in tissues that contain complex mixtures of cell types, additivity of GTPase activation may be an indication that the distinct receptors in question reside in different cells.

4.3. Agonist Stimulation of [^{35}S]GTPγS Binding and of GTP Analog Photoaffinity Labeling

Because receptor activation promotes the rate of nucleotide exchange on a G protein, it should be possible to use agonist activation of the rate of [^{35}S]GTPγS binding to detect receptor–G protein interactions. Since this analog will not be hydrolyzed by the endogenous GTPase activity of the G protein, then once bound, the label will not be removed. Agonist promotion of the rate of binding of [^{35}S]GTPγS has indeed been observed in a number of situations. This approach can be used in a similar fashion to others described above to define if the relevant G protein is a substrate for pertussis toxin, since receptor stimulation of the rate of binding of the nucleotide will be prevented by prior treatment of the cell or tissue with pertussis toxin if the relevant G protein is a substrate for ADP-ribosylation catalyzed by this toxin. The principal limitation of this approach is that it does not provide a covalent modification of the G protein, such that the identity of the G protein can be demonstrated following SDS-PAGE and immunoblotting or following selective immunoprecipitation.

A modified strategy that may provide a more direct approach is to use a radiolabeled, photoreactive GTP analog as the nucleotide. To date, this has only been explored to a limited degree. Offermans et al. (1990) have incubated membranes of HL-60 cells with various chemotactic agents in the presence of the photoreactive GTP analog [^{32}P]GTP azidoanilide. Following subsequent exposure to UV light and SDS-PAGE, they noted an increased labeling of a 40-kDa polypeptide in assays that included the chemotactic agents, and this was dependent on the concentration of agonist used. Although it has not yet been recorded, it should be easy to immunoprecipitate the labeled polypeptide(s) with specific antisera and, hence, define the G protein(s) linked to these receptors. Although such an approach has not yet been applied to neural systems, it should be equally applicable.

5. Direct Activation of G Proteins and Interactions of G Proteins and Effectors

5.1. Analogs of Guanine Nucleotides and Caged Compounds

The binding of a poorly hydrolyzed analog of GTP to a G protein will produce activation of the G protein. As such, analogs, including GTPγS and GMP-PNP, have been widely used to examine G protein regulation of effector systems in the absence of receptor activation both in membrane preparations and in cells permeabilized, for example, either by electrical discharge or with low concentrations of detergent (*see* chapter by Taylor et al.). Such nucleotides have proven to be useful in studies on the kinetics of activation of adenylyl cyclase, phosphoinositidase C, and a range of ion channels. Of particular use in a number of electrophysiological experiments have been "caged" analogs of GTP (*see* Dolphin [1990] for review; Parker, this vol.), where the analog can be introduced into the cell in an inactive form and only released following photoactivation. One limitation of the use of such caged compounds is that illumination is generally able to release a small fraction of the active nucleotide, and as such, dose response curves must be either carefully calibrated or treated with considerable caution.

5.2. Differential Affinities of GTP Analogs for G_s and "G_i"

One potential difficulty with the use of guanine nucleotides and the various poorly hydrolyzed analogs in examining G protein regulation of effector function is that, at high levels, any analog of GTP would be anticipated to activate all of the G pro-

teins in the membrane fraction or cell. As such, it may be difficult to eliminate the possibility of interactions (crosstalk) between
distinct signaling pathways being responsible for the observations noted. In relation to the adenylyl cyclase cascade, with both
stimulatory and inhibitory control being exerted over the same
effector by separate G proteins, problems relating to activation
of both G proteins are particularly relevant. However, it is fortu-
nate that the affinities of the individual G proteins (G_s and "G_i")
for both GTP and for various poorly hydrolyzed analogs of the
nucleotide are markedly different. Concentration curves for GTP
in the regulation of adenylyl cyclase show marked activation up
to concentrations close to $1 \times 10^{-7}M$. However, at higher concentrations, reduced adenylyl cyclase activities are noted. This is
taken to imply that G_s has a higher affinity for GTP than "G_i".
The loss of the inhibitory phase of GTP regulation of adenylyl
cyclase in adipocytes isolated from streptozotocin-induced dia-
betic rats (Strassheim et al., 1990) is consistent with this pattern,
since it is known that guanine nucleotide regulation of G_i is
attenuated in this disease model (Gawler et al., 1987). The poorly
hydrolyzed analog of GTP, GMP-PNP, is the most easily available and, hence, most commonly used nucleotide that displays
the reverse pattern of affinity for the G proteins of the adenylyl
cyclase cascade. As such, low concentrations (usually close to
1 nM) can be used to activate "G_i" selectively and, hence, allow
study of G protein control of the inhibitory arm of adenylyl
cyclase with minimal contribution from stimulatory input.

5.3. Direct Activators, Inhibitors of G Proteins

Activation of G proteins can be achieved in the absence of
added guanine nucleotides. The most common strategy is to use
fluoride ions. The rationale behind the use of fluoride ions is
that, owing to the intrinsic GTPase activity of the G protein α
subunit and the facilitating role of agonist-stimulated receptors
in promoting the rate of release of guanine nucleotides from the
nucleotide binding site, in the basal state, each G protein will
have a molecule of GDP tightly associated. Fluoride ions (the
actual active species appears to be AlF_4^-) appear to bind along-
side GDP and, by mimicking the terminal phosphate of GTP

(Bigay et al., 1985), promote hydrolysis-resistant activation of the G protein. The use of AlF_4^- is also appropriate in many situations, because it can enter cells relatively easily and, hence, produce activation of G proteins in whole cells, which obviates the necessity of permeabilizing cells with electrical discharge, detergent treatment, and so forth, which would be required to allow guanine nucleotides access to their site of action.

The limitations with the use of AlF_4^- are that it is even less selective than the various guanine nucleotide analogs inasmuch as all G proteins appear to have similar affinities for the anion and, in some circumstances, the other known activities of fluoride ions, e.g., inhibition of phosphatase activities, may provide unwanted side effects that would compromise interpretation of an experiment.

6. Reconstitution Assays

6.1. $G_s \cdot Cyc^-$ Membranes, Cholera Toxin "Stripped" Membranes

The cyc^- mutant of the S49 lymphoma cell line might be assumed from its name to lack functional adenylyl cyclase. This was demonstrated to be incorrect, however, since stimulation of cAMP production can be noted in the presence of Mn^{2+}. By contrast, both receptor stimulation and AlF_4^- stimulation of adenylyl cyclase are lacking, and it is well established that neither mRNA coding for, or protein corresponding to, $G_s\alpha$ can be detected within these cells. Membranes from these cells provide an extremely sensitive acceptor system to assess the presence of exogenously provided G_s, and indeed the availability of such membranes was crucial in the purification of G_s (*see* Section 2.). In the bulk of reconstitution protocols, the differential ability of AlF_4^- to stimulate cAMP production in cyc^- membranes in the presence and absence of a putative source of detergent solubilized G_s is measured. Although the lack of expression of $G_s\alpha$ makes membranes from cyc^- cells the most appropriate reconstitution system, it is possible to use membranes from certain cells that have been treated with high concentrations of cholera toxin as an acceptor system. Owing to the downregulation of $G_s\alpha$ produced by this treatment (Chang and Bourne, 1989; Milligan et al., 1989), such membranes, which have been "stripped" of virtually all $G_s\alpha$, can use exogenously provided G_s to restore AlF_4^- stimulation of adenylyl cyclase. Such an approach is limited to cells in which levels of G_s represent, or become, the limiting component for transduction of hormonal stimulation of adenylyl cyclase. As such, membranes from cholera toxin-treated NG108-15 cells represent an appropriate reconstitution system (MacLeod and Milligan, 1990) (Table 3), but it is unlikely that membranes from cholera toxin-treated pituitary GH_3 cells would, since in these cells, loss of some 90% of immunologically detectable $G_s\alpha$ does not reduce forskolin-stimulated adenylyl cyclase activity (Chang and Bourne, 1989).

6.2. Pertussis Toxin-Sensitive G Proteins

The lack of a suitable reconstitution system for "G_i" in purification studies implied that the functional activity of purified pertussis toxin substrates could not be assessed adequately. The earliest attempts at reconstitution involved the addition of "G_i," or its isolated subunits, purified from rabbit liver, to pertussis toxin-treated platelet membranes to attempt to reconstitute α_2-adrenergic receptor- and guanine nucleotide-mediated inhibition of adenylyl cyclase. This was successful, as was addition of purified "G_i" from bovine brain to membranes of pertussis toxin-treated NG108-15 cells in reconstituting opioid peptide inhibition of adenylyl cyclase (Milligan and Klee, 1985). However, in these studies, fractions enriched in either G_i1 or G_o displayed similar abilities to reconstitute opioid receptor inhibition of adenylyl cyclase activity, suggesting that both proteins could function as "G_i." As an alternate strategy, Florio and Sternweis (1985) partially purified bovine brain muscarinic acetylcholine receptors and reconstituted these with purified bovine brain "G_i" or G_o, using a variety of protocols to colocalize the receptors and G proteins in phospholipid vesicles. Each G protein was able to enhance the affinity of agonist binding to the receptor some 10–20-fold, indicating that each G protein was able to interact functionally with the receptor. Despite the potential difficulties that complex mixtures of muscarinic receptor subtypes might

Table 3
Use of Membranes of NG108-15 Cells
Following Cholera Toxin-Treatment of the Cells
as an Acceptor System for Reconstitution of $G_s\alpha$[a]

	Adenylyl cyclase activity, pmol cAMP/min·mg acceptor protein
Basal	94.4 ± 0.6
Basal + sodium fluoride (10 mM)	62.4 ± 1.3
Basal + extract untreated NG108-15 cells	104.7 ± 1.2
Basal + extract untreated NG108-15 cells + sodium fluoride (10 mM)	301.3 ± 1.6

[a]Following cholera toxin treatment (100 ng/mL, 16 h) of NG108-15 cells, immunologically detectable levels of $G_s\alpha$ are reduced to virtually undetectable levels. In the presence of fluoride ions, a marked reduction in levels of adenylyl cyclase activity is noted, presumably owing to the activation of G_i. Membranes that have thus been "stripped" of $G_s\alpha$ by cholera toxin treatment can be used in reconstitution assays as an acceptor system to monitor exogenously provided $G_s\alpha$.

have added to the interpretation of these results, a range of conceptually similar experiments using a number of neurotransmitter receptors in varying states of purity have indicated (disappointingly) little selectivity of receptors for the various pertussis toxin-sensitive G proteins, and even interactions between G_i and β-adrenergic receptors, which normally function to stimulate adenylyl cyclase, have been noted (Asano et al., 1984). However, selectivity of reconstitution between G_i and G_o with the kyotorphin (tyrosine-arginine) receptor has recently been reported (Ueda et al., 1989). Furthermore, Senogles et al. (1990) have provided evidence that the dopamine D_2 receptor from anterior pituitary is able to interact selectively with G_i2 rather than other G proteins in reconstitution systems. Given the overall primary sequence similarity and probable structural similarity of the pertussis toxin-sensitive G proteins, it is possible that nonphysiologically relevant interactions might take place among receptors and G proteins in such artificial reconstitution systems, either because of the physical orientation that the proteins are able to adopt in such conditions or because the relative proportion of receptors and G proteins is inappropriate. The second possibility seems somewhat unlikely given the apparent vast molar excess of pertussis toxin-sensitive G proteins over receptors in brain. However, there may be physical constraints on G protein availability in the native state, and some evidence would suggest a role for cytoskeletal components in controlling the availability of G proteins (Wang et al., 1990). With the considerations outlined above in mind, experiments have recently been performed to attempt to address whether receptor–G protein–effector specificity is the normal situation in membranes and in whole cells (*see below*).

7. Immunological Identification of G Proteins

7.1. Historical Perspectives

It was the availability of a polyclonal antiserum generated against purified bovine rod transducin that demonstrated that the two major pertussis toxin-sensitive polypeptides purified from bovine brain were not likely to be derived from one another. This antiserum (CW6) identified the 41-kDa polypeptide ("G_i"), but not the 39-kDa polypeptide (G_o) (Milligan and Klee, 1985; Pines et al., 1985). Since both of these G proteins had been purified on the basis of being substrates for pertussis toxin-catalyzed ADP-ribosylation, the 2 kDa apparent mobility difference between the two polypeptides could not result from a proteolytic cleavage of the C-terminus, because this would have removed the cysteine residue, which is the ADP-ribose acceptor site. Also, since antiserum CW6 had been epitope mapped to a site close to the C-terminus of the α subunit of rod transducin, it was unlikely that it would identify a site at the extreme N-terminus of "G_i" such that proteolytic cleavage in this region would generate a 39 kDa species (G_o), which did not show immunological crossreactivity with "G_i." These data suggested (with hindsight, correctly) that "G_i" would prove to be more closely related to rod transducin than would G_o. Confirmation that G_o was not derived from "G_i" was obtained by the production of polyclonal antisera (e.g., antiserum RV3), which identified brain G_o, but not "G_i" on immunoblots (Gierschik et al.,1986). It was apparent that further diversity in the family of pertussis toxin-sensitive G proteins expressed in brain must exist, because rat

C6 glioma cells expressed high levels of a pertussis toxin substrate that was not identified by either antiserum CW6 or RV3 (Milligan et al., 1986).

Since antiserum CW6 could be seen to display weak reactivity against G_o and because the use of mixtures of brain "G_i" and G_o as antigen in rabbits had been ineffective in generating antibodies against "G_i," then an alternative strategy was employed. As noted above, antiserum CW6 was generated against bovine rod transducin and identified an epitope(s) within the C-terminal 5-kDa peptide of the α subunit of this polypeptide. Because this antibody also identified brain "G_i," and rod transducin was the only G protein for which a corresponding cDNA clone was available at the time, Spiegel, Unson, and Milligan produced a synthetic peptide corresponding to the C-terminal decapeptide of rod transducin and used this as an antigen to attempt to generate an antiserum that would identify "G_i," but not G_o. This approach was highly successful (Goldsmith et al., 1987). The antiserum produced (AS7) identified brain "G_i," but not G_o. More surprisingly, this antiserum also identified the major pertussis toxin-sensitive polypeptide of rat C6 glioma cells, which had not been identified by antiserum CW6 (Milligan et al., 1986). As such, the "G_i"s of bovine brain and rat glioma were demonstrated to be related, but distinct. This immunological difference was not a reflection of species variation in "G_i." Subsequent isolation of cDNAs corresponding to "G_i" indicated that they could be divided into two categories, G_i1, which corresponded to the brain type, and G_i2, which corresponded to the glioma form and was isolated from a series of cells derived from hemopoietic stem cells (*see* Lochrie and Simon, 1988; Kaziro, 1990 for review). The cDNAs corresponding to these two forms of "G_i" predicted that they would have identical C-terminal decapeptides and that this sequence differed by a single conservative substitution from the equivalent region of rod transducin. The ability of antiserum AS7 to identify each of these forms was thus explained. cDNAs corresponding to G_o indicated alterations of amino acids in five of the 10 positions of the C-terminal deca-peptide of this pertussis toxin-sensitive G protein in comparison to rod transducin and provided a rationale for the inability of antiserum AS7 to identify G_o.

7.2. Recent Studies

Discrimination in mobility between G_i1 and G_i2 can be achieved on Western blots of SDS-polyacrylamide gels either by alterations in the gel conditions to lower the concentrations of bisacrylamide and, hence, reduce the degree of crosslinking in the resolving gel (Mitchell et al., 1989) or by the inclusion of $4M$ deionized urea in the resolving gel *(see* for example Mullaney and Milligan, 1990b). Further, antipeptide antisera against sections of the primary sequences of G_i1 and G_i2, which are distinct to each protein, have been used to generate selective antisera (Goldsmith et al., 1987). Antisera against such divergent internal sequences of G_i1 and G_i2 have in general not produced antisera of such high titer as those directed against the C-terminal sequence, but have provided highly discriminatory probes.

Antisera directed against the C-terminal regions of G proteins have also recently proved to be useful tools in a range of functional studies *(see* Section 7.4.). The isolation of cDNAs corresponding to a third "G_i-like" G protein (G_i3) (Jones and Reed, 1987) has required the reassessment of immunological crossreactivities of previously generated antisera (since G_i1 and G_i3 are 94% identical at the primary sequence level, antisera able to discriminate between these gene products might be anticipated to be difficult to generate); however, antisera generated against synthetic peptides corresponding to amino acids 159–168 of $G_i1\alpha$ and $G_i3\alpha$ appear to provide selective probes (Goldsmith et al., 1988b). It further appears that antisera directed against the C-terminal decapeptide of transducin, which identify G_i1, display little affinity for G_i3. By contrast, antisera generated against the C-terminal decapeptide of G_o do crossreact with G_i3 (Spiegel, personal communication; Milligan, 1990a). This may be largely a result of the fact that the C-terminal amino acid of G_o and G_i3 is a tyrosine, whereas the equivalent amino acid in G_i1, G_i2, and transducin is a phenylalanine. G_i3 immunoreactivity can be shown to migrate between G_i1 and G_i2 in SDS-polyacrylamide gels produced in the presence of low bisacrylamide concentrations (Mitchell et al., 1989). In comparison to $G_i1\alpha$ and $G_i2\alpha$, which are highly expressed, relatively low levels of $G_i3\alpha$ immunoreactivity have been detected in brain.

To add further complexity to the immunological identification of individual pertussis toxin-sensitive G proteins in brain, it has recently become clear that "G_o" is not comprised of a single molecular entity. Two distinct forms have been isolated (G_o and G_o^*) from bovine brain (Goldsmith et al., 1988a), largely resulting from differences in their chromatography on high-resolution ion exchange columns, such as TSK-toyopearl and mono-Q. Both of these proteins can be considered immunologically to be forms of G_o, because a range of antipeptide antisera directed against discrete regions of the primary sequence of G_o, as deduced from cDNA clones, identify both. Resolution of NG108-15 neuroblastoma × glioma hybrid cell membranes by two-dimensional electrophoresis also allows resolution of two forms of G_o, which have isoelectric points of 5.5 and 5.8 (Mullaney and Milligan, 1990a,b). However, the recently identified second form of G_o in NG108-15 membranes may not be identical with brain G_o^*. Resolution of the two forms of G_o from NG108-15 membranes in one-dimensional SDS-PAGE cannot be achieved unless $4M$ deionized urea is included in the resolving gel (Fig. 4). Under such conditions, the more acidic form migrates somewhat more slowly through the gel than the more basic form. G_o^* from rat brain migrates considerably more slowly in such SDS-urea PAGE gels than either of the forms present in NG108-15 membranes (Mullaney and Milligan, 1990b). Adding further to the apparent complexity, Katada and coworkers (Kobayashi et al., 1989) have recently reported on the resolution of multiple forms of G_o by high-resolution ion exchange chromatography of detergent extracts from brain. The rationale for these apparent multiple isoforms is yet to be examined.

A number of ontogenic studies have been performed to assess levels of pertussis toxin-sensitive G proteins in rat brain during development (*see* Milligan et al., 1987a, for example). These have confirmed G_o to be the most prevalent of the pertussis toxin-sensitive G proteins in brain. It increases in amount from some 30 pmol/mg protein in the neonate to 120 pmol/mg protein in

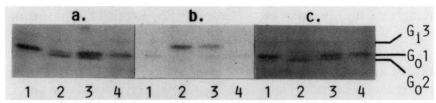

Fig. 4. Resolution of two distinct isoforms of $G_o\alpha$ in one-dimensional SDS-urea-PAGE. Membranes of (1) rat brain (5 µg), (2) untreated NG108-15 cells (50 µg), (3) dibutyryl cAMP-differentiated (1 mM, 6 d) NG108-15 cells (50 µg), and (4) mouse brain (5 µg) were resolved by one-dimensional SDS-urea-PAGE and immunoblotted with two distinct antipeptide antisera, which identify $G_o\alpha$ (a,c), and with an antipeptide antiserum against $G_i3\alpha$ (b) as primary reagents. Data are adapted from Mullaney and Milligan (1990b) and demonstrate that NG108-15 cells express two distinct forms of G_o. The more acidic form of G_o, which is present in higher levels following differentiation of the cells (compare lane 3 and lane 2), migrates more slowly in these gel conditions and corresponds exactly to G_o from both rat and mouse brain. The more rapidly migrating form of G_o in NG108-15 cells has no obvious equivalent in rat or mouse brain and, hence, is neither $G_i3\alpha$ nor G_o*, as defined by Goldsmith et al. (1988a).

the adult. G_i1 and G_i2 are some fivefold lower in concentration and alter in amount less markedly during development (Table 4).

Relatively few studies have used immunocytochemistry to identify pertussis toxin-sensitive G proteins in brain slices (however, *see* Worley et al., 1986). *In situ* hybridization experiments have, however, been used to demonstrate that the distribution of mRNA corresponding to $G_s\alpha$ and $G_i2\alpha$ are very similar in slices of rat brain, although the distribution patterns of $G_o\alpha$ and $G_i1\alpha$ mRNA are different (Brann et al., 1987). The similarities of G_s and G_i2 mRNA distribution might indicate that both G proteins impinge on the same effector, which, given the known role of G_s, would suggest G_i2 as the adenylyl cyclase inhibitor (more direct evidence to support this idea is discussed in Sections 3.2. and 7.4.). However, it should be noted that a similar distribution in brain of $G_o\alpha$, as assessed by specific antibodies, and protein kinase C, as assessed by the binding of [3H] phorbol ester binding, led to the suggestion that G_o might act to regulate receptor control of phosphatidylinositol 4,5-bisphosphate hydrolysis (Worley et al., 1986). However, receptor control of

Table 4
Ontogenic Development of Levels of G_i1 and G_o in Rat Brain[a]

Age, d	$G_i1\alpha$	$G_o\alpha$
	pmol/mg membrane protein	
5	7.5	30.0
30	18.5	117.5

[a]Levels of each of G_i1a and G_oa were measured in membranes derived from forebrains of female Osborne-Mendel rats by immunoblotting with specific antisera. Data are adapted from Milligan et al. (1987a).

phosphoinositidase C in brain is generally insensitive to treatment with pertussis toxin, implying that G_o cannot function as the transducer protein.

7.3. Methodological Considerations for the Immunological Identification of G Proteins

1. The continued identification of cDNAs corresponding to novel G proteins requires reassessment of the specificity/ selectivity of antisera. For example, antisera directed against the C-terminal region of $G_o\alpha$ do not crossreact with $G_i1\alpha$ or G_i2. They do, however, display a degree of crossreactivity with $G_i3\alpha$. Antibodies to the equivalent region of G_i3 also crossreact with G_o (Milligan, 1990a) (Fig. 5).
2. Each animal producing an antiserum, even against the same conjugate, must be analyzed separately. Crossreactivities and titers can vary markedly, even from bleed to bleed.
3. Addition of ADP-ribose to pertussis toxin-sensitive G proteins retards their mobility in SDS-PAGE (Goldsmith et al., 1987). Other covalent modifications, such as phosphorylation, might do likewise. Identification should thus not rely exclusively on the mobility of a protein in a gel.

7.4. Immunological Approaches to the Identification of Interactions Between Receptors and G Proteins

The functional consequence of pertussis toxin-catalyzed ADP-ribosylation of its G protein substrates is to prevent contact between receptors and the G protein. Since ADP-ribosylation

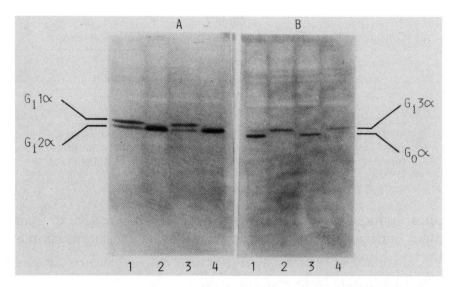

Fig. 5. Crossreactivity of antisera directed against the C-terminal decapeptides of $G_o\alpha$ and $G_i3\alpha$. Neither identify $G_i1\alpha$ or $G_i2\alpha$. Membranes of rat glioma C6 BU1 cells (2, 100 µg; 4, 50 µg), which express high levels of G_i2 and substantial quantities of G_i3, and membranes of rat cerebral cortex (1, 100 µg; 3, 50 µg), which have high levels of G_i1 and G_o and lower levels of G_i2 and G_i3, were resolved on SDS-PAGE using conditions able to resolve the various pertussis toxin-sensitive G proteins. Samples were immunoblotted with either antiserum SG1 (panel A) (directed against the C-terminal decapeptide of $G_i1\alpha$ and $G_i2\alpha$) or with antiserum I3B (directed against the C-terminal decapeptide of $G_i3\alpha$) (panel B). Antiserum I3B identified a 41-kDa polypeptide in C6 BU1 cells ($G_i3\alpha$), but in cerebral cortex, although low levels of G_i3 were noted (panel B, lane 1), the antiserum also identified a 39-kDa polypeptide ($G_o\alpha$), which was not detected in C6 BU1 membranes. The crossreactivity of this antiserum between G_i3 and G_o, but not with G_i1 or G_i2, strongly suggests that the C-terminal amino acid (phenylalanine for G_i1 and G_i2, tyrosine for G_i3 and G_o) plays a key role in epitope identification for antisera directed against the extreme C-terminal region of the G proteins. Data are taken from Milligan (1990a).

occurs on a cysteine residue, which is only four amino acids from the C-terminus of the α subunit of these G proteins, it was apparent that the C-terminal region of a G protein would likely represent a key domain for the interaction of the G protein with receptors.

Recognition that the molecular defect in the *unc* mutation of the murine S49 lymphoma cell line was a single base substitution in the $G_s\alpha$ gene such that a proline residue was substituted for an arginine in $G_s\alpha$ of the wild type, and that this substitution occurred six amino acids from the C-terminus (Sullivan et al., 1987) provided further evidence for the importance of the C-terminus in receptor–G protein interactions. It was thus possible that antisera directed against the C-terminal regions of the various pertussis toxin-sensitive G proteins would be able to interfere with such contacts and allow identification of which species of pertussis toxin-sensitive G protein interacted with particular receptors. Interaction of rhodopsin and rod transducin can be attenuated with the antirod transducin C-terminal antiserum AS7 (Cerione et al., 1988) or with synthetic peptides that form part of the primary sequence of the α subunit (Hamm et al., 1988).

Because multiple species of pertussis toxin-sensitive G proteins may be expressed by a single cell, it is important to define the cellular complement of G proteins prior to using such antisera as probes of receptor–G protein interactions. The neuroblastoma × glioma hybrid cell line, NG108-15, has been widely used as a model system to study signaling cascades in nervous tissue. This cell appears to express at least three distinct pertussis toxin-sensitive G proteins. These have been immunologically identified as the products of the G_i2, G_i3, and G_o genes (Milligan et al., 1990; McKenzie and Milligan, 1990). Both receptor-mediated inhibition of adenylyl cyclase and receptor-mediated inhibition of Ca^{2+} currents are attenuated by pretreatment of these cells with pertussis toxin, demonstrating that the relevant G protein(s) is (are) a substrate for toxin-catalyzed ADP-ribosylation.

The interaction of a receptor with a G protein can be conveniently assessed using each of three distinct assays. The first of these is dependent upon the observation that the affinity of binding of receptor and hormone or neurotransmitter is reduced when the experiment is performed in the presence of analogs of GTP. This is a reflection that the receptor–G protein complex displays higher affinity for the agonist than does the receptor in isolation. Such GTP shifts are not observed when an antagonist is used as the ligand (*see* Section 4.1.). As such, attenuation of receptor–G protein interaction can be assayed in ligand binding studies using either [^{3}H]agonist binding or agonist displacement of [^{3}H]antagonist binding.

When membranes of NG108-15 cells were incubated with an IgG fraction from either normal rabbit serum or antiserum AS7 prior to performing a saturation binding study with the enkephalin agonist [^{3}H]DADLE as a ligand for the δ opioid receptor on these cells, in both cases, Scatchard analysis of the specific binding demonstrated an apparent single class of receptor sites. In each case, the total number of sites was the same. However, the affinity of [^{3}H]DADLE binding was two- to threefold lower in samples preincubated with the IgG fraction from antiserum AS7 than with the IgG fraction from normal rabbit serum (McKenzie and Milligan, 1990; Milligan et al., 1990) (Fig. 6). The alteration in affinity for [^{3}H]DADLE was equivalent to that seen when comparing either the effects of the poorly hydrolyzed analog of GTP, GMP-PNP (100 μM), or pretreatment of the cells with pertussis toxin prior to making membranes. As such, the pertussis toxin-sensitive G protein linked to the opioid receptor must be identified by antiserum AS7. Antiserum AS7 cannot discriminate between G_i1 and G_i2, since these two G proteins have identical C-terminal decapeptides. However, as noted above, we have been unable to detect the expression of either the G_i1 polypeptide or mRNA corresponding to this protein in NG108-15 cells (McKenzie and Milligan, 1990). By contrast, G_i2 is expressed in substantial amounts. G_i2 was thus demonstrated to interact directly with the δ opioid receptor in these cells. Preincubation of NG108-15 membranes with IgG fractions from antisera directed against the C-terminal decapeptides of either G_o (antiserum OC1) or G_i3 (antiserum I3B) did not modulate the affinity of specific [^{3}H]DADLE binding to these membranes, suggesting that the opioid receptor interacted selectively with G_i2 rather than the other pertussis toxin-sensitive G proteins, which are expressed in these cells and, hence, should potentially be available to interact with the receptor (McKenzie and Milligan, 1990).

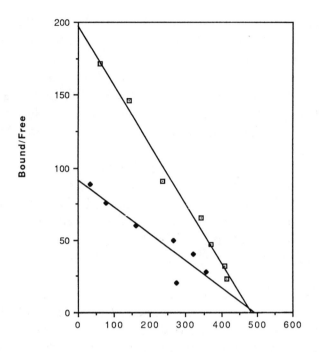

Bound (fmol/mg protein)

Fig. 6. Antibodies against the C-terminal region of $G_i2\alpha$ uncouple the δ opioid receptor of NG108-15 cells from the G protein signaling system. Membranes of NG108-15 cells were preincubated at 37°C for 60 min with an IgG fraction from either normal rabbit serum (□) or antiserum AS7 (which in these cells identifies only $G_i2\alpha$) (◆). The specific bindings of a range of concentrations of [^{3}H]DADLE were then assessed. Binding parameters in this experiment were B_{max}, 481 ± 12 fmol/mg protein; K_d, 2.4 nM for membranes incubated with normal rabbit serum; and B_{max}, 510 ± 23 fmol/mg protein; K_d, 5.6 nM for membranes incubated with antiserum AS7. Since the antiserum against $G_i2\alpha$ converts the opioid receptor from a form with high affinity for [^{3}H]DADLE to one with a lower affinity, this experiment defines that the opioid receptor interacts directly with G_i2. Exactly the same results were produced by incubation of the membranes with GMP-PNP. Data are adapted from McKenzie and Milligan (1990).

Although it was noted above that the NG108-15 cell is widely used as a model system to examine neuronal signaling cascades, it should be pointed out that the major form of "G_i" expressed in brain is G_i1, which is not expressed by these cells. It is well established that the δ opioid receptor of NG108-15 cells can function in a pertussis toxin-sensitive manner to inhibit adenylyl cyclase activity. Preincubation of membranes of these cells with IgG from antiserum AS7, but not with the IgG fraction from normal rabbit serum, prevented DADLE inhibition of adenylyl cyclase, thus defining G_i2 as the true inhibitory G protein of the adenylyl cyclase cascade ("G_i") (McKenzie and Milligan, 1990; Milligan et al., 1990a).

G_i2 appears to be expressed universally in mammalian tissues, whereas G_i1 is expressed in a more limited range of tissues (Milligan, 1990b) and appears to be particularly associated with neuronal tissues. This is not exclusively so, however, since, for example, it is present in high concentrations in white adipose tissue (Mitchell et al., 1989; Green et al., 1990). The function of G_i1 remains unknown.

The third convenient assay for interactions among receptors and G proteins is based on agonist stimulation of GTPase activity (Section 4.2.). In the resting state, the G protein α subunit has GDP in the nucleotide binding site. Occupation of a G protein-linked receptor by an agonist promotes the release of GDP from the associated G protein and its exchange by GTP. It is the GTP bound form of the G protein that is active in regulating the activity of effector systems, and a GTPase activity inherent to the α subunit of all G proteins functions to limit temporally the active state of the G protein. If this GTPase activity is reduced or inactivated, then the G protein will be constitutively active, as occurs for example to the stimulatory G protein of the adenylyl cyclase cascade (G_s) following ADP-ribosylation catalyzed by cholera toxin. Because it is GDP–GTP exchange that is the rate-limiting step in G protein activation, enhanced exchange, produced as a consequence of agonist activation of a receptor, will be reflected in a hormone or neurotransmitter-dependent increase in GTPase activity, which can be measured using $\gamma[^{32}P]GTP$ as substrate. Assays of this nature have been particularly suited to studies of receptor interactions with pertussis toxin-sensitive G proteins. In the case of G proteins that are not substrates for this toxin, it is frequently difficult, if not impossible, to measure receptor stimulation of high-affinity GTPase activity above the background "noise" of the system (*see* Section 4.2.).

When the IgG fractions of each of antisera AS7, OC1 I3B, and CS1 (which identifies the C-terminal decapeptide of forms of $G_s\alpha$) and normal rabbit serum were incubated with membranes of NG108-15 cells prior to measurement of basal and opioid peptide stimulation of high-affinity GTPase activity, it was only the IgG fraction from antiserum AS7 that was able to prevent DADLE stimulation of GTPase activity (McKenzie and Milligan, 1989,1990). Interestingly, the IgG fraction from antiserum AS7 did not have any effect on the basal GTPase, indicating that its only effect was to prevent interaction between receptor and G protein.

In a number of systems, pertussis toxin pretreatment produces a reduction in "basal" GTPase activity in membranes derived from the pretreated cells in comparison to untreated cells. Since pertussis toxin is believed only to prevent receptor interaction with a G protein and not to alter the intrinsic GTPase activity of that G protein, such observations require explanation. It has been suggested that this effect may be produced by activation of the G protein by unoccupied receptors (Costa et al., 1990). Limited direct evidence to support such a model has been presented, but it has been noted that opioid antagonists, such as ICI 174864, that display "negative intrinsic activity" (Costa and Herz, 1989) produce a consistent inhibitory effect on basal high-affinity GTPase activity, and this has been suggested to represent a reversal of the spontaneous association between opioid receptors and G proteins (Costa and Herz, 1989; Costa et al., 1990). It would be anticipated that an antibody directed against the C-terminal region of the relevant G protein would both attenuate agonist stimulation of GTPase activity and also reduce the apparent "basal" GTPase activity in such systems. Such studies remain to be performed.

7.5. Methodological Considerations

The specificity of an antibody on a western blot may not be identical with that in immunoprecipitation or in functional assays. It is important to use either affinity-purified antibodies or IgG fractions of antisera rather than the crude antisera for such studies to minimize potential artifacts. We have noted in some experiments with crude antisera that apparent effects of the antibodies could not be titrated out with dilutions of the antiserum that would be anticipated to make the antibody concentration limiting.

8. Physical Location of G Proteins Within the Cell: Implications for Function

Because G proteins function to allow communication between proteins that are either transmembrane or are associated with the internal face of the plasma membrane, it would be anticipated that they be located at the cytoplasmic face of the plasma membrane. With the exception of the two forms of transducin, G proteins generally require detergent treatment to release them from the membrane. How they remain in association with the plasma membrane is a matter of considerable debate, but considerable elucidation of this point has been achieved recently. None of the three subunits contain a sequence or sequences consistent with it representing a transmembrane spanning element. The α subunit of each of the pertussis toxin-sensitive G proteins has an N-terminal sequence (MetGlyXXXSer, where Met is the initiator amino acid and is removed posttranslationally) that is a consensus sequence for high-affinity amide-linked myristoylation (Grand, 1989). Indeed, a number of the pertussis toxin-sensitive G protein α subunits have been shown to contain covalently linked myristic acid (Buss et al., 1987). It should be noted, however, that G_s also requires detergent treatment to remove it from the membrane, and although it also has an N-terminal glycine residue, it does not have a serine at position 5. No evidence suggests that $G_s\alpha$ has N-terminal myristic acid, but the polypeptide may contain other fatty acyl groups, such as palmitic acid, which provide a similar anchorage function. Although the concept that N-terminal myristate acts as an anchor to the plasma membrane for the pertussis toxin-sensitive G proteins is attractive and is supported indirectly by the observation that the tryptic cleavage of the N-terminal 2 kDa peptides of these proteins converts them into soluble, nonmembrane associated forms (Eide et al., 1987), this question will only be answered conclusively by the expression of forms of the G proteins that have the N-terminal glycine converted to another amino acid, such as alanine, by site-directed mutagenesis and that can be demonstrated to be able to interact with βγ subunits. Such studies are in progress in a number of laboratories. Indeed, Jones et al. (1990) have demonstrated that myristoylation of the N-terminal glycine residue of $G_i1\alpha$ is essential for membrane anchorage of this polypeptide using exactly the strategy outlined above. Although the expressed alanine mutant of $G_i1\alpha$ appeared to be able to interact with βγ subunits on the basis that the addition of βγ subunits promoted pertussis toxin-catalyzed ADP-ribosylation of both the mutant and normal $G_i1\alpha$, the mutant was unable to incorporate [^{3}H]myristate and did not associate with the plasma membrane.

It is also possible that the βγ subunits, which are more hydrophobic than the α subunits, play a key role in anchoring the α subunit to the membrane. The interaction of α with the βγ complex has been demonstrated to be dependent on the integrity of the N-terminus of the α subunit (Neer and Clapham, 1988; Neer et al., 1988). Tryptic hydrolysis of the extreme N-terminal region of $G_o\alpha$ prevents interaction with βγ. However, the role of βγ in α subunit anchorage cannot be all important since guanine nucleotide-induced dissociation of the G protein αβγ holomer does not result in the rapid release of G protein α subunits from the membrane. It has been noted, however, that analogs of GTP, but not of GDP or ATP, will cause a slow, time-dependent release of the α, but not the βγ subunit of pertussis toxin-sensitive G proteins from membranes of both NG108-15 and glioma C6 cells. The mechanism of this GTP-dependent release process is not known, and it remains to be determined if the released α subunit still contains myristate. This guanine nucleotide-dependent release does not require proteolytic cleavage of the G proteins, since the released forms are the same size in SDS-PAGE as the membrane attached forms. Further, antisera directed against both the extreme N- and C-terminal regions of

$G_o\alpha$ identified both the membrane bound and released forms (McArdle et al., 1988). A particular role for the γ subunit in membrane anchorage appears to be likely in light of observations that the extreme C-terminal region of γ subunits consists of a CAAX (-Cys Ala Ala X) sequence. In a range of proteins, such a sequence is a signal for a complex process that involves removal of the three C-terminal amino acids and subsequent isoprenylation (farnesylation?) and carboxymethylation of the now C-terminal cysteine residue (*see* Glomset et al., 1990 for review). Subsequent palmitolylation of upstream cysteine residues are also frequently observed in such proteins.

In ultrastructural studies on the localization of $G_o\alpha$ in cultured fetal and adult murine neurons, a fraction of the immunological reactivity corresponding to $G_o\alpha$ was detected in the cytoplasmic matrix as well as on the cytoplasmic face of the plasma membrane (Gabrion et al., 1989). It is possible that localization of $G_o\alpha$ immunoreactivity to the cytoplasmic matrix represents endosome-associated G protein. The association of this polypeptide with endosomes has been suggested to be owing to persistent anchorage of the protein to the membrane during membrane recycling. It is also worth noting that $G_i\alpha$ and β subunit have been shown to be associated with endocytic vesicles in liver (Ali et al., 1989) and that in polarized cells, such as hepatocytes (Ali et al., 1989) and intestinal epithelia, immunoreactivity corresponding to various pertussis toxin-sensitive G proteins is unequally distributed in different domains of the plasma membrane.

9. G Proteins and Ion Channels

9.1. Basic Aspects

As noted above, in contrast to G_i2 and G_i3, which appear to be present in virtually all mammalian cells and tissues, immunoreactivity corresponding to G_i1 and G_o is limited in distribution. It is also particularly noticeable that these two pertussis toxin-sensitive polypeptides tend to be present in high levels in neuronal tissue. On such a basis, it was tempting to speculate that these G proteins might allow receptor-regulation of ion channel function.

In a considerable number of cases, it has been demonstrated that either receptor- or guanine nucleotide-regulation of ion channel activity is not dependent upon prior generation of an intracellular messenger and, hence, is unlikely to be the consequence of a phosphorylation event (*see* Dolphin, 1990 for review). One of the most studied systems to examine direct interactions of G proteins with an ion channel has involved preparation of cell-free patches from atrial myocytes. Here, the inwardly rectifying K^+ current can be activated by application of a poorly hydrolyzed analog of GTP to the cytoplasmic side of the patch. Similarly, addition of a muscarinic agonist to the external face of the membrane and GTP + Mg^{2+} to the cytoplasmic face produced equivalent results (Kurachi et al., 1986). The effect of agonist could be prevented by treatment with the active enzymatic subunit of pertussis toxin, defining a role for a pertussis toxin-sensitive G protein as the transducer.

Reconstitution of this effect to patches of pertussis toxin-pretreated myocytes was achieved by the addition of a pertussis toxin-sensitive G protein isolated from human erythrocytes, which was thus named G_k (Codina et al., 1987). It was shown subsequently that the sequence of G_k was entirely consistent with that predicted from a cDNA corresponding to $G_i3\alpha$ (Codina et al., 1988). The conclusion that $G_i3\alpha$ is able to regulate K^+ channels specifically must, however, be viewed carefully in light of subsequent experiments from the same group in which proteins expressed from cDNAs corresponding to each of $G_i1\alpha$, $G_i2\alpha$, and $G_i3\alpha$ were shown to be essentially equipotent in reconstitution of channel activity (Yatani et al., 1988). Such crosstalk among G proteins and effectors is inconsistent with the hypothesis that distinct G proteins have specific functions and interact specifically with receptors and effector systems. Potential difficulties with the interpretation of receptor–G protein and G protein–effector reconstitution studies have been discussed above (Section 6.2.) and, dependent upon the experimental protocol, may indicate potential interactions under essentially pharmacological, rather than physiological, conditions.

In both central and peripheral neuronal tissue, many agonists, including those that act at $GABA_B$, dopamine D_2, somatostatin, and both μ and δ opioid receptors, produce a hyper-polarization by increasing K^+ conductance (Nicoll, 1988). These effects are transduced by a pertussis toxin-sensitive G protein, although the identity of this G protein has not yet been defined. However, in inside-out patches of cultured hippocampal neurons, K^+ channel activity was induced by the addition of guanine nucleotide-activated G_o or activated, recombinant $G_o\alpha$ (Van Dongen et al., 1988). Multiple classes of K^+ channel were activated as assessed by variations in single-channel conductance and by the presence or absence of inward rectification. Unde-fined pertussis toxin-sensitive G proteins have also been impli-cated in hippocampal long-term potentiation (Goh and Pennefather, 1989) and in the generation of inhibitory postsyn-aptic potentials in hippocampal CA3 neurons after activation of mossy fibers (Thalmann, 1988), a conductance that is mediated by $GABA_B$ receptors. It is no great surprise that pertussis toxin-sensitive G proteins, which make up some 2% of total brain pro-tein, should be involved in the regulation of many ion channel activities in neuronal tissue; however, studies in this area now require the use of more elegant techniques and discriminatory probes to define the roles of the individual pertussis toxin-sensi-tive G proteins. Treatment with pertussis toxin and addition of analogs of GTP are unlikely to provide further insights.

One approach has been to make use of single identified neurons of invertebrates. Histamine, acetylcholine, and dopam-ine each regulate a K^+ current via activation of a pertussis toxin-sensitive G protein(s) in isolated neurons of *Aplysia* (Sasaki and Sato, 1987). Although this G protein(s) has yet to be identified, a combination of pertussis toxin-catalyzed ADP-ribosylation and immunological studies should allow the relevant G protein to be identified.

An approach of this nature was adopted by Harris-Warnick et al. (1988) using snail neurons. They demonstrated that dop-aminergic inhibition of voltage-sensitive Ca^{2+} channels (VSCC) is mediated by a pertussis toxin-sensitive G protein. Pertussis toxin-catalyzed ADP-ribosylation identified what appeared to be a single substrate for the toxin, which was identified by a polyclonal antiserum generated against $G_o\alpha$ purified from bo-vine brain. The snail G protein is not identical to mammalian brain $G_o\alpha$, however, since it migrates slightly more slowly in SDS-PAGE. However, when introduced into the neurons, the anti-$G_o\alpha$ antibodies were able to block the effect of dopamine on VSCC and the effects of pertussis toxin treatment were overcome by the introduction of purified mammalian $G_o\alpha$. How specific the reconstitution was for G_o, however, was not tested since parallel experiments with other purified pertussis toxin-sensitive G proteins were not performed. Control experiments of this type would have been particularly useful in relation to previous studies in which δ opioid receptor-regulation of neuronal VSCC was examined in neuroblastoma × glioma hybrid, NG108-15, cells. Opioid peptides produce an inhibition of Ca^{2+} currents in such cells, if the cells have been caused to "differentiate" by treatment with agents that elevate intracellular concentrations of cAMP. Opioid inhibition of VSCC was attenuated by pretreatment of the cells with pertussis toxin, but was reconstituted some tenfold more effectively by brain G_o than by G_i (Hescheler et al., 1987). It is not clear if the (weaker) effect of the G_i preparation (largely G_i1?) was because G_i1 could indeed function to couple receptors to VSCC or whether a degree of contamination of the G_i preparation was responsible for the effect. Although unknown at the time, G_o^* is likely to copurify with G_i1 in all conditions currently used for G protein purification, apart from high-resolution ion exchange chromatography (Section 2.).

Although such reconstitution experiments provide insights into potential coupling mechanisms, it would be more appropriate to be able to interfere with the transduction cascade in a selective manner or to delete components of the cascade. The first of these objectives has recently been achieved in experiments examining α_2-adrenergic inhibition of VSCC in prostaglandin E_1-differentiated NG108-15 cells (McFadzean et al., 1989). As noted in Section 7.4., these cells express at least three distinct pertussis toxin-sensitive G proteins, G_i2, G_i3, and G_o. Antibodies directed against the C-terminal decapeptides of each of these G proteins were introduced into the cells. To ensure that the cells could be loaded adequately with the antibodies, and that the antibodies would have free access to all parts of the cell, following injection of a number of cells, they were fixed and treated with fluorescein-coupled immunoglobulin raised against rabbit immunoglobulins and viewed under UV light. The entire area of the cell, including the neurite-like processes, displayed positive fluorescein staining. In cells injected with a $G_o\alpha$ C-terminal antibody (OC1), noradrenaline-inhibition of VSCC was almost obliterated. In contrast, in cells injected with either the $G_i2\alpha$ (AS7) or $G_i3\alpha$ (I3B) antisera, noradrenaline was as effective as in cells injected with antibodies from normal rabbit serum. These data provide strong evidence to support the notion that a (the) function of G_o in neuronal tissues is to regulate VSCC. The differentiation procedure that is required in these cells before receptor control of VSCC can be recorded increases substantially levels of $G_o\alpha$ in these cells (Mullaney and Milligan, 1989). As noted above, two distinct isoforms of $G_o\alpha$ can be identified in membranes of these cells using a series of anti$G_o\alpha$ antisera, but to date no antiserum has been useful in discriminating between them. Particularly noticeable increases in the more acidic isoform are associated with cAMP-induced differentiation (Mullaney and Milligan, 1990b), but it is yet to be ascertained if it is only this form that can interact with VSCC. Further evidence intimating a role for G_o in transducing the effects of neurotransmitters on inhibition of Ca^{2+} currents has been provided by Miller and coworkers (Ewald et al., 1988) in experiments in which exogenously added G proteins or isolated α subunits were able to reconstitute neurotransmitter sensitivity to pertussis toxin-treated cells (*see* Rosenthal et al., 1988 for review).

9.2. Methodological Considerations

It is difficult to assess whether G protein-mediated regulation of an ion channel is a direct effect or consequent on the generation of an intracellular messenger when using whole cell configurations. In many systems, the current may run down before suitable equilibration of the antibodies can be achieved within the cell.

10. Regulation of G Protein Expression

Differential regulation of expression of mRNA and protein corresponding to $G_s\alpha$ and $G_i\alpha$ in rat brain cortex has been noted following chronic treatment with corticosterone (Saito et al., 1989).

Elevated levels of $G_s\alpha$ have also been noted in a pituitary cell line following exposure to dexamethasone (Chang and Bourne, 1987). These results imply that glucocorticoid hormones may be able to regulate expression of G protein genes. The recent isolation of a genomic clone for $G_i2\alpha$ (Weinstein et al., 1988) has indicated the presence of consensus DNA binding domains for AP1 and AP2, factors that appear to mediate the transcriptional effects of phorbol esters and cAMP.

The control of regulation of expression of G protein genes, the understanding of which is currently in its infancy, and how this may be altered in pathological states are likely to be studied actively in the near future.

11. Future Trends

The availability of cDNA clones corresponding to the various pertussis toxin-sensitive G proteins is likely to lead to a rapid expansion in our understanding of the function of these proteins. These advances are likely to be produced by a number of distinct approaches.

11.1. Expression of Recombinant Protein

The availability of considerable quantities of genetically defined G proteins will assist in allocation of function to the various pertussis toxin substrates. Protein produced in this manner will eliminate one major problem that has hindered reconstitution experiments, which is that it has been difficult to negate the possibility of artifactual results produced by low-level contamination by related G proteins. This potential problem has produced considerable debate regarding whether atrial K^+ channels are regulated by the α subunit of G_k (G_i3?) or by $\beta\gamma$ subunits (Neer and Clapham, 1988; Birnbaumer et al., 1990). It remains impossible to purify homogenous preparations of the individual G proteins by conventional chromatographic techniques. It should be remembered, however, that one of the first reports on a reconstitution system using recombinant forms of $G_i1\alpha$, $G_i2\alpha$, and $G_i3\alpha$ indicated that all three proteins were able to function as G_k (Yatani et al., 1988). It is not clear if these results have physiological meaning, but atrial cells express a considerable number of distinct pertussis toxin-sensitive G proteins. Given the remarkable conservation of the primary sequence of each of these proteins in a range of mammalian species, it must be concluded that strong evolutionary pressure exists to maintain the exact sequence and structure. Such considerations suggest distinct and specific roles for each member of this family.

11.2. Site-Directed Mutagenesis

As noted above, information on the importance of myristoylation for membrane attachment is likely to be produced by alteration of the *N*-terminal glycine residue, which all of the pertussis toxin-sensitive G proteins contain (Jones et al., 1990). Alteration of Cys 186 in $p21^{N\text{-}ras}$, which was believed to represent the site for the attachment of palmitate prior to recent studies on isoprenylation of CAAX motifs (*see* Section 8.), in this polypeptide has previously indicated both that fatty acylation is crucial for interaction of the protein with the plasma membrane and that this polypeptide may play a crucial role in growth factor control of mitogenesis (Hancock et al., 1988). Alterations in the arginine residue, which is the site for cholera toxin-catalyzed ADP-ribosylation in $G_s\alpha$, produces constitutive activation of this protein. Each of the pertussis toxin-sensitive G proteins has an arginine residue in the equivalent position. Mutation of this amino acid might then be expected to produce consitutive activation of these proteins, since this amino acid plays a key role in the GTPase activity of G protein α subunits. Studies of this nature should assist in defining the effector functions of the individual G proteins.

11.3. The Construction of Genetic Chimeras

Considerable evidence indicates that the extreme C-terminal region of G protein α subunits is of prime importance in the recognition and interaction with receptors (Section 7.4.). Much less is currently known about how interactions with effector systems are controlled. A region of considerable diversity of sequence between the individual pertussis toxin-sensitive G proteins is between amino acids 95–125. As such, this area was considered a likely region to define the effector specificity of the G proteins.

However, the construction of a chimeric α_i/α_s cDNA consisting of the N-terminal 212 amino acids of murine $G_i2\alpha$ and the C-terminal 160 amino acids of murine $G_s\alpha$ and its expression in S49 cyc⁻ cells (which do not express $G_s\alpha$) indicated that effector recognition was located in the C-terminal region, since this chimera functioned to activate adenylyl cyclase (Masters et al., 1988). Similar constructs may help to define the effector site more clearly.

11.4. Isolation of G Protein Mutants

The identification of genes encoding G protein α subunits from *Drosophila*, some of which are preferentially expressed in the nervous system (*see*, for example, Quan et al., 1989), suggests that genetic manipulation can be used in this system to investigate the role of G proteins in complex multicellular processes. This is of particular interest since known mutations in *Drosophila*, which produce alterations in the levels of cAMP (the rutabaga gene controls regulation of adenylyl cyclase, whereas the product of the dunce gene corresponds to a cAMP phosphodiesterase) cause both developmental and learning defects. It is likely that alterations in either the G protein genes or the level of expression of these genes will be associated with known mutants.

11.5. Use of G Protein Antisera in Electrophysiological Experimentation

Antisera directed against both the receptor and effector interaction sites of the various pertussis toxin-sensitive G proteins are likely to be used on a more widespread basis in electrophysiological experimentation. Probes of this nature will be useful both in whole cell configurations and in single-channel recordings, and should help to define the specificity (or otherwise) of direct G protein interactions with ion channels.

Acknowledgments

Experiments in GM's laboratory are supported by the Medical Research Council and the Agricultural and Food Research Council.

Abbreviations

cAMP	adenosine 3',5' cyclic monophosphate
DADLE	[D-Ala^2-D-Leu^5] enkephalin
FMLP	N-formyl-methionyl-leucyl-phenylalanine
GDP	guanosine 5'-diphosphate
GMP-PNP	guanosine 5'-[βγ-imido]triphosphate
GTP	guanosine 5'-triphosphate
GTPγS	guanosine 5'-[γ-thio]triphosphate
VSCC	voltage-sensitive Ca^{2+} channel

References

Ali N., Milligan G., and Evans W. H. (1989) Distribution of G-proteins in rat liver plasma-membrane domains and endocytic pathways. *Biochem. J.* **261,** 905–912.

Asano T., Katada T., Gilman A. G., and Ross E. M. (1984) Activation of the inhibitory GTP-binding protein of adenylyl cyclase, G$_i$, by β-adrenergic receptors in reconstituted phospholipid vesicles. *J. Biol. Chem.* **259,** 9351–9354.

Banga H. S., Walker R. K., Winberry L. K., and Rittenhouse S. E. (1987) Pertussis toxin can activate human platelets. Comparative effects of holotoxin and its ADP-ribosylating S1 subunit. *J. Biol. Chem.* **262,** 14871–14874.

Bigay J., Deterre P., Pfister C., and Chabre M. (1985) Fluoroaluminates activate transducin-GDP by mimicking the δ-phosphate of GTP at its binding site. *FEBS Lett.* **191,** 181–185.

Birdsall N. J. M., Hulme E. C., and Burgen A. (1980) The character of the muscarinic receptors in different regions of the rat brain. *Proc. R. Soc. Lond. [Biol.]* **207,** 1–12.

Birnbaumer L., Abramowitz J., and Brown A. M. (1990) Receptor–effector coupling by G proteins. *Biochim. Biophys. Acta* **1031,** 163–226.

Brann M. R., Collins R. M., and Spiegel A. (1987) Localization of mRNAs encoding the α subunits of signal-transducing G-proteins within rat brain and among peripheral tissues. *FEBS Lett.* **222,** 191–198.

Buss J. E., Mumby S. M., Casey P. J., Gilman A. G., and Sefton B. M. (1987) Myristoylated α subunits of guanine nucleotide binding regulatory proteins. *Proc. Natl. Acad. Sci. USA* **84,** 7493–7497.

Cassel D. and Selinger Z. (1976) Catecholamine-stimulated GTPase activity in turkey erthyrocyte membranes. *Biochim. Biophys. Acta* **452**, 538–551.

Cerione R. A., Kroll S., Rajaram R., Unson C., Goldsmith P., and Spiegel A. M. (1988) An antibody directed against the carboxyl terminal decapeptide of the α subunit of the retinal GTP-binding protein, transducin. Effects on transducin function. *J. Biol. Chem.* **263**, 9345–9352.

Chang, F.-H. and Bourne H. R. (1987) Dexamethasone increases adenylyl cyclase activity and expression of the α subunit of Gs in GH3 cells. *Endocrinology* **121**, 1711–1715.

Chang, F.-H. and Bourne H. R. (1989) Cholera toxin induces cAMP-independent degradation of G_s. *J. Biol. Chem.* **264**, 5352–5357.

Codina J., Olate J., Abramowitz J., Mattera R., Cook R. G., and Birnbaumer L. (1988) $α_i3$ cDNA encodes the α subunit of G_k, the stimulatory G-protein of receptor-regulated K^+ channels. *J. Biol. Chem.* **263**, 6746–6750.

Codina J., Yatani A., Grenet D., Brown A. M., and Birnbaumer L. (1987) The α subunit of the GTP binding protein G_k opens atrial potassium channels. *Science* **236**, 442–445.

Costa T. and Herz A. (1989) Antagonists with negative intrinsic activity at δ opioid receptors coupled to GTP-binding proteins. *Proc. Natl. Acad. Sci. USA* **86**, 7321–7325.

Costa T., Lang J., Gless C., and Herz A. (1990) Spontaneous association between opioid receptors and GTP-binding regulatory proteins in native membranes: Specific regulation by antagonists and sodium ions. *Mol. Pharmacol.* **37**, 383–394.

Dolphin A. C. (1990) G-proteins and the regulation of ion channels, in *G-Proteins as Mediators of Cellular Signalling Processes* (Houslay M. D. and Milligan G., eds.), Wiley and Sons, Chichester, pp. 125–150.

Eide B., Gierschik P., Milligan G., Mullaney I., Unson C., Goldsmith P., and Spiegel A. (1987) GTP-binding proteins in brain and neutrophil are tethered to the plasma membrane via their amino termini. *Biochem. Biophys. Res. Commun.* **148**, 1398–1405.

Ewald D. A., Sternweis P. C., and Miller R. J. (1988) Guanine nucleotide-binding protein G_o-induced coupling of neuropeptide Y receptors to Ca^{2+} channels in sensory neurons. *Proc. Natl. Acad. Sci. USA* **85**, 3633–3637.

Florio V. A. and Sternweis P. C. (1985) Reconstitution of resolved muscarinic cholinergic receptors with purified GTP-binding proteins. *J. Biol. Chem.* **260**, 3477–3483.

Freissmuth M. and Gilman A. G. (1989) Mutations of $G_sα$ designed to alter the reactivity of the protein with bacterial toxins. Substitutions at Arg[187] result in loss of GTPase activity. *J. Biol. Chem.* **264**, 21907–21914.

Gabrion J., Brabet Ph., Dao B. N. T., Homberger V., Dumuis A., Sebben M., Rouot B., and Bockaert J. (1989) Ultrastructural localization of the GTP-binding protein G_o in neurons. *Cell. Signalling* **1**, 107–123.

Gao B., Gilman A. G., and Robishaw J. D. (1987) A second form of the β-subunit of signal transducing G-proteins. *Proc. Natl. Acad. Sci. USA* **84**, 6122–6125.

Gautam N., Baetscher M., Aebersold R., and Simon M. I. (1989) A G-protein gamma subunit shares homology with *ras* proteins. *Science* **244**, 971–974.

Gawler D., Milligan G., Spiegel A. M., Unson C. G., and Houslay M. D. (1987) Abolition of the expression of inhibitory guanine nucleotide regulatory protein G_i activity in diabetes. *Nature* **327**, 229–232.

Gierschik P. and Jakobs, K.-H. (1987) Receptor mediated ADP-ribosylation of a phospholipase C-stimulating G-protein. *FEBS Lett.* **224**, 219–223.

Gierschik P., Milligan G., Pines M., Goldsmith P., Codina J., Klee W., and Spiegel A. (1986) Use of specific antibodies to quantitate the guanine nucleotide binding protein G_o in brain. *Proc. Natl. Acad. Sci. USA* **83**, 2258–2262.

Gierschik P., Sidiropoulos, D., and Jakobs, K.-H. (1989) Two distinct G_i-proteins mediate formyl peptide receptor signal transduction in human leukemia (HL-60) cells. *J. Biol. Chem.* **264**, 21470–21473.

Gilman A. G. (1987) G-proteins: Transducers of receptor generated signals. *Annu. Rev. Biochem.* **56**, 615–649.

Glomset J. A., Gelb M. H., and Farnsworth C. C. (1990) Prenyl proteins in eukaryotic cells: A new type of membrane anchor. *T.I.B.S.* **15**, 139–142.

Goh J. W. and Pennefather P. S. (1989) A pertussis toxin-sensitive G-protein in hippocampal long-term potentiation. *Science* **244**, 980–983.

Goldsmith P., Backlund P. S. Jr., Rossiter K., Carter A., Milligan G., Unson C. G., and Spiegel A. (1988a) Purification of heterotrimeric GTP-binding proteins from brain: Identification of a novel form of G_o. *Biochemistry*, **27**, 7085–7090.

Goldsmith P., Gierschik P., Milligan G., Unson C. G., Vinitsky R., Malech H. L., and Spiegel A. M. (1987) Antibodies directed against synthetic peptides distinguish between GTP-binding proteins in neutrophil and brain. *J. Biol. Chem.* **262**, 14683–14688.

Goldsmith P., Rossiter K., Carter A., Simonds W., Unson C. G., Vinitsky R., and Spiegel A. M. (1988b) Identification of the GTP-binding protein encoded by G_{i3} complementary DNA. *J. Biol. Chem.* **263**, 6476–6479.

Grand R. J. A. (1989) Acylation of viral and eukaryotic proteins. *Biochem. J.* **258**, 625–638.

Graziano M. P. and Gilman A. G. (1989) Synthesis in *Escherichia coli* of GTPase-deficient mutants of $G_s\alpha$. *J. Biol. Chem.* **264**, 15475–15482.

Green A., Johnson J. L., and Milligan G. (1990) Down-regulation of G_i subtypes by prolonged incubation of adipocytes with an A1 adenosine receptor agonist. *J. Biol. Chem.* **265**, 5206–5210.

Hamm H. E., Deretic D., Arendt A., Hargrave P. A., Koenig B., and Hofmann K. P. (1988) Site of G-protein binding to rhodopsin mapped with synthetic peptides to the α subunit. *Science* **241**, 832–834.

Hancock J. F., Marshall C. J., McKay I. A., Gardner S., Houslay M. D., Hall A., and Wakelam M. J. O. (1988) Mutant but not normal p21 *ras* elevates inositol phospholipid breakdown in two different cell systems. *Oncogene* **3**, 187–193.

Harris B. A., Robishaw J. D., Mumby S. M., and Gilman A. G. (1985) *Science* **229**, 1274–1277.

Harris-Warrick R., Hammond C., Paupardin-Tritsch D., Homberger V., Rouot B., Bockaert J., and Gerschenfeld H. M. (1988) An α40 subunit of a GTP-binding protein immunologically related to G_o mediates a dopamine-induced decrease of Ca^{2+} current in snail neurons. *Neuron* **1**, 27–32.

Helper J. R. and Harden T. K. (1986) Guanine nucleotide-dependent pertussis toxin insensitive stimulation of inositol phosphate formation by carbachol in a membrane preparation from human astrocytoma cells. *Biochem. J.* **239**, 141–146.

Hescheler J., Rosenthal W., Trautwein W., and Schultz G. (1987) The GTP-binding protein, G_o regulates neuronal calcium channels. *Nature* **325**, 445–447.

Houslay M. D., Bojanic D., Gawler D., O'Hagan S., and Wilson A. (1986) Thrombin, unlike vasopressin, appears to stimulate two distinct guanine nucleotide regulatory proteins in human platelets. *Biochem. J.* **238**, 109–113.

Hsia J. A., Moss J., Hewlitt E. L., and Vaughan M. (1984) ADP-ribosylation of adenylyl cyclase by pertussis toxin. Effects on inhibitory agonist binding. *J. Biol. Chem.* **259**, 1086–1090.

Huff R. M. and Neer E. J. (1986) Subunit interactions of native and ADP-ribosylated α39 and α41, two guanine nucleotide-binding proteins from bovine cerebral cortex. *J. Biol. Chem.* **261**, 1105–1110.

Iiri T., Tohkin M., Morishima N., Ohoka Y., Ui M., and Katada T. (1989) Chemotactic peptide receptor-supported ADP-ribosylation of a pertussis toxin substrate GTP-binding protein by cholera toxin in neutrophil-type HL-60 cells. *J. Biol. Chem.* **264**, 21394–21400.

Jones D. T. and Reed R. R. (1987) Molecular cloning of five GTP-binding protein cDNA species from rat olfactory neuroepithelium. *J. Biol. Chem.* **262**, 14241–14249.

Jones T. L. Z., Simonds W. F., Merendino J. J., Brann M. R., and Spiegel A. M. (1990) Myristoylation of an inhibitory GTP-binding protein α subunit is essential for its membrane attachment. *Proc. Natl. Acad. Sci. USA* **87**, 568–572.

Kaziro Y. (1990) Molecular biology of G-protein, in *G-Proteins as Mediators of Cellular Signalling Processes* (Houslay M. D. and Milligan G., eds.) Wiley and Sons, Chichester, pp. 47–66.

Kobayashi I., Shibasaki H., Takahashi K., Kikkawa S., Ui M., and Katada T. (1989) Purification of GTP-binding proteins from bovine brain membranes. Identification of heterogeneity of the α subunit of G_o proteins. *FEBS Lett.* **257**, 177–180.

Koski G. and Klee W. A. (1981) Opiates inhibit adenylyl cyclase by stimulating GTP hydrolysis. *Proc. Natl. Acad. Sci. USA* **78**, 4185–4189.

Kurachi Y., Nakajima T., and Sugimoto T. (1986) Role of intracellular Mg^{2+} in the activation of muscarinic K^+ channels in cardiac atrial cell membrane. *Pflugers Arch.* **407,** 572–574.

Kurose H., Katada T., Amano, T., and Ui M. (1983) Specific uncoupling by islet activating protein, pertussis toxin, of negative signal transduction via α-adrenergic, cholinergic and opiate receptors in neuroblastoma × glioma hybrid cells. *J. Biol. Chem.* **258,** 4870–4875.

Landis C. A., Masters S. B., Spada A., Pace A. M., Bourne H. R., and Vallar L. (1989) GTPase inhibiting mutations activate the α chain of G_s and stimulate adenylyl cyclase in human pituitary tumours. *Nature* **340,** 692–696.

Lochrie M. A. and Simon M. I. (1988) G-protein multiplicity in eukaryotic signal transduction systems. *Biochemistry* **27,** 4958–4965.

Macleod K. G. and Milligan G. (1990) Bipahasic regulation of adenylyl cyclase by cholera toxin in neuroblastoma × glioma hybrid cells is due to the activation and subsequent loss of the α subunit of the stimulatory GTP binding protein (G_s) *Cell. Signalling* **2,** 139–151.

Maguire M. E., Van Arsdale P. M., and Gilman A. G. (1976) An agonist specific effect of guanine nucleotides on binding to the β adrenergic receptor. *Mol. Pharmacol.* **12,** 335–339.

Martin T. F. J., Lucas D. O., Bajjalieh S. M., and Kowalchyk J. A. (1986) Thyrotropin-releasing hormone activates a Ca^{2+}-dependent polyphosphoinositide phosphodiesterase in permeable GH3 cells. GTPγS potentiation by a cholera and pertussis toxin-insensitive mechanism. *J. Biol. Chem.* **261,** 2918–2927.

Masters S. B., Miller R. T., Chi, M.-H., Chang, F.-H., Beiderman B., Lopez N. G., and Bourne H. R. (1989) Mutations in the GTP-binding site of $G_s\alpha$ alter stimulation of adenylyl cyclase. *J. Biol. Chem.* **264,** 15467–15474.

Masters S. B., Stroud R. M., and Bourne H. R. (1986) Family of G-protein α chains: Amphipathic analysis and predicted structure of functional domains. *Protein Eng.* **1,** 47–54.

Masters S. B., Sullivan K. A., Miller R. T., Beiderman B., Lopez N. G., Ramachandran J., and Bourne H. R. (1988) Carboxyl terminal domain of $G_s\alpha$ specifies coupling of receptors to stimulation of adenylyl cyclase. *Science* **241,** 448–451.

McArdle H., Mullaney I., Magee A., Unson C., and Milligan G. (1988) GTP analogues cause release of the alpha subunit of the GTP binding protein, G_o, from the plasma membrane of NG108-15 cells. *Biochem. Biophys. Res. Commun.* **152,** 243–251.

McFadzean I., Mullaney I., Brown D. A., and Milligan G. (1989) Antibodies to the GTP binding protein, G_o, antagonize noradrenaline-induced calcium current inhibition in NG108-15 hybrid cells. *Neuron* **3,** 177–182.

McKenzie F. R., Kelly E. C. H., Unson C. G., Spiegel A. M., and Milligan G. (1988) Antibodies which recognise the C-terminus of the inhibitory guanine nucleotide binding protein (G_i) demonstrate that opioid peptides and foetal calf serum stimulate the GTPase activity of two separate pertussis toxin substrates. *Biochem. J.* **249,** 653–659.

McKenzie F. R. and Milligan G. (1989) The use of specific antisera to locate functional domains of guanine nucleotide binding proteins, in *Receptors, Membrane Transport and Signal Transduction. NATO ASI Series H: Cell Biology,* vol. 29 (Evangelopoulos A. E., Changeux J. P., Packer L., Sotiroudis T. G., and Wirtz K. W. A., eds.), Springer-Verlag, Berlin, pp. 65–74.

McKenzie F. R. and Milligan G. (1990) δ opioid-receptor mediated inhibition of adenylyl cyclase is transduced specifically by the guanine nucleotide binding protein G$_i$2. *Biochem. J.* **267,** 391–398.

Milligan G. (1988) Techniques used in the identification and analysis of function of pertussis toxin-sensitive guanine nucleotide binding proteins. *Biochem. J.* **255,** 1–13.

Milligan G. (1989) Foetal calf serum enhances cholera toxin-catalysed ADP-ribosylation of the pertussis toxin-sensitive guanine nucleotide binding protein, G$_i$2, in rat glioma C6BU1 cells. *Cell. Signalling* **1,** 65–74.

Milligan G. (1990a) Immunological probes and the identification of guanine nucleotide binding proteins, in *G-Proteins as Mediators of Cellular Signalling Processes* (Houslay M. D. and Milligan G., eds.), Wiley and Sons, Chichester, pp. 31–46.

Milligan G. (1990b) Tissue distribution and subcellular location of guanine nucleotide binding proteins: Implications for cellular signalling. *Cell. Signalling* **1,** 411–419.

Milligan G., Gierschik P., Spiegel A. M., and Klee W. A. (1986) The GTP binding regulatory proteins of neuroblastoma × glioma, NG108-15, and glioma, C6, cells. Immunochemical evidence of a pertussis toxin substrate that is neither Ni nor No. *FEBS Lett.* **195,** 225–230.

Milligan G., Gierschik P., Unson C. G., and Spiegel A. M. (1987b) The use of specific antisera to study the developmental regulation of guanine nucleotide binding proteins. *Protides Biol. Fluids* **35,** 415–418.

Milligan G. and Klee W. A. (1985) The inhibitory guanine nucleotide binding protein (Ni) purified from bovine brain is a high affinity GTPase. *J. Biol. Chem.* **260,** 2057–2063.

Milligan G. and McKenzie F. R. (1988) Opioid peptides promote cholera toxin-catalysed ADP-ribosylation of the inhibitory guanine nucleotide binding protein (G$_i$) in membranes of neuroblastoma × glioma hybrid cells. *Biochem. J.* **252,** 369–373.

Milligan G., Mitchell F. M., Mullaney I., McClue S. J., and McKenzie, F. R. (1990) The role and specificity of guanine nucleotide binding proteins in receptor–effector coupling, in *Hormone Perception and Signal Transduction in Animals and Plants* (Roberts J., Venis M., and Kirk C., eds.), Society of Biologists, London (157–172).

Milligan G., Streaty R. A., Gierschik P., Spiegel A. M., and Klee W. A. (1987a) Development of opiate receptors and GTP-binding regulatory proteins in neonatal rat brain. *J. Biol. Chem.* **262,** 8626–8630.

Milligan G., Unson C. G. and Wakelam M. J. O. (1989) Cholera toxin treatment produces down-regulation of the α subunit of the stimulatory guanine nucleotide binding protein (G_s) *Biochem. J.* **262**, 643–649.

Mitchell F. M., Griffiths S. L., Saggerson E. D., Houslay M. D., Knowler J. T., and Milligan G. (1989) Guanine nucleotide binding proteins expressed in rat white adipose tissue. Identification of both mRNAs and proteins corresponding to G_i1, G_i2 and G_i3. *Biochem. J.* **262**, 403–408.

Mullaney I. and Milligan G. (1989) Elevated levels of the guanine nucleotide binding protein, G_o, are associated with differentiation of neuroblastoma × glioma hybrid cells. *FEBS Lett.* **244**,113–118.

Mullaney I. and Milligan G. (1990a) Identification and analysis of two distinct isoforms of the guanine nucleotide-binding protein G_o in NG108-15 cells. *Biochem. Soc. Trans.* **18**, 396–399.

Mullaney I. and Milligan G. (1990b) Identification of two distinct isoforms of the guanine nucleotide binding protein, G_o, in neuroblastoma × glioma hybrid cells. Independent regulation during cyclic AMP-induced differentiation. *J. Neurochem.* **55**, 1890–1898.

Murphy P. M., Eide B., Goldsmith P., Brann M., Gierschik P., Spiegel A., and Malech H. L. (1987) Detection of multiple forms of $G_i\alpha$ in HL 60 cells. *FEBS Lett.* **221**, 81–86.

Nakamura T. and Ui M. (1985) Simultaneous inhibitons of inositol phospholipid breakdown, arachidonic acid release and histamine secretion in mast cells by islet activating protein, pertussis toxin. A possible involvement of the toxin-specific substrate in the Ca^{2+} mobilizing receptor-mediated biosignaling system. *J. Biol. Chem.* **260**, 3584–3593.

Neer E. J. and Clapham D. E. (1988) Roles of G protein subunits in transmembrane signalling. *Nature* **333**, 129–134.

Neer E. J., Lok J. M., and Wolf L. G. (1984) Purification and properties of the inhibitory guanine nucleotide regulatory unit of brain adenylate cyclase. *J. Biol. Chem.* **259**, 14222–14229.

Neer E. J., Pulsifer, L., and Wolf L. G. (1988) The amino terminus of G protein α subunits is required for interaction with β/γ. *J. Biol. Chem.* **263**, 8996–9000.

Nicoll R. A. (1988) The coupling of neurotransmitter receptors to ion channels in the brain. *Science* **241**, 545–551.

Offermans S., Schafer R., Hoffman B., Bombien E., Spicher K., Hinsch, K.-D., Schultz G., and Rosenthal W. (1990) Agonist-sensitive binding of a photoreactive GTP analog to a G-protein α subunit in membranes of HL-60 cells. *FEBS Lett.* **260**, 14–18.

Ohta H., Okajima F., and Ui M. (1985) Inhibition by islet-activating protein of a chemotactic peptide-induced early breakdown of inositol phospholipids and Ca^{2+} mobilization in guinea pig neutrophils. *J. Biol. Chem.* **260**, 15771–15780.

Pines M., Gierschik P., Milligan G., Klee, W., and Spiegel A. (1985) Antibodies against the carboxy-terminal 5-kDa peptide of the a subunit of transducin crossreact with the 40-kDa but not the 39-kDa guanine nucleotide binding protein from brain. *Proc. Natl. Acad. Sci. USA* **82,** 4095–4099.

Quan F., Wolfgang W. J., and Forte M. A. (1989) The *Drosophila* gene coding for the α subunit of a stimulatory G-protein is preferentially expressed in the nervous system. *Proc. Natl. Acad. Sci. USA* **86,** 4321–4325.

Rodbell M. (1980) The role of hormone receptors and GTP-regulatory proteins in membrane transduction. *Nature* **284,** 17–22.

Rodbell M., Krans H. M. J., Pohl S., and Birnbaumer L. (1971) The glucacon-sensitive adenyl cyclase system in plasma membranes of rat liver. IV. Effects of guanyl nucleotides on binding of ^{125}I-glucagon. *J. Biol. Chem.* **246,** 1872–1876.

Rosenthal W., Hescheler J., Trautwein, W., and Schultz G. (1988) Control of voltage-dependent Ca^{2+} channels by G-protein-coupled receptors. *FASEB J.* **2,** 2784–2790.

Saito N., Guitart X., Hayward M., Tallman J. F., Duman R. S., and Nestler E. J. (1989) Corticosterone differentially regulates the expression of $G_s\alpha$ and $G_i\alpha$ messenger RNA and protein in rat cerebral cortex. *Proc. Natl. Acad. Sci. USA* **86,** 3906–3910.

Sasaki K. and Sato M. (1987) A single GTP-binding protein regulates K^+-channels coupled with dopamine, histamine and acetylcholine receptors. *Nature* **325,** 259–262.

Senogles S. E., Spiegel A. M., Padrell E., Iyengar R., and Caron M. G. (1990) Specificity of receptor–G-protein interactions. Discrimination of G_i subtypes by the D_2 dopamine receptor in a reconstituted system. *J. Biol. Chem.* **265,** 4507–4514.

Spiegel A. M. (1990) Structure and identification of G-proteins: Isolation and purification, in *G-Proteins as Mediators of Cellular Signalling Processes* (Houslay M. D. and Milligan G., eds.), Wiley and Sons, Chichester, pp. 15–30.

Sternweis P. C. and Robishaw J. D. (1984) Isolation of two proteins with high affinity for guanine nucleotides from membranes of bovine brain. *J. Biol. Chem.* **259,** 13806–13813.

Strassheim D., Milligan G., and Houslay M. D. (1990) Diabetes abolishes the GTP-dependent, but not the receptor-dependent inhibitory function of the inhibitory guanine-nucleotide-binding regulatory protein (G_i) on adipocyte adenylyl cyclase function. *Biochem. J.* **266,** 521–526.

Sullivan K. A., Miller R. T., Masters S. B., Beiderman B., Heideman W., and Bourne H. R. (1987) Identification of receptor contact site involved in receptor–G-protein coupling. *Nature* **330,** 758,759.

Thalmann R. H. (1988) Evidence that guanosine triphosphate (GTP) bind-
    ing proteins contol a synaptic response in brain. Effect of pertussis
    toxin and GTPγS on the late inhibitory postsynaptic potential of hip-
    pocampal CA3 neurons. *J. Neurosci.* **8,** 4589–4602.
Ueda H., Yoshihara Y., Misawa H., Fukushima N., Katada T., Ui M., Takagi
    H., and Satoh M. (1989) The kyotorphin (tyrosine-arginine) receptor
    and a selective reconstitution with purified G_i, measured with GTPase
    and phospholipase C assays. *J. Biol. Chem.* **264,** 3732–3741.
VanDongen A. J. M., Codina J., Olate J., Mattera R., Joho R., Birnbaumer L.,
    and Brown A. M. (1988) Newly identified brain potassium channels
    gated by the guanine nucleotide binding protein G_o. *Science* **242,**
    1433–1436.
Wang N., Yan K., and Rasenick M. M. (1990) Tubulin binds specifically to
    the signal-transducing proteins, $G_s\alpha$ and $G_i\alpha1$. *J. Biol. Chem.* **265,**
    1239–1242.
Weinstein L. S., Spiegel A. M., and Carter A. D. (1988) Cloning and charac-
    terization of the human gene for the α subunit of G_i2, a GTP-binding
    signal transduction protein. *FEBS Lett.* **232,** 333–340.
Worley P. F., Baraban J. M., Van Dop C., Neer E. J., and Snyder S. H. (1986)
    G_o, a guanine nucleotide-binding protein: Immunohistochemical
    localization in rat brain resembles distribution of second messnger sys-
    tems. *Proc. Natl. Acad. Sci. USA* **83,** 4561–4565.
Yatani A., Mattera R., Codina J., Graf R., Okabe K., Padrell E., Iyengar R.,
    Brown A. M., and Birnbaumer L. (1988) The G-protein-gated atrial
    K^+ channel is stimulated by three distinct $G_i\alpha$ subunits. *Nature* **336,**
    680–682.

Methods for the Analysis of Phosphoinositides and Inositol Phosphates

Philip P. Godfrey

1. Introduction

There is now compelling evidence that stimulation of inositol phospholipid metabolism via phosphoinositidase C (PIC) is the signal transduction pathway for a wide variety of receptors in the mammalian nervous system. A by no means comprehensive list is shown in Table 1. Receptors are thought to be coupled to PIC through a guanine nucleotide-binding protein (G protein—*see* Chapter 1 by Milligan) termed G_p, by a mechanism analogous to coupling of receptors with adenylyl cyclase, though as yet the identity of this G protein has not been established.[*] (For general reviews, *see* Hawthorne and Pickard, 1979; Downes, 1986; Fisher and Agranoff, 1987; Watson and Godfrey, 1988; Berridge and Irvine, 1989; Nahorski and Potter, 1989). The initial receptor-coupled event is a PIC-mediated breakdown of phosphatidylinositol 4,5-bisphosphate (PtdIns4,5P_2) to give the two intracellular messengers inositol 1,4,5-trisphosphate (Ins(1,4,5)P)$_3$, which mobilizes intracellular Ca^{2+} (*see* Chapter 3 by Taylor et al.), and 1,2-diacylglycerol (DG), which activates protein kinase C (*see* Chapter 6 by Burgess). Ins(1,4,5)P_3 is then either phosphorylated to inositol 1,3,4,5-tetrakisphosphate (Ins(1,3,4,5)P_4) or dephosphorylated to Ins(1,4)P_2, the reactions catalyzed by a 3-kinase or 5-phosphatase, respectively. Ins(1,3,4,5)P_4 is subsequently dephosphorylated to

[*]G proteins of the G_q family are now thought to couple receptors to activation of PIC.

From: *Neuromethods, Vol. 20: Intracellular Messengers*
Eds: A. Boulton, G. Baker, and C. Taylor © 1992 The Humana Press Inc.

Table 1
Receptors Known To Be Coupled to PIC in the CNS or Related Cell Lines

Receptor	Subtype(s)	Tissue
Muscarinic	m1, m3, m5	Brain slices
		Neuronal cell lines
		Neuronal and glial primary cultures
		Transfected cell lines
Adrenergic	α_1	Brain slices
		Synaptoneurosomes
		Primary cultures
		Transfected cell lines
Histaminergic	H_1	Brain slices
		Neuronal cell lines
Serotonergic	5-HT$_{1C}$, 5-HT$_2$	Brain slices
		Choroid plexus
		Neuronal cell lines
		Transfected cell lines
Glutamatergic	Quis/metabotropic	Brain slices
		Cerebellar granule cells
		Primary neuronal culture
Tachykinin	NK1, NK2, NK3	Brain slices
		Astrocytoma cell lines
Neurotensin		Brain slices
		Neuronal cell lines
Bradykinin		Neuronal cell lines
CCK		Brain slices
Vasopressin		Brain slices
		SCG[a]
Depolarization		Brain slice
		SCG[a]

[a]SCG—superior cervical ganglion.

Ins(1,3,4)P_3 by the same 5-phosphatase that degrades Ins(1,4,5)P_3, and is then further metabolized to a variety of inositol bis- and monophosphates (*see* Shears, 1989 for a detailed review of inositol phosphate metabolism). A general scheme for the metabolic pathways of the inositides is shown in Fig. 1. One feature of particular note in the catabolic pathways for Ins(1,4,5)P_3 is that two key enzymes, inositol polyphosphate 1-phosphatase and inositol monophosphatase, are specifically inhibited by Li$^+$ in the mil-

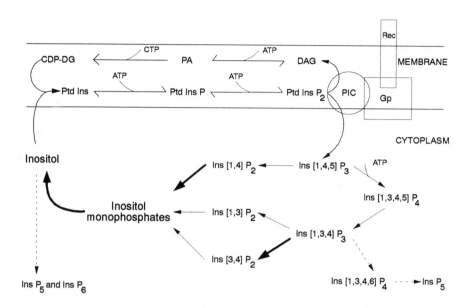

Fig. 1. The metabolic pathways of the inositol phospholipids and inositol phosphates. This is a relatively current understanding of the pathways of metabolism of the inositol phosphates and inositol phospholipids. I have concentrated in particular on those pathways activated following receptor stimulation. Most abbreviations are defined in the text; $InsP_2$—inositol bisphosphates, $InsP_3$—inositol trisphosphates, $InsP_4$—inositol tetrakis-phosphates, $InsP_5$ and $InsP_6$—inositol pentakisphosphate and inositol hexakis-phosphate, respectively; numbers indicate known positional iso-mers. Rec—receptor, PLC—phospholipase C (-PIC), Gp—putative G pro-tein coupling to PIC, DAG—diacylglycerol, PA—phosphatidic acid. Bold arrows indicate pathways that are inhibited by Li⁺, and dashed arrow indi-cate pathways that are not yet fully defined.

limolar range (Hallcher and Sherman, 1980; Majerus et al., 1988). Stimulation of cells in the presence of Li⁺ results in a substantial buildup of inositol monophosphates, and this has been used to provide the basis for a simple assay of phosphoinositide turn-over (Berridge et al., 1982; *see later*). It has also been proposed that this effect of Li⁺ is the biochemical event underlying its action in the therapy of manic-depressive illness (Berridge et al., 1982,1989; Godfrey et al., 1989).

In this chapter, I will provide detailed methodological information on the different techniques we use for the extraction, separation, and analysis of both inositol lipids and inositol phosphates. There are several other techniques which we have not used ourselves that readers may find useful, and I will provide the appropriate references for these methods. This chapter complements that by Hajra, Fisher, and Agranoff (1988) in an earlier *Neuromethods* volume, and I will allude to, but not go over in detail, many of the methods they described. Other reviews of methods for analysis of inositol phosphates have recently appeared (Dean and Beaven, 1989; Palmer and Wakelam, 1989), and I would recommend these to readers.

The type of analysis procedure used will depend very much on the type of question that the investigator wishes to ask. For example, if the experiment is to correlate changes in $Ins(1,4,5)P_3$ or $Ins(1,3,4,5)P_4$ levels with increases in cytoplasmic $[Ca^{2+}]$, then analysis of the mass of these metabolites would be most appropriate. If effects of receptor stimulation on metabolic interconversions of inositol phosphates are being studied, then separation using dowex column chromatography or HPLC will be necessary. When a simple measure of the extent of receptor activation is required (e.g., when investigating the pharmacology of a particular receptor coupled to PtdIns hydrolysis), the assay of a total inositol phosphate (InsPs) fraction in the presence of LiCl provides a more accurate estimation of the extent of receptor activation than does measurement of a single metabolite.

2. Tissue Preparation

For most of our studies, we have used brain slices or clonal cell lines, though other workers have also successfully used synaptosomes (Brammer et al., 1988), brain membrane preparations (Litosch, 1987; Claro et al., 1990), nerve ending membranes (Van Rooijen et al., 1983), or primary cultures of neurons (Weiss et al., 1988). Although tissue preparation/buffer conditions will vary depending on individual investigator's requirements, the procedures for studying agonist-stimulated inositide turnover are essentially similar.

Most studies involve prelabeling cells with either $^{32}PO_4^{3-}$ or [^{3}H]-inositol, though for mass measurements of inositol phosphates, this is unnecessary. In general, with brain slices, cells are labeled for a short period with [^{3}H]-inositol, and then stimulated with agonist either in the continued presence of label or following removal of the extracellular label by washing. We have found both procedures to give essentially similar results. One disadvantage to using tissue preparations is that it is impossible to maintain cellular viability long enough to obtain labeling of inositides to isotopic equilibrium. This may then cause problems when trying to relate changes in radioactivity to changes in mass. A way around this is to use cultured cells and label them (usually with [^{3}H]-inositol) for 24–48 h. This will label PtdIns$P$ and PtdInsP_2 to equilibrium, and has been used to estimate inositol lipid and phosphate mass following stimulation (*see* Cook et al., 1990). However, there is evidence that some inositol phosphates, particularly the more highly phosphorylated ones, are not labeled to equilibrium even under these conditions (Cook et al., 1990).

2.1. Preparation and Stimulation of Brain Slices

Rats or mice were killed by decapitation. The brain was removed and placed on a block at room temperature, and cross chopped slices (350 × 350 μm) were prepared from cerebral cortex (or other brain region as required) using a McIlwain tissue chopper. Slices were then washed 4 × with 25 vol of a Krebs-Ringer-HEPES buffer (composition [mM]: NaCl 130, KCl 5, MgSO$_4$ 1.2, CaCl$_2$ 1.2, HEPES 20, Na$_2$HPO$_4$ 1.2, and glucose 10; pH 7.4) equilibrated with 100% O$_2$ at 37°C. The slices were then suspended in 50 vol of buffer (w/v), gassed, and incubated in a shaking water bath at 37°C for 60 min with changes of buffer at 20 and 40 min. This preincubation is essential for resynthesis of the polyphosphoinositides, which become depleted during the tissue preparation; agonist responses are significantly smaller if it is omitted.

The slices (approx 400 mg tissue) are then allowed to settle under gravity and incubated in 5 mL gassed buffer in a 25-mL flask with [^{3}H]-inositol. When analyzing a total lipid or Ins$P$s fraction, 10 μCi of label is used; when it is necessary to separate individual inositol phosphates, up to 100 μCi of [³H]-inositol may be required, depending on the resolution needed. The flask is gassed, capped, and incubated with shaking for 45–60 min. Slices are then washed with a further 4 × 25 mL buffer and allowed to settle. Then 50 μL of packed slices are added to a flat-bottomed minivial (e.g., Beckman Bio-vial) containing 250 μL buffer. When a total InsPs fraction is wanted, this buffer should contain 10 mM LiCl (*see* Berridge et al., 1982 and Introduction), though when individual inositol phosphates are being studied, it should be left out. Incubations are continued for the required time, usually 30–60 min for a total InsPs experiment, and then stopped and extracted using one of the methods described in Section 3.

2.2. Stimulation of Cell Lines

Cells are split and grown for 24–48 h in normal medium containing serum in Costar 6-well tissue culture clusters. The medium is then removed and replaced with inositol-free medium without serum, but containing [³H]-inositol (2 μCi/mL for total InsPs experiments, or more if a detailed inositol phosphate analysis is required), and grown for a further 24 h. Cells are then washed with phosphate-buffered saline and incubated in 1 mL of Krebs-Ringer-HEPES (containing 10 mM LiCl if necessary) for 15 min before addition of agonist. To stop the reaction, the medium is aspirated and 0.75 mL of ice-cold methanol is added to the cells, which are then scraped off the plates into glass test tubes. The plates are then washed with 0.75 mL methanol, which is added to the first extract. Water (0.5 mL) and chloroform (0.5 mL) are then added, and a single phase should ensue. Samples are allowed to extract for 30 min, and the phases are then split by addition of a further 0.5 mL water and chloroform. Following vortex mixing and configuration (500 g for 5 min) the upper phase is removed for inositol phosphate analysis.

3. Extraction of Inositides from Tissues

Following the pioneering work of Folch in the 1940s, a number of methods, based on extraction of phosphoinositides into organic solvents, have been developed (*see* Hajra et al., 1988).

The most common one now in use involves the quantitative extraction of all inositol lipid species using acidified chloroform/ methanol (Wells and Dittmer, 1965). For extraction of the water-soluble inositol phosphates from tissues, several methods are regularly used, including chloroform/methanol, chloroform/ methanol/HCl (as for the lipids), perchloric acid (PCA) followed by neutralization by KOH or by FREON/octylamine, and tri-chloroacetic acid (TCA) with neutralization by diethyl ether. We have used all of these, with consistent results, in our lab. For extraction of acid-labile cyclic inositol phosphates, a mixture of phenol, chloroform, and methanol can be used (*see* Hawkins et al., 1987 for details).

It should be pointed out that no one extraction procedure will be totally satisfactory in all respects and that the procedure of choice will depend on the measurements required. It is always advisable to check efficacy of extraction by following recoveries of radioactive standards added to tissue extracts following termination of the incubation.

3.1. Extraction of Lipids

When we are analyzing inositol lipids from [^{3}H]-inositol-labeled brain slices, the incubations are terminated by the addition of 2.25 mL chloroform/methanol (1:2, v/v) containing 20 μL of 6M HCl and 300 μL buffer. Samples are then allowed to extract for 30 min before addition of 0.75 mL chloroform and 0.75 mL 2M KCl. Initially, the samples should be in a single phase, which then splits into an upper (aqueous) and lower (organic) phase upon addition of chloroform and KCl. The exact volumes used are not important, though the ratios of chloroform to methanol to water should be consistent. Extracts are then vortexed and centrifuged, and the lower phase is removed. The upper phase is then washed with 1.4 mL chloroform, and the samples washed and recentrifuged. The lower phase is then combined with the original extract, and this is then washed with methanol/1M HCl (1:1, v/v) and dried down.

This procedure provides a consistent and reproducible extraction of all the inositol lipids. Neutral chloroform/methanol will extract PtdIns, but it will not provide quantitative extraction of the polyphosphoinositides.

3.2. Chloroform/Methanol Extraction and Analysis of a Total InsPs Fraction

This procedure is based on that originally described by Berridge et al. (1982). When measuring a total InsPs fraction, the incubations (310 μL) are terminated with 0.94 mL of chloroform/methanol (1:2, v/v), and the phases split with 0.31 mL water and 0.31 mL chloroform. After vortexing and centrifuging, 0.75 mL of the upper layer is removed and added to 2.25 mL water in a 15-mL conical test tube. Dowex AG1-X8 (0.5 mL) (200–400 mesh) anion-exchange resin (formate form) is then added and the samples vortexed. The gel is allowed to settle and the supernatant is poured off. The gel is then washed 3–5 × with 3 mL water (5 mM *myo*-inositol can also be used) to remove contaminating [^{3}H]-inositol (counting samples of the washes will determine how many are required). The InsPs are then extracted twice with 0.5 mL each time of 1 M ammonium formate/0.1 M formic acid. Scintillant (10 mL) is then added to each combined extract, and the samples counted for radioactivity in the liquid phase (efficiency is about 30%). Chloroform/methanol will not quantitatively extract the highly phosphorylated InsPs (InsP$_3$, InsP$_4$, and so forth), though since these generally make up <5% of a total InsPs fraction, this does not significantly affect results. Acidification of the chloroform/methanol with HCl at the initial extraction step will provide a more reproducible extraction of the higher InsPs, though samples will then need to be neutralized prior to anion-exchange chromatography.

3.3. Extraction with Perchloric Acid

Two distinct methods have been developed that employ perchloric acid (PCA) as an extracting agent. It has been suggested that, to ensure good recoveries of radiolabeled inositol phosphates, a small amount of phytic acid hydrolysate (Wreggett and Irvine, 1987) should be added with the PCA.

In the first method, 0.6 mL of ice-cold 4.5% PCA is added to 0.31 mL of sample, and the samples left to extract for 10 min on ice. These are then vortexed and centrifuged. A portion (0.7 mL) is then removed and neutralized with a sufficient volume of $0.5M$ KOH/9 mM $Na_2B_4O_7$ to give a pH of 8–9. The samples are then centrifuged to remove precipitated potassium perchlorate and the supernatants used for inositol phosphate analysis. In the second method, the incubation (0.31 mL) is stopped with 0.31 mL ice-cold 10% PCA, 20 µL of 1% EDTA is added, and the samples left on ice for 10 min. Samples are then centrifuged, and 0.5 mL of supernatant is transferred to an Eppendorf tube; 0.6 mL of 1,1,2-trichlorotrifluoroethane (FREON)/tri-*n*-octylamine (1:1 mixture freshly made up) is then added, the samples vortexed vigorously for at least 15 s, and then centrifuged. The mixture should resolve into three phases: a bottom phase containing FREON and unreacted octylamine, a middle phase consisting of octylamine perchlorate, and an upper aqueous phase containing the neutral inositol phosphate extract. This upper extract can then be analyzed. Caution should be exercised when using these extracts for mass analysis of Ins(1,4,5)P_3 (*see* Section 5.), since it has been reported that direct addition of these extracts to the binding assay may result in denaturation of the binding protein (Cook et al., 1990). We have never noticed this problem, though we do make up our Ins(1,4,5)P_3 standards for the binding assay in buffer that has been through the extraction procedure to ensure compatibility.

3.4. Extraction with Trichloroacetic Acid

The reactions (250 µL) are terminated with 50 µL of ice-cold 30% trichloroacetic acid (TCA). Phytate hydrolysate (about 25 µg phosphate) is added if necessary, and the samples are transferred to Eppendorf tubes, vortexed, and centrifuged. Supernatant (250 µL) is then removed and added to 1.0 mL of water-saturated diethyl-ether, and the samples vortexed. The tubes are then placed in a bath of methanol/dry ice, which rapidly freezes the aqueous samples while leaving the upper ether phase liquid. This can then be easily poured off. The samples are washed a further 3 × with ether and then dried down *in vacuo.* Extracts can then be resuspended in 10 mM $NaHCO_3$/5 mM EDTA or other buffer as required.

4. Separation and Analysis of Phosphoinositides

A variety of techniques are available for separation of the lipid intermediates of the phosphoinositide cycle, including paper or column chromatography (*see* Hajra et al., 1988; Dean and Beaven, 1989), thin-layer chromatography (TLC), and deacylation followed by separation of the water-soluble products. Those techniques we have used are detailed below. I will also describe a simple method we have developed for the measurement of CDP-diacylglycerol (CDP-DG) generated during stimulation of PtdIns cycle metabolism.

4.1. Thin-Layer Chromatography

We have generally used TLC for the separation of ^{32}P-labeled lipids from ^{32}PO$_4{}^{3-}$-labeled cells, because a wide variety of phospholipids can be measured with relative convenience. We have found that we could never adequately separate all phospholipid species, including poylphosphoinositides, on a single TLC system, and so we usually split our samples into two, running one plate for polyphosphoinositides and one for the other lipids. A wide variety of TLC systems have been used over the years (*see* Hajra et al., 1988). I will describe the ones we have used most successfully.

We have used 0.25-mm thick E. Merck silica gel 60 plates, which have been oxalate (1% potassium oxalate) impregnated (to reduce streaking). An aliquot of dried lipid is resuspended in chloroform/methanol (9:1) and spotted on the plate. For separation of polyphosphoinositides, a one-dimensional system of chloroform/methanol/conc. ammonia/water (90:90:7:20) is used. For other phospholipids, we have used a variety of solvent systems including:

1. Chloroform/methanol/acetone/acetic acid/water (40:15: 13:12:8 by vol);
2. A double solvent system, developed in the same direction of chloroform/methanol/conc. ammonia (400:10:1) followed by chloroform/methanol/conc. ammonia/water (65:35:2:3); and

3. A double solvent system, developed in the same direction of chloroform/methanol/acetic acid/water (first, 40:10:10:1, then, 120:46:19:3).

The R_f values differ for each solvent system, so nonradioactive standards should be run to verify the positions of each of the lipids. We normally visualize the lipids using the Sigma molybdenum spray reagent. The positions of the radioactive spots can be easily determined by autoradiography, and the appropriate lipids can then be scraped and counted for radioactivity.

Lipid samples can also be taken for estimation of PO_4^{3-}. We scrape the lipids into borosilicate test tubes and digest for 2 h in 0.2 mL 70% (v/v) perchloric acid at 200°C. The samples are then made up to 2mL and centrifuged to remove the silica. The supernatant can then be removed and assayed for PO_4^{3-} (Godfrey and Putney, 1984).

4.2. Deacylation and Separation of Water-Soluble Products

For separation of [^{3}H]-inositol lipids, we have used the deacylation followed by anion-exchange chromatography protocol described by Creba et al. (1983). Radioactivity in all three inositol lipids can be easily quantified. Dried lipids are dissolved in 1 mL of chloroform to which is added 0.2 mL methanol and 0.2 mL of 1M NaOH in methanol/water (19:1). The samples are allowed to digest at room temperature for 20 min. Then 0.6 mL of methanol and water are added, and the tubes are vortexed and centrifuged. Upper phase (1 mL) is then removed and neutralized with boric acid (0.25M). Samples are then diluted to 5 mL with sufficient ammonium formate and sodium tetraborate to give final concentrations of 0.18M and 5 mM, respectively. The mixture is then loaded onto a 1-mL dowex AG1-X8 (200–400 mesh) anion-exchange column and eluted with a further 15 mL 0.18M ammonium formate/5 mM sodium tetraborate. The eluate is combined with that from the original column loading, and a sample is taken for scintillation counting. This eluate contains labeled glycerophosphoinositol, the deacylated product of phosphatidylinositol. Glycerophosphoinositol phosphate, the product of phosphatidylinositol 4-phosphate, is then eluted with 20 mL of $0.4M$ ammonium formate/$0.1M$ formic acid, and glycerophosphoinositol bisphosphate (from PtdInsP_2) is then eluted with 20 mL of $1.0M$ ammonium formate/$0.1M$ formic acid. Columns are regenerated by washing with $2M$ ammonium formate/$0.1M$ formic acid and water, and are used 4–5 times. Samples from each fraction are added to scintillation fluid and counted. Recently, another inositol lipid, PtdIns3P, has been identified in some cell types, though its precise function is still unclear. It can be separated from PtdIns4P by HPLC *(see below)* following deacylation of the lipids. In most cells, this lipid makes up only a small proportion (<10%) of the total PtdIns$P$ pool.

4.3. Assay of CDP-Diacylglycerol (CDP-DG)

When cells are stimulated in the presence of Li$^+$, there is a reduction in intracellular inositol levels, and this is reflected in a reduction in PtdIns resynthesis and a compensatory accumulation of the other precursor of PtdIns, namely CDP-DG *(see* Fig. 1). We have developed a simple assay for CDP-DG involving prelabeling of endogenous CTP with cytidine, which then accumulates as CDP-DG upon stimulation in the presence of LiCl. The rise in CDP-DG can be reversed by preincubation with *myo*-inositol and can provide a useful index of cellular inositol concentrations (Godfrey, 1989). The advantage of the technique is that CDP-DG is the only lipid that labels with cytidine; samples can therefore be analyzed without the need for time-consuming TLC. A disadvantage is that it could also reflect other diacylglycerol-producing phospholipase activity, though this will only be important if it subsequently cycles to PtdIns.

Brain slices are prepared and preincubated as normal (Section 2.1.). Then 50 μL of packed slices are incubated in flat-bottom vials in 0.25 mL buffer containing [^{14}C]- or [^{3}H]-cytidine (0.2 μCi/mL or 2 μCi/mL, respectively) and 10 mM LiCl. Samples are then incubated for 15 min at 37°C prior to addition of agonist. The incubations are continued for a further 60 min and then stopped with 0.94 mL chloroform/methanol (1:2). Phases are split with 0.31 mL each of chloroform and water, and the samples are centrifuged. A 0.45-mL portion of the bottom layer is removed, washed with 1 mL methanol/1M HCl, and then dried down and counted for radioactivity.

5. Analysis of Inositol Phosphates

There are a wide variety of methods for estimation of mass levels of inositol phosphates, of varying complexity (*see* Hajra et al., 1988; Cook et al., 1990; Dean and Beaven, 1989). The most common method involves analysis of mass levels of Ins(1,4,5)P_3 or Ins(1,3,4,5)P_4 by the use of specific and selective binding proteins, and this will be described below. Mass levels of inositol phosphates can also be measured by gas chromatography (Leavitt and Sherman, 1982). This method is most useful for assay of the inositol monophosphates (InsP_1) where sensitivities of under 1 pmol can be obtained. It is also one of the few techniques that can directly measure endogenous levels of inositol phosphates in small tissue samples in vivo. We have some experience in this technique, and I will detail our method. Radiolabeled inositol phosphates are usually separated by anion-exchange chromatography or HPLC (*see below*); paper chromatography can also be used (*see* Dean and Beaven, 1989), though this has now largely been superseded by the ion-exchange techniques.

Some techniques can be applied to all inositol phosphates (this usually involves the use of radiolabeled precursors), whereas some will measure only a few. The method of choice depends on a number of factors including:

1. The inositol phosphate of interest;
2. The number of samples that need to be analyzed;
3. The availability of material for analysis; and
4. Initial outlay for equipment and running costs.

5.1. Separation of Radiolabeled Inositol Phosphates

Analysis of a total InsPs fraction (*see* Section 3.2.) is simple, convenient, and cheap, and a large number of samples can be assayed (60 samples a day with relative ease). To separate individual inositol polyphosphate species, anion-exchange chromatography is normally employed. This techniue is again simple and, although a little more time-consuming than the Ins*P*s method, reasonably large numbers of samples can be processed. It also provides a more detailed profile of the inositol phosphates. The aqueous tissue extract (*see* Section 3.) is diluted to 10 mL with water and then put on a 1-mL dowex AG1-X8 (formate form, 200–400 mesh) column. Free inositol is eluted with a further 20 mL water. Then glycerophosphoinositol is eluted with 16 mL of 60 m*M* ammonium formate/5 m*M* sodium tetraborate. Ins*P* is then eluted with 16 mL 0.2*M* ammonium formate/0.1*M* formic acid, Ins*P*$_2$ with 16 mL of 0.4*M* ammonium formate/0.1*M* formic acid, Ins*P*$_3$ with 8 mL of 0.8*M* ammonium formate/0.1*M* formic acid, and Ins*P*$_4$ with 8 mL of 1.2*M* ammonium formate/ 0.1*M* formic acid. The elution profiles should be checked with radioactive standards. A sample of each eluate can then be taken for scintillation counting. With the higher salt concentrations, the scintillation fluid will not always form a single phase with the sample; under these circumstances either add a little methanol to the mixture or predilute the eluate with water. The major disadvantage of this technique is that it does not separate the isomers of inositol phosphates.

A simple enzymic method for separation of [^{3}H]-Ins(1,4,5)P_3 and [^{3}H]-Ins(1,3,4)P_3 has recently been developed (Kennedy et al., 1989), based on the principle that the Ins(1,4,5)P_3 5-phosphatase is an Mg^{2+}-dependent enzyme, whereas the Ins(1,3,4)P_3 4-phosphatase is Mg^{2+}-independent. Thus, in the presence of EDTA, Ins (1,3,4)P_3 is selectively degraded, and the remaining radioactivity can then be attributed to Ins(1,4,5)P_3. A crude preparation of rat brain cytosol is used as an enzyme source; a rat brain is homogenized (two 15 s bursts with a polytron, setting 6) in 250 m*M* HEPES/2 m*M* MgCl$_2$, pH 7.4, at a concentration of 25% (w/v), and then centrifuged at 4°C for 90 min at 100,000*g*. The supernatant is used as enzyme source. Top layer (300 μL) from a FREON/octylamine tissue extraction (*see* Section 3.3.) is then added to 50 μL 250 m*M* HEPES/5 m*M* EDTA and 17.5 μL of brain supernatant, and the incubation continued at 37°C for 30–60 min. The reaction is stopped with 370 μL of 10% PCA, and the samples are then reextracted with FREON/octylamine.

Ins(1,4,5)P_3 can then be separated from Ins(1,3)P_2, the product of Mg^{2+}-independent dephosphorylation of Ins(1,3,4)P_3, using conventional dowex column chromatography *(see above)*. Ins(1,3,4)P_3 content can be assumed to be the remaining counts in a total InsP_3 fraction after subtraction of the Ins(1,4,5)P_3 counts.

When individual inositol phosphate isomers need to be analyzed, for example, when the metabolic interconversions of inositol phosphates are being studied, then HPLC should be the method of choice. This is the only method that can resolve all the different inositol phosphates. Its disadvantages are that it is very time-consuming (each run can take up to 2 h), it is expensive (in terms of equipment and scintillation fluid), and only a small number of samples can be analyzed easily. Dean and Beaven (1989) have described HPLC methodology in some detail, and Batty et al. (1989) described HPLC methods to separate individual inositol phosphate isomers prepared from carbachol-stimulated rat cerebral cortex slices. These papers should provide a useful starting point for people wishing to set up the technique.

5.2. Mass Analysis of Inositol Phosphates

5.2.1. Specific Binding Assays

In the past 2–3 yr, specific binding assays for the measurement of both Ins(1,4,5)P_3 (Challiss et al., 1988; Palmer et al., 1989) and InsP_4 (Bradford and Irvine, 1987; Donie and Reiser, 1989; Challiss and Nahorski, 1990) mass have been developed, based on displacement of [^{3}H]-Ins(1,4,5)P_3 or [^{32}P]-Ins(1,3,4,5)P_4 from specific binding proteins in bovine adrenal cortex or rat cerebellum, respectively. The assays are easy, and large numbers can be processed simultaneously. The only drawbacks are the costs of the radiolabeled inositol phosphates and the time it takes to prepare the binding proteins. Time can be saved by preparing the latter in bulk and freezing aliquots of membranes; we have found the adrenal protein to be stable for at least 6 mo at –70°C.

Ins(1,4,5)P_3 binding is done using a bovine adrenal binding protein (Palmer et al., 1989), which is basically a crude P2 pellet of bovine adrenal cortex membranes. Bovine adrenal cortex is dissected on ice and homogenized in 20 vol of ice-cold 20 mM

NaHCO$_3$ using a polytron (setting 6, 3 bursts of 15 s). The preparations are then centrifuged at 1000g for 10 min at 4°C; the supernatant is removed, and the pellet rehomogenized and centrifuged again. The combined supernatants are then centrifuged at 20,000 rpm (38,000g) at 4°C for 20 min, and the membranes resuspended to 5–10 mg/mL protein in incubation buffer (20 mM NaCl, 100 mM KCl, 1 mM EDTA, 1 mg/mL BSA, and 20 mM Tris-HCl, pH 8.3). Aliquots of this can be stored frozen. The binding assays are done on ice in a final vol of 250 μL, in Eppendorf tubes. Each assay contains 0.25–0.5 nM [^{3}H]-Ins(1,4,5)P_3 (25 μL), and either cold Ins(1,4,5)P_3 (for displacement curves) or sample as appropriate (25 μL). Nonspecific binding is defined by 10 μM cold Ins(1,4,5)P_3 (we use the Sigma inositol trisphosphate preparation). Reactions are started by addition of binding protein (150 μL), and samples are vortexed and incubated for 10 min. Incubations are terminated by centrifugation in a microfuge for 10 min; the supernatant is then poured off and the pellet washed twice with distilled water. The pellet is solubilized overnight with 100 μL tissue solubilizer. Then 1 mL of scintillant is added and the samples counted. Tissue samples are prepared by one of the methods described earlier and Ins(1,4,5)P_3 levels are determined by measuring displacement of label and comparison to a displacement curve (0.1–100 nM) with standard Ins(1,4,5)P_3.

The ability of ATP and many other normal intracellular components (Nunn and Taylor, 1990) to displace Ins(1,4,5)P_3 from its binding site can pose considerable problems for measurement of Ins(1,4,5)P_3 in cell extracts by radio receptor assay. Rigorous controls are essential and have been described (Challiss et al., 1990).

Ins(1,3,4,5)P_4 binding is done using a rat cerebellar binding protein (Challiss and Nahorski, 1990), which is basically a crude preparation of cerebellar membranes. Cerebella are removed from rats following decapitation and homogenized in 20 vol of ice-cold 20 mM NaHCO$_3$/1 mM dithiothreitol. The preparations are then centrifuged twice at 20,000 rpm (38,000g) at 4°C for 20 min, and the membranes resuspended to 5–10 mg/mL protein. Aliquots of this can be frozen. The binding assays are done on ice in a final vol of 160 μL in Eppendorf tubes. Each assay contains 25 mM sodium acetate, 25 mM K$_2$HPO$_4$, 2 mM EDTA (pH 5.0), approx 15,000 dpm of [^{32}P]-Ins(1,3,4,5)P_4, and either cold Ins(1,3,4,5)P_4 (for displacement curves) or sample as appropriate. Nonspecific binding is defined by 10 μM cold Ins(1,3,4,5)P_4. Reactions are started by addition of binding protein, and samples are vortexed occasionally during the 30-min incubation period. Incubations are stopped by rapid filtering through Whatman GF/B filters, and free ligand is removed by washing 3 × with 3 mL of 25 mM sodium acetate, 25 mM K$_2$HPO$_4$, 5 mM NaHCO$_3$, and 1 mM EDTA (pH 5.0). Radioactivity remaining on the filters is determined by scintillation counting. Tissue samples are prepared by one of the methods described earlier, and Ins(1,3,4,5)P_4 levels are determined by measuring displacement of label and comparison to a displacement curve with standard Ins(1,3,4,5)P_4.

5.2.2. Gas Chromatography

Gas chromatography (GC) has been used to measure free *myo*-inositol (Allison and Blisner, 1975), inositol monophosphates (Leavitt and Sherman, 1982; Hirvonen et al., 1988), and InsP_3 (Agranoff and Seguin, 1974; Rittenhouse and Sasson, 1985) in extracts of tissues. We have used GC to analyze Ins$P$ levels in small samples of brain tissue, and our method is based on that described by Leavitt and Sherman (1982). The experiments are very slow and time-consuming, though they will provide the most sensitive measures of mass inositol phosphate levels from in vivo tissue samples.

We normally do our experiments in the presence of LiCl, which significantly enhances inositol phosphate levels and can potentiate agonist-stimulated increases in Ins$P$ levels (*see* Sherman et al., 1986). LiCl or NaCl (3 mEq/kg) is administered subcutaneously to rats, and 24 h later, the compound of interest is administered. After the appropriate period, rats are then decapitated, and the head immediately placed in liquid nitrogen. When frozen (approx 30 s), the heads can be removed into a –70°C freezer for storage (InsP_1 levels are stable for several months at –70°C). The head usually splits down the midline upon freezing, and samples of frozen brain tissue can usually be removed fairly easily. We usually remove small samples of frontal cortex. This is done in a cryostat at –15°C; samples should not be allowed to thaw. When very selective brain regions are needed, it may be easier to remove the brain from the skull prior to freezing (Hirvonen et al., 1988); however, this can result in elevated basal $InsP_1$ concentrations. Tissue samples are then lyophilized, in Eppendorf tubes, at –35°C for 24–48 h and then at room temperature for a further 24 h. The dried samples are then weighed, and 3–5 mg of tissue is transferred to reactivials (Pierce) and derivatized in a 1:1 mixture of *N,O-bis*(trimethylsilyl) trifluoro-acetamide containing 10% trimethylchlorosilane:dry pyridine. We use 1 μL of derivatizing agent/20 μg tissue; this provides a reasonable $InsP_1$ concentration for later GC analysis, without leaving the derivatizing liquid extract too viscous for easy handling. We derivatize brain tissue samples directly, without any further purification of the inositol phosphates. The samples are derivatized for 24–48 h at room temperature and are kept in a dry atmosphere (stored over silica gel) at all times. Derivatized extracts are stable indefinitely at –70°C. Ins1*P*, Ins2*P*, and glucose 6-phosphate standards are derivatized in the same way. Known concentrations of Ins2*P* can be added to derivatized tissue extracts to act as internal standards for these samples, since Ins2*P* is absent from biological materials and will separate easily from other inositol monophosphates on GC.

All our analyses are done using a Varian model 3400 GC with a flame photometric detector in phosphorus-specific mode. The columns we use are either 2 m × 4 mm id glass columns packed with 3% OV–17 on chromosorb W-HP, 80/100 mesh (each new batch should be checked for inertness, separating ability, and sensitivity by doing dilution curves with the Ins2*P* standard; a linear dilution curve down to the sensitivity limit of the machine should be obtained), or megabore columns (20 mm DB-1, J & W Scientific). Silanized glass wool is used to plug the injector end of the column lightly, which helps prevent particulate matter from the derivatized tissue extracts blocking the columns and altering the running characteristics. Argon is used as the carrier gas. The operating conditions vary depending on the columns being used, though normally the injector temperature is

220°C, detector temperature 250°C, and oven temperature 200°C. Both temperature and flow rate of carrier gas need to be optimized to get the best running conditions (flow rates in particular will depend on the type of column being used). Columns are primed at the beginning of each set of runs with five injections of brain extract, to ensure good linearity and prevent adsorptive losses. Standards are run after every 4–5 samples to check for potential adsorption problems. Ins1P is present at higher concentrations in brain tissue than the other monophosphates, and we usually measure only this isomer. Basal Ins$P$ levels are normally around 0.2 nmol/mg dry wt of tissue (Sherman et al., 1986), and this is usually increased four- to fivefold with 3 mmol/kg LiCl, and to an even greater extent with agonist plus LiCl.

6. Conclusions

In this chapter, I have attempted to cover all of the commonly used methods for the preparation and analysis of phosphoinositides and inositol phosphates in tissue, particularly neuronal tissue, samples. I have described methods for a wide variety of analyses and, hopefully, have provided adequate references for methods that I have not covered in detail. As I said in the Introduction, the type of analytical procedure to be used will depend greatly on the type of question that needs to be answered; I trust that the information provided in this chapter will help people decide which is the most useful technique for their particular needs.

Acknowledgments

I would like to thank S. Arkinstall for critically reading the manuscript, and all my technicians who helped with the experiments, particularly S. J. McClue, Z. Taghavi, and S. Neidhart.

References

Agranoff B. W. and Seguin E. B. (1974) Preparation of inositol trisphosphate from brain: GLC of trimethylsilyl derivative. *Prep. Biochem.* **4,** 359–366.
Allison J. H. and Blisner M. E. (1975) Inhibition of the effect of lithium on brain inositol by atropine and scopolamine. *Biochem. Biophys. Res. Commun.* **68,** 1332–1338.

Batty I., Letcher A. J., and Nahorski, S. R. (1989) Accumulation of inositol polyphosphate isomers in agonist-stimulated cerebral-cortex slices. *Biochem. J.* **258,** 23–32.

Berridge M. J. and Irvine R. F. (1989) Inositol phosphates and cell signalling. *Nature* **341,** 197–205.

Berridge, M. J., Downes, C. P., and Hanley, M. R. (1982) Lithium amplifies the agonist dependent phosphatidylinositol responses in brain and salivary glands. *Biochem. J.* **206,** 587–595.

Berridge, M. J., Downes, C. P., and Hanley, M. R. (1989) Neural and developmental actions of lithium: a unifying hypothesis. *Cell* **59,** 411–419.

Bradford P. G. and Irvine R. F. (1987) Specific binding sites for [^{3}H]-inositol 1,3,4,5-tetrakisphosphate on membranes of HL60 cells. *Biochem. Biophys. Res. Commun.* **149,** 680–685.

Brammer, M. J., Hajimohammadreza I., Sardiwal S., and Weaver K. (1988) Is inositol bisphosphate the product of A23187 and carbachol-mediated polyphosphoinositide breakdown in synaptosomes? *J. Neurochem.* **51,** 514–521.

Challiss R. A. J. and Nahorski S. R. (1990) Neurotransmitter and depolarisation-stimulated accumulation of inositol 1,3,4,5-tetrakisphosphate mass in rat cerebral cortex slices. *J. Neurochem.* **54,** 2138–2141.

Challiss R. A. J., Chilvers E. R., Willcocks A. L., and Nahorski S. R. (1990) Heterogeneity of [^{3}H] inositol 1,4,5-trisphosphate binding site in adrenal cortical membranes. Characterization and validation of a radio-receptor assay. *Biochem. J.* **265,** 421–427.

Challiss R. A. J., Batty I., and Nahorski S. R. (1988) Mass measurements of ino, itol 1,4,5-trisphosphate in rat cerebral cortex slices using a radio-receptor assay: Effects of neurotransmitters and depolarisation. *Biochem. Biophys. Res. Commun.* **157,** 684–691.

Claro E., Wallace M. A., and Fain J. N. (1990) Dual effect of fluoride on phosphoinositide metabolism in rat brain cortex. *Biochem. J.* **268,** 733–737.

Cook et al. (1990) Mass measurement of inositol 1,4,5-triphosphate and sn-1,2-diacylglycerol in bombesin-stimulated Swiss 3T3 mouse fibroblasts. *Biochem. J.* **265,** 617–620.

Creba J. A., Downes C. P., Hawkins P. T., Brewster G., Michell R. H., and Kirk C. J. (1983) Rapid breakdown of phosphatidylinositol 4-phosphate and phosphatidylinositol 4,5-bisphosphate in rat hepatocytes stimulated by vasopressin and other Ca^{2+}-mobilising hormones. *Biochem. J.* **212,** 733–747.

Dean N. M. and Beaven M. A. (1989) Methods for the analysis of inositol phosphates. *Anal. Biochem.* **183,** 199–209.

Donie F. and Reiser G. (1989) A novel specific binding assay for quantitation of intracellular inositol 1,3,4,5-tetrakisphosphate using a high-affinity InsP$_4$ receptor from cerebellum. *FEBS Lett.* **254,** 155–158.

Downes C. P. (1986) Agonist-stimulated phosphatidylinositol 4,5-bisphosphate metabolism in the nervous system. *Neurochem. Int.* **9,** 211–230.

Fisher S. K. and Agranoff B. W. (1987) Receptor activation and inositol lipid hydrolysis in neural tissues. *J. Neurochem.* **48,** 999–1017.

Godfrey P. P. (1989) Potentiation by lithium of CMP-phosphatidate formation in carbachol-stimulated rat cerebral-cortical slices and its reversal by *myo*-inositol. *Biochem. J.* **258**, 621–624.

Godfrey P. P., McClue S. J., White A. M., Wood A. J., and Grahame-Smith D. G. (1989) Subacute and chronic in vivo lithium treatment inhibits agonist- and sodium fluoride-stimulated inositol phosphate production in rat cortex. *J. Neurochem.* **52**, 498–506.

Godfrey P. P. and Putney J. W. (1984) Receptor-mediated metabolism of the phosphoinositides and phosphatidic acid in rat lacrimal acinar cells. *Biochem. J.* **218**, 187–195.

Hajra A. K., Fisher S. K., and Agranoff B. W. (1988) Isolation, separation, and analysis of phosphoinositides from biological sources, in *Neuromethods* vol. 7 (Boulton A. A., Baker G. B., and Horrocks L. A., eds.), Humana, Clifton, NJ, pp. 211–225.

Hallcher L. M. and Sherman W. R. (1980) The effects of lithium and other agents on the activity of *myo*-inositol 1-phosphatase from bovine brain. *J. Biol. Chem.* **255**, 10896–10901.

Hawkins P. T., Berrie C. P., Morris A. J., and Downes C. P. (1987) Inositol 1,2-cyclic 4,5-trisphosphate is not a product of muscarinic receptor-stimulated phosphatidylinositol 4,5-bisphosphate hydrolysis in rat parotid glands. *Biochem. J.* **243**, 211–218.

Hawthorne J. N. and Pickard M. R. (1979) Phospholipids in synaptic function. *J. Neurochem.* **32**, 5–14.

Hirvonen M.-R., Lihtamo H., and Savolainen K. (1988) A gas chromatographic method for the determination of inositol monophosphates in rat brain. *Neurochem. Res.* **13**, 957–962.

Kennedy E. D., Batty I. H., Chilvers E. R., and Nahorski S. R. (1989) A simple enzymic method to to separate [^{3}H]-1,4,5- and 1,3,4-inositol trisphosphate isomers in tissue extracts. *Biochem. J.* **260**, 283–286.

Leavitt A. L. and Sherman W. R. (1982) Resolution of DL-*myo*-inositol 1-phosphate and other sugar enantiomers by gas chromatography. *Methods Enzymol.* **89**, 3–18.

Litosch I. (1987) Guanine nucleotide and NaF stimulation of phospholipase C activity in rat cerebral-cortical membranes. *Biochem. J.* **244**, 35–40.

Majerus P. W., Connolly T. M., Bansal V. S., Inhorn R., Ross T. S., and Lips D. L. (1988) Inositol phosphates: Synthesis and degradation. *J. Biol. Chem.* **263**, 3051–3054.

Nahorski S. R. and Potter B. V. L. (1989) Molecular recognition of inositol polyphosphates by intracellular receptors and metabolic enzymes. *Trends Pharmacol. Sci.* **10**, 139–144.

Nunn D. L. and Taylor C. W. (1990) Liver inositol 1,4,5-trisphosphate-binding sites are the calcium-mobilizing receptors. *Biochem. J.* **270**, 227–232.

Palmer S., Hughes K. T., Lee D. Y., and Wakelam M. J. O. (1989) Development of a novel Ins(1,4,5)P$_3$ specific binding assay. *Cell. Signal.* **1**, 147–153.

Palmer S. and Wakelam M. J. O. (1989) Mass measurements of inositol phosphates. *Biochim. Biophys. Acta* **1014**, 239–246.

Rittenhouse S. E. and Sasson J. P. (1985) Mass changes in myoinositol tris-phosphate in human platelets stimulated by thrombin. *J. Biol. Chem.* **260**, 8657–8660.

Shears S. B. (1989) Metabolism of the inositol phosphates produced upon receptor activation. *Biochem. J.* **260**, 313–324.

Sherman W. R., Gish B. G., Honchar M. P., and Munsell L. Y. (1986) Studies on the effects of lithium on phosphoinositide metabolism in vivo. *Fed. Proc.* **45**, 2639–2646.

Van Rooijen, L. A. A., Seguin E. B., and Agranoff B. W. (1983) Phosphodie-steratic breakdown of endogenous polyphosphoinositides in nerve ending membranes. *Biochem. Biophys. Res. Commun.* **112**, 919–926.

Watson S. P. and Godfrey P. P. (1988) The role of receptor-stimulated inosi-tol phospholipid hydrolysis in the autonomic nervous system. *Pharmacol. Ther.* **38**, 387–417.

Weiss S., Schmidt B. H., Sebben M., Kemp D. E., Bockaert J., and Sladeczek F. (1988) Neurotransmitter-induced inositol phosphate formation in neurons in primary culture. *J. Neurochem.* **50**, 1425–1433.

Wells M. A. and Dittmer J. C. (1965) The quantitative extraction and analy-sis of brain polyphosphoinositides. *Biochemistry* **4**, 2459–2468.

Wreggett K. A. and Irvine R. F. (1987) A rapid separation method for inosi-tol phosphates and their isomers. *Biochem. J.* **245**, 655–660.

Inositol Trisphosphate Receptors and Intracellular Calcium

Experimental Approaches

Colin W. Taylor, Jennifer M. Bond, David L. Nunn, and Katherine A. Oldershaw

1. Introduction

1.1. Formation of Ins(1,4,5)P₃

Inositol 1,4,5-trisphosphate (Ins(1,4,5)P_3) was first shown to stimulate mobilization of intracellular Ca^{2+} from permeabilized pancreatic acinar cells (Streb et al., 1983) and has since become recognized as an intracellular messenger formed after receptor activation and then responsible for mobilizing Ca^{2+} from the intracellular stores of many different cell types. The interactions between most receptors and the phosphoinositidase C (PIC) that catalyzes formation of Ins(1,4,5)P_3 and 1,2-diacylglycerol (DG) have much in common with the more completely understood interactions between receptors and adenylyl cyclase (Gilman, 1987; Taylor, 1990b; Fig. 1). In both cases, agonist-occupied receptors catalyze activation of a specific guanine nucleotide dependent regulatory protein(s) (G protein) by allowing it to lose its tightly bound GDP and replace it with GTP. The active GTP-bound G protein, which has yet to be identified for the phosphoinositide pathway[*], then regulates the activity of an

---

[*]G proteins belonging to the G_q family are the pertossis toxin-insensitive G proteins that couple receptors to activation of PIC.

From: *Neuromethods, Vol. 20: Intracellular Messengers*
Eds: A. Boulton, G. Baker, and C. Taylor © 1992 The Humana Press Inc.

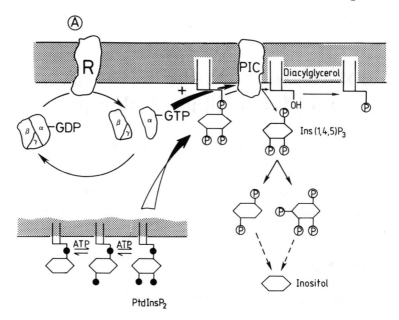

Fig. 1. Receptor-regulated formation of Ins(1,4,5)P_3. The complex of receptor (R) and agonist (A) catalyzes the exchange of GDP for GTP on an unidentified G protein. Binding of GTP to the α subunit of the oligomeric G protein probably promotes its dissociation into α-GTP and βγ subunits. The former probably directly stimulates PIC activity until its intrinsic GTPase activity hydrolyzes the bound GTP, inactivating it and allowing it to reassociate with the βγ complex to form the complete G protein. Active PIC catalyzes hydrolysis of PtdInsP$_2$ to Ins(1,4,5)P_3 and diacyglycerol. The latter can activate certain proteins kinase C before its metabolism by a specific kinase to phosphatidic acid or by specific lipases that remove the fatty acid residues. Ins(1,4,5)P_3 enters the cytosol and regulates Ca^{2+} mobilization from intracellular stores before it is metabolized by a specific 5-phosphatase or 3-kinase. The Ins(1,4)P_2 and Ins(1,3,4,5)P_4 formed are the first substrates of a complex series of phosphorylation and dephosphorylation reactions (not shown) that can eventually lead to inositol that can be recycled to the membrane phosphoinositides.

intracellular effector, for example, adenylyl cyclase or PIC. Stimulation of the latter causes increased hydrolysis of the membrane phospholipid, phosphatidylinositol 4,5-bisphosphate (PtdInsP$_2$), and the formation of Ins(1,4,5)P_3 and DG. The G protein is inactivated, probably within a few seconds, when its intrinsic GTPase activity hydrolyzes the bound GTP and restores the G protein to its inactive state. Further details of G protein function and of the methods used to study G proteins are included in the chapter by Milligan (chapter 1).

A distinguishing feature of the polyphosphoinositide signaling pathway is the complexity of the metabolism of each of the intracellular messengers formed by hydrolysis of $PtdInsP_2$. $Ins(1,4,5)P_3$ may be either dephosphorylated by $Ins(1,4,5)P_3$ 5-phosphatase to inositol 1,4-bisphosphate or phosphorylated by $Ins(1,4,5)P_3$ 3-kinase to inositol 1,3,4,5-tetrakisphosphate (Ins $(1,3,4,5)P_4$) (Berridge and Irvine, 1989; Shears, 1989). Both pathways curtail the ability of $Ins(1,4,5)P_3$ to mobilize intracellular Ca^{2+} stores, and both may ultimately allow inositol to be returned to the plasma-membrane phosphoinositides, but the pathways may also lead to formation of additional intracellular messengers (Berridge and Irvine, 1989). An additional complexity has recently become apparent with the demonstration that $Ins(1,3,4,5)P_4$ may be recycled to $Ins(1,4,5)P_3$ by a specific 3-phosphatase that may be subject to regulation (Hodgson and Shears, 1990; McIntosh and McIntosh, 1990). The methods that allow inositol phosphate metabolism to be analyzed are discussed in the chapter by Godfrey (chapter 2). The complex and rapid metabolism of $Ins(1,4,5)P_3$, and the possibility that it may lead to additional intracellular messengers, present considerable experimental problems in analyzing the intracellular actions of $Ins(1,4,5)P_3$. These problems are further discussed below.

The other product of $PtdInsP_2$ hydrolysis is 1,2-diacylglycerol (DG), which in concert with Ca^{2+} and phosphatidylserine stimulates the activities of a family of protein kinase C enzymes (Nishizuka, 1988). This limb of the phosphoinositide signaling pathway is discussed in the chapter by Burgess (chapter 6), and its possible role in contributing to the complex intracellular Ca^{2+} signals evoked by extracellular stimuli are discussed in Section 4.

The essential features of the phosphoinositide signaling pathway are shown in Fig. 1. Several reviews and books provide more complete coverage of the subject (Michell, 1975; Berridge and Irvine, 1984,1989; Downes and Michell, 1985; Putney, 1986a; Taylor and Merritt, 1986; Berridge and Michell, 1988; Downes et al., 1988; Michell et al., 1989).

1.2. Ins(1,4,5)P$_3$-Sensitive Ca^{2+} Stores

After years of uncertainty, largely because of the experimental problems that bedevil attempts to measure intracellular Ca^{2+} distribution free of its artifactual redistribution, it is now generally accepted that mitochondria can accumulate Ca^{2+} only when the cytoplasmic [Ca^{2+}] is high (>1 μM). Mitochondria may therefore protect the cell from very high and potentially damaging increases in cytoplasmic [Ca^{2+}], but they are not important intracellular Ca^{2+} regulators in unstimulated cells, in which the free [Ca^{2+}] is typically about 100 nM (Burgess et al., 1983; Somlyo et al., 1985). Ins(1,4,5)P$_3$, if it is to mediate the effects of cell-surface receptors on intracellular Ca^{2+} stores, must therefore be able to stimulate Ca^{2+} mobilization from nonmitochondrial stores. This prediction was confirmed by the first demonstration of the effects of Ins(1,4,5)P$_3$ on the intracellular Ca^{2+} stores of pancreatic acinar cells (Streb et al., 1983) and has subsequently been confirmed in numerous other studies (reviewed in Berridge and Irvine, 1984).

Although it is clear that Ins(1,4,5)P$_3$ causes release of Ca^{2+} from a vesicular organelle that becomes loaded with Ca^{2+} only if ATP is available, the nature and localization of these Ins(1,4,5)P$_3$-sensitive Ca^{2+} stores are still unclear. In most cells stimulated with agonists that evoke formation of Ins(1,4,5)P$_3$ (Burgess et al., 1984), and in permeabilized cells stimulated directly with Ins(1,4,5)P$_3$ (Taylor and Putney, 1985), only a fraction of the intracellular Ca^{2+} stores are released. Under appropriate conditions, the released Ca^{2+} can then be resequestered into other nonmitochondrial stores (Dawson and Irvine, 1984). Typically less than half the stores are Ins(1,4,5)P$_3$-sensitive, but there is considerable variation between cell types (Joseph and Williamson, 1989) and other regulators, for example GTP (Mullaney et al., 1988), influence the size of the Ins(1,4,5)P$_3$-sensitive stores. Recent studies (Muallem et al., 1989; Taylor and Potter, 1990; Section 1.3.) suggest an additional complexity because the size of the Ins(1,4,5)P$_3$-sensitive store itself seems to depend on the Ins(1,4,5)P$_3$ concentration.

The heterogeneity among intracellular Ca^{2+} stores may not be restricted to differential distribution of the receptor that mediates the effects of $Ins(1,4,5)P_3$, because the Ca^{2+}-accumulating mechanisms of the different stores may also differ. The Ca^{2+}-ATPases of the $Ins(1,4,5)P_3$-sensitive and $Ins(1,4,5)P_3$-insensitive stores may differ in size (Burgoyne et al., 1989), and perhaps in their sensitivity to 2,5-di-(*tert*-butyl)-1,4-benzohydroquinone (tBu BHQ) (Kass et al., 1989; Oldershaw and Taylor, 1990; Fig. 2), though probably not in their sensitivity to thapsigargin (Thastrup et al., 1990). In pancreatic acinar cells, an even clearer difference between the stores is suggested by the demonstration that $Ins(1,4,5)P_3$-sensitive stores may load by H^+–Ca^{2+} exchange and the $Ins(1,4,5)P_3$-insensitive stores by a Ca^{2+}-ATPase (Thévenod et al., 1989). However, these results are inconsistent with studies of other cells, hepatocytes, for example, in which the $Ins(1,4,5) P_3$-sensitive stores accumulate Ca^{2+} by a mechanism that is inhibited by vanadate and insensitive to the protonophore 2,4-dinitrophenol (Burgess et al., 1984; Taylor and Putney, 1985). Understanding the mechanisms of Ca^{2+} sequestration by different intracellular stores is important if we are to establish the interplay among them and may also provide for rational development of tools that allow the stores to be discriminated experimentally. However, our present understanding is very limited.

Release of Ca^{2+} induced by Ca^{2+} is an established feature of some of the intracellular Ca^{2+} stores of muscle and perhaps neurons and, at least in muscle, is known to be mediated by the ryanodine receptor (Lai et al., 1988). In addition to the endogenous regulators, Ca^{2+}, Mg^{2+}, calmodulin, and ATP, these stores are sensitive to caffeine, ryanodine, and ruthenium red. More general interest in Ca^{2+}-induced Ca^{2+} release has been provoked by the suggestion that it may play a part in generating oscillatory changes in intracellular Ca^{2+} (Berridge and Galione, 1989; Section 4.2.). It now seems likely that caffeine-sensitive stores are more widespread than first supposed and that in several, though not all, tissues the stores released by $Ins(1,4,5)P_3$ are distinct from those released by caffeine (Malgoroli et al., 1990; Matsumoto, 1990). The distinction is particularly clear in adrenal chromaffin cells, in which the two stores are clearly located in different parts of the cell (Burgoyne et al., 1989).

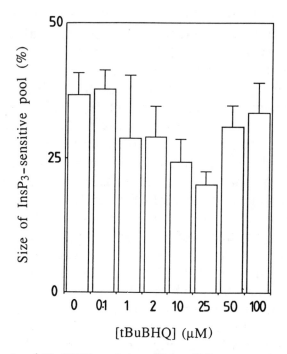

Fig. 2. Effects of tBuBHQ on intracellular Ca^{2+} stores of permeabilized hepatocytes. Permeabilized hepatocytes loaded to steady-state with $^{45}Ca^{2+}$ were stimulated for 2 min with various concentrations of tBuBHQ and then for a further 30 s with a maximal concentration of $Ins(1,4,5)P_3$ (10 µM). The figure shows the fraction of the $^{45}Ca^{2+}$ pool remaining after tBuBHQ treatment that was released by $Ins(1,4,5)P_3$. The effects of tBuBHQ alone, which released up to 54% of the stores (Oldershaw and Taylor, 1990), are not shown, but it is clear that low concentrations of tBuBHQ ($\leq$25 µM) have some selectivity for $Ins(1,4,5)P_3$-sensitive Ca^{2+} stores.

It is clear that only a discrete fraction of the intracellular Ca^{2+} stores of unstimulated cells are sensitive to $Ins(1,4,5)P_3$ and that the remaining stores may be subject to independent regulation (reviewed in Meldolesi et al., 1990). However, the morphological identities of these different stores are far less clear.

Subcellular fractionation studies have suggested that $Ins(1,4,5)P_3$-sensitive Ca^{2+} stores copurify with fractions enriched in endoplasmic reticulum enzyme markers (Streb et al., 1984), with plasma-membrane markers (Guillemette et al., 1988), or with neither (Volpe et al., 1988). Immunocytochemical studies of the cerebellar Ins(1,4,5)P$_3$ receptor with antibodies to the purified protein are equally inconclusive, with reports of its localization to rough, and to a lesser extent smooth, endoplasmic reticulum, nuclear membrane, subplasmalemmal cisternae, and plasma membrane (Maeda et al., 1989; Mignery et al., 1989; Ross et al., 1989).

Cell fractionation combined with immunocytochemical analyses of many intact cells have identified discrete small (50–200 nm in diameter) intracellular vesicular organelles that contain a Ca^{2+}-ATPase related to that of sarcoplasmic reticulum and large amounts of the low-affinity high-capacity Ca^{2+}-binding protein, calreticulin. These organelles have been termed calciosomes by Volpe and his colleagues (1988), and were initially proposed to be the Ins(1,4,5)P$_3$-sensitive Ca^{2+} stores. More recent studies (reviewed in Meldolesi et al., 1990) suggest that the situation is more complex. Calreticulin has been identified in endoplasmic reticulum, the Ca^{2+}-ATPase of the Ins(1,4,5)P$_3$-sensitive stores of adrenal chromaffin cells is clearly distinct from that present in calciosomes (Burgoyne et al., 1989), and even in the initial study, the association between Ins(1,4,5)P$_3$ sensitivity and calciosome markers was inconclusive (Volpe et al., 1988).

In skeletal muscle, Ca^{2+} uptake occurs throughout the sarcoplasmic reticulum, but Ca^{2+} release occurs only at the terminal cisternae, where the transverse tubules and sarcoplasmic reticulum are closely apposed and the ryanodine receptors are concentrated. It is worth considering the possibility that such regional specialization may also occur within Ins(1,4,5)P$_3$-sensitive stores. Such a situation could explain the difficulty that many groups have experienced in observing Ins(1,4,5)P$_3$-induced Ca^{2+} mobilization from subcellular fractions, because fragmentation of the stores could produce vesicles enriched in only Ca^{2+}-ATPase or Ins(1,4,5)P$_3$ receptors, but not both. Evidence from a variety of sources suggests that the situation may be even more complex with very dynamic interactions between different Ca^{2+} stores. This could explain the otherwise inconsistent reports of the localization of Ins(1,4,5)P$_3$ receptors, and is more directly supported by the effects of GTP on intracellular Ca^{2+} stores, perhaps reflecting the involvement of a small GTP-binding protein in mediating fusion between discrete Ca^{2+} stores (Mullaney et al., 1988). The possibility that the cytoskeleton may play a part in maintaining the subcellular organization of $Ins(1,4,5)P_3$-sensitive stores must also be considered, because in liver, the copurification of specific $Ins(1,4,5)P_3$-binding sites with plasma membrane markers is disrupted by cytochalasin B (Rossier et al. 1991). There is no doubt that intracellular Ca^{2+} stores are heterogenous, and there is growing evidence (Section 4.) that specific interactions between them are an important feature of the response of intact cells to extracellular stimuli. There is an urgent need for tools that will reliably discriminate between the different stores, locate them within intact cells, and allow a better understanding of the dynamic interactions between them.

1.3. Mechanism of Ins(1,4,5)P₃-Induced Ca²⁺ Mobilization

Acceptance of the important role of $Ins(1,4,5)P_3$ in regulating intracellular Ca^{2+} stores has focused attention on its mechanism of action. These studies have received additional impetus from the recent purifications of cerebellar and smooth-muscle $Ins(1,4,5)P_3$-binding sites (Section 3.3.). In this section, we will consider only the properties of the $Ins(1,4,5)P_3$ receptor in its native membrane environment.

Studies of many different tissues have established several key features of $Ins(1,4,5)P_3$ action:

1. $Ins(1,4,5)P_3$ stimulates Ca^{2+} efflux rather than inhibiting a Ca^{2+}-uptake pathway. ATP is not required for the response (Prentki et al., 1984; Taylor and Putney, 1985, ruling out earlier suggestions that $Ins(1,4,5)P_3$ might act via a specific kinase.
2. Ca^{2+} release is little affected by temperature (Smith et al., 1985; Joseph and Williamson, 1986), and $Ins(1,4,5)P_3$ can facilitate Ca^{2+} transfer in either direction across membrane vesicles according to the electrochemical gradient for Ca^{2+} (Muallem et al., 1985). These results provided the first indication that $Ins(1,4,5)P_3$ opens a Ca^{2+} channel.
3. The charge movements resulting from $Ins(1,4,5)P_3$-induced electrogenic Ca^{2+} efflux are compensated for by an influx of permeant monovalent cations (probably K^+ under physio-

logical conditions) or an efflux of permeant anions. Removal of permeant cations, blockade of K$^+$ channels, or addition of high concentrations of permeant anions therefore reduce or totally inhibit Ins(1,4,5)P$_3$-induced Ca^{2+} mobilization (Muallem et al., 1985; Joseph and Williamson, 1986; Shah and Pant, 1988).

4. Stopped flow measurements of permeabilized cells (Champeil et al., 1989; Meyer et al., 1990) and flash photolysis of caged Ins(1,4,5)P$_3$ in intact cells (Ogden et al., 1990; Parker, chapter 7) have demonstrated that a maximal concentration of Ins(1,4,5)P$_3$ stimulates Ca^{2+} mobilization within 100 ms of its addition, again consistent with rapid opening of a Ca^{2+} channel.

5. Opening of a Ca^{2+} channel by Ins(1,4,5)P$_3$ is steeply cooperative (Meyer et al., 1988, 1990; Nunn and Taylor, 1990), though binding of Ins(1,4,5)P$_3$ to its receptor is not (Section 3.1.). This cooperativity may amplify small changes in cytoplasmic Ins(1,4,5)P$_3$ concentration and could explain the long latency (up to 1 s) between the addition of Ins(1,4,5)P$_3$ and channel opening at low Ins(1,4,5)P$_3$ concentrations or after stimulation with submaximal concentrations of extracellular hormone (Meyer et al., 1990; Ogden et al., 1990; Miledi and Parker, 1989).

6. Binding of Ins(1,4,5)P$_3$ to its receptor and Ins(1,4,5)P$_3$-induced Ca^{2+} mobilization are optimal at alkaline pH (typically pH 8) (Worley et al., 1987; Joseph et al., 1989; Meyer et al., 1990). This effect may be physiologically important, because protein kinase C, by activating Na$^+$–H$^+$ exchange, can increase cytoplasmic pH, and this has been implicated in the action of many growth factors.

7. Ins(1,4,5)P$_3$ receptor does not become desensitized (Taylor et al., 1989). This is important because a sustained response to extracellular stimuli, which requires Ca^{2+} entry, is proposed to require an active Ins(1,4,5)P$_3$ receptor (Section 4.1.).

8. Ins(1,4,5)P$_3$-induced Ca^{2+} mobilization is unaffected by agents (ruthenium red, caffeine, ATP, voltage-sensitive Ca^{2+}-channel blockers, dantrolene, Co^{2+}, Ni^{2+}, Mn^{2+}, Cd^{2+}) that affect other Ca^{2+} fluxes (Berridge and Irvine, 1984, 1989; Joseph and Williamson, 1986).

Although Ins(1,4,5)P_3-induced Ca^{2+} mobilization is not subject to homologous desensitization, it does appear to be regulated, though the full physiological significance has yet to be established. In some cells (e.g., neuroblastoma cell lines, pancreatic acinar cells, and cerebellum) (Cheuh and Gill, 1986; Jean and Klee, 1986; Willems et al., 1990; Joseph et al., 1989), but not in others (e.g., hepatocytes) (Burgess et al., 1984), Ins(1,4,5)P_3-induced Ca^{2+} mobilization is inhibited by an increase in cytoplasmic [Ca^{2+}]. This negative feedback is a key feature of one model for intra-cellular Ca^{2+} spiking (Section 4.2.). The effects of Ca^{2+} on the interaction between Ins(1,4,5)P_3 and its receptor have been most thoroughly examined in cerebellum, where the purified receptor is unaffected by Ca^{2+}, but Ca^{2+}-sensitivity is conferred by a distinct 300-kDa protein, calmedin (Danoff et al., 1988). Preliminary studies suggest that although calmedin is widely distributed in central nervous system, it is far less abundant, and perhaps absent from many peripheral tissues (Danoff et al., 1988). This aspect of Ins(1,4,5)P_3-induced Ca^{2+} mobilization, the effects of cytoplasmic Ca^{2+}, and particularly the role of accessory proteins, clearly deserves to be examined more closely in peripheral tissues.

The cerebellar Ins(1,4,5)P_3 receptor is phosphorylated by cAMP-dependent protein kinase, but the phosphorylation has no effect on its affinity for Ins(1,4,5)P_3 and only a very modest inhibitory effect on the ability of Ins(1,4,5)P_3 to induce Ca^{2+} mobilization (Supattapone et al., 1988a).

Other factors, such as GTP, perhaps Ins(1,3,4,5)P_4 (Irvine, 1989), and protein kinase C (Burgess, chapter 6), may indirectly regulate the effects of Ins(1,4,5)P_3 by altering the amount of Ca^{2+} in the Ins(1,4,5)P_3-sensitive stores rather than by direct interaction with the Ins(1,4,5)P_3 receptor.

The ability of many normal intracellular constituents (e.g., ATP, GTP, 2,3-bisphosphoglycerate) to compete with Ins(1,4,5)P_3 for its receptor has important implications, because these competitive antagonists are present in sufficient amounts to reduce the apparent affinity of the receptor for Ins(1,4,5)P_3 substantially (Nunn and Taylor, 1990; Guillemette et al., 1990). Within the cell, therefore, the concentrations of Ins(1,4,5)P_3 required to evoke Ca^{2+} mobilization may be substantially higher than predicted from experiments in vitro. Consistent with this suggestion is the observation that careful measurements of the cytoplasmic levels of Ins(1,4,5)P$_3$ in resting and stimulated cells are substantially higher (2 and 25 μM) (Horstman et al., 1988) than expected from the concentration-dependence of Ins(1,4,5)P$_3$-induced Ca^{2+} mobilization in vitro (typical EC$_{50}$ = 0.2 μM).

Another regulatory influence on Ins(1,4,5)P$_3$-induced Ca^{2+} mobilization has recently come to light, but the mechanisms underlying it are unclear. We (Taylor and Potter, 1990), and others (Muallem et al., 1989) have shown that even when Ca^{2+} sequestration is inhibited (e.g., by removal of ATP) submaximal concentrations of Ins(1,4,5)P$_3$ release only a fraction of the Ins(1,4,5)P$_3$-sensitive stores, whereas a maximal concentration wholly empties them. This "fractional release" of Ca^{2+} stores is not a consequence of desensitization of the Ins(1,4,5)P$_3$ receptor (Taylor and Potter, 1990), but may reflect a synergistic interaction between cytoplasmic Ins(1,4,5)P$_3$ and luminal Ca^{2+} in regulating the Ca^{2+} channel (Irvine, 1990). A maximal concentration of Ins(1,4,5)P$_3$ may, even in the absence of luminal Ca^{2+}, be sufficient to open the Ca^{2+} channel, whereas lower concentrations of Ins(1,4,5)P$_3$ would require higher luminal Ca^{2+} contents to open it. Under these conditions, the effects of low concentrations of Ins(1,4,5)P$_3$ would become limited by the Ca^{2+} content of the stores, thereby preventing their complete emptying. This model for "fractional Ca^{2+} release" is presently being tested, because it could provide a very attractive mechanism for receptor-regulated Ca^{2+} entry (Section 4.1.).

2. Ins(1,4,5)P$_3$ and Ca^{2+} Mobilization: Experimental Approaches

2.1. Cell Permeabilization

The negative charges of Ins(1,4,5)P$_3$ prevent it from penetrating the plasma membrane, thereby ensuring that it is not lost from the cell and can effectively function as an intracellular messenger. The effects of directly applied Ins(1,4,5)P$_3$ can therefore be examined only after its microinjection into intact cells or after application to cells in which the plasma membrane no longer presents a barrier.

Advantages of microinjection, especially when applied to large cells, are the minimal disruption of the intracellular environment and the opportunity to examine the effects of injected messengers on the final cell response, for example, secretion (Burgoyne, chapter 11). The large size of *Limulus* photoreceptors and *Xenopus* oocytes, the latter >1 mm in diameter, make them particularly amenable to microinjection of Ins(1,4,5)P_3. Such studies have shown that the Ins(1,4,5)P_3-sensitive Ca^{2+} stores are highly localized to discrete regions, the light-sensitive rhabdomeral lobe of the photoreceptor (Payne and Fein, 1987), and the animal pole of the oocyte (Berridge, 1988). Even within the animal pole, the effects of Ins(1,4,5)P_3 injection are spatially restricted so that only the immediately surrounding Ca^{2+} stores are depleted, with neighboring stores retaining their sensitivity to subsequent injections of Ins(1,4,5)P_3 (Berridge, 1989). Microinjection of Ins(1,4,5)P_3 thus provided some of the first evidence for spatial organization of intracellular Ca^{2+} signals, and that evidence is now supported by studies with single-cell imaging of cells loaded with Ca^{2+} indicators (Berridge et al., 1988; Cheek, 1989; Moreton, chapter 5; Poenie, chapter 4).

Although microinjection of Ins(1,4,5)P_3 is a powerful technique, it can be easily applied only to small numbers of relatively large cells and is inappropriate for examining the properties of Ins(1,4,5)P_3-induced Ca^{2+} mobilization that require stringent control of the receptor environment. Such experiments are best performed with subcellular fractions or, more usually and certainly more easily, with permeabilized cells.

A number of techniques, including ATP^{4-}-permeabilization of mast cells (Cockcroft and Gomperts, 1979) and transformed fibroblasts (Rozengurt and Heppel, 1975) and permeabilization of pancreatic acinar cells by Ca^{2+}-free media (Streb and Schulz, 1983), have been successfully applied to specific cell types, but are not generally applicable. Agents that form a complex with membrane-bound cholesterol, certain bacterial toxins, and physical methods (such as intense electric fields or osmotic shocks) are more generally applicable cell-permeabilization techniques.

Saponin (Glauert et al., 1962), a saponin-ester β-escin (Kobayashi et al., 1989), digitonin (Gogelein and Huby, 1984), and the structurally unrelated polyene antibiotics (e.g., filipin) (Norman et al., 1972) interact with cholesterol in plasma membranes to produce holes about 8–10 nm in diameter. These agents are relatively selective in permeabilizing only the plasma membrane and not the membranes of intracellular organelles, because the latter contain negligible amounts of cholesterol. We permeabilize rat hepatocytes by incubating cells at a final density of about 5×10^5 cells/mL in a Ca^{2+}-free, cytosol-like medium (140 mM KCl, 20 mM NaCl, 2 mM MgCl$_2$, 1 mM EGTA, 20 mM PIPES; pH 6.8) for 10 min at 37°C with saponin (75 µg/mL). After permeabilization, the cells are washed twice in the same medium without saponin, and then resuspended in an appropriate cytosol-like medium. With this treatment, >95% of cells become permeable to trypan blue and remain permeable after removal of saponin, but variable amounts of cytosolic proteins are lost. One solution to this problem is to permeabilize cells in very dense suspensions (10^7 cells/mL) with more saponin (200 µg/mL for 10 min, with additional ATP [4.5 mM]) and then to use the cells without washing away the cytosolic proteins that leak to the medium. With such a protocol, we find that hepatocytes accumulate $^{45}Ca^{2+}$ and respond normally to the stable analog of Ins(1,4,5) P_3, Ins(1,4,5)P_3[S]$_3$, despite the continued presence of saponin and the unusually high cell density (Oldershaw and Taylor, unpublished observation).

Intracellular Ca^{2+} stores behave relatively normally after saponin permeabilization (Burgess et al., 1983), but receptor regulation of Ins(1,4,5)P_3 formation and, presumably, the interaction of other receptors with their G proteins are usually disrupted. A recent study of receptor- and Ins(1,4,5)P_3-stimulated contraction of permeabilized smooth-muscle cells suggests one solution to this problem, because β-escin permeabilized the cells without abolishing functional interactions between receptors and their effector systems (Kobayashi et al., 1989). Therefore, β-escin may be preferable to saponin when functional interactions between receptors and their signaling pathways are important considerations.

Of the many microbial cytolysins (Thelestam and Mollby, 1979), only streptolysin-O from *Streptococcus pyogenes* and α-toxin from *Staphylococcus aureus* have been widely used to permeabilize cells. The pores formed by α-toxin are too small to allow passage of the toxin itself and, although the much larger pores formed by streptolysin-O could allow the toxin access to intracellular membranes, it is simple to devise protocols that restrict its action to the plasma membrane (Ahnert-Hilger and Gratzl, 1988). Both toxins can therefore provide selective permeabilization of the plasma membrane: α-toxin by creating pores that allow passage of small molecules (up to 1000 dalton) and streptolysin-O by creating much larger pores that allow passage of large protein molecules. Both toxins, saponin, and β-escin are available from Sigma or Calbiochem.

Exposure to intense electric fields (typically about 2 kV/cm, $\tau = 200$ μs) causes localized breakdown of the plasma membrane, leaving pores of about 4 nm in diameter. Knight (1981) has described both the practical and the theoretical aspects of the method and has successfully applied the technique to studies of the intracellular factors that control exocytosis (*see* Burgoyne, chapter 11). In our experience, it is an effective means of preparing permeabilized cells that retain functional interactions between receptors and phosphoinositidase C (Merritt et al., 1986), but cells that have been chemically permeabilized are generally better able to accumulate Ca^{2+} in intracellular stores and then to respond to $Ins(1,4,5)P_3$.

A variety of techniques allow cells to be reversibly permeabilized with large, otherwise impermeant, molecules becoming trapped in the cell when the plasma membrane reseals. These procedures cannot offer the temporal resolution needed to examine the intracellular effects of such active signal molecules as $Ins(1,4,5)P_3$. They do, however, allow intracellular loading of inactive precursors of the signal molecules, caged $Ins(1,4,5)P_3$ for example, which can later be instantaneously photoactivated after the cells have resealed (*see* chapter 7 by Parker).

2.2. Ca²⁺ Mobilization

A number of techniques are commonly applied to examine the effects of $Ins(1,4,5)P_3$ in permeabilized cells and subcellular fractions. Because each has its advantages and limitations there is much to be said for combining a variety of approaches.

The free [Ca^{2+}] of the medium surrounding permeabilized cells may be monitored using Ca^{2+}-selective electrodes and Ca^{2+}-sensitive indicator dyes. When the intracellular organelles of the cells are the major Ca^{2+} buffers, Ca^{2+} handling by the stores may be monitored. Permeabilized cells are suspended in a cytosol-like medium without EGTA and the free [Ca^{2+}] of the medium is then recorded as ATP is added, allowing the cells to sequester Ca^{2+} and so reduce the free [Ca^{2+}] of the medium. This situation, in which Ca^{2+} transport by intracellular stores causes the free [Ca^{2+}] of the surrounding medium to change, more closely resembles the situation in intact cells than do the conditions usually used for $^{45}Ca^{2+}$ flux studies, in which the free [Ca^{2+}] is fixed by EGTA. The free [Ca^{2+}] of medium prepared without addition of EGTA may exceed 10 µM and this may be too high for the cells to reduce it substantially. Under these conditions, the medium may need to have Ca^{2+} removed by treatment with a Ca^{2+}-chelating resin (Meyer et al., 1990) or a small amount of a weak Ca^{2+} buffer (e.g., ATP^{4-}) may be added to bring the free [Ca^{2+}] into an appropriate range (<1 µM) (Dawson, 1985).

Electrode and indicator methods allow the effects of Ins(1,4,5)P$_3$ to be observed immediately, which can be useful when sequential additions of drugs are under investigation. The principal disadvantage shared by these methods is that the change in free [Ca^{2+}] is the assay of cellular activity; contamination of drugs (e.g., Ins(1,4,5)P$_3$) with Ca^{2+} can therefore cause problems and the effects of variations in the free [Ca^{2+}] of the medium are difficult to examine. The methods also examine only the net flux of Ca^{2+} between the stores and medium; unidirectional flux measurements require tracer methods.

The preparation and calibration of Ca^{2+}-selective macroelectrodes, the characteristics of the Ca^{2+}-ionophores used to make them (Thomas, 1982; Ammann, 1986), and their application to studies of Ca^{2+} stores in permeabilized cells and subcellular fractions (Dawson, 1985; Joseph et al., 1984; Thévenod et al., 1989) are described elsewhere. A major limitation is the poor temporal resolution (response time of up to 1 s). Fluorescent Ca^{2+} indicators overcome this problem, but with affinities for Ca^{2+} of <600 nM they are less suited to measurements of high free [Ca^{2+}] (>5 µM) than are electrodes. Nevertheless, for most purposes Ca^{2+} indicators will be the first choice.

Luminescent Ca^{2+}-sensitive photoproteins (e.g., aequorin) are used to monitor Ca^{2+} changes in intact cells and, although more difficult to use than fluorescent indicators, they have the advantage of responding to Ca^{2+} concentrations that would saturate the presently available fluorescent dyes (Campbell, 1983; Cobbold and Rink, 1987). The photoproteins, however, are not suitable for studies of permeabilized cells.

The most commonly used fluorescent Ca^{2+} indicators (e.g., quin2, fura-2, fluo-3, indo-1) are all derivatives of the selective Ca^{2+}-chelator, 1,2-bis-(2-aminophenoxy)ethane-$N,N,N'N',$-tetracetic acid (BAPTA; a relative of EGTA) to which various fluorophores have been attached (Cobbold and Rink, 1987; Tsien, 1989; Poenie, chapter 5). The indicators are extensively used to load intact cells because the lipophilic acetoxymethyl esters of the dyes are membrane-permeant, but once inside the cell they are cleaved by nonspecific esterases and the Ca^{2+}-sensitive hydrophilic indicator is then trapped within the cell. The methods used to monitor intracellular Ca^{2+} in intact cells with these indicators are described in chapters 4 and 5 by Moreton and Poenie. The membrane-impermeant, free-acid forms of the dyes have also found widespread use as monitors of Ca^{2+} handling by permeabilized cells.

We use fluo-3 to measure Ca^{2+} fluxes in permeabilized cells because its relatively low affinity for Ca^{2+} (K_d= 864 nM at 37°C) (Merritt et al., 1990) provides greater sensitivity in the relevant range of free [Ca^{2+}] than fura-2 (K_d= 224 nM). Permeabilized hepatocytes (10^7/mL) are suspended in nominally Ca^{2+}-free cytosol-like medium with mitochondrial inhibitors (Section 2.1.) but without added EGTA. Fluorescence (λ_{ex} = 503 nm, λ_{em} = 530 nm) is measured with a Perkin-Elmer LS50 luminescence spectrometer. Autofluorescence is recorded before addition of fluo-3 free acid (2 µM) and Ca^{2+} uptake is then initiated by addition of ATP (1.5 mM) and a regenerating system (creatine kinase 5 U/mL; creatine phosphate 5 mM). The permeabilized cells then reduce the free [Ca^{2+}] from about 400 nM to a new steady-state of about 200 nM and the effects of added Ins(1,4,5)P_3 can then be examined (Fig. 3). Traces are calibrated by addition of EGTA (2 µM) to obtain the minimum fluorescence signal, F_{min}, followed by Ca^{2+}

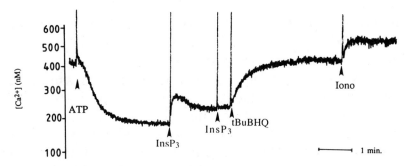

Fig. 3. Handling of Ca^{2+} by permeabilized hepatocytes monitored with fluo-3. Permeabilized hepatocytes rapidly sequester Ca^{2+} after addition of ATP, causing the free $[Ca^{2+}]$ of the medium, monitored with fluo-3 fluorescence, to fall. Addition of $Ins(1,4,5)P_3$ (10 µM) stimulates release of a fraction of this sequestered Ca^{2+}, some of which is then resequestered. A second addition of $Ins(1,4,5)P_3$ evokes no further release, indicating that reuptake is not into $Ins(1,4,5)P_3$-sensitive stores. Addition of tBuBHQ (25 µM) releases most of the Ca^{2+} not released by $Ins(1,4,5)P_3$, and the remainder is discharged by the Ca^{2+} ionophore ionomycin (1 µM). Minor Ca^{2+}-contamination of drugs accounts for the slightly greater fluorescence signal after ionomycin relative to that before addition of ATP.

(15 µM) to obtain F_{max}. The free $[Ca^{2+}]$ is then computed from

$$[Ca^{2+}] = K_d (F - F_{min})/(F_{max} - F)$$

where K_d = 864 nM (Merritt et al., 1990).

A clear advantage of the fluorescent indicators is their applicability to measuring rapid changes in free $[Ca^{2+}]$. Stopped-flow measurements combined with techniques similar to those described above have exploited this potential and now provide opportunities to measure the effects of $Ins(1,4,5)P_3$ on intracellular Ca^{2+} stores over a time-scale of tens of milliseconds (Champeil et al., 1989; Meyer et al., 1990; Section 1.3.).

The *bis*-azo indicators, aresenazo III and antipyralazo III (Thomas, 1982), can be used with subcellular fractions, but their use is limited to situations in which their low affinities for Ca^{2+} (K_ds between 15 and 500 µM) offer considerable advantages (Palade, 1987). In most situations the poor temporal resolution that results from the need to monitor the ratio of two absorbances, the elaborate spectrophotometers that such measurements require, and the considerable Ca^{2+} fluxes required to give detectable signals are serious disadvantages.

In contrast to the methods so far described, measurements of $^{45}Ca^{2+}$ flux allow the behavior of intracellular Ca^{2+} stores to be monitored without changes in the free $[Ca^{2+}]$ of the incubation medium, and they can readily be adapted to allow unidirectional flux measurements. We incubate permeabilized hepatocytes (8.5×10^5 cells/mL) in the cytosol-like medium with mitochondrial inhibitors and $^{45}CaCl_2$ (2–3 µCi/mL). Samples (200 µL) are stopped by dilution into 5 mL of cold isotonic sucrose (310 mM)/ EGTA (1 mM) and rapidly filtered through Whatman glass-fiber GF/C filters, and the filters are washed with a further 5 mL rinse of the sample tube with sucrose/EGTA. Our earlier experiments included ^3H_2O in the first sucrose/EGTA addition to correct for trapped volume, but with much-improved filtration apparatus the results are now sufficiently reproducible without this correction. The filters are counted, and ^{45}Ca uptake and the effects of drugs (e.g., $Ins(1,4,5)P_3$) are then expressed as a fraction of ATP-dependent ^{45}Ca accumulation. In our experiments, permeabilized hepatocytes load to steady-state with ^{45}Ca within 10 min of addition of ATP and the regenerating system, and maintain the same ^{45}Ca content for at least 20 min.

Another approach, now much neglected, is provided by the tetracycline antibiotic chlortetracycline. In aqueous solution, chlortetracycline has very low affinity for Ca^{2+} (K_d = 440 µM) and is therefore insensitive to the usual changes in free $[Ca^{2+}]$ in cytoplasm or in the medium surrounding permeabilized cells. Binding of Ca^{2+} (or Mg^{2+}) slightly increases the fluorescence of chlortetracycline, but more important, it becomes more lipophilic and associates with membranes, where its fluorescence (λ_{ex} = 380 nm; λ_{em} = 520 nm) is substantially increased (Caswell, 1979). The indicator therefore reports the free $[Ca^{2+}]$ in close proximity to membranes and, by virtue of its low affinity for Ca^{2+}, it is sensitive only to high $[Ca^{2+}]$ close to membranes. In permeabilized cells, and in intact cells from which extracellular indicator has been removed (Marcotte et al., 1990), the indicator therefore provides a monitor of the free Ca^{2+} content of intracellular stores. Chlortetracycline has recently been successfully used to examine intracellular Ca^{2+} mobilization in basophilic leukemia cells (Marcotte et al., 1990). Practical details of the methods can be found in that report.

The many methods available for measuring Ca^{2+} handling by intracellular stores provide opportunities to measure unidirectional fluxes, net Ca^{2+} transport between the stores and medium, and perhaps also the free $[Ca^{2+}]$ of the stores. Each method has its limitations, but together they provide complementary approaches to our understanding of Ca^{2+} handling by permeabilized cells and subcellular fractions.

3. Ins(1,4,5)P_3 Receptor: Experimental Approaches

3.1. Identification of Ins(1,4,5)P$_3$ Receptors

The existence of specific intracellular $Ins(1,4,5)P_3$ receptors was initially inferred from the structural specificity of Ca^{2+} mobilization induced by inositol phosphate (Burgess et al., 1984). Synthetic inositol phosphates and their analogs (Potter, chapter 8) now provide a substantial battery of ligands with which the receptor may be better characterized.

Structure–activity studies with inositol phosphates and the analogs that are presently available have identified only full agonists with a wide range of affinities. The natural messenger, D-$Ins(1,4,5)P_3$, is the most potent. The only antagonists presently available (heparin, ATP, and perhaps GTP and 2,3-bisphosphoglycerate) (Nunn and Taylor, 1990; Guillemette et al., 1990) bind with relatively low affinity and bind equally well to many other intracellular targets. Heparin, the most potent antagonist (K_d = about 0.5 μM) (Worley et al., 1987), has been used to purify $Ins(1,4,5)P_3$ receptors (Section 3.3.), but there is urgent need for more specific antagonists.

From the limited data available, the following structural features of $Ins(1,4,5)P_3$ appear to contribute to its activity at the Ca^{2+}-mobilizing receptor (Fig. 4) (substituents are numbered as D-*myo* inositol derivatives [further details in Berridge and Irvine, 1989]):

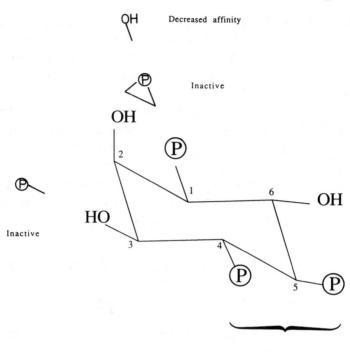

Fig. 4. Structural features of Ins(1,4,5)P_3 essential for Ca²⁺ mobilization. The effects of substituting various groups for the 1-phosphate, and the 2- and 3-hydroxyls of Ins(1,4,5)P_3 are shown.

1. Recognition is stereospecific: L-Ins(1,4,5)P_3 (= D-Ins(3,5,6)P_3) is at least 1000-fold less active than D-Ins(1,4,5)P_3 (Taylor et al., 1989).
2. Adjacent phosphates at the 4 and 5 positions are essential, but their replacement by phosphorothioate groups reduces activity only slightly.
3. Removal of the 1-phosphate substantially reduces affinity (>300-fold). The affinity of Ins(2,4,5)P_3 is only about 40-fold less than that of Ins(1,4,5)P_3 (Nunn and Taylor, 1990), implying that the 1- or 2-phosphate can increase affinity. Removal of the 2-hydroxyl has little effect (Hirata et al., 1989).
4. Attachment of even very large additional substituents to the 1-phosphate decreases affinity by only 10- to 100-fold

(Henne et al., 1988), and substitution of the 2-hydroxyl decreases affinity by less than tenfold (Hirata et al., 1989). These modifications, by which groups that are more reactive can be attached to the 1- or 2-positions without loss of activity, offer great promise because they may allow synthesis of derivatives suitable for many applications, including affinity chromatography and photoaffinity labeling. Some progress has already been made in this direction (Hirata et al., 1990).

5. Other modifications that cause substantial loss of activity include removal of the 2-, 3-, and 6-hydroxyl groups (Polokoff et al., 1988) and methylation or removal of the 6-hydroxyl (Polokoff et al., 1988; Nahorski and Potter, 1989).

6. Other naturally occurring inositol phosphates, Ins(1,3,4)P$_3$, Ins(1,3,4,5)P$_4$, and Ins(cyclic1:2,4,5)P$_3$, are probably inactive. Earlier reports of activity are probably attributable to contamination with inositol phosphates that are more active (Nahorski and Potter, 1989).

Clearly, these structure–activity analyses are beginning to provide indications of the interactions between Ins(1,4,5)P$_3$ and its receptor that allow such specific recognition (Fig. 4). The increasing availability of synthetic ligands and the opportunities for site-directed mutagenesis provided by the cDNA sequence of the cerebellar receptor (Section 3.4.) will surely allow these interactions to be more fully characterized in the near future.

Structure–activity studies show unequivocally the interactions between ligands and their physiological receptors. Radioligand-binding studies must be more cautiously interpreted until the binding site can be shown to be the recognition site of the physiological receptor. Specific intracellular Ins(1,4,5)P$_3$ binding sites were first identified in permeabilized hepatocytes and neutrophils (Spät et al., 1986) and subsequently in many other tissues (reviewed in Nahorski and Potter, 1989; Joseph and Williamson, 1989; Fig. 5). In general, the rank order of potency of ligands in displacing specifically bound Ins(1,4,5)P$_3$ correlates well to their potency in stimulating Ca^{2+} mobilization. However,

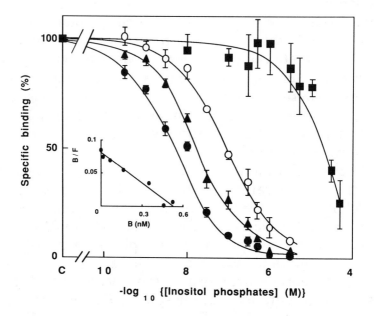

Fig. 5. Inhibition of specific Ins(1,4,[^{32}P]5)P_3 binding by Ins(1,4,5)P_3 analogs in permeabilized hepatocytes: The specificity of the Ins(1,4,5)P_3 binding site for inositol trisphosphates is demonstrated. Inhibition (%) of specific Ins(1,4,[^{32}P$_3$]5)P_3 binding by D-Ins(1,4,5)P_3 (●), DL-Ins(1,4,5)P_3 (▲), DL-Ins(1,4,5)P_3[S]$_3$ (○), and L-Ins(1,4,5)P_3 (■) is plotted (mean ± SEM, n = 4–8) as a function of increasing concentration of unlabeled competitor. The data for Ins(1,4,5)P_3 is redrawn in the form of a Scatchard plot to demonstrate the simple bimolecular nature of the interaction of Ins(1,4,5)P_3 with its binding site. Reproduced from Nunn and Taylor (1990) with permission of the Biochemical Society.

the binding sites typically have much higher affinity than expected from Ca^{2+}-mobilization studies, casting some doubt on their physiological significance. Another problem is the difficulty of distinguishing Ins(1,4,5)P_3 binding to its Ca^{2+}-mobilizing receptor from its binding to the enzymes that metabolize it. Recent studies from our laboratory address these problems.

Freeze-dried saponin-permeabilized hepatocytes reconstituted (0.2 mg cell protein/mL) in cytosol-like medium (Section 2.1.) are incubated in microfuge tubes in a final vol of 0.5 mL at 4°C with labeled Ins(1,4,5)P_3 (0.02 μCi [^{3}H]Ins(1,4,5)P_3, 34 Ci/

mmol, or 0.02 μCi Ins(1,4,[^{32}P]5)P$_3$ >1000 Ci/mmol) and competing drugs. Specific binding reaches equilibrium within 5 min. Bound and free ligands are separated by rapid centrifugation (20,000g at 4°C) and the pellet rinsed superficially with 200 μL of ice-cold buffer. Experiments in which specific Ins(1,4,[^{32}P]5)P$_3$ binding was measured without the final rinse, but with ^{3}H$_2$O included to correct for trapped volume, established that the rinse caused no loss of specific binding. The pellets are resuspended in 100 μL of distilled water by sonication and counted in the microfuge tubes in 1 mL of scintillation fluid. Under the conditions of these experiments, in which the incubations were on ice and both the period of incubation and the cell-protein concentration were minimized, degradation of added Ins(1,4,[^{32}P]5)P$_3$ never exceeded 5%, and the total amount bound was about 3%.

The affinity (K_d) and number of binding sites (B_{max}) are determined by nonlinear least-squares curve fitting by the iterative computer program LIGAND (Munson and Rodbard, 1980), which determines nonspecific binding as the limiting slope (at infinite free-ligand concentration) of the plot of total vs free ligand concentration. This analysis avoids the need for a single concentration of unlabeled Ins(1,4,5)P$_3$ to be arbitrarily selected to measure nonspecific binding because it allows the entire competition curve to be used to define it. Similar values of K_d and B_{max} are obtained when nonspecific binding is defined as binding in the presence of 300 nM Ins(1,4,5)P$_3$; this value was used to provide an initial estimate for the more rigorous curve-fitting analysis.

In permeabilized hepatocytes, Ins(1,4,[^{32}P]5)P$_3$ binds to a single class of high-affinity sites (K_d = 7.8 ± 1.1 nM) with a Hill coefficient of 1.02 ± 0.09, and the number of sites (B_{max} = 1.2 ± 0.3 pmol/mg protein) (Fig. 5) increases linearly over a tenfold range of protein concentrations. Association of Ins(1,4,[^{32}P]5)P$_3$ is rapid, and once equilibrium is attained, addition of excess unlabeled Ins(1,4,5)P$_3$ rapidly displaces bound Ins(1,4,[^{32}P]5)P$_3$ (dissociation rate constant = 0.03 min^{-1}). Similar methods have been applied to many tissues with essentially similar results (reviewed in Nahorski and Potter, 1989; Joseph and Williamson, 1989). The cerebellum is noteworthy for its unusually large number of binding sites (20 pmol/mg protein) (Worley et al., 1987), a feature that aided the purification of the receptor protein from that source (Section 3.3.).

We have provided direct evidence that the high-affinity $Ins(1,4,5)P_3$-binding sites of permeabilized hepatocytes are the receptors that mediate the effects of $Ins(1,4,5)P_3$ on intracellular Ca^{2+} stores: first by demonstrating that the binding is to neither of the enzymes that metabolize $Ins(1,4,5)P_3$, and second by measuring $Ins(1,4,5)P_3$ binding and Ca^{2+} mobilization under identical conditions (Nunn et al., 1990; Nunn and Taylor, 1990).

When measured under conditions identical to those used for $Ins(1,4,5)P_3$ binding, the K_m of $Ins(1,4,5)P_3$ 5-phosphatase for $Ins(1,4,5)P_3$ is 17 μM. It is therefore unlikely to be the protein to which $Ins(1,4,5)P_3$ binds with an affinity of 7.8 nM. Furthermore, radiation-inactivation target size analysis of freeze-dried hepatocytes suggests that whereas the binding site has a molecular target size of 260 kDa, the enzyme has a target size of 190 kDa (Nunn et al., 1990).

A stable analog of $Ins(1,4,5)P_3$, inositol 1,4,5-trisphosphorothioate ($Ins(1,4,5)P_3[S]_3$) (Potter, chapter 8), has allowed us to eliminate the possibility that the $Ins(1,4,5)P_3$ binding site is the substrate binding site of $Ins(1,4,5)P_3$ 3-kinase. Although $Ins(1,4,5)P_3[S]_3$ stimulates Ca^{2+} release from the intracellular stores of a variety of cells, including hepatocytes, it is not a substrate of either the 3-kinase or 5-phosphatase enzymes (Taylor et al., 1989; Potter, chapter 8). More important for the present discussion, a substantial excess of $Ins(1,4,5)P_3[S]_3$ does not affect the rate of $Ins(1,4,5)P_3$ phosphorylation by the 3-kinase, implying that it is not recognized by the enzyme. However, $Ins(1,4,5)P_3[S]_3$ does displace specifically bound $Ins(1,4,5)P_3$ from its binding site, and compared with other ligands, its rank order of potency is similar in Ca^{2+}-mobilization and binding assays (Figs. 5 and 6).

These results confirm that the specific high-affinity $Ins(1,4,5)P_3$-binding sites measured in permeabilized hepatocytes are not the substrate-binding sites of either of the enzymes known to metabolize $Ins(1,4,5)P_3$. Furthermore, the metabolic stability of $Ins(1,4,5)P_3[S]_3$ and its ability to discriminate between the receptor and 3-kinase suggest that when it becomes avail-

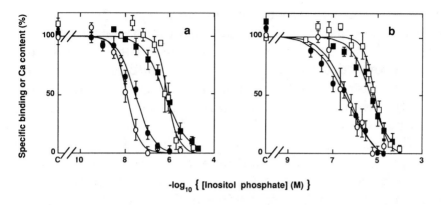

Fig. 6. Ins(1,4,5)P_3 binding and Ca^{2+} mobilization measured under identical conditions. When Ins(1,4,5)P_3 binding and the functional response are measured under identical conditions in permeabilized hepatocytes, the concentrations of ligand that cause half-maximal Ca^{2+} mobilization (EC$_{50}$) (open symbols) are very similar to the concentrations that cause half-maximal inhibition of specific Ins(1,4,[^{32}P]5)P_3 binding (= K_D in these experiments, in which the radioligand is present at very low concentration) (solid symbols). Results are shown as means ± SEM of 6–8 independent experiments. a: Ins(1,4,5)P_3 (O ●), Ins(2,4,5)P_3 (□ ■) b: DL-Ins(1,4,5)P_3[S]$_3$ (O ●), Ins(4,5)P_2 (□ ■). Reproduced from Nunn and Taylor (1990) with permission.

able labeled with ^{35}S it may be the ligand of choice for studies of the Ins(1,4,5)P_3 receptor.

To examine Ins(1,4,5)P_3 binding and Ca^{2+} mobilization under identical conditions, saponin-permeabilized hepatocytes (Section 2.1.) were resuspended at a very high cell density (8.5 × 10^6 cells/mL) in cytosol-like medium with the free [Ca^{2+}] adjusted to about 120 nM and with antimycin (10 μM), oligomycin (10 μM) and ^{45}CaCl$_2$ (2 μCi/mL) added. After 3 min, ATP (1.5 mM), creatine phosphate (5 mM), and creatine kinase (1 U/mL) were added to stimulate Ca^{2+} uptake into nonmitochondrial stores. Ten minutes later, when cells had loaded to steady-state with ^{45}Ca^{2+}, they were simultaneously cooled and added to test compounds by rapid dilution of cells (40 μL) into medium (360 μL) at 0°C. After 5 min, the reactions were terminated and the ^{45}Ca^{2+} contents of the cells determined as described in Section 2.3. Inhibition of specifically bound Ins(1,4,[^{32}P]5)P_3

was measured under identical conditions, except that $^{45}Ca^{2+}$ was omitted and cells were diluted into medium containing Ins(1,4, [^{32}P]5)P_3 (0.06 µCi/tube) and ^3H_2O (0.6 µCi/tube, to allow subsequent correction for trapped volume) in addition to test compounds. After 5 min, the reactions were terminated by rapid centrifugation, and binding parameters were established as described above.

The results of these experiments (Fig. 6) show that when care is taken to measure binding and the functional response under identical conditions, the concentrations of inositol phosphates that cause half-maximal Ca^{2+} mobilization (EC$_{50}$) are very similar to their affinities for the specific Ins(1,4,5)P_3-binding site (K_d). This is further evidence that the high-affinity Ins(1,4,5)P_3-binding sites of permeabilized hepatocytes are the ligand-recognition sites of the receptors that mediate the effects of Ins(1,4,5)P_3 on intracellular Ca^{2+} stores (Nunn and Taylor, 1990).

The Ins(1,4,5)P_3 receptor, the existence of which was initially inferred from structure–activity studies, can now be identified by radioligand binding, which has been instrumental in allowing its purification and functional reconstitution (Section 3.3.). In addition, the specificity of Ins(1,4,5)P_3-binding sites can now be exploited to measure the mass of Ins(1,4,5)P_3 in various tissues by radioreceptor assay (e.g., Challiss et al., 1990; Godfrey chapter 2).

3.2. Receptor Solubilization

The first step in purifying any integral membrane protein, including the Ins(1,4,5)P_3 receptor, is to remove it from its native membrane without destroying the protein. Detergents are usually used to solubilize receptors, but organic solvents have also been used, particularly when the proteins have very tightly bound lipids or when detergent removal presents problems (Boyan and Clement-Cormier, 1984). The choice of detergent and optimization of conditions for receptor solubilization remain matters of trial-and-error. In establishing a suitable method, a variety of detergents, protein/detergent ratios (0.1–10), ionic strengths, and incubation times must be tested (Hjelmeland and Chrambrach, 1984). Finally, optimal conditions for protein solubilization need to be matched with optimal conditions for subsequent analysis of the solubilized protein, radioligand binding or removal of detergent, and functional reconstitution, for example.

Rat cerebellar membranes, which have the highest known density of $Ins(1,4,5)P_3$-binding sites (20 pmol/mg protein), were the first tissue from which $Ins(1,4,5)P_3$-binding sites were successfully solubilized (Supattapone et al., 1988b). Our methods are modified from those of earlier reports. Rat cerebella (1 g) are homogenized in buffered saline (50 mL; $1M$ NaCl, 1 mM EDTA, 50 mM Tris-HCl, pH 8.3 at 4°C) using an Ultraturrax® T-25 homogenizer (9500 rpm, 20 s). After centrifugation (35,000g, 20 min), the pellet is resuspended in homogenization buffer without NaCl and the centrifugation repeated three times. Membranes are then resuspended (50 mg wet wt/mL) in NaCl-free homogenization buffer with the nonionic detergent Triton X-100 (1%, w/v) (final protein/detergent = 0.1). The suspension is incubated on ice, with stirring, for 30 min and the soluble fraction recovered as the supernatant after centrifugation (105,000g, 60 min). Our recovery of solubilized $Ins(1,4,5)P_3$-binding sites is typically 80%.

Solubilization of $Ins(1,4,5)P_3$-binding sites from other tissues has proved more difficult, although smooth-muscle $Ins(1,4,5)P_3$-binding sites have recently been solubilized, purified, and characterized (Chadwick et al., 1990). Our studies have concentrated on rat liver $Ins(1,4,5)P_3$ receptors because hepatocytes are one of the cell types in which hormonal regulation of intracellular Ca^{2+} and the behavior of the $Ins(1,4,5)P_3$ receptor in its native environment are best understood.

A crude membrane fraction is prepared by perfusing a rat liver *in situ* with Ca^{2+}-free medium (50 mL; 118 mM NaCl, 5.4 mM KCl, 0.8 mM $MgSO_4$, 0.96 mM NaH_2PO_4, 1 mM EGTA, 25 mM $NaHCO_3$, 95:5 O_2:CO_2; pH 7.4 at 4°C), homogenizing it in buffered sucrose (50 mL; 250 mM sucrose, 1 mM EGTA, 5 mM HEPES; pH 7.4 at 4°C) with 10 passes of a loose-fitting plunger followed by three passes with a tight-fitting plunger in a Dounce homogenizer. After centrifugation (1000g, 5 min), the supernatant is recentrifuged (105,000g, 60 min). The pellet is resuspended (20 mg protein/mL) in buffer (50 mM Tris-HCl, 1 mM EDTA, 0.1 mM PMSF; pH 8.3). Membranes (5 mg protein/ mL) are then solubilized with 3-[3-cholamidopropyl dimethyl- ammonio]-1 propanesulfonate (CHAPS, 1% w/v) in buffered KCl (100 mM KCl, 1 mM EDTA, 50 mM Tris-HCl; pH 8.3) on ice and stirred for 75 min. Solubilized proteins are recovered as the supernatant after centrifugation (105,000g, 60 min). This method allows about 30% solubilization of hepatocyte Ins(1,4,5)P_3-bind- ing sites.

The filtration or centrifugation methods routinely used to separate bound from free ligand in radioligand-binding assays to receptors in their native membranes cannot be applied to sol- ubilized receptors. Several appropriate methods have been described (El-Rafei, 1984), but we will concentrate on the two methods in use in our laboratory, spun-column chromatogra- phy and polyethylene glycol (PEG) precipitation followed by centrifugation.

Spun-column chromatography separates bound and free Ins(1,4,5)P_3 by gel filtration, a technique that exploits the differ- ence in size between the receptor-bound and free forms of the ligand. Free ligand remains within the column matrix; ligand– receptor complexes are completely excluded and are collected in the void volume. Samples are incubated as usual with [³H]Ins (1,4,5)P_3 (about 10,000 cpm; 200 μL final vol) and then applied to a 1-mL BioGel P10 gel-filtration column and centri- fuged (1000g, 6 min) (Supattapone et al., 1988b). The void volume (= bound [³H]Ins(1,4,5)P_3) is collected and counted for radio- activity. The technique is reproducible, but time-consuming and too expensive for routine use.

PEG can be used to precipitate receptor–ligand complexes in the presence of a carrier protein (bovine γ-globulin) and the precipitate of bound radioligand recovered by centrifugation or filtration. The sample (final vol, 0.5 mL) is incubated with [³H]Ins(1,4,5)P_3 in buffer (50 mM Tris-HCl, 1 mM EDTA; pH 8.3) with detergent (0.1% w/v) at 4°C for 5 min. PEG 6000 (0.5 mL, 30%, w/v) and bovine γ-globulin (10 μL, 25 mg/mL) are added, the samples placed on ice for 5 min, and then centrifuged (20,000g, 5 min). The supernatant is removed, and the pellet superficially washed with PEG 6000 (15%, w/v) and then resuspended in water (1 mL) by sonication before counting for radioactivity. PEG precipitation provides a rapid, economical, and convenient method for routine analysis of solubilized Ins(1,4,5)P$_3$ receptors.

For both methods, specific binding is analyzed by the methods described earlier (Section 3.1.). Our results suggest that both methods give very similar results.

3.3. Receptor Purification

Size, charge, hydrophobicity, and more specific properties (e.g., affinity for particular ligands) are all routinely used to separate proteins. The hydrophobic nature and special conditions required for solubilization of integral membrane proteins (Section 3.2.) create particular problems when applying purification methods more commonly used for hydrophilic proteins.

A number of strategies have been applied to purification of Ins(1,4,5)P$_3$-binding sites (Fig. 7). Anion exchange (DEAE and DE52) has been applied with some success to purification of cerebellar Ins(1,4,5)P$_3$-binding sites, but in other tissues, in which the concentration of binding sites is considerably lower, affinity chromatography is more likely to be successful. Affinity chromatography has the advantage of speed, a critical factor in optimizing recovery of small amounts of a protein that may not be very stable after solubilization.

Heparin inhibits Ins(1,4,5)P$_3$ binding to its receptor (Worley et al., 1987; Tones et al., 1989; Chilvers et al., 1990; Section 3.1.), and is presently the step in affinity chromatography common to all successful purifications of Ins(1,4,5)P$_3$-binding sites (Supattapone et al., 1988b; Chadwick et al., 1990; Fig. 7). Ins(1,4,5) P$_3$-binding sites adhere to lectin columns, suggesting that the receptor is a glycoprotein; indeed the cDNA sequence of the cloned receptor has revealed multiple consensus sequences for N-linked glycosylation (Furuichi et al., 1989). This glycosylation allows another purification step that has been widely applied, lectin affinity chromatography. The receptor binds quantitatively to concanavalin A (Con A) affinity columns, but there are reports of difficulty in elution (Maeda et al., 1990). Lentil

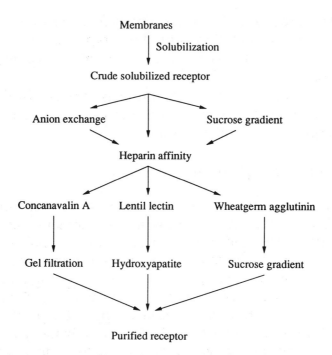

Fig. 7. Purification of Ins(1,4,5)P_3 receptors. Strategies adopted in purification of Ins(1,4,5)P_3 receptors include separation on the basis of charge (anion exchange and hydroxyapatite chromatography), size (gel-filtration chromatography and sucrose-gradient centrifugation), and specificity (general: lectin affinity chromatography with Con A, lentil lectin, and wheatgerm agglutinin; specific: heparin affinity chromatography).

lectin and wheatgerm agglutinin affinity resins do not bind the receptor as efficiently, but reported recoveries are generally better (Maeda et al., 1990; Chadwick et al., 1990).

The high M_r of Ins(1,4,5)P_3-binding sites (225,000–260,000) (Supattapone et al., 1988b; Chadwick et al., 1990; Nunn et al., 1990) makes techniques that exploit size fractionation (e.g., gel filtration, sucrose-gradient centrifugation) very attractive. However, gel filtration in the presence of detergents needs to be approached with caution. Column buffers should contain detergent near or above its CMC (the concentration above which detergents form micelles) to prevent protein aggregation or precipitation; to avoid irreversible adsorption to the resin, a particular problem with dextran-based media, resins that are more inert, such as polyacrylamide, should be used. Sucrose-gradient centrifugation, which has been successfully used to purify the ryanodine receptor (M_r = 400,000) (Lai et al., 1988), has more recently been applied to smooth muscle Ins(1,4,5)P_3-binding sites (Chadwick et al. 1990).

Hydroxylapatite, a type of calcium phosphate that binds to negatively charged groups in proteins, has also been used as a final step in purification of Ins(1,4,5)P_3 receptors from cerebellar membranes (Maeda et al., 1990). Binding of the protein to hydroxylapatite may be more than a nonspecific interaction between the column material and negative groups on the protein, and may result from a specific interaction with the parts of the receptor that recognize the Ca^{2+} that passes through the channel.

Ins(1,4,5)P_3 binding to rat liver microsomes is inhibited by triazine dyes (Cibacron Blue and patent blue) (Bootman et al., 1990). These dyes, which have been widely used as affinity-chromatography media for purification of other proteins (Thompson et al., 1975; Atkinson et al., 1981), may therefore provide another approach to purifying Ins(1,4,5)P_3 receptors. The methods described have allowed purification of Ins-(1,4,5)P_3 receptors from cerebellum and smooth muscle, but simpler purification procedures, perhaps involving affinity columns prepared from modified inositol phosphates (Hirata et al., 1990; Section 3.1.), will be very useful.

3.4. Functional Properties of Ins(1,4,5)P₃ Receptors

A number of approaches, including covalent modification of the receptor protein, functional reconstitution of the purified receptor, single-channel recording, and molecular biology techniques, are beginning to provide information about the molecular basis of the interactions between Ins(1,4,5)P_3 and the Ca^{2+} channel it regulates.

Sulfhydryl groups appear to be implicated in Ins(1,4,5)P_3 binding and Ca^{2+} release: AgCl, *p*-chloromercuribenzoate, and *p*-hydroxymercuribenzoate inhibit Ins(1,4,5)P_3-induced Ca^{2+} release in platelets and permeabilized insulinoma cells (Adunyah and Dean, 1986; Föhr et al., 1989); N-ethyl maleimide and iodo-acetamide inhibit Ins(1,4,5)P_3 binding to bovine adrenal-cortex microsomes (Guillemette and Segui, 1988); and p-chloromercuric benzoyl sulfonate blocks Ins(1,4,5)P_3 binding to cerebellar membranes (Supattapone et al., 1988b). Dithiothre-itol (DTT) pretreatment prevents most of these effects; indeed DTT and β-mercaptoethanol slightly stimulate Ins(1,4,5)P_3 binding to cerebellar membranes (Supattapone et al., 1988b), though the opposite effect was reported for adrenal cortex (Guillemette and Segui, 1988).

The substantial negative charge of all known ligands for the Ins(1,4,5)P_3 receptor suggests that basic residues of the receptor (Lys or Arg) may be involved in ligand recognition. The high pK_a (12.5) of Arg guanidine groups gives them a positive charge over the physiological pH range. X-ray crystallography has shown that the guanidyl groups of guanidine form hydrogen bonds to phosphate oxygen atoms (Cotton et al., 1973), whereas Arg residues provide hydrogen-bonding sites with nucleotide phosphate groups (Riordan, 1979, Raess et al., 1985). Treatment of platelet membranes with p-hydroxyphenylglyoxal, an Arg-modifying reagent, inhibited [^{3}H]Ins(1,4,5)P_3 binding (O'Rourke and Feinstein, 1990). Our preliminary studies of rat liver membranes suggest that Lys residues may also be important for Ins(1,4,5)P_3 binding because citraconic anhydride, a Lys-modifying reagent, reversibly blocks Ins(1,4,5)P_3 binding (Bond and Taylor, unpublished observation).

Carboxy-group-modifying reagents, such as N,N'-dicyclo-hexylcarbodimide and 1-ethyl-3-(3-dimethylaminopropyl) carbodimide, inhibit Ins(1,4,5)P_3-induced Ca^{2+} release from platelet membrane vesicles without affecting Ins(1,4,5)P_3 binding, suggesting that the Ca^{2+} channel of the receptor may have been modified (O'Rourke and Feinstein, 1990).

Chemical modifications combined with sequence analysis will eventually reveal the specific amino acid residues involved in receptor function, and site-directed mutagenesis will provide the ultimate confirmation of these results. Recent work with P$_{400}$ protein *(see below)* should allow rapid progress in this area.

P$_{400}$, a protein initially studied because it is absent from mutant mice that lack cerebellar Purkinje cells, is now known to be the cerebellar Ins(1,4,5)P$_3$ receptor (Furuichi et al., 1989). P$_{400}$ cDNA has been cloned and, from the sequence, the structure of the cerebellar Ins(1,4,5)P$_3$ receptor has been predicted to be a protein of 2749 amino acid residues with an M_r of 313,000. A hydropathy profile of the primary sequence reveals the existence of several stretches of hydrophobic amino acids that may represent multiple membrane-spanning sequences. The large hydrophilic *N*-terminal domain is cytoplasmic; the short *C*-terminal stretch although initially proposed to be luminal is now thought to be cytoplasmic. An interesting feature of the P$_{400}$ cDNA sequence is its fragmentary sequence similarity to that of the ryanodine receptor, the channel responsible for Ca^{2+} release from skeletal and cardiac muscle sarcoplasmic reticulum (Furuichi et al., 1989). In addition, both proteins have, or are predicted to have, large cytoplasmic *N*-terminal domains and short *C*-terminal regions, and both form tetrameric homo-oligomers that probably represent the functional receptor-channel complex. P$_{400}$ has no clear sequence similarity with other Ca^{2+}-binding sites, such as the E–F hand of Ca^{2+}-binding proteins, the hypothetical Ca^{2+}-binding site in the dihydropyridine receptor, or the putative Ca^{2+}-binding region of Ca^{2+}-ATPases.

Functional reconstitution of both native and engineered Ins (1,4,5)P$_3$ receptors is crucial to further studies of the molecular basis of Ins(1,4,5)P$_3$ binding and action. The purified cerebellar Ins(1,4,5)P$_3$ receptor has been reconstituted into phospholipid vesicles by dialyzing the protein–detergent–phospholipid mixtures against buffer to remove the detergent and effect vesicle formation (Ferris et al., 1989). The vesicles then show Ins(1,4,5)P$_3$-sensitive Ca^{2+} fluxes, but although the response has the appropriate pharmacological specificity, it is small, and this presently severely limits opportunities for more detailed analyses.

We have reconstituted partially purified cerebellar Ins $(1,4,5)P_3$ receptors into phosphatidylcholine (PtdCho) vesicles according to the method of Horne et al. (1986). A sample (400 µL) in Triton X-100 (0.1%, w/v) in buffer (50 mM Tris-HCl, 1 mM EDTA; pH 8.3) is mixed with sonicated PtdCho (100 µL, 20 mg/mL, Sigma, Type III-S). After incubation on ice for 2 min, $MgCl_2$ (50 µL, 100 mM) is added and the mixture incubated for a further 30 min. The sample is then applied to an ExtractiGel™ column (1 mL, Pierce-Warriner, Chester) prepared exactly as described by Horne et al. (1986), using Tris-HCl (50 mM, pH 8.3) as the buffering system, and eluted with 1.5 mL of their final equilibrating buffer. Ins $(1,4,5)P_3$ binding to the vesicles is determined by incubation with [^{3}H]Ins(1,4,5)P_3 for 5 min at 4°C in a total incubation volume of 0.5 mL followed by rapid filtration through Whatman GF/B filters with a 2-mL wash of ice-cold incubation buffer. These methods, which allow solubilized Ins(1,4,5)P_3 receptors to be reconstituted into phospholipid vesicles, are now being used to examine in more detail the interactions between Ins(1,4,5)P_3 and its receptor.

Single-channel recordings from Ins(1,4,5)P_3-sensitive Ca^{2+} channels have been made from excised patches of plasma membrane from T-lymphocytes (Kuno and Gardner, 1987) and plant vacuoles (Lassalles and Kado, 1990), and from planar lipid bilayers after their fusion with vesicles prepared from smooth-muscle sarcoplasmic reticulum (Ehrlich and Watras, 1988) or skeletal muscle transverse tubules (Vilven and Coronado, 1988). No single-channel recordings have yet been reported for the purified Ins(1,4,5)P_3 receptor.[*] Such recordings are needed because the characteristics of the smooth-muscle sarcoplasmic reticulum Ca^{2+} channel are the only ones presently consistent with the known properties of the Ins(1,4,5)P_3 receptor of the intracellular Ca^{2+} stores.

---

[*]After submission of this manuscript Maeda et al. (1991) reported single-channel recordings from purified cerebellar Ins(1,4,5)P_3 receptors reconstituted into plural lipid bilayer.

4. Complex Changes in Intracellular Ca^{2+}: The Role of $Ins(1,4,5)P_3$

4.1. Ca^{2+} Entry

In most cells stimulated by agonists that evoke PtdInsP_2 hydrolysis, two phases of the Ca^{2+} response can be distinguished. The initial response is typically independent of extracellular Ca^{2+} and reflects mobilization of intracellular Ca^{2+} stores by $Ins(1,4,5)P_3$ (Section 1.). The sustained response requires Ca^{2+} entry from the extracellular space. Although the mechanisms underlying this phase of the response are not yet clear, current hypotheses all suggest that $Ins(1,4,5)P_3$, directly or indirectly, plays a central role.

The "capacitative Ca^{2+} entry" model (Putney, 1986b) proposed that empty $Ins(1,4,5)P_3$-sensitive Ca^{2+} stores provide the signal that stimulates Ca^{2+} entry across the plasma membrane (Fig. 8). Early versions of the model assumed that the intracellular stores were closely associated with the plasma membrane and that Ca^{2+} passed directly from the extracellular space into the stores and from there, in the continued presence of $Ins(1,4,5)P_3$, through the stores and into the cytosol (Putney, 1986b; Merritt and Rink, 1987). More recent evidence (reviewed in Taylor, 1990a) suggests that although the Ca^{2+} content of the intracellular stores is the key regulator of Ca^{2+} entry, entry occurs directly into the cytosol and not through the stores. The outstanding difficulty is the nature of the signal that communicates the state of the intracellular stores to the plasma membrane.

Irvine and his colleagues (Irvine and Moor, 1986; Morris et al., 1987; reviewed by Irvine [1989]) have presented persuasive evidence that in some cells $Ins(1,3,4,5)P_4$ in concert with Ins $(1,4,5)P_3$ regulates Ca^{2+} entry. These results were reconciled with the capacitative model by suggesting that $Ins(1,3,4,5)P_4$ facilitated communication between the $Ins(1,4,5)P_3$-sensitive stores and the plasma membrane, thereby allowing empty stores to activate the Ca^{2+}-entry pathway (Irvine, 1989). However, Ins $(1,3,4,5)P_4$ appears not to be a universal regulator, because refilling of empty Ca^{2+} stores can occur long after $Ins(1,3,4,5)P_4$

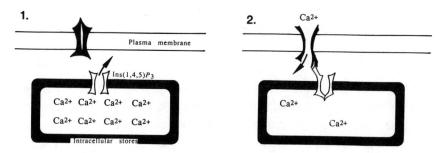

Fig. 8. Receptor-regulated Ca²⁺ entry. One possible model is shown, whereby the Ca²⁺ content of the Ins(1,4,5)P_3-sensitive Ca²⁺ stores directly regulates Ca²⁺ entry from the extracellular space across the plasma membrane and into the cytosol (1) Ins(1,4,5)P_3 binds to its receptor in the membrane of the intracellular Ca²⁺ stores, opens the integral ion channel, and Ca²⁺ begins to leave the stores and enter the cytosol. The Ca²⁺ entry pathway of the plasma membane remains closed. (2) As the Ca²⁺ stores empty, their decreased Ca²⁺ content is proposed to alter the conformation of the Ins(1,4,5)P_3 receptor with two important consequences. First, the sensitivity of Ins(1,4,5)P_3-induced Ca²⁺ mobilization to Ins(1,4,5)P_3 is reduced. This may account for the inability of submaximal concentrations of Ins(1,4,5)P_3 to release the Ins(1,4,5) P_3-sensitive stores fully (Section 1.3.) because the Ca²⁺ channel closes once there is no longer sufficient intraluminal Ca²⁺ and cytoplasmic Ins(1,4,5)P_3 to keep it open. Second, the conformational change in the Ins(1,4,5)P_3 receptor is proposed to be detected by the plasma membrane Ca²⁺ entry pathway leading to increased Ca²⁺ entry into the cytosol.

has been degraded (Hughes et al., 1988; Takemura and Putney, 1989). The role of Ins(1,3,4,5)P_4 in regulating Ca²⁺ entry therefore remains unclear.

The homologous structures of the Ins(1,4,5)P_3 and ryanodine receptors (Section 3.4.), and a proposal that luminal Ca²⁺ may regulate the responsiveness of the Ins(1,4,5)P_3 receptor (Section 1.3.) have recently provided the basis for a revised capacitative Ca²⁺-entry model (Irvine, 1990). The ryanodine receptor of the sarcoplasmic reticulum and the dihydropyridine receptor of the sarcolemma are assumed to be linked to form the junctional "feet." This link may allow the voltage change across the sarcolemma to be relayed to the intracellular Ca²⁺ stores from which Ca²⁺ is released to the cytosol (Gill, 1989). Irvine (1990) has proposed a similar organization in nonmuscle cells with a direct physical link between the plasma membrane proteins responsible for Ca^{2+} entry and the $Ins(1,4,5)P_3$ receptor allowing communication between them. The ability of the latter to monitor cytoplasmic $Ins(1,4,5)P_3$ and its proposed ability to monitor luminal Ca^{2+} could then provide a mechanism whereby empty intracellular stores signal their state to the plasma membrane and thereby stimulate Ca^{2+} entry (Fig. 8). This very attractive model has little direct experimental support, but it does provide a framework for future experiments.

4.2. Organization of Intracellular Ca^{2+} Signals

The pioneering studies of Cobbold and his colleagues (Woods et al., 1986) and the introduction of Ca^{2+} indicators that allow the cytoplasmic $[Ca^{2+}]$ to be measured in single cells (Tsien, 1989; Poenie, chapter 5) have transformed our understanding of Ca^{2+} signaling. It is now clear that the intracellular Ca^{2+} signal is generally neither uniformly distributed across the cell nor its temporally organization as simple as first appeared from measurements of populations of cells. Furthermore, it is becoming increasingly clear that this precise spatial and temporal organization of intracellular Ca^{2+} signals has important effects on the final cell response (Cheek, 1989). In bovine adrenal chromaffin cells, for example, stimuli (e.g., nicotinic agonists, membrane depolarization) that evoke Ca^{2+} entry, but not mobilization of intracellular stores, uniformly increase the free cytoplasmic $[Ca^{2+}]$ around the cell periphery and cause substantial exocytosis. By contrast, stimuli (e.g., muscarinic agonists, bradykinin) that cause polyphosphoinositide hydrolysis, mobilization of intracellular Ca^{2+} stores, and less Ca^{2+} entry produce a more localized increase in cytoplasmic $[Ca^{2+}]$ and are much weaker stimulants of exocytosis (O'Sullivan et al., 1989; Burgoyne, chapter 11).

In many cells, the initial, localized increase in cytoplasmic $[Ca^{2+}]$ that follows activation of PIC propagates across the cell as a Ca^{2+} wave, the characteristics of which are strikingly similar in different cells (Berridge and Galione, 1988). This feature may be particularly important in cells in which detection of the extracellular stimulus is necessarily polarized, but the entire cell must respond. In hepatocytes, for example, glycogenolysis is a coordinated response of the entire cell, yet the transmembrane signaling machinery may be restricted to the sinusoidal membranes, where they have access to blood-borne hormones (Rooney et al., 1990). Ca^{2+} waves may be the means whereby an initial, very localized response is rapidly propagated throughout the cell.

The mechanisms underlying regenerative Ca^{2+} waves are unknown. Current models suggest that the initial Ca^{2+} rise stimulates a further Ca^{2+} response either by directly activating PIC and thereby generating further $Ins(1,4,5)P_3$ or by first overloading a distinct Ca^{2+} store from which Ca^{2+} is later released by a Ca^{2+}-induced Ca^{2+}-release mechanism (Section 1.2.) (Berridge and Galione, 1988; Berridge et al., 1988; Berridge and Irvine, 1989; Rooney et al., 1990).

Intracellular Ca^{2+} signals evoked by receptors that stimulate PIC often consist of a series of Ca^{2+} spikes with the free $[Ca^{2+}]$ typically exceeding 1 μM for a few seconds before returning to the resting level for periods of up to several minutes (Woods et al.,1986; Berridge and Galione, 1988; Jacob et al., 1988; Jacob, 1990). In many situations, the frequency of these Ca^{2+} spikes increases with increasing concentration of extracellular stimulus. This, and the observation that the shape of individual spikes may be characteristic for each agonist (Berridge et al., 1988), has provoked the suggestion that Ca^{2+} spikes may provide a digitally encoded intracellular Ca^{2+} signal in which the frequency of the spikes conveys the intensity of the extracellular stimulus, and the shape of each conveys its nature.

Variations of two basic models have been proposed to account for the spiking behavior of intracellular Ca^{2+}. Neither is yet supported by conclusive evidence. The first model suggests that cyclical changes in intracellular $Ins(1,4,5)P_3$ concentration underlie the cyclical release of intracellular Ca^{2+} stores, and a variety of feedback loops (Cobbold et al., 1990) and theoretical models (Meyer and Stryer, 1988) have been proposed in support of this model. Perhaps the major evidence against it is the demonstration that the stable phosphorothioate analog of Ins(1,4,5)P$_3$, which cannot undergo cycles of formation and degradation, evokes Ca^{2+} spikes in pancreatic acinar cells (Wakui et al., 1989). However, even this evidence cannot wholly exclude the possibility that Ins(1,4,5)P$_3$[S]$_3$ stimulates Ca^{2+} mobilization and that this then stimulates PIC, initiating the cycles of Ins(1,4,5)P$_3$ formation and degradation.

The second class of models assumes that the intracellular concentration of Ins(1,4,5)P$_3$ remains elevated throughout the period of stimulation, and that periodic Ca^{2+} release is then a consequence of periodic changes in the sensitivity of the Ins(1,4,5)P$_3$ receptor or of Ca^{2+} redistribution among intracellular stores. As discussed earlier (Section 1.3., Taylor et al.,1989), there is no evidence to suggest an inherent ability of the Ins(1,4,5)P$_3$ receptor to undergo homologous desensitization, but in some tissues Ca^{2+} does inhibit Ins(1,4,5)P$_3$-induced Ca^{2+} mobilization. This inhibitory effect of Ca^{2+} could allow Ca^{2+} to be resequestered despite the continued presence of Ins(1,4,5)P$_3$ until the free [Ca^{2+}] again fell to a level that allowed Ins(1,4,5)P$_3$ to discharge the stores. A problem with this model is that in some tissues in which Ca^{2+} spikes have been reported (e.g., hepatocytes) there is clearly no effect of Ca^{2+} on Ins(1,4,5)P$_3$-induced Ca^{2+} mobilization.

Another variation of this second class of model has been proposed by Berridge and his colleagues (Berridge and Galione, 1988; Goldbeter et al., 1990). The Ins(1,4,5)P$_3$-sensitive stores are proposed to remain empty, or partially empty, for as long as the cytoplasmic Ins(1,4,5)P$_3$ level remains elevated. This both inactivates a substantial fraction of the normal intracellular Ca^{2+} sequestering capacity and stimulates Ca^{2+} entry (Section 4.1.). The cells therefore face an increased cytoplasmic Ca^{2+} load that is sequestered by Ins(1,4,5)P$_3$-insensitive Ca^{2+} stores. The latter are proposed to share features with the Ca^{2+}-induced Ca^{2+} release stores of sarcoplasmic reticulum and to discharge their accumulated Ca^{2+} as they become overloaded and the cytoplasmic [Ca^{2+}] continues to rise. Although this model attractively accounts for many of the features of Ca²⁺ spiking behavior, there is presently no wholly convincing evidence to support the existence of Ca^{2+}-induced Ca^{2+} release stores in the nonexcitable cells that show Ca^{2+} spiking. The complexity of the Ca^{2+} responses of intact cells to extracellular stimuli is now a focus of enormous interest, from which is likely to emerge a better understanding of the nature of the responses, their functional significance, and, in combination with the approaches described in this chapter, their control.

5. Concluding Remarks

The emphasis throughout this chapter has been methodological, in that we have attempted to provide an overview of some of the methods currently used to examine the effects of $Ins(1,4,5)P_3$ on intracellular Ca^{2+} stores and to consider in more detail the methods with which we have first-hand experience. It therefore seems appropriate to stress the crucial part played by improved methods in the recent transformation of our understanding of intracellular Ca^{2+} signaling. The fluorescent Ca^{2+} indicators developed by Tsien (reviewed in Tsien, 1989) and his colleagues provide the clearest example, but other methodological developments have also been important: methods of permeabilizing cells without disrupting intracellular organelles, the introduction of caged intracellular messengers, and simple methods of measuring inositol phosphates are a few examples. It seems likely that future strides in our understanding of intracellular Ca^{2+} signaling will also be led by improved methods; perhaps those methods will eventually provide techniques that will allow other steps in the signaling pathway to be measured with the temporal and spatial resolution that are presently available only for Ca^{2+}.

Acknowledgments

Work in the authors' laboratory is supported by grants from the Wellcome Trust, the Royal Society, and the Medical Research Council.

Abbreviations

CMC	Critical micelle concentration
DG	1,2,-Diacylglycerol
$Ins(1,4,5)P_3$	D-*myo* inositol 1,4,5-trisphosphate. Other inositol phosphates are abbreviated in accord with the same convention.
PC	Phosphatidylcholine
PIC	Phosphoinositidase C (= polyphosphoinositide-specific phospholipase C)
PMSF	Phenylmethylsulfonylfluoride
tBuBHQ	2,5-Di-(*tert*-butyl)-1,4-benzohydroquinone

References

Ahnert-Hilger G. and Gratzl M. (1988) Controlled manipulation of the cell interior by pore-forming proteins. *Trends Pharmacol. Sci.* **9,** 195–197.

Adunyah S. E. and Dean W. L. (1986) Effects of sulfhydryl reagents and other inhibitors on Ca^{2+} transport and inositol trisphosphate-induced Ca^{2+} release from human platelet membranes. *J. Biol. Chem.* **261,** 13,071–13,075.

Ammann D. (1986) *Ion-Selective Electrodes Principles Design and Application,* Springer-Verlag, Heidelberg.

Atkinson T., Hammond P. M., Hartwell R. D., Hughes P., Scawen M. D., Sherwood R. F., Small D. A. P., Bruton C. J., Harvey M. J., and Lowe C. R. (1981) Triazine dye affinity chromatography. *Biochem. Soc. Trans.* **9,** 290–293.

Berridge M. J. (1988) Inositol trisphosphate-induced membrane potential oscillations in *Xenopus* oocytes. *J. Physiol.* **403,** 589–599.

Berridge M. J. (1989) Inositol trisphosphate-induced calcium mobilization is localized in *Xenopus* oocytes. *Proc. R. Soc. Lond. Biol.* **238,** 235–243.

Berridge M. J. and Galione A. (1988) Cytosolic calcium oscillators. *FASEB J.* **2,** 3074–3082.

Berridge M. J. and Irvine R. F. (1984) Inositol trisphosphate, a novel second messenger in cellular signal transduction. *Nature* **312,** 315–321.

Berridge M. J. and Irvine R. F. (1989) Inositol phosphates and cell signalling. *Nature* **341,** 197–205.

Berridge M. J. and Michell R. H. (1988) *Inositol Lipids and Transmembrane Signalling* (The Royal Society, London), p. 200.

Berridge M. J., Cobbold P. H., and Cuthbertson K. S. R. (1988) Spatial and temporal aspects of cell signalling. *Philos. Trans. R. Soc. Lond. Biol.* **320**, 325–343.

Bootman M. D., Pay G. F, Rick C. E., and Tones M. A. (1990) Two sulphonated dye compounds which compete for inositol 1,4,5-trisphosphate binding to rat liver microsomes: effects on 5'phosphatase activity. *Biochem. Biophys. Res. Commun.* **166**, 1334–1339.

Boyan B. D. and Clement-Cormier Y. (1984) Organic solvent extraction of membrane proteins, in *Receptor Biochemistry and Methodology* vol. 1, *Membranes, Detergents and Receptor Solubilization* (Venter J. C. and Harrison L. C., eds.), Liss, NY, pp. 47–64.

Burgess G. M., Irvine R. F., Berridge M. J., McKinney J. S., and Putney J. W. (1984) Actions of inositol phosphates on Ca^{2+} pools in guinea-pig hepatocytes. *Biochem. J.* **224**, 741–746.

Burgess G. M., McKinney J. S., Fabiato A., Leslie B. A., and Putney J. W. (1983) Calcium pools in saponin-permeabilized guinea-pig hepatocytes. *J. Biol. Chem.* **258**, 15,336–15,345.

Burgoyne R. D., Cheek T. R., Morgan A., O'Sullivan A. J., Moreton R. B., Berridge M. J., Mata A. M., Colyer J., Lee A. G., and East J. M. (1989) Distribution of two distinct Ca^{2+}-ATPase-like proteins and their relationships to the agonist-sensitive calcium store in adrenal chromaffin cells. *Nature* **342**, 72-74.

Campbell A. K. (1983) *Intracellular Calcium. Its Universal Role as Regulator.* (John Wiley, Colchester).

Caswell A. H. (1979) Methods of measuring intracellular calcium. *Int. Rev. Cytol.* **56**, 145–165.

Chadwick C. C., Saito A., and Fleischer S. (1990) Isolation and characterization of the inositol trisphosphate receptor from smooth muscle. *Proc. Natl. Acad. Sci. USA* **87**, 2132–2136.

Challis R. A. J., Chilvers E. R., Willcocks A. L., and Nahorski S. R. (1990) Heterogeneity of [^{3}H]inositol 1,4,5-trisphosphate binding sites in adrenal cortical membranes. Characterization and validation of a radioreceptor assay. *Biochem. J.* **265**, 421–427.

Champeil P., Combettes L., Berthon B., Doucet E., Orlowski S., and Claret M. (1989) Fast kinetics of calcium release induced by *myo*-inositol trisphosphate in permeabilized rat hepatocytes. *J. Biol. Chem.* **264**, 17,665–17,673.

Cheek T. R. (1989) Spatial aspects of calcium signalling. *J. Cell Sci.* **93**, 211–216.

Cheuh S. -H. and Gill D. L. (1986) Inositol 1,4,5-trisphosphate and guanine nucleotides activate calcium release from endoplasmic reticulum via distinct mechanism. *J. Biol. Chem.* **261**, 13,883–13,886.

Chilvers E. R., Challiss R. A. J., Willcocks A. L., Potter B. V. L., Barnes P. J., and Nahorski S. R. (1990) Characterization of stereospecific binding sites for inositol 1,4,5-trisphosphate in airway smooth muscle. *Br. J. Pharmacol.* **99**, 297–302.

Cobbold P. H. and Rink T. J. (1987) Fluorescence and bioluminescence measurement of cytoplasmic free calcium. *Biochem. J.* **248**, 313–328.

Cobbold P., Dixon J., Sanchez–Bueno A., Woods N., Daly M., and Cuthbertson K. (1990) Receptor control of calcium transients, in *Transmembrane Signalling. Intracellular Messengers and Implications for Drug Development* (Nahorski S. R., ed.), John Wiley, Chichester, UK, pp. 185–206.

Cockcroft S. and Gomperts B. D. (1979) ATP induces nucleotide permeability in rat mast cells. *Nature* **279**, 541,542.

Cotton F. A., Day V. W. Hazen E. E., and Larsen S. (1973) Structure of methylguanidium dihydrogenorthophosphate. A model compound for arginine-phosphate hydrogen bonding. *J. Am. Chem. Soc.* **95**, 4834–4840.

Danoff S. K., Supattapone S., and Snyder S. H. (1988) Characterization of membrane protein from brain mediating the inhibition of inositol 1,4,5-trisphosphate receptor binding by calcium. *Biochem. J.* **254**, 701–705.

Dawson A. P. (1985) GTP enhances inositol trisphosphate-stimulated Ca^{2+} release from liver microsomes. *FEBS Lett.* **185**, 147–150

Dawson A. P. and Irvine R. F. (1984) Inositol(1,4,5)trisphosphate-promoted Ca^{2+} release from microsomal fractions of rat liver. *Biochem. Biophys. Res. Commun.* **120**, 858–864

Downes C. P. and Michell R. H. (1985) Inositol phospholipid breakdown as a receptor-controlled generator of second messengers, in *Molecular Mechanisms of Transmembrane Signalling* (Cohen P. and Houslay M. D., eds.), Elsevier Science, Amsterdam, pp. 3–56.

Downes C. P., Berrie C. P., Hawkins P. T., Stephens L., Boyer J. L., and Harden T. K. (1988) Receptor and G-protein-dependent regulation of turkey erythrocyte phosphoinositidase C. *Philos. Trans. R. Soc. Lond. Biol.* **320**, 267–280.

Ehrlich B. E. and Watras J. (1988) Inositol 1,4,5-trisphosphate activates a channel from smooth muscle sarcoplasmic reticulum. *Nature* **336**, 583–586

El-Rafei M. F. (1984) Assay of soluble receptors, in *Receptor Biochemistry and Methodology* vol. 1, *Membranes, Detergents and Receptor Solubilization* (Venter J. C. and Harrison L. C., eds.), Alan R. Liss, NY, pp. 99–108.

Ferris C. D., Huganir R. L., Supattapone S., and Snyder S. H. (1989) Purified inositol 1,4,5-trisphosphate receptor mediates calcium flux in reconstituted lipid vesicles. *Nature* **342**, 87–89.

Föhr K. J., Scott J., Ahnert-Hilger G., and Gratzl M. (1989) Characterization of the inositol 1,4,5-trisphosphate-induced calcium release from permeabilized endocrine cells and its inhibition by decavanadate and p-hydroxymercuribenzoate. *Biochem. J.* **262**, 83–89.

Furuichi T., Yoshikawa S., Miyawaki A., Wada K., Maeda N., and Mikoshiba K. (1989) Primary structure and functional expression of the inositol 1,4,5-trisphosphate-binding protein P_{400}. *Nature* **342**, 32–38.

Gill D. L. (1989) Receptor kinships revealed. *Nature* **342**, 16–18.

Gilman A. G. (1987) G proteins: Transducers of receptor-generated signals. *Annu. Rev. Biochem.* **56,** 615–649.

Glauert A. M., Dingle J. T., and Lucy J. A. (1962) Action of saponin on biological cell membranes, *Nature* **196,** 952–955.

Gogelein H. and Huby A. (1984) Interactions of saponin and digitonin with black lipid membranes and lipid monolayers. *Biochim. Biophys. Acta* **773,** 32–38.

Goldbeter A., Dupont G., and Berridge M. J. (1990) Minimal model for signal-induced Ca^{2+} oscillations and for their frequency encoding through protein phosphorylation. *Proc. Natl. Acad. Sci. USA* **87,** 1461–1465.

Guillemette G. and Segui J. A. (1988) Effects of pH, reducing and alkylating reagents on the binding and Ca^{2+} release activities of inositol 1,4,5-triphosphate in the bovine adrenal cortex. *Mol. Endocrinol.* **2,** 1249–1255

Guillemette G., Balla,T., Baukal, A. J. and Catt, K. J. (1988) Characterization of inositol 1,4,5-trisphosphate receptors and calcium mobilization in hepatic plasma membrane fractions. *J. Biol. Chem.* **263,** 4541–4548.

Guillemette G., Favreau I., Lamontagne S., and Boulay G. (1990) 2,3-diphosphoglycerate is a nonselective inhibitor of inositol 1,4,5-trisphosphate action and metabolism. *Eur. J. Pharmacol.* **188,** 251–260.

Henne V., Mayr G. W., Grabowski B., Koppitz B., and Söling H.-D. (1988) Semisynthetic derivatives of inositol 1,4,5-trisphosphate substituted at the 1-phosphate group. *Eur. J. Biochem.* **174,** 95–101.

Hirata M., Watanabe Y., Ishimatsu T., Yanaga F., Koga T., and Ozaki S. (1990) Inositol 1,4,5-trisphosphate affinity chromatography. *Biochem. Biophys. Res. Commun.* **168,** 379–386.

Hirata M., Watanabe Y., Ishimatsu T., Ikebe T., Kimura Y., Yamaguchi K., Ozaki S., and Koga T. (1989) Synthetic inositol trisphosphate analogs and their effects on phosphatase, kinase, and the release of Ca^{2+}. *J. Biol. Chem.* **264,** 20,303–20,308.

Hjelmeland L. M. and Chrambrach A. (1984) Solubilization of functional membrane bound receptors, in *Receptor Biochemistry and Methodology,* vol. 1, *Membranes, Detergents and Receptor Solubilization* (Venter J. C. and Harrison L. C., eds.), Alan R. Liss, NY, pp. 35–46.

Hodgson M. E. and Shears S. B. (1990) Rat liver contains a potent endogenous inhibitor of inositol 1,3,4,5-tetrakisphosphate 3-phosphatase. *Biochem. J.* **267,** 831–834.

Horne W. A., Weiland G. A., Oswald R. E., and Cerione R. A. (1986) Rapid incorporation of the solubilized dihydropyridine receptor into phospholipid vesicles. *Biochim. Biophys. Acta* **863,** 205–212.

Horstman D. A., Takemura H., and Putney J. W. (1988) Formation and metabolism of [^{3}H]inositol phosphates in AR42J pancreatoma cells. *J. Biol. Chem.* **263,** 15,297–15,303.

Hughes A. R, Takemura H., and Putney J. W. (1988) Kinetics of inositol 1,4,5-trisphosphate and inositol cyclic 1:2,4,5-trisphosphate metabolism in intact rat parotid acinar cells. *J. Biol. Chem.* **263,** 10,314–10,319.

Irvine R. F. (1989) Functions of inositol phosphates, in *Inositol Lipids in Cell Signalling* (Michell R. H., Drummond A. H., and Downes C. P., eds.), Academic, London, pp. 135–161.

Irvine R. F. (1990) "Quantal" Ca^{2+} release and the control of Ca^{2+} entry by inositol phosphates—a possible mechanism. *FEBS Lett.* 263, 5–9.

Irvine R. F. and Moor R. M. (1986) Micro-injection of inositol 1,3,4,5-tetrakisphosphate activates sea urchin eggs by a mechanism dependent on extrernal Ca^{2+}. *Biochem. J.* 240, 917–920.

Jacob R. (1990) Calcium oscillations in electrically nonexcitable cells. *Biochim. Biophys. Acta* 1052, 427–438.

Jacob R., Merritt J. E., Hallam T. J., and Rink T. J. (1988) Repetitive spikes in cytoplasmic calcium evoked by histamine in human endothelial cells. *Nature* 335, 40–45.

Jean T. and Klee C. B. (1986) Calcium modulation of inositol 1,4,5-trisphosphate-induced calcium release from neuroblastoma x glioma hybrid (NG108-15) microsomes. *J. Biol. Chem.* 261, 16,414–16,420.

Joseph S. K. and Williamson J. R. (1986) Characteristics of inositol trisphosphate-mediated Ca^{2+} release from permeabilized hepatocytes. *J. Biol. Chem.* 261, 14,658–14,664.

Joseph S. K. and Williamson J. R. (1989) Inositol polyphosphates and intracellular calcium release. *Arch. Biochem. Biophys.* 273, 1–15.

Joseph S. K., Rice H. L., and Williamson J. R. (1989) The effect of external calcium and pH on inositol trisphosphate-mediated calcium release from cerebellum microsomal fractions. *Biochem. J.* 258, 261–265.

Joseph S. K., Thomas A. P., Williams R. J., Irvine R. F., and Williamson J. R. (1984) *myo*-Inositol 1,4,5-trisphosphate. A second messenger for the hormonal mobilization of intracellular Ca^{2+} in liver. *J. Biol. Chem.* 259, 3077–3081.

Kass G. E. N., Duddy S. K., Moore G. A., and Orrenius S. (1989) 2.5-Di(*tert*-butyl)1,4-benzohydroquinone rapidly elevates cytosolic Ca^{2+} concentration by mobilizing the inositol 1,4,5-trisphosphate-sensitive Ca^{2+} pool. *J. Biol. Chem.* 264, 15,192–15,198.

Knight D. E. (1981) Rendering cells permeable by exposure to electric fields, in *Techniques in Cellular Physiology*, P113 Elsevier/North Holland Scientific, Amsterdam, pp. 1–20.

Kobayashi S., Kitazawa T., Somlyo A. V., and Somlyo, A. P. (1989) Cytosolic heparin inhibits muscarinic and α-adrenergic Ca^{2+} release in smooth muscle. *J. Biol. Chem.* 264, 17,997–18,004.

Kuno M. and Gardner P. (1987) Ion channels activated by inositol 1,4,5-trisphosphate in plasma membrane of human T-lymphocytes. *Nature* 326, 301–304.

Lai F. A., Erickson H. P., Rousseau E., Liu Q.-Y., and Meissner G. (1988) Purification and reconstitution of the calcium release channel from skeletal muscle. *Nature* 331, 315–319.

Lassalles A. J. P. and Kado R. T. (1990) Opening of Ca^{2+} channels in isolated red beet root vacuole by inositol 1,4,5-trisphosphate. *Nature* 343, 567–570.

McIntosh R. P. and McIntosh J. E. A. (1990) Metabolism of the biologically active inositol phosphates Ins(1,4,5)P_3 and Ins(1,3,4,5)P_4 by ovarian follicles of *Xenopus laevis. Biochem. J.* **268,** 141–145.

Maeda N., Kawasaki, T., Nakade, S., Yokota, N. Taguchi, T., Kasai, M., and Mikoshiba, K. (1991) Structural and functional characterization of inositol 1,4,5-trisphosphate receptor channel from mouse cerebellum. *J. Biol. Chem.* **266,** 1109-1116.

Maeda N., Niinobe M., and Mikoshiba K. (1990) A cerebellar Purkinje cell marker P_{400} protein is an inositol 1,4,5-trisphosphate (InsP_3) receptor protein. Purification and characterisation of InsP_3 receptor. *EMBO J.* **9,** 61–68.

Maeda N., Niinobe M., Inoue Y., and Mikoshiba K. (1989) Developmental expression and intracellular location of P_{400} protein characteristic of Purkinje cells in the mouse. *Dev. Biol.* **133,** 67–76.

Malgoroli A., Fesce R., and Meldolesi J. (1990) Spontaneous [Ca^{2+}]$_i$ fluctuations in rat chromaffin cells do not require inositol 1,4,5-trisphosphate elevations but are generated by a caffeine- and ryanodine-sensitive intracellular Ca^{2+} store. *J. Biol. Chem.* **265,** 3005–3008.

Marcotte G. V., Millard P. J., and Fewtrell C. (1990) Release of calcium from intracellular stores in rat basophilic leukemia cells monitored with the fluorescent probe chlortetracycline. *J. Cell Physiol.* **142,** 78–88.

Matsumoto T., Kanaide H., Shogakiuchi Y., and Nakamura M. (1990) Characteristics of the histamine-sensitive calcium stores of vascular smooth muscle. Comparison with norepinephrine- or caffeine-sensitive stores. *J. Biol. Chem.* **265,** 5610–5616.

Meldolesi J., Madeddu L., and Pozzan T. (1990) Intracellular Ca^{2+} storage organelles in non muscle cells: Heterogeneity and functional assignment. *Biochim. Biophys. Acta* **1055,** 130-140.

Merritt J. E. and Rink T. J. (1987) Regulation of cytosolic free calcium in fura-2-loaded rat parotid acinar cells. *J. Biol. Chem.* **262,** 17,362–17,369.

Merritt J. E., McCarthy S. A., Davies M. P. A., and Moores K. E. (1990) Use of fluo-3 to measure cytosolic Ca^{2+} in platelets and neutrophils: Loading cells with the dye; calibration of traces; measurements in the presence of plasma; buffering of cytosolic Ca^{2+}. *Biochem. J.* **269,** 513-519.

Merritt J. E., Taylor C. W., Rubin R. P., and Putney J. W. (1986) Evidence suggesting that a novel guanine nucleotide-dependent regulatory protein couples receptors to phospholipase C in exocrine pancreas. *Biochem. J.* **232,** 435–438.

Meyer T. and Stryer L. (1988) Molecular model for receptor-stimulated calcium spiking. *Proc. Natl. Acad. Sci. USA* **85,** 5051–5055.

Meyer T., Holowka D., and Stryer L. (1988) Highly cooperative opening of calcium channels by inositol 1,4,5-trisphosphate. *Science* **240,** 653–656.

Meyer T., Wensel T., and Stryer L. (1990) Kinetics of calcium channel opening by inositol 1,4,5-trisphosphate. *Biochemistry* **29,** 32–37.

Michell R. H. (1975) Inositol phospholipids and cell surface receptor function. *Biochim. Biophys. Acta* **415,** 81–147.

Michell R. H., Drummond A. H., and Downes C. P. (1989) *Inositol Lipids in Cell Signalling,* Academic, London, pp. 534.

Mignery G. A., Südhof T. C., Takei K., and Camilli P. D. (1989) Putative receptor for inositol 1,4,5-trisphosphate similar to ryanodine receptor. *Nature* **342,** 192–195.

Miledi R. and Parker I. (1989) Latencies of membrane currents evoked in *Xenopus* oocytes by receptor activation, inositol trisphosphate and calcium. *J. Physiol.* **415,** 189–210.

Morris A. P., Gallacher D. V., Irvine R. F., and Petersen O. H. (1987) Synergism of inositol trisphosphate and tetrakisphosphate in activating Ca²⁺-dependent K⁺ channels. *Nature* **330,** 653–655.

Muallem S., Pandol S. J., and Beeker T. G. (1989) Hormone-evoked calcium release is a quantal process. *J. Biol. Chem.* **264,** 205–212.

Muallem S., Schoeffield M., Pandol S., and Sachs G. (1985) Inositol trisphosphate modification of ion transport in rough endoplasmic reticulum. *Proc. Natl. Acad. Sci. USA* **82,** 4433–4437.

Mullaney J. M., Yu M., Ghosh T. K., and Gill D. L. (1988) Calcium entry into the inositol 1,4,5-trisphosphate-releasable calcium pool is mediated by a GTP-regulatory mechanism. *Proc. Natl. Acad. Sci. USA* **85,** 2499–2503.

Munson P. J. and Rodbard D. (1980) LIGAND: A versatile computerized approach for characterization of ligand-binding systems. *Anal. Biochem.* **107,** 220–239.

Nahorski S. R. and Potter B. V. L. (1989) Molecular recognition of inositol polyphosphates by intracellular receptors and metabolic enzymes. *Trends Pharmacol. Sci.* **10,** 139–144.

Nishizuka Y. (1988) The molecular heterogeneity of protein kinase C and its implications for cellular regulation. *Nature* **334,** 661–665.

Norman A. W., Demel R. A., DeKruyff B., Geurts Van Kessel W. S. M., and Van Deenen L. L. M (1972) Studies on the biological properties of polyene antibiotics: Comparison of other polyenes with filipin in their ability to interact specifically with sterol. *Biochim. Biophys. Acta* **290,** 1–14.

Nunn D. L. and Taylor C. W. (1990) Liver inositol 1,4,5-trisphosphate-binding sites are the calcium-mobilizing receptors. *Biochem. J.* (in press).

Nunn D. L., Potter B. V. L., and Taylor C. W. (1990) Molecular target size of inositol trisphosphate receptors in cerebellum and liver. *Biochem. J.* **266,** 189–194.

Ogden D. C., Capiod T., Walker J. W., and Trentham D. R. (1990) Kinetics of the conductance evoked by noradrenaline, inositol trisphosphate or Ca²⁺ in guinea-pig isolated hepatocytes. *J. Physiol.* **422,** 585–602.

Oldershaw K. A. and Taylor C. W. (1990) 2,5-Di(*tert*-butyl)-1,4-benzohydroquione mobilizes inositol 1,4,5-trisphosphate-sensitive and insensitive Ca²⁺ stores. *Febs Lett.* **274,** 214–216.

O'Rourke F. and Feinstein B. (1990) The inositol 1,4,5-trisphosphate receptor binding sites of platelet membranes. *Biochem. J.* **267,** 297–302.

O'Sullivan A. J., Cheek T. R., Moreton R. B., Berridge M. J., and Burgoyne R. D. (1989) Localization and heterogeneity of agonist-induced changes in cytosolic calcium concentration in single bovine adrenal chromaffin cells from video imaging of fura-2. *EMBO J.* **8**, 401–411.

Palade P. (1987) Drug-induced Ca^{2+} release from isolated sarcoplasmic reticulum. *J. Biol. Chem.* **262**, 6135–6141.

Payne R. and Fein A. (1987) Inositol 1,4,5-trisphosphate releases calcium from specialized sites within *Limulus* photoreceptors. *J. Cell Biol.* **104**, 933–937.

Polokoff M. A., Bencen G. H., Vacca J. P., deSolms S. J., Young S. D., and Huff J. R. (1988) Metabolism of synthetic inositol trisphosphate analogs. *J. Biol. Chem.* **263**, 11,922–11,927.

Prentki M., Wollheim C. B., and Lew P. D. (1984) Ca^{2+} homeostasis in permeabilized human neutrophils. Characterization of Ca^{2+}-sequestering pools and the action of inositol 1,4,5-trisphosphate. *J. Biol. Chem.* **259**, 13,777–13,782.

Putney J. W. (1986a) *Receptor Biochemistry and Methodology*, vol. 7: *Phosphoinositides and Receptor Mechanisms.* Alan R. Liss, NY.

Putney J. W. (1986b) A model for receptor-regulated calcium entry. *Cell Calcium* **7**, 1–12.

Raess B. U., Record D. M., and Tunnicliff G. (1985) Interaction of phenylglyoxal with the human erythrocyte (Ca^{2+} + Mg^{2+})-ATPase. *Mol. Pharmacol.* **27**, 444–450.

Riordan F. (1979) Arginyl residues and anion binding sites in proteins. *Mol. Cell. Biochem.* **26**, 71–92.

Rooney T. A., Sass E. J., and Thomas A. P. (1990) Agonist-induced cytosolic calcium oscillations originating from a specific locus in single hepatocytes. *J. Biol. Chem.* **265**, 10,792-10,796.

Ross C. A., Meldolesi J., Milner T. A., Satah T., Supattapone S., and Snyder S. H. (1989) Inositol 1,4,5-trisphosphate receptor localized to endoplasmic reticulum in cerebellar Purkinje neurons. *Nature* **339**, 468–470.

Rossier, M. F., Bird, G. St. J., and Putney, J. W. (1991) Subcellular distribution of the calcium-storing inositol 1,4,5-trisphosphate-sensitive organelle in rat liver. *Biochem. J.* **274**, 643-650.

Rozengurt E. and Heppel L. A. (1975) A specific effect of external ATP on the permeability of transformed 3T3 cells. *Biochem. Biophys Res. Commun.* **67**, 1581–1588.

Shah J. and Pant H. C. (1988) Potassium channel blockers inhibit inositol trisphosphate-induced calcium release in the microsomal fraction isolated from rat brain. *Biochem. J.* **250**, 617–620.

Shears S. B. (1989) Metabolism of the inositol phosphates produced upon recep-tor activation. *Biochem. J.* **260**, 313–324

Smith J. B., Smith L., and Higgins B. L. (1985) Temperature and nucleotide dependence of calcium release by *myo*-inositol 1,4,5-trisphosphate in cultured vascular smooth muscle cells. *J. Biol. Chem.* **260**, 14,413–14,416.

Somlyo A. P., Bond M., and Somlyo A. V. (1985) Calcium content of mito-chondria and endoplasmic reticulum in liver rapidly frozen in vivo. *Nature* **314**, 622–625.

Spät A., Bradford P. G., McKinney J. S., Rubin R. P., and Putney J. W. (1986) A saturable receptor for ^{32}P-inositol-1,4,5-trisphosphate in hepatocytes and neutrophils. *Nature* **319**, 514–516.

Streb H. and Schulz I. (1983) Regulation of cytosolic free Ca^{2+} concentration in acinar cells of rat pancreas. *Am. J. Physiol.* **245**, G347–G357.

Streb H., Irvine R. F., Berridge M. J., and Schulz I. (1983) Release of calcium from nonmitochondrial stores in pancreatic acinar cells by inositol-1,4,5-trisphosphate. *Nature* **306**, 67–69.

Streb H., Bayerdörffer E., Haase W., Irvine R. F., and Schulz I. (1984) Effect of inositol-1,4,5-trisphosphate on isolated subcellular fractions of rat pancreas. *J. Membr. Biol.* **81**, 241–253.

Supattapone S., Danoff S. K., Theibert A., Joseph S. K., Steiner J., and Snyder S. H. (1988a) Cyclic AMP-dependent phosphorylation of a brain inositol trisphosphate receptor decreases its release of calcium. *Proc. Natl. Acad. Sci. USA* **85**, 8747–8750.

Supattapone S., Worley P. F., Baraban J. M., and Snyder S. H. (1988b) Solubilization, purification, and characterization of an inositol trisphosphate receptor. *J. Biol. Chem.* **263**, 1530–1534.

Takemura H. and Putney J. W. (1989) Capacitative calcium entry in parotid acinar cells. *Biochem. J.* **258**, 409–412.

Taylor C. W. (1990a) Receptor-regulated Ca^{2+} entry: Secret pathway or secret messenger. *Trends Pharmacol. Sci.* **11**, 269-271.

Taylor C. W. (1990b) The role of G proteins in transmembrane signalling. *Biochem. J.* **272**, 1-13.

Taylor C. W. and Merritt J. E. (1986) Receptor coupling to polyphosphoinositide turnover: a parallel with the adenylate cyclase system. *Trends Pharmacol. Sci.* **7**, 238–242.

Taylor C. W. and Potter B. V. L. (1990) The size of inositol 1,4,5-trisphosphate-sensitive Ca^{2+} pools depends on inositol trisphosphate concentration. *Biochem. J.* **266**, 189–194.

Taylor C. W. and Putney J. W. (1985) Size of the inositol trisphosphate-sensitive calcium pool in guinea-pig hepatocytes. *Biochem. J.* **232**, 435–438.

Taylor C. W., Berridge M. J., Cooke A. M., and Potter B. V. L. (1989) Inositol 1,4,5-trisphosphorothioate, a stable analogue of inositol trisphosphate which mobilizes intracellular calcium. *Biochem. J.* **259**, 645–650.

Thastrup O., Cullen P. J., Drøbak B. K., Hanley M. R., and Dawson A. P. (1990) Thapsigargin, a tumor promoter, discharges intracellular Ca^{2+} stores by specific inhibition of the endoplasmic reticulum Ca^{2+}-ATPase. *Proc. Natl. Acad. Sci. USA* **87**, 2466–2470.

Thelestam M. and Mollby R. (1979) Classification of microbial, plant and animal cytolysins based on their membrane-damaging effects on human fibroblasts. *Biochim. Biophys. Acta* **557**, 156–169.

Thévenod R., Dehlinger-Kremer M., Kemmer T. P., Christian A.-L., Potter B. V. L., and Schulz I. (1989) Characterization of inositol 1,4,5-trisphosphate-sensitive (IsCaP) and -insensitive (IisCaP) nonmitochondrial Ca^{2+} pools in rat pancreatic acinar cells. *J. Membr. Biol.* **109**, 173–186.

Thomas M. V. (1982) *Techniques in Calcium Research.* Academic, London, p. 214.

Thompson S. T., Cass K. H., and Stellwagen E. (1975) Blue dextran-Sepharose: An affinity column for the dinucleotide fold in proteins. *Proc. Natl. Acad. Sci. USA* **72**, 669–672.

Tones M. A., Bootman M. D., Higgins B. F., Lane D. A., Pay G. F., and Lindahi U. (1989) The effect of heparin on the inositol 1,4,5 trisphosphate receptor in rat liver microsomes: Dependence on sulphate content and chain length. *FEBS Lett.* **252**, 105–108.

Tsien R. Y. (1989) Fluorescent probes of cell signaling. *Annu. Rev. Neurosci.* **12**, 227–253.

Vilven J. and Coronado R. (1988) Opening of dihydropyridine calcium channels in skeletal muscle membranes by inositol trisphosphate. *Nature* **336**, 587–589.

Volpe P., Krause K.-H., Hashimoto S., Zorzato F., Pozzan T., Meldolesi J., and Lew D. P. (1988) "Calciosome," a cytoplasmic organelle: The inositol 1,4,5-trisphosphate-sensitive Ca^{2+} stores of nonmuscle cells? *Proc. Natl. Acad. Sci. USA* **85**, 1091–1095.

Wakui M., Potter B. V. L., and Petersen O. H. (1989) Pulsatile intracellular calcium release does not depend on fluctuations in inositol trisphosphate concentration. *Nature* **339**, 317–320.

Willems P. H. G. M., DeJong M. D., DePont J. J. H. H. M., and van Os C. H. (1990) Ca^{2+}-sensitivity of inositol 1,4,5-trisphosphate-mediated Ca^{2+} release in permeabilized pancreatic acinar cells. *Biochem. J.* **265**, 681–687.

Woods N. M., Cuthbertson K. S. R., and Cobbold P. H. (1986) Repetitive transient rises in cytoplasmic free calcium in hormone-stimulated hepatocytes. *Nature* **319**, 600–602.

Worley P. F., Baraban J. M., Supattapone S., Wilson V. S., and Snyder S. H. (1987) Characterization of inositol trisphosphate receptor binding in brain. *J. Biol. Chem.* **262**, 12,132–12,136.

Measurement of Intracellular Calcium with Fluorescent Calcium Indicators

Martin Poenie

1. Introduction

Changes in cytosolic free calcium ion concentration ($[Ca^{2+}]_i$) accompany many cellular transitions and stimulation of cell surface receptors. Both release of Ca^{2+} from internal stores and Ca^{2+} entry through plasma membrane ion channels can contribute to cellular Ca^{2+} transients. In the past, direct measurement of $[Ca^{2+}]_i$ was limited mainly to cells that could be penetrated with Ca^{2+} electrodes or injected with membrane impermeant indicators, such as aequorin. Alternatives, such as $^{45}Ca^{2+}$ flux measurements, reduction of extracellular Ca^{2+}, or artificial elevation of intracellular Ca^{2+} using ionophores, provided indirect evidence for the role of Ca^{2+} in particular cell processes. However, the clues provided by indirect approaches are not always reliable. Inward and outward Ca^{2+} fluxes can be triggered by depolarizing sea urchin eggs without measurable changes in $[Ca^{2+}]_i$ (M. Poenie, unpublished) or the activity of Ca^{2+}-dependent enzymes (Schmidt et al.,1982). Although cells can transduce a change in membrane potential or $[Ca^{2+}]_i$ into a meaningful cell signal, the same cannot can be said for a Ca^{2+} current (Jaffe, 1986). Calcium currents are significant when they alter $[Ca^{2+}]_i$ somewhere inside the cell, and change the proportion of free and Ca^{2+}-bound forms of Ca^{2+}-binding proteins. Thus, understanding the significance of Ca^{2+} currents will ultimately require knowledge of how they alter $[Ca^{2+}]_i$.

From: Neuromethods, Vol. 20: Intracellular Messengers
Eds: A. Boulton, G. Baker, and C. Taylor © 1992 The Humana Press Inc.

Changes in cellular activities upon adding Ca^{2+} ionophores, such as A23187 or ionomycin, have provided insight into the role of Ca^{2+} in cell function. However, addition of an ionophore does not guarantee a physiologically relevant change in $[Ca^{2+}]_i$. The effectiveness of the Ca^{2+} ionophores varies with pH, ionic strength, temperature, and cell type. One could as easily raise $[Ca^{2+}]_i$ to extraordinarily high and toxic levels or not change $[Ca^{2+}]_i$ at all. Thus, although a Ca^{2+} ionophore can inhibit cell motility at one concentration, at other concentrations it can also stimulate cell movements. This leads to disparate interpretations of the role of Ca^{2+} in cell motility.

Chelators of extracellular Ca^{2+} are widely used to inhibit the entry of extracellular Ca^{2+} and can expose events that depend on the presence of extracellular Ca^{2+}. However, some cells may tolerate low extracellular Ca^{2+} (ca. $10^{-7}M$) only for short periods. Incubations with EGTA can slowly deplete Ca^{2+} stores, so that cellular responses may vary with the duration of exposure to Ca^{2+}-free medium. Furthermore, EGTA can deplete other metals, such as Zn^{2+}. Once again, apart from measurement of $[Ca^{2+}]_i$, one does not know how the treatment affects the Ca^{2+} signal. Thus, chelators, ionophores, and tracers should be considered as important tools, but not as substitutes for direct measurements of $[Ca^{2+}]_i$.

The fluorescent Ca^{2+} indicators and their acetoxymethyl (AM) ester derivatives provide important and convenient tools for studying cellular Ca^{2+} signals. With commercialization of the technology, studies using these indicators have become commonplace. Yet the availability of good, user-friendly instruments and commercial sources of the indicators does not automatically guarantee meaningful biological results. One needs an understanding of the properties of the indicators in their various forms and of how they interact with the cell in order to interpret the data accurately. Verifying the quality of dye loading, spectral response, and calibration procedures constitutes an important part of any study using these indicators. The following discussion aims at acquainting those new to this field with some of the problems commonly encountered in using these indicators and how they can be avoided.

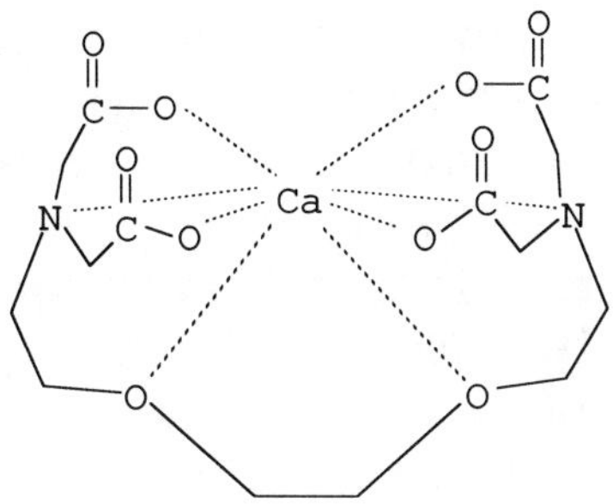

EGTA

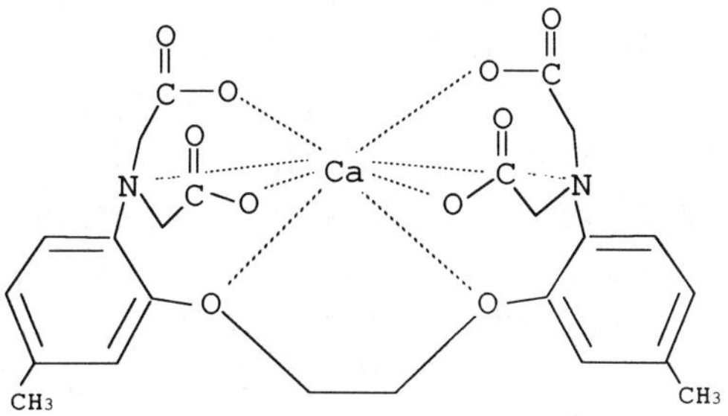

Bapta

Fig. 1. The structure of EGTA (top) is compared to that of bapta (bottom). Both compounds exhibit a single octa-coordinate binding site for Ca²⁺. One oxygen from each of the four carboxyl groups, the two tertiary amino groups, and the two ether-linked oxygens all contribute to Ca²⁺ binding. Bapta differs from EGTA in that the amino groups of bapta are aromatic and are not significantly protonated at pH 7.0.

2. Properties of Fluorescent Ca²⁺ Indicators

2.1. Chelation Properties

The fluorescent Ca²⁺ indicators including fura-2, indo-1, and fluo-3, derive their Ca²⁺-binding properties and ion selectivity from bapta (Tsien, 1980; Grynkiewicz et al., 1985; Tsien and Minta, 1989). Bapta is a tetracarboxylate analog of EGTA (*see* Fig. 1), cleverly designed by Roger Tsien to be pH-insensitive in the physiological pH range while retaining high affinity and selectivity for Ca²⁺ over Mg²⁺. Bapta contains a single high-affinity binding site for Ca²⁺, which involves coordination of one Ca²⁺

ion to eight ligands on the bapta molecule. Here, Ca^{2+} binds to one oxygen on each of the four carboxyl groups, the two aromatic amino groups, and the two phenoxyethane oxygens (Gerig et al., 1987). This 1:1 stoichiometry provides for simple interpretation of spectral changes associated with the binding of Ca^{2+} to bapta or other members in this family. At physiologic pH, there are only two forms of the dye, free and Ca^{2+}-bound. These two forms typically have distinct spectral properties. Spectra of a mixture of both forms of the chelator represent the contributions from free and the Ca^{2+}-bound forms in proportion to their concentration. For many metallochromic indicators, the stoichiometry of Ca^{2+} binding is not simply 1:1, and relating spectral changes to Ca^{2+} binding is much more complex.

Bapta differs from EGTA in that *o*-aminophenoxy groups are substituted for the aminoethoxy groups on EGTA (*see* Fig. 1). This switch from alkyl to aromatic amino groups results in a number of improved features in bapta. In EGTA, the pK_as of the two amino groups are 9.58 and 8.96 (Martell and Smith, 1974), respectively, and are largely protonated over the physiologic pH range. Protonation of these nitrogen groups greatly reduces the affinity of EGTA for Ca^{2+}. As a result, the affinity of EGTA for Ca^{2+} is highly pH-dependent. The aromatic amino groups of bapta have much lower pK_as (pK_a = 5.47 and 6.36; Tsien, 1980) and are not significantly protonated over the physiologic pH range. Because of these differences in amine pK_as, bapta is not sensitive to minor variations in pH and binds Ca^{2+} much more rapidly than EGTA. For EGTA to bind Ca^{2+}, protons bound at the amino groups must first dissociate. The binding is essentially a Ca^{2+}–H^+ exchange reaction, in which dissociation of protons is rate-limiting. This exchange reaction slows the rate at which EGTA can bind to Ca^{2+}. Since bapta amino groups are not protonated to begin with, there is no rate-limiting proton dissociation step (Hellam and Podolsky, 1969; Tsien, 1980). The faster kinetics of Ca^{2+} binding by bapta is significant when attempting to buffer or block fast Ca^{2+} transients in neurons and muscle cells.

Bapta has good selectivity for Ca^{2+} over Mg^{2+}. The Mg^{2+} ion is smaller than Ca^{2+} and, therefore, cannot fully coordinate with all the available ligands. Instead, Mg^{2+} appears to use only half of the bapta Ca^{2+}-binding pocket. At very high Mg^{2+} concentrations, two Mg^{2+} ions will bind to bapta. This has led to the development of fluorescent indicators selective for Mg^{2+}, such as mag-fura (furaptra), which is essentially the fura-2 chromophore linked to half of the bapta molecule (Levy et al., 1988; Raju et al., 1989; Molecular Probes Catalog). These indicators have dissociation constants for Mg^{2+} in the millimolar range.

The affinity of bapta for Ca^{2+} is strongly influenced by the presence and position of electron-withdrawing substituents on the aromatic rings. Consequently, a variety of bapta-based buffers have been generated with Ca^{2+} dissociation constants ranging from $10^{-7}M$ to $10^{-4}M$ (Tsien, 1980; Speksnuder et al., 1989).

2.2. Spectral and Fluorescence Properties

When a fluorescent chromophore is electronically conjugated to bapta, the effects of Ca^{2+} binding are relayed to the fluorophore and influence its fluorescence. The changes in fluorescence spectra associated with Ca^{2+} binding differ among the various fluorescent indicators. For quin2, fluo-3, and rhod-2, Ca^{2+} binding mainly increases fluorescence intensity (Fig. 2). Quin2 does have an isoexcitation point, but the fluorescence is very weak at the isoexcitation wavelength, so that it is not generally used as a ratiometric indicator. With fura-2, Ca^{2+} binding shifts the preferred excitation wavelength from 362 to 335 nm. Indo-1 exhibits similar shifts in excitation, but also exhibits shifts in the emission spectrum as well. The shift in either excitation or emission spectra upon Ca^{2+} binding permits the use of fura-2 (Fig. 3) and indo-1 (Fig. 4) as ratiometric indicators where $[Ca^{2+}]_i$ measurements depend on the ratio of fluorescence at two excitation or emission wavelengths, rather than simply the intensity at one wavelength. Unfortunately, the presently available indicators that exhibit shifts in wavelength (fura-2, indo-1) require UV excitation (Grynkiewicz et al., 1985).

Recently, long-wavelength indicators based on fluorescein (fluo-3) and rhodamine (rhod-2) have become available, but these do not exhibit enough shift in wavelength to make them useful as ratiometric indicators (Minta et al., 1989). Even so, the longer excitation and emission wavelengths of fluo-3 and its 40-fold

A Excitation spectra for fluo-3, Ca titration
emission 530 nm, 0.1M KCl, pH 7.0, 22°

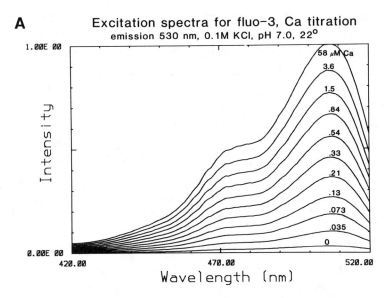

B Emission spectra for fluo-3, Ca titration
excitation 490 nm, 0.1M KCl, pH 7.0, 22°

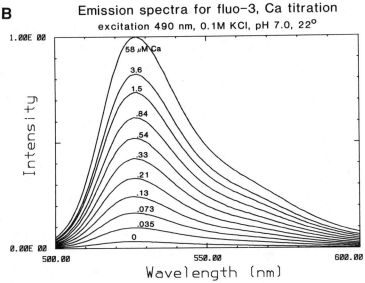

Fig. 2. Fluorescence excitation and emission spectra of fluo-3. **A.** Excitation spectra (530 nm emission), recorded at 22°C, were recorded as fluo-3 was titrated incrementally with Ca^{2+}. The zero $[Ca^{2+}]$ spectrum was obtained from 15 μM fluo-3 in 100 mM KCl, 10 mM K-MOPS, 10 mM K_2H_2EGTA, pH 7.0, and then $[Ca^{2+}]$ was increased by additions of K_2CaEGTA to give values indicated in the figure. **B.** Emission spectra (490 nm excitation) were recorded for fluo-3 as in A. (From Minta et al., 1989.)

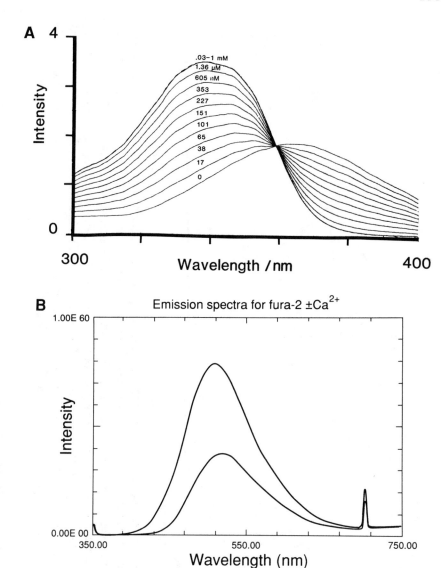

Fig. 3. Fluorescence excitation and emission spectra for fura-2. **A.** Excitation spectra of fura-2, titrated incrementally with Ca^{2+}, were recorded between 300–400 nm with the emission set at 505 nm. The preferred excitation wavelength shifts from 365 nm when $[Ca^{2+}]$ is zero to 335 nm at saturating Ca^{2+}. (From Grynkiewicz et al., 1985.) **B.** The emission spectra of fura-2 was recorded at zero and saturating $[Ca^{2+}]$ while exciting at 351 nm. The small shift in fura-2 emission wavelengths with Ca^{2+} is generally considered insignificant. (Spectra provided by R. Tsien.)

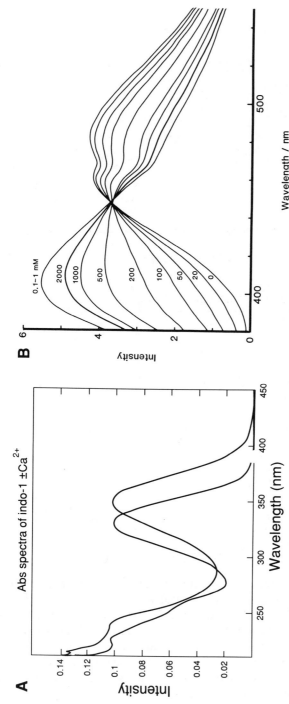

Fig. 4. Absorbance and fluorescence emission spectra of indo-1. Indo-1 exhibits a shift in both excitation and emission spectra with Ca^{2+}. A. The absorbance spectra of indo-1 are given for zero and saturating [Ca^{2+}]. Here absorbance rather than excitation spectra are given because the absorbance spectrum does not depend on choice of a particular emission wavelength. B. Emission spectra for indo-1 (355 nm excitation) were recorded as Ca^{2+} was incrementally increased from zero to 1 mM [Ca^{2+}]. The zero [Ca^{2+}] spectrum was obtained at 37°C from 6 μM indo-1 in 115 mM KCl, 20 mM NaCl, 10 mM MOPS, and 1.115 mM K$_2$H$_2$EGTA brought to pH 7.05 with KOH. Aliquots of K$_2$CaEGTA were then added to increase [Ca^{2+}] to the values indicated in the figure. Unless specified otherwise, all [Ca^{2+}] are expressed as nM. (From Grynkiewicz et al., 1985.)

Table 1
Spectral Properties of the Fluorescent Calcium Indicators

	bapta[b]	quin2[b]	fura-2[c]	indo-1[c]	fluo-3[d]	rhod-2[d]
K_{dCa}, nM	107[a]	126	224, 135[a]	250	450	1000
K_{dMg}, mM	17	1	5.6		9.0	
Abs + Ca, nm		352	335	331	506	553
Abs − Ca, nm			362	349	506	553
Em + Ca, nm		492	505	410	526	578
Em − Ca, nm		492	512	485	526	578

[a]Determined at 22°C; unless indicated otherwise, dissociation constants (K_d) are measured at 37°C.
[b]Tsien, 1980.
[c]Grynkiewicz et al., 1985.
[d]Minta et al., 1989.

increase in brightness upon binding Ca²⁺ have led to a steady increase in its use. A comparison of the properties of the different indicators is given in Table 1.

The data given in Table 1 apply only to the indicators in their salt or free acid form. However, most experiments use the AM esters of these indicators because they provide an easy means of loading them into the cell (*see below*). The properties of indicator/AM esters are quite different from those of the indicator itself. This is illustrated in the comparison of fura-2 and fura-2/AM given in Table 2.

2.3. Calcium Buffering

One of the important concerns raised in using fluorescent Ca²⁺ indicators is the issue of Ca²⁺ buffering. With quin2, dye concentration inside the cell after loading often approached a concentration of 1 mM. One might question the possibility of measuring 1 μM free Ca²⁺ inside a cell containing millimolar concentrations of indicator. However, this is similar to pH measurements using pH indicator dyes, where the indicator concentration greatly exceeds [H⁺]. This works because a pH buffer is included in the solution, which makes the concentration of the pH indicator insignificant. Measurement of Ca²⁺ using fluorescent Ca²⁺ indicators operates on the same principle and assumes the pres-

Table 2
A Comparison of Fura-2 and Fura-2/AM [a]

Fura-2 Penta-Anion	Fura-2/AM
Brightly fluorescent	Brightly fluorescent
Soluble in aqueous media	Soluble in organic solvents
Binds Ca^{2+}	Does not bind Ca^{2+}
Excitation shifts when Ca^{2+} is added	Excitation does not shift when Ca^{2+} is added
Fluorescence quenched by Mn^{2+}	Fluorescence is not quenched by Mn^{2+}
Does not bind to membranes	Binds to membranes
Impermeable to cells	Permeable through the cell membrane
Uses:	Use:
microinjection into cells, calibration standards	noninvasive loading of fura-2 into cells

[a] Data compiled from Grynkiewicz et al. (1985); Roe et al. (1990); Poenie (1990); Popov et al. (1988).

ence of an intracellular Ca^{2+} buffer. Most studies indicate that cells contain almost 1 mM of Ca^{2+} buffering power/L of cell water. These estimates have been obtained in different ways. In neutrophils, the Ca^{2+} buffering capacity was estimated to be 760 μM with 20% of the available sites occupied under resting conditions (von Tscharner et al., 1986). In these studies, quin2 was introduced into cells at different concentrations, and rates of Ca^{2+} entry were measured after addition of ionophore. The contribution of the cytoplasmic buffer affected the apparent rate of Ca^{2+} entry, depending on the amount of quin2 present. Using this approach, the cytoplasmic buffer component could be mathematically isolated and quantified. In a similar study using platelets, Johansson and Haynes (1988) reported the total concentration of cytoplasmic buffer as 730 μM with a K_d of 140 nM and attributed much of it to calmodulin. Using the data of White et al. (1981), who reported 0.16 mM as the intracellular calmodulin concentration and four Ca^{2+}-binding sites per calmodulin, one arrives at 0.64 mM of total Ca^{2+} buffer owing to calmodulin alone. One apparent problem here is that the K_d of calmodulin for Ca^{2+} is reportedly in the micromolar range. However, when calmodulin is bound to other cell components, the K_d can drop to $10^{-8}M$ (Kohse and Heilmeyer, 1981).

Estimates of the Ca^{2+} buffer in smooth muscle can also be derived from the electron probe data of Bond et al. (1984). Their measurements indicate the presence of 188 μM bound Ca^{2+} in resting muscle cytoplasm, which drops to 94 μM in the presence of extracellular EGTA. Given a resting $[Ca^{2+}]_i$ of $10^{-7}M$, these data would again suggest a substantial concentration of high-affinity cytoplasmic Ca^{2+} buffer. In stimulated smooth muscle cells, the total Ca^{2+} rises by an additional 236 μM, but the free only rises from 10^{-7} to a few micromolar. Using a single dissociation constant of 144 nM, the minimum total Ca^{2+} buffer that could account for their data is 550 μM. However, many cellular Ca^{2+}-binding proteins have affinities much weaker than 144 nM, and total concentration of buffer would need to be much greater than 550 μM.

Cells loaded with fura-2 typically contain between 50–100 μM of the indicator. Given a K_d of 224 nM, 50 μM intracellular fura-2, and a resting Ca^{2+} level of $10^{-7}M$, approx 15 μM of fura-2 will be bound to Ca^{2+} and 35 μM will be free. Based on the estimates of bound Ca^{2+} and cytosolic buffer, the presence of fura-2 would expand the pool of bound cytosolic Ca^{2+} by about 10%. The remaining free fura-2 would represent only 5–10% of the total free cytosolic sites available for Ca^{2+} binding. For typical fura-2 loadings, this would appear to be a relatively small perturbation.

2.4. Kinetics of Ca²⁺ Binding

Calcium studies in nerve and muscle cells, where transients occur over a period of milliseconds, require indicators with suitable time constants. The kinetics of Ca^{2+} binding has been studied by Kao and Tsien (1988) using temperature-jump relation kinetics and by Jackson et al. (1987) using stopped-flow techniques. Both studies gave similar results. For fura-2 at 20°C and 0.561 μM free Ca^{2+}, Kao and Tsien obtained an association rate constant (k_1) of $6.02 \times 10^8 M^{-1}.s^{-1}$ and a dissociation rate constant (k_{-1}) of $96.7 s^{-1}$ and at this Ca^{2+} concentration, a relaxation time of approx 2.5 ms. Here relaxation time (τ) is related to k_1 and k_{-1} by the equation

$$1/\tau = k_1[Ca^{2+}] + k_{-1} \tag{1}$$

The relaxation time is faster at higher Ca^{2+} concentrations, so that at 1 μM Ca^{2+}, the relaxation rate falls to 1.4 ms. Kao and Tsien suggest that an indicator should be able to track a Ca^{2+} transient on a time scale 3–4 × slower than the relaxation time. Thus, in theory, fura-2 should be able to track Ca^{2+} transients on a time scale of 5–10 ms. In practice, Ca^{2+} transients have been observed in neurons with rise times in the range of 7–10 ms (Lev-Ram and Grinvald, 1987). During this period, the authors estimated a change from 10 to 90% saturation of fura-2 with Ca^{2+}.

In contrast to results described above, several authors indicate that once fura-2 is inside the cell, its rate of Ca^{2+} binding becomes much slower. They claim, for example, that fura-2 does not accurately track muscle Ca^{2+} transients as compared to the signal recorded using antipyrlazo III. In the studies of Hollingworth and Baylor (1987), interaction of fura-2 with the cytoplasm influenced the kinetics of Ca^{2+} binding. Reportedly, fura-2 diffused through the myoplasm threefold slower than expected for a freely diffusible small molecule, whereas anisotropy measurements indicated a restriction in fura-2 mobility. They also suggested that the fura-2 response behaved as if the dissociation rate were three- to fourfold slower than that reported by Kao and Tsien. The results of Klein et al. (1988) reported even greater kinetic delays in the response of fura-2 when compared to that of antipyrlazo III. They calculated on and off rates for Ca^{2+} binding that were approximately tenfold slower than those determined by Kao and Tsien. Taken together, these studies indicate that, under ideal conditions, the kinetics of Ca^{2+} binding are close to those needed for measurement of fast neuronal or muscle Ca^{2+} transients, but in the cytoplasm of some cells, these conditions are not always met.

3. Application of Fluorescent Ca²⁺ Indicators

3.1. Indicator Quality

Good biological results using fluorescent indicators depend on the quality of the indicator. Most investigators obtain the indicators from commercial sources and depend on the supplier's quality control. The quality of product from commercial suppliers is generally good, but problems arise occasionally. We have experienced periods in which cells lost their responses to any stimulus. In some circumstances, this seemed to relate to the health of the cells, whereas in other cases, the problem could be traced to a particular batch of indicators.

Two problems arise on occasion and may be difficult to detect without titration of the indicator or careful analytical separations, such as HPLC. One problem is incomplete hydrolysis of the ethyl or t-butyl ester precursors prior to preparation of the AM ester. Another problem is where the indicator does not have the full complement of carboxyl groups either because of incomplete alkylation or spontaneous decarboxylation. Both problems can show up during titrations as the presence of a low-affinity indicator.

Several simple analytical tests can be performed routinely to test the purity and authenticity of the dye preparation. The AM ester should be chromatographically pure and migrate as a single spot by thin-layer chromatography (TLC) or, better, as a single peak in a good reverse-phase HPLC separation. Samples of the AM ester can be spotted onto silica plates and developed in ethyl acetate:hexane 1:1, ethyl acetate:toluene 1:1, or chloroform:methanol 3:1. Using HPLC, samples can be chromatographed on a C_{18} reverse-phase column using an isocratic mobile phase of 70% methanol:water. The salts of the indicator dyes can be readily analyzed with reverse-phase TLC plates (we use Whatman $MKC_{18}F$) developed in 70% water:methanol. Here, the penta-anion of fura-2 is the fastest moving spot, whereas unhydrolyzed ester or decarboxylated forms of the dye move more slowly.

Hydrolysis of the dye AM esters, which occurs readily with KOH (1M) solution at 37°C, provides another check on the quality of the dye. The products of hydrolysis should have the same properties as an authentic sample of the indicator salt. Ethyl esters take harsher conditions or longer to hydrolyze than the AM ester, and persist after all the AM esters are hydrolyzed. With fluo-3 and rhod-2, t-butyl esters are used, instead of ethyl esters, during synthesis to protect the carboxyl groups. These are base-stable and are normally removed with trifluoroacetic acid. It should be noted that problems requiring this type of analysis are unusual, but not without precedent.

Titration of the indicator dye is a useful method for determining the purity and quality of the chelator. For fura-2, the Scatchard plot of $\Delta F/[Ca^{2+}]$ vs ΔF, at either 340 or 380 nm, should yield a straight line. Fura-2 should saturate at around 30 μM $[Ca^{2+}]$, so that no additional change in fluorescence is seen when Ca^{2+} is incremented in the millimolar range. Additional fluorescence change through millimolar increments of Ca^{2+} indicates the presence of low-affinity indicator. Contaminants or impurities are typically seen as fluorescence at 380 nm that does not decrease as Ca^{2+} levels are increased. One can use the 340:380 nm fluorescence ratio at saturating $[Ca^{2+}]$ as a sensitive standard for the presence of fluorescent contaminants.

3.2. Indicator Loading

3.2.1. General Considerations

Perhaps the most critical aspect of using the fluorescent Ca^{2+} indicators concerns the method of loading dye into the cell. Where feasible, injection of the chelator salt provides the best assurance of good dye loading in the cell. However, microinjection finds application mainly for single cell studies. The primary alternative is the use of AM ester derivatives (Tsien, 1981). The AM ester protects and masks the negatively charged carboxyl groups on the indicator, so that it can easily cross the plasma membranes of cells. The AM ester derivative is different from many other simple esters in that it is readily hydrolyzed once inside the cell, liberating the original indicator.

Cells differ in their ability to accumulate and fully hydro-lyze the AM ester derivatives. Most plants, fungi, and many marine invertebrates metabolize the AM esters too slowly to be loaded by this approach. In the worst cases, the AM ester is the major fluorescent component in the cell after loading. Unhydrolyzed dye in the cell is especially serious when using ratiometric indicators, the results being underestimation of [Ca^{2+}] and small changes in ratio during [Ca^{2+}]$_i$ transients *(see* Almers and Neher, 1985; chapter by Moreton in this vol.). Where the loading and hydrolysis are good, residual AM ester constitutes <5% of the total indicator fluorescence in the cell. Achieving a good degree of hydrolysis is therefore a prerequisite to any quantitative use of the fluorescent Ca^{2+} indicators. Experimentation can sometimes lead to significant improvements in loading because cells with good potential for loading can, under some conditions, load poorly.

$$R\text{-}\overset{O}{\overset{||}{C}}\text{-O-C-O-}\overset{O}{\overset{||}{C}}\text{-CH}_3 \quad \xrightarrow[\text{esterases}]{\text{cellular}} \quad R\text{-}\overset{O}{\overset{||}{C}}\text{-O}^- \;+\; H\text{-}\overset{O}{\overset{||}{C}}\text{-H} \;+\; CH_3\overset{O}{\overset{||}{C}}\text{-O}^- \;+\; H^+$$

Acetoxymethyl Ester   Free dye   Formaldehyde   Acetate   (2)

3.2.2. Formaldehyde as a Byproduct of Acetoxymethyl Ester Hydrolysis

As can be seen above, hydrolysis of the AM ester group yields a free dye-carboxylate group, acetate, H^+, and formaldehyde. Formaldehyde is known to be toxic and is therefore of some concern. To date, the most serious examples of formaldehyde toxicity relate to the use of quin2 or fura-2 in retinal rod outer segments, which are especially sensitive. One of the important effects of formaldehyde is the inhibition of glycolysis, which seems to explain its toxicity to retinal rod outer segments (Winkler, 1981). Problems stemming from formaldehyde toxicity have been avoided, even in rod outer segments, by using enriched media and loading cells in a relatively large volume of solution so as to dilute out the formaldehyde (Korenbrot et al., 1986). Pyruvate and ascorbate are two components in enriched media that appear to protect cells from formaldehyde (Garcia-

Sancho, 1984; Tsien and Pozzan, 1989). We routinely include 1 mM pyruvate in culture media used for loading the indicators. However, it is important, especially in studies in which cells are studied over several hours after loading, to control for the effects of formaldehyde. As such a control, one can use methylene diacetate, the AM ester of anisidine diacetate (AM-o-aminophenoxymethane-N,N-diacetate), which represents half of the bapta/AM molecule, or perhaps even the AM ester of the pH indicator BCECF (Rink et al., 1982). In the latter case, one could easily measure the concentration of BCECF loading so that the amount of formaldehyde released inside the cell can be calculated. In the case of fura-2, where much lower concentrations of cytosolic dye are required to obtain good signals, we have not detected any obvious toxic effects that could be attributable to formaldehyde. Comparisons of cytotoxic T lymphocyte killing efficiencies revealed no significant differences between cells incubated in various concentrations of fura-2/AM and unloaded controls (Poenie et al., 1987).

3.2.3. Indicator Solubility

One hindrance to good dye loading relates to the insolubility and hydrophobicity of the AM ester derivatives of the Ca^{2+} indicators. An ideal ester of fura-2 or other indicator dye would dissolve readily in water, quickly cross the plasma membrane, and hydrolyze to completion in the cytosol with residual unhydrolyzed dye easily washed away. Unfortunately, we do not yet have dye esters with these properties, and one has to contend with the solubility and hydrophobicity problems. The AM esters of fura-2 and indo-1 are only slightly soluble in water, but dissolve readily in DMSO and other organic solvents. When a stock solution of fura-2 in DMSO is added to water, the dye tends to come out of solution and form particulates even at concentrations of a few micromolar. This problem is more severe with marine organisms because of the high salt in sea water and the colder marine environment. One can readily demonstrate this by comparing the fluorescence of an aqueous fura-2 solution before and after filtration through a 0.2 μm-filter.

The hydrophobicity of the AM ester causes several problems in loading cells with the indicator. Formation of dye particulates or precipitates tends to defeat attempts to increase dye loading by increasing the concentration of AM ester, unless agents that help solubilize the dye are also present. This does not mean that it is impossible to load indicator into cells when the dye forms particulates. The particulates can stick to the cell membrane in which the dye dissolves and is then transferred to the cytoplasm. This is similar to the behavior of A23187 in sea water, where the particulate form of the indicator can mediate egg activation. On the other hand, some aspects of dye loading become worse as the concentration of AM ester is increased. The particulates of fura-2 tend to stick to the outside of the cells and make it difficult to wash off the AM ester at the end of the loading procedure. Partitioning of AM ester into hydrophobic compartments becomes worse as AM ester concentration is increased, so that unhydrolyzed dye ends up as the major form of the indicator in the cell. For many cell types, 0.1–2.0 μM fura-2 gives adequate loading.

Although one normally avoids using high concentrations of AM ester (5–10 μM), it may be necessary, in order to circumvent other problems, such as rapid dye leakage or compartmentation of indicator after it has been hydrolyzed. The strategy here is to both load cells and make measurements as rapidly as possible. Loading of PtK cells improved when incubated with high concentrations (15 μM) of fura-2/AM and a solubilizing agent for short periods (15–30 min). Bovine serum albumin, pluronic F–127 (Cohen et al., 1974; Poenie et al., 1986), and cyclodextrins have all been used as fura-2/AM-solubilizing agents.

As mentioned above, residual unhydrolyzed ester must be minimized since it greatly degrades ratiometric Ca^{2+} measurements. Measurement of unhydrolyzed AM ester after loading a batch of cells provides useful information in developing a good loading protocol. One approach is to compare fluorescence spectra from aliquots of cells loaded with indicator, an aliquot of unloaded cells, and an aliquot of cells lysed in the presence of 1–2 mM $MnSO_4$. The spectra of unloaded cells serve as a background and are subtracted from the remaining two spectra. Manganese ions quench the fluorescence of fully hydrolyzed fura-2, quin2, and indo-1, so that residual indicator fluorescence is attributed to unhydrolyzed dye. One should compare intensities at an isoexcitation or isoemission wavelength, wavelengths at which the fluorescence intensity does not depend on [Ca^{2+}].

Alternatively, one can hydrolyze residual uncleaved AM ester in dye-loaded cells with KOH. Comparison of fluorescence scans before and after hydrolysis, both in the presence of saturating Ca^{2+}, reveals a change owing to AM ester hydrolysis. With fura-2, for example, an excitation spectrum of a 2.5-mL suspension of loaded cells is recorded between 300–400 nm. This suspension should be in a normal physiological saline containing 1 mM Ca^{2+}. Subsequently, 20 μL of 10M KOH is added, and the suspension incubated at 37°C for approx 10 min. The pH of the suspension is neutralized, and the excitation spectrum is recorded again. Comparison of the two spectra will show a decrease in fluorescence at 380 nm and an increase in fluorescence at 340 nm. This results from the base-catalyzed conversion of the fura2/AM to fura-2 penta-anion, which then binds Ca^{2+}. One can estimate the percent of unhydrolyzed ester as the change in fluorescence at 340 nm over the total fluorescence at 340 nm.

3.2.4. Determining Indicator Concentration Inside Cells

Knowing the cytoplasmic concentration of indicator after loading addresses the important issue of Ca^{2+} buffering. One can determine the total dye concentration inside the cell relatively easily using a cell suspension and a fluorimeter if one knows the cell density and the average cell volume, and has a solution of known concentration of the indicator. Here, intensity measurements either must use an isoexcitation (or isoemission) wavelength, or [Ca^{2+}] must be fixed for all measurements. Experimentally, one simply lyses a known number of cells in a known volume of HBSS (available, e.g., from GIBCO catalog). After obtaining a spectrum from the lysed suspension, an aliquot of indicator standard is added to give a known final concentration in the cuvet. Preferably, one adds an amount of reference indicator that roughly matches that already present in the cuvet. This will cause an increase in fluorescence proportional to the concentration of the added reference indicator. One should avoid adding so much additional indicator that the intensity goes far off scale or out of the intensity range suitable for the photomultiplier. A proportion can then be used to determine the original concentration of indicator in the cuvet after cell lysis.

$$\frac{\text{conc. dye released from cells}}{\text{flourescence of dye from cells}} = \frac{\text{conc. reference standard added to cuvet}}{\text{increase in fluorescence after addition}} \quad (3)$$

Once the concentration of indicator in the cuvet is known, this value is converted to moles and then divided by the total volume of cytoplasm to give the cytoplasmic concentration.

Time will be saved if one knows ahead of time how much of the reference solution to add, so that the fluorescence remains on scale. A starting point can be derived from estimates of typical cell loadings. For example, a 10^6 cells/mL suspension of typical blood cells represents roughly 2 μL of cytoplasm. Using 50 μM as a typical cytoplasmic dye concentration after loading, there will be 0.1 nmol of total indicator. This gives a concentration of approx 40 nM indicator in the 2.5 mL cuvet.

3.2.5. Strategies for Loading Cells with Indicator AM Esters

First attempts at ester loading should utilize approaches that, through experience, have worked for many cells, and then variations should be tried. Furthermore, one needs to assess the quality of loading quantitatively, in order to determine the nature of problems that might arise for particular cell types. One should determine total dye content and the degree of hydrolysis. As a starting point, two loading conditions are suggested.

Loading conditions for blood cells:

1. Suspend cells in RPMI 1640 (available, e.g., from GIBCO) containing 1 mM pyruvate and 1 μM fura-2/AM
2. Incubate at 37°C for 30 min
3. Wash cells by gentle centrifugation

4. Resuspend cells in HEPES-buffered HBSS without phenol red, pH 7.2
5. Incubate for an additional 15 min at room temperature before starting measurements

Loading conditions for fibroblasts and adherent cells:

1. Prepare a solution containing 2.5 µL of 1 µM fura-2/AM in DMSO and 2.5 µL 25% Pluronic F127 in DMSO
2. Place cells adhered to cover slip in a small plastic Petri dish containing 1 mL of HEPES-buffered RPMI containing 1 mM added pyruvate
3. Quickly add the fura-2 solution to the Petri dish and mix
4. Place on an orbital shaker for 30 min at room temperature
5. Wash the loading solution off the cells using fresh medium

The above suggestions only represent a beginning, and numerous variations can follow. One can vary fura-2 concentration, loading temperature, dispersing agents, such as Pluronic, loading time, and incubation time after loading. The use of Pluronic provides a way to help disperse the indicator and prevent formation of precipitates. Alternatively, one can use BSA as a dispersant or a 0.2-µm filter to remove particulates before adding the RPMI–indicator mixture to the cells.

3.2.6. Leakage of Indicator After Loading

Temperature affects dye solubility, rate of ester hydrolysis, and the rate at which cells compartmentalize and extrude the dyes. Although increased dye solubility and ester hydrolysis favor good loading, the compartmentalization and extrusion of the dye are detrimental. For T and B lymphocytes one can obtain good loading at 37°C in RPMI containing 1% serum and 1 µM fura-2/AM for 30 min. The same conditions used to load many fibroblast cell types often result in patches of compartmentalized dye with little fluorescence in the cytosol. Loading fibroblasts at room temperature gives much more uniform loading. However, indicator that starts out uniformly dispersed in the cytosol will progressively leak out of the cell or become concentrated in intracellular organelles. In experiments using cell suspensions, this loss of dye to the medium is seen as an apparent slow steady increase in baseline [Ca^{2+}]$_i$. As fura-2 leaks out of the cell, it moves from a low Ca^{2+} environment (cytosol) to the exterior where $[Ca^{2+}]$ is high. Addition of 1 mM Mn^{2+} ion should instantly quench this extracellular dye revealing the true baseline.

3.2.7. Compartmentalization of Indicator into Organelles

The hydrophobic AM esters can concentrate in hydrophobic compartments within a cell. A different process can lead to compartmentalization of fully hydrolyzed indicator. For example, the potassium salt of fura-2, when injected into sea urchin eggs, becomes compartmentalized over a period of an hour. After compartmentalization occurs, the indicator cannot be fully released by rupturing the egg. In another case, iontophoresis of the fura-2 potassium salt into plant cells was followed by a rapid transfer of dye from cytosol to vacuole (Hepler and Callahan, 1987).

Both compartmentalization and leakage of the fully hydrolyzed indicator of the AM esters can be inhibited by compounds, such as probenecid and sulfinpyrazone, that inhibit anion transport (DiVirglio et al., 1988,1990). This suggests that compartmentalization of fura-2 is an active, enzyme-mediated transport that recognizes the fura-2 penta-anion. This process is slowed somewhat when cells are cooled from 37°C to room temperature. It has also been reported that prechilling cells to 4°C for 5–15 min before loading at room temperature with the AM ester retards the processes of compartmentalization and leakage for up to 2 h (Roe et al., 1990).

For some studies, especially when Ca^{2+} gradients or spatial variations of Ca^{2+} are of interest, it is important to determine the amount of dye compartmentalization exhibited by the cells being utilized for Ca^{2+} measurements so that one knows how much of the intracellular indicator is actually cytosolic. Two problems can be encountered. The presence of unhydrolyzed indicator associated with hydrophobic compartments can produce regions of the cytoplasm where Ca^{2+} appears artificially low. Alternatively, accumulation of fully cleaved dye into membrane-bound compartments, such as the endoplasmic reticulum, gives the appearance of local regions with high Ca^{2+}. Thus, artifactual

Ca^{2+} gradients can easily be generated. We have employed several methods to determine the degree of compartmentation of intracellular fura-2. One method relies on selective lysis of the plasma membrane by digitonin, which at threshold concentrations leaves intracellular membranes intact. The appropriate digitonin concentration varies with cell type, and with whether cells are adhered or in suspension. We obtain better results using adherent cells or cells attached to coverslips with poly-L-lysine. Polylysine-coated coverslips are prepared from new coverslips coated with a film of 1 mg/mL poly-L-lysine (100,000–300,000 mol wt) in $1M$ KCl. While still wet, the coverslips are rinsed twice in distilled water and then air-dried. They should be used within a few hours. Sometimes the bond formed with poly-L-lysine is too rough for especially fragile cells. An alternative, Cell Tak™, seems to be more gentle and is preferred in these cases. To assess compartmentation, we compare the amount of fura-2 released along with the cytosol (digitonin lysis) to the total fura-2 released using Triton X-100. Here, lactate dehydrogenase (LDH) is used as a marker enzyme for the cytosol. In addition acridine orange (or 9-aminoacridine), which partitions into acidic compartments, such as lysosomes, is used as an indicator for integrity of membrane-bound compartments (Zeitz et al.; 1980, Lee et al., 1982).

In this procedure, we first determine a digitonin concentration that maximizes the release of LDH from the cell, but gives minimal release of acridine orange. Aliquots of cells are adhered either to plastic Petri dishes or to polylysine-coated coverslips. Cells are then incubated with $5–10 \times 10^{-6}M$ acridine orange for 15 min and washed 3 × with HBSS. Acridine orange may continue to leach out slowly, so the lysis that follows should be carried out immediately. In the first step, an aliquot of cells is treated with digitonin for 5 min. An effective concentration of digitonin is typically between 25–100 μM. The supernatant is then gently removed and replaced with HBSS containing 1% Triton X-100. The digitonin supernatant was first assayed for LDH activity and then scanned to determine the amount of acridine orange present. The assay for LDH utilized the conversion of NAD (nonfluorescent) to NADH (fluorescent). A reaction mixture is prepared containing 100 mM KCl, 10 mM lactate, 1 mM MgCl$_2$, 1 mM NAD, and 10 mM HEPES, pH 7.2. An aliquot of the digitonin supernatant is added to the cuvet, and the conversion of NAD to NADH monitored at 340 nm excitation and 440 nm emission for a few minutes to obtain a good slope. After obtaining the data from the digitonin treatment, the Triton X-100 supernatant is removed and the assays repeated. From these data, one can estimate the amount of LDH released by digitonin compared to the residual LDH released by Triton X-100. One obtains the same information regarding the acridine orange. After determining the optimal concentration of digitonin to use, the detergent treatments are then repeated on cells loaded with fura-2, but not with acridine orange. The degree of hydrolysis and concentration of fura-2 in each fraction can then be determined as described above.

3.2.8. Photobleaching

Photobleaching is the loss or alteration of the fluorophore, usually because of its reactivity in the excited state. Photobleaching can cause a simple loss of fluorescence or convert the indicator to forms that remain fluorescent, but whose properties differ. This can complicate the measurement of intracellular Ca²⁺. Quin2 is relatively sensitive to photobleaching, which is seen as a loss of fluorescence. Fura-2 is relatively resistant, and photobleaching, when it occurs, does not cause a simple loss of fluorescence (Becker and Fay, 1988). Under strong illumination, photobleaching causes accumulation of a Ca²⁺-insensitive form of the indicator, which is preferentially excited at the longer excitation wavelength. The buildup of fluorescent breakdown products in the cell causes an underestimation of [Ca²⁺]$_i$. In their discussion, Becker and Fay offer several useful suggestions for minimizing the photobleaching of fura-2.

Suggestions for minimizing photobleaching (adapted from Becker and Fay [1988]):

1. Minimize exposure to light—excite only while collecting data. We typically employ a shutter to block the excitation beam when not collecting data. Keep loaded cells (and fura-2 stocks) out of light. Attenuate excitation intensity. When

monochromator sources are used, one can narrow the slits. Otherwise, a diaphragm can be used to cut down the excitation light.

2. Improve emission collection efficiency—use highest numerical aperture objective available. Objectives with numerical apertures of 1.3–1.4 are available. Replace prism-type mirrors with front-faced mirrors. Use wide bandpass emission filters. Fura-2 exhibits a relatively large stokes shift, so that emission and excitation spectra show little spectral overlap. Except where cellular autofluorescence is a serious problem, we prefer to use narrow excitation bands and a broad emission bandpass 50–60 nm or even a yellow UV cutoff filter for the emitted light.

3. Improve emission signal to noise (background). Use higher dye loadings when possible. Reduce the amount of dye in extracellular fluid (and use media with low intrinsic fluorescence). Prevent stray light contamination (for example, work in a darkened room).

4. Calibration and Quantitation

The issue of calibration is complicated by the different methods and lengths individual laboratories go to in correcting for various envionmental variables. At the heart of it, calibration involves subtracting background autofluorescence, in some cases correcting for unhydrolyzed dye, and determining the fluorescence end points for the indicator. The experimental procedures differ somewhat depending on whether the measurements are done in a cuvet or on single cells. Here, a general discussion of the spectral properties of the dye will be presented first, followed by a more specific discussion devoted to each type of measurement.

4.1. Calibration Parameters

4.1.1. Calibration
of Nonratiometric Measurements

Calibration of fluorescent indicators refers to relating fluorescence values obtained from the dye-loaded specimen to concentrations of bound (C_b) and free (C_f) indicator. The fraction bound/free is related to [Ca²⁺] by the K_d such that $C_b/C_f =$ [Ca²⁺]/K_d. For nonratiometric indicators, such as quin2 and fluo-3, the fraction C_b/C_f is equivalent to $(F - F_{min})/(F_{max} - F)$ provided that certain conditions are met, namely, that at the low concentrations of dye inside the cell, fluorescence is proportional to concentration, and the dye binds with a 1:1 stoichiometry. For quin2, the concentration of Ca²⁺-bound indicator varies from zero where fluorescence is at a minimum (F_{min}) to 100% where fluorescence is at a maximum (F_{max}). Ideally, fluorescence will change linearly with the concentration of bound indicator. Fluorescence properties of the indicator, illumination brightness, pathlength, total dye concentration, illumination and emission wavelength settings, instrument light collecting efficiency, and sensitivity must all remain unchanged between determinations of F, F_{min}, and F_{max}. Most of these parameters can be maintained when using a fluorometer where the cuvet maintains a fixed pathlength. F_{min} and F_{max} are usually obtained at the end of the experiment using a detergent, such as digitonin, to release the trapped indicator into the medium. In most extracellular media, the [Ca²⁺] is 1–2 mM, so that after addition of detergent, one obtains F_{max}. Following lysis, excess EGTA and Tris base are added in a 1:2 ratio to give R_{min}. The pH needs to be raised to 8.5 or above to increase the binding affinity of EGTA for Ca²⁺. If the pH is raised too high, however, hydroxides of Ca²⁺ or Mg²⁺ will precipitate, causing a great increase in light scattering.

Experience using quin2 has led to the development of working calibration equations that incorporate convenient modifications to the calibration procedures (Tsien et al., 1982; Rink et al., 1983; Hesketh et al., 1983; Rink and Pozzan, 1985; Tsien and Pozzan, 1989). The cellular autofluorescence of quin2-loaded cells can be readily determined using Mn²⁺ to quench the fluorescence of quin2 (or fura-2), giving the autofluorescence + unhydrolyzed ester (F_{Mn}). The value F_{min} can be obtained from the following relationship:

$$F_{min} = F_{Mn} + 0.167 \, (F_{max} - F_{Mn}) \qquad (4)$$

As given here, the equation applies only to quin2, since it incorporates the sixfold increase in brightness of quin2 upon binding to Ca^{2+}. Procedurally, one obtains F_{max} after cell lysis and then adds $MnSO_4$ to obtain F_{Mn}, from which F_{min} can be calculated (Hesketh et al., 1983; Cobbold and Rink, 1987).

4.1.2. Calibration of Ratiometric Indicators

Ratiometric indicators provide several advantages over single wavelength indicators, but have some additional complications as well. When using fluorescence ratios, fluorescence measurements are taken at two different wavelengths over closely spaced intervals. During the interval between measurement at one wavelength until measurement is completed at the second wavelength, acquisition parameters must remain constant. However, after a ratio is taken, several factors, such as pathlength, total dye concentration, illumination intensity, collection efficiency, and sensitivity, cancel. This permits direct comparison of ratios obtained at different times, even though some parameters may have changed with time. This is of particular importance in single cell measurements and imaging. The fluorescence ratio obtained in one region of a cell can be compared to the ratio in a different region of the cell, even though pathlength and dye concentration may differ. On the other hand, comparison of ratios still requires constancy of some parameters.

Calibration of ratiometric indicators also aims at relating fluorescence ratios to the concentration of Ca^{2+}-bound/free indicator (C_b/C_f) and, in turn, relating C_b/C_f to $[Ca^{2+}]$. For fura-2 and indo-1, the fluorescence at each excitation or emission wavelength contains contributions from both the Ca^{2+}-bound and free forms of the dye. Using the conventions of Grynkiewicz et al. (1985), the fluorescence ($F1$ and $F2$) at wavelengths $\lambda 1$ and $\lambda 2$, respectively, are given by:

$$F1 = S_{f1}C_f + S_{b1}C_b \qquad (5)$$
$$F2 = S_{f2}C_f + S_{b2}C_b$$

Here C_f and C_b refer to the concentration of free and bound forms of the indicator, and $S_{f1}, S_{f2}, S_{b1},$ and S_{b2} are four proportionality coefficients that relate the concentration of a particular form of the dye (b = bound or f = free) at a particular wavelength (1 or 2) to its particular fluorescence intensity. The factor S incorporates both properties of the indicator, including its extinction coefficient and quantum efficiency, as well as pathlength, illumination intensity, and instrument detection efficiency. It should be noted that some components of S_{f1}, S_{f2}, S_{b1}, and S_{b2} are environmentally sensitive and are constant only for a particular set of environmental conditions. A fluorescence ratio R is obtained by dividing $F1/F2$ so that:

$$R = F1/F2 = (S_{f1}C_f + S_{b1}C_b)/(S_{f2}C_f + S_{b2}C_b) \tag{6}$$

The $[Ca^{2+}]$ can be related to C_b and C_f by the equation:

$$C_b = C_f[Ca^{2+}]K_d \tag{7}$$

which when substituted into the equation for a fluorescence ratio gives:

$$R = \{S_{f1} + (S_{b1}[Ca^{2+}]/K_d)\}/\{S_{f2} + (S_{b2}[Ca^{2+}]/K_d)\} \tag{8}$$

This equation is now solved for Ca^{2+}:

$$R\{S_{f2} + (S_{b2}[Ca^{2+}]/K_d)\} = \{S_{f1} + (S_{b1}[Ca^{2+}]/K_d)\}$$

$$RS_{f2} - S_{f1} = (S_{b1}[Ca^{2+}]/K_d) - R(S_{b2}[Ca^{2+}]/K_d)$$

$$K_d(RS_{f2} - S_{f1}) = [Ca^{2+}](S_{b1} - RS_{b2})$$

$$K_d(RS_{f2} - S_{f1})/(S_{b1} - RS_{b2}) = [Ca^{2+}] \tag{9}$$

Dividing both sides by S_{f2}/S_{b2} gives

$$K_d(R - S_{f1}/S_{f2})/(S_{b1}/S_{b2} - R) = [Ca^{2+}]/(S_{f2}/S_{b2})$$

$$K_d\{(R - S_{f1}/S_{f2})/(S_{b1}/S_{b2} - R)\}(S_{f2}/S_{b2}) = [Ca^{2+}] \tag{10}$$

Under conditions where the concentration of bound form (C_b) is zero, $F1/F2$ becomes $S_{f1}C_f/S_{f2}C_f$ or simply S_{f1}/S_{f2}. Likewise, when the concentration of free form (C_f) of the indicator is zero, $F1/F2$

becomes $S_{b1}C_b/S_{b2}C_b$, or simply S_{b1}/S_{b2}. These two parameters constitute R_{min} and R_{max}, respectively (*see* Fig. 5). The equation therefore can be restated as:

$$[Ca^{2+}] = K_d\{(R - R_{min})/(R_{max} - R)\}(S_{f2}/S_{b2}) \qquad (11)$$

The parameter S_{f2}/S_{b2}, referred to by some as β, is required when one of the wavelengths used in ratioing is not set to the isoexcitation or isoemission point. With fura-2, for example, setting wavelength 2 at the isoexcitation wavelength makes $S_{f2} = S_{b2}$, and therefore, β drops out of the equation. Most investigators working with fura-2 use 380 nm for wavelength 2 where S_{f2}/S_{b2} is given as 15.3 (Grynkiewicz et al., 1985).

The values R_{min} and R_{max} can be obtained from two fura-2 solutions, one set to zero Ca^{2+} and the other set to saturating Ca^{2+} levels. In a cuvet, they can be obtained by lysing cells after a Ca^{2+} measurement, and setting the solution to saturating high and then low Ca^{2+} levels. For single cell measurements on a microscope stage, they can be obtained by using thin films of high and low Ca^{2+} calibration solutions sandwiched between coverslips. However, using two separate solutions for determining S_{f2}/S_{b2} is dangerous, because here path length does not cancel out and path length might easily differ between two separate dye-sandwich preparations. We prefer to consider S_{f2}/S_{b2} as an intrinsic property of the dye much the same as K_d. It is typically determined carefully and then used repeatedly.

4.1.3. Determination of K_d

The dissociation constant for the indicators is determined from the ratio C_b/C_f for a range of known Ca^{2+} concentrations. The dissociation constant can then be readily determined from a Scatchard plot. Solutions of known $[Ca^{2+}]$ are normally prepared using EGTA as the Ca^{2+} buffer. The prospect of preparing EGTA buffers is foreboding to some, and complicated by the various literature values given for the different EGTA constants and methods to obtain these values (Anderegg, 1964; Boyd et al., 1965; Bers, 1982; Harafuji and Ogawa, 1980; Martell and Smith, 1974; Moisescu and Pusch, 1975; Portzehl et al., 1964; Smith and Miller, 1984, 1985; Swarzenbach, 1957). However, if one is willing to

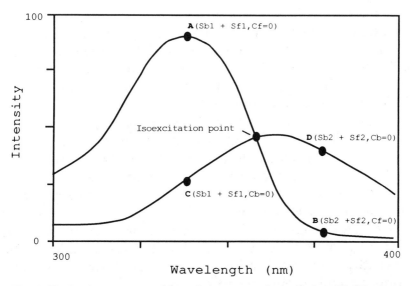

Fig. 5. Excitation spectra of fura-2 at zero and saturating [Ca²⁺] are given to illustrate the wavelengths frequently used for measurement and calibration. Here $A/B = R_{max}$, $C/D = R_{min}$, and $D/B = Sf_2/Sb_2$ or β. Since the two traces shown here represent spectra of fura-2 at high- and low-Ca²⁺ end points, each trace should represent the spectra of only one form of the dye. Thus, point A would represent the relative contributions from both forms of the dye ($S_{b1} + S_{f1}$) at any intermediate Ca²⁺ concentration. Here, however, the spectra are obtained from Ca²⁺ saturated fura-2, so that the concentration of free fura-2 (Cf) is zero. The isoexcitation point near 360 nm is useful as a measure of indicator concentration, because fluorescence intensity at this excitation wavelength does not change with Ca²⁺. The wavelengths used for ratioing are usually chosen to maximize the range from R_{min} to R_{max}, while still ensuring that one has adequate signal at each wavelength for high or low [Ca²⁺]. Some prefer to use the isoexcitation wavelength as one of the wavelengths for ratioing.

accept a given apparent K_d for a given pH value, then the procedure is fairly simple. The Ca²⁺ for an EGTA buffer is given by:

$$[Ca^{2+}] = K_d [CaEGTA]/[free\ EGTA] \qquad (12)$$

Here, K_d is the apparent K_d, that which is defined for a particular pH, temperature, and ionic strength. This effective K_d is calculated from the proton ionization constants for EGTA and the absolute dissociation constants of the EGTA⁴⁻ and EGTA³⁻ forms.

However, tables of the apparent dissociation constants for EGTA over a wide range of conditions have been published (Tsien and Pozzan, 1988). Given an apparent dissociation constant, one can readily generate Ca^{2+} buffers covering a range of $[Ca^{2+}]$ levels. A simple and reliable procedure is to start by preparing a CaEGTA solution (Moisescu and Pusch, 1975; Tsien and Pozzan, 1989). This has to be done carefully, but once prepared, one need only to mix CaEGTA:free EGTA at various ratios to generate a buffer. The ratio is determined by $[Ca^{2+}]/K_d = [CaEGTA]/[free\ EGTA]$.

As an alternative to manual calculations, one can use a microcomputer program that calculates all the necessary parameters. One such program, "MaxChelator," has been adapted from the calculator program of Fabiato and Fabiato (1979), and runs on PC-AT-compatible computers. MaxChelator is available from Chris Patton, Hopkins Marine Station, Pacific Grove, CA, 93950, for only a small shipping charge.

4.1.4. Titrations

Titrations usually involve recording excitation or emission spectra for the indicator over a suitable range of Ca^{2+} concentrations. Although simple in theory, minor differences in dye concentration between samples can cause large errors in the titration. The best approach to take is one that minimizes the effect of small pipeting errors. One example of such an approach is given by Williams and Fay (1990).

4.2. Variables Affecting Dye Calibration Parameters

A number of factors can influence the parameters R_{min}, R_{max}, and K_d, and therefore influence the results obtained from intracellular measurements. Below is a list of parameters that are known to alter either K_d or the fluorescence properties of the indicator.

Factors that influence K_d of fura-2 for Ca^{2+} [*]

Factor	Effect
1. Temperature	K_d decreases as temperature decreases
2. Ionic strength	K_d increases with increasing salt

[*] Data compiled from Grynkiewicz et al., 1985; Roe et al., 1990; Poenie, 1990; Popov et al., 1988.

3. Viscosity   Disputed, some propose that increasing viscosity increases apparent K_d

4. Dye-protein binding   Increases apparent K_d

5. pH   Apparent K_d increases as pH drops significantly below 7.0.

Factors that influence spectral characteristics of fura-2[*]

Factor	Effect
1. Temperature	Decrease in temperature enhances fluorescence at the longer excitation wavelengths preferentially
2. Polarity	Erratic effects, mostly enhances fluorescence at longer excitation wavelengths of fura-2
3. Viscosity	Enhances fluorescence at long excitation wavelengths.
4. Binding to other metals	Variable, Mn^{2+} quenches fluorescence of the free anion, Zn^{2+} acts similar to calcium, Mg^{2+} preferentially enhances the fluorescence at the longer excitation wavelengths of fura-2

4.3. Methods of Calibration for Microscopy

4.3.1. Calibration of Fluo-3

Fluo-3, although not a ratiometric indicator, is attractive because of its long excitation and emission wavelengths, and the 40-fold increase in brightness when it binds to Ca^{2+}. Thus, although not ideal, it is nevertheless being used for single cell measurements. The calibration regimen is similar to the meth-

---

[*] The effects of viscosity and polarity on indo-1 emission spectra differ in that they preferentially enhance the shorter emission wavelengths. Without correction viscosity effects cause an underestimation of $[Ca^{2+}]_i$ using fura-2 and an overestimation of $[Ca^{2+}]_i$ using indo-1.

odology employing Mn^{2+} for calibrating quin2. However, a saturating concentration of Mn^{2+} does not quench fluo-3, but simulates the fluorescence of the same amount of fluo-3 at 20% Ca^{2+} saturation. This effect has been used by Kao et al. (1989) to obtain F_{min} and F_{max} values from individual fluo-3-loaded cells. Here the background autofluorescence (F_{bkg}) is determined, and Mn^{2+} is introduced into the cells with ionomycin to give F_{Mn}. F_{max} is then calculated from the equation:

$$F_{max} = \{(F_{Mn} - F_{bkg})/0.2\} + F_{bkg} \tag{13}$$

After F_{max} is obtained, F_{min} can be calculated from the relationship:

$$F_{min} = \{(F_{max} - F_{bkg})/40\} + F_{bkg} \tag{14}$$

Finally, having determined F_{max} and F_{min} $[Ca^{2+}]_i$ can be determined using the standard equation:

$$[Ca^{2+}]_i = K_d(F - F_{min})/(F_{max} - F) \tag{15}$$

4.3.2. Calibration of Ratiometric Indicators

Calibration of fura-2 measurements on single cells requires determination of R_{min} and R_{max} using the same microscope illumination system and optics as used for the experimental measurements. For simplicity of discussion, it was suggested that R_{min} and R_{max} can be determined using two thin films of dye sandwiched between coverslips, one set to saturating Ca^{2+} and the other to zero Ca^{2+}. There are actually two different schools of thought as to how best to determine R_{min} and R_{max}. One approach, as described above, uses the dye-sandwich approach (in vitro calibration), whereas the other attempts to force the fura-2 trapped inside cells sequentially to R_{min} and then R_{max} (in vivo calibration). The latter utilizes high concentrations of ionomycin to permeabilize the cell to Ca^{2+}. Each approach has advantages and pitfalls.

4.3.3. In Vitro Determination of R_{min} and R_{max}

Using thin films of dye is a simple, rapid means of obtaining R_{min} and R_{max}. It is relatively easy to control the concentration of Ca^{2+} in the calibration solutions by simply preparing one solution containing 1 mM Ca^{2+} (100 mM KCl, 20 mM HEPES, pH 7.2) and the other containing EGTA with no added Ca^{2+}. Typically, we use large concentrations of EGTA (50 mM K_2H_2EGTA, 20 mM HEPES, pH 7.2) in these solutions, because the volume of solution trapped between the coverslips is small and Ca^{2+} leaches out from the glass coverslips. One can reduce, but not totally eliminate this effect by soaking the coverslips in EDTA. If one is attempting to determine S_{f2}/S_{b2}, the concentrations of indicator in both solutions must be identical. We typically put a mark near the middle of the lower coverslip, which is in contact with the objective lens, using a permanent black marker. After drying, this dot will be used for focusing on the coverslip surface. The lower cover slip, a 25 mm circle, is attached to a plastic Petri dish with the center removed and mounted onto the microscope stage of the inverted microscope. The microscope is focused so that the black marker dot is in view, and then the microscope field is positioned to a clear area adjacent to the dot. Next, 2 µL of calibration solution is placed in the middle of the coverslip and an 18 mm circular coverslip is added so that it rests directly on the larger coverslip, causing the dye droplet to flatten out. The signals are then recorded. For the low Ca^{2+} solution, all operations after adding the droplet of calibration solution must be done quickly to avoid problems owing to leaching of Ca^{2+} from the glass. The same procedures can be repeated using the high Ca^{2+} calibration solution. Background fluorescence is determined for a sample of calibration solution that does not contain the indicator dye.

4.3.4. In Vivo Determination of R_{min} and R_{max}

An alternative means of calibration utilizes Ca^{2+} ionophore to permeabilize the cell to Ca^{2+}. In its simplest form one can simply add high concentrations of ionophore to cells in high and low [Ca^{2+}] solutions to achieve R_{max} and R_{min}, respectively. Monensin and nigericin are sometimes added to dissipate Na^+ and K^+ gradients, respectively, so as to block Na^+-dependent Ca^{2+} transport (Williams and Fay, 1990). One has to assume that cellular Ca^{2+} equilibrates completely with that of the external solution. The procedure of Williams and Fay (1990) is somewhat more elaborate. A suspension of cells is equilibrated in a solution buffered to a known [Ca^{2+}] containing monensin and nigericin. Sufficient ionomycin (>1 μM) or 4-bromo-A23187 is then added so that measured ratios achieve an end point close (within 5%) to that of a fura-2 solution buffered to the same [Ca^{2+}].

The virtue of this approach is that it compensates for cytoplasmic alterations in fluorescence properties of the indicator, even if one does not know how the cytoplasm affects the dye. In theory, this might seem to be an ideal approach, but there are practical problems that make it more questionable. An indicator, such as fura-2, trapped inside the cell is a fairly good competitor with EGTA, unless the pH of the EGTA solution is raised to a nonphysiological pH. Changing the pH should not have any great impact on the indicator dye, but may have effects on the cytoplasm. Furthermore, in the studies of Bond et al. (1984), EGTA did not remove all the Ca^{2+} inside the cell. The rate of diffusion of Ca^{2+} across the membrane is governed largely by the concentration gradient. As the Ca^{2+} concentration approaches equilibrium, the rate of diffusion slows. One cannot really independently verify that complete equilibration has occurred. The use of an external calibration solution for reference does automatically provide a solution to this problem. If the cellular fura-2 ratios are really shifted to lower values for any given Ca^{2+} concentration because of viscosity or other factors, matching cellular ratios to that of an external reference solution stops short of Ca^{2+} equilibration. In order to avoid this, the reference solution must match the cellular cytoplasm in terms of viscosity, ionic strength, and other factors. The accuracy of this approach will ultimately depend on how closely the reference solution mimics the cytoplasm, which makes this similar to the strategy of the in vitro approach. It should be added that cells load heterogeneously and may contain various amounts of unhydrolyzed ester. With time, intracellular fura-2 can leak out of cells, leaving behind unhydrolyzed ester as an increasing proportion of the total signal. The presence of unhydrolyzed ester can cause serious underestimation of R_{max}, so that its presence needs to be quantified. Scanlon et al. (1987) have described a procedure that permits measurement and subtraction of a spectrum of unhydrolyzed ester. The procedure depends on modeling fura-2 spectra and normalizing the cellular fluorescence signal to one in which there is no unhydrolyzed dye.

4.4. Corrections for In Vitro Calibrations

One of the main objections against this approach to determining R_{min} and R_{max} is that it does not take into account alterations in fura-2 fluorescence properties that may occur inside the cell. It is also clear that the fluorescence properties of fura-2 can be altered in the intracellular environment. Using simple aqueous calibration solutions does in fact give R_{min} and R_{max} values that are too high, and as a result, the calculated $[Ca^{2+}]_i$ values are too low.

The simple aqueous calibration solutions containing physiological concentrations of K^+ and Mg^{2+} do not fully simulate the cytoplasm of cells. The fura-2 R_{min} and R_{max} values obtained from simple calibration solutions are somewhat higher than would be obtained if the calibration solution fully simulated the cytoplasm of the cell. For indo-1, the situation is reversed so that the calibration solution gives R_{min} and R_{max} values that are lower than if the calibration solution simulated the cytoplasm. As a result, $[Ca^{2+}]_i$ is often underestimated with fura-2 and overestimated with indo-1. The largest source of this error appears to be an alteration in the spectral properties of the indicator in the cytoplasm that may arise either from restricted mobility of the dye in the cytoplasm (viscosity) or perhaps because of differences in the polarity of the dye environment.

Evidence that an alteration in spectral properties of fura-2 takes place in the cytoplasm can be seen as a drop in the fluorescence intensity at both excitation wavelengths when suspensions of fura-2 loaded cells are lysed with digitonin. When the cells are lysed and the dye moves from cytoplasm to extracellular environment, one might expect to see changes in fluorescence owing to differences in $[Ca^{2+}]_i$ and $[Ca^{2+}]_o$. However, changes in Ca^{2+} changes always send signals at the two excitation wavelengths in opposite directions. A loss in signal at both excitation wavelengths cannot be explained as a change in $[Ca^{2+}]$. Furthermore, the percentage change is greater at 380 nm than at 340 nm, as if the fluorescence were preferentially enhanced at 380 nm excitation. Upon lysis, this enhancement is lost and the fluorescence at both excitation wavelengths decreases, but more so at 380 nm excitation (Fig. 7).

Immobilizing the dialkylamino arms of fura-2 could produce an enhancement of fluorescence in a manner similar to that described by Drexhage (1973). To determine if viscosity was a suitable explanation for the alteration in fluorescence properties of fura-2 in the cytoplasm, measurements were made to determine if fura-2 exhibited a greater restriction of mobility or rotational diffusion inside the cell relative to fura-2 in free solution. It was found that the rotational diffusion (measured as steady-state fluorescence polarization) of fura-2 in the cytoplasm was significantly reduced (Poenie, 1990). Other studies have shown a similar restriction of rotational diffusion using different approaches (Keating et al., 1989; Popov et al., 1988; Konishi et al., 1988). When a solution of fura-2 was made viscous so that the fluorescence polarization values equal that found for cytoplasmic fura-2, a marked enhancement of the fluorescence at 380 nm excitation was observed. A much smaller enhancement was seen at 340 nm excitation (*see* Fig. 6).

Another way to look at the effect of viscosity is to monitor the fluorescence change at 380 nm relative to that at 340 nm during a Ca^{2+} transient. The parameters S_{f1}, S_{f2}, S_{b1}, and S_{b2} reflect a relationship between the relative fluorescence of the two forms of the indicator such that when $[Ca^{2+}]_i$ changes, the fluorescence at $\lambda 1$ changes in fixed proportion to the change at $\lambda 2$. Comparison of this relationship in vitro and in vivo reveals a difference between the behavior of fura-2 in free solution and inside the cell. For example, if one is measuring resting Ca^{2+} levels in cells and stimulates an increase in $[Ca^{2+}]_i$, the fura-2 fluorescence at 340 nm will rise, while that at 380 nm will fall. The rise at 340 nm will be in fixed proportion to the fall at 380 nm. Fura-2 solutions at high and low $[Ca^{2+}]$ can be used to obtain the same information for the indicator in simple aqueous media. Differences between the indicator inside the cell and in free solution reflect a change in one of the S factors. In effect, what can be determined is how the S factors are altered inside the cell. Recalling a component of the calibration equation, we note that:

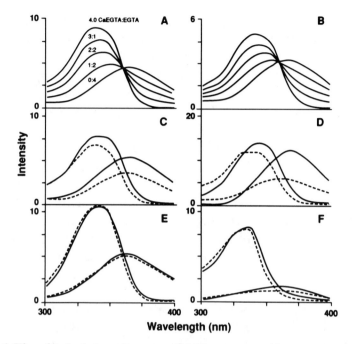

Fig. 6. The effect of viscosity on excitation spectra of fura-2. (A–B) Fura-2 was titrated with Ca²⁺ in the absence (A) or presence (B) of sucrose. In B, the fura-2 stock solutions used in A were mixed with sucrose to give a final concentration of 2M sucrose. Hence, the final concentration of fura-2 is lower in B than in A, and fluorescence intensity is reduced. Otherwise, all parameters were identical. The inclusion of sucrose shifts the relative 340 nm:380 nm ratios, so that for any given [Ca²⁺] the ratio is lower when sucrose is present. The K_d relative to that of EGTA appears to be the same in the presence or absence of sucrose. In panel C, fura-2 solutions contain either 0 Ca²⁺, 1 mM EGTA, or 1 mM Ca²⁺ in the presence (—) or absence (- - -) of 20% PVP. Panel D is similar to panel C, except that 10% Kodak™ gelatin was used as the thickener. In panel E, 10% agarose was used as the thickening agent. Panel F shows 0 Ca²⁺, 1 mM EGTA, or 1 mM Ca²⁺ solutions of quin2 in the presence (—) and absence (- - -) of 2M sucrose. As can be seen, the effect of viscosity on quin2 is similar to that of fura-2.

$$F1 = S_{f1}C_f + S_{b1}C_b$$
$$F2 = S_{f2}C_f + S_{b2}C_b \quad\quad (16)$$

and that the ratio:

$$R = F1/F2 = S_{f1}C_f + S_{b1}C_b/S_{f2}C_f + S_{b2}C_b \quad\quad (17)$$

Furthermore, for a change in Ca^{2+} (ΔCa) the change in concentrations of bound form (ΔC_b) of the dye equals an opposite change in concentration of the free form (ΔC_f) of the dye so that $\Delta C_b = -\Delta C_f$. Thus, during a Ca^{2+} transient, the change in signal at 340 nm relative to the change in signal at 380 nm is governed by the ratio $S_{f1} + S_{b1}/S_{f2} + S_{b2}$. During a Ca^{2+} transient the change in fluorescence at 340 nm ($\Delta F340$ nm) relative to the change in fluorescence at 380 nm ($\Delta F380$ nm) reflects $S_{f1} + S_{b1}/S_{f2} + S_{b2}$. One can compare this to the $\Delta F340$ nm and $\Delta F380$ nm obtained from fura-2 in the high and low Ca^{2+} calibration solution provided that dye concentration, path length, and all instrumental parameters are identical when measuring the two solutions. Using this comparison, one readily finds that this proportion is not the same for fura-2 inside the cell and fura-2 in free solution, and reflects a simple 15–30% enhancement of fluorescence at 380 nm relative to that at 340 nm.

These observations provide a simple means for correcting fura-2 calibrations for the effect of viscosity. One can either prepare calibration standards in viscous media, or take the experimentally measured R_{min} and R_{max} and multiply them by the value:

$$\frac{\Delta F340/\Delta F380 \qquad \text{measured in the cell}}{\Delta F340/\Delta F380 \qquad \text{obtained from calibration stds} \pm Ca^{2+}} \qquad (18)$$

5. Choosing a Fluorescent Calcium Indicator

The choice of a fluorescent Ca^{2+} indicator depends on several factors, including the type of measurement and the available equipment. One can readily exclude quin2 for most cases owing to the high concentrations of dye that must be loaded into cells, leading to substantial Ca^{2+} buffering. However, quin2 is valuable as a Ca^{2+} buffer, and its fluorescence is an added bonus because one can readily determine how much quin2 was loaded into the cell. Whether to use a ratiometric indicator or a single-wavelength indicator, such as fluo-3, depends in part on availability of a suitable detection system. Since fluo-3 is not a ratiometric indicator, its use is similar to that of quin2. For studies using a fluorimeter with populations of cells, fluo-3 offers a large dynamic range because of a 40-fold increase in fluorescence between the Ca^{2+}-bound and Ca^{2+}-free forms of the dye. The background autofluorescence is less of a factor in quantitation of single-wavelength measurements, unless the autofluorescence is itself altered by Ca^{2+}. Less is currently known regarding how fluo-3 fluorescence is affected by the intracellular environment, but viscosity effects have thus far been more pronounced using dual wavelengths with the ratiometric indicators. Offsetting these advantages are the susceptibility of single-wavelength measurements to fluorescence artifacts.

We prefer to use indo-1 even in cuvet measurements. The fact that the emission wavelengths change in opposite directions with changes in Ca^{2+} provides confidence that a change in ratio reflects a change in Ca^{2+}. One can assume that a change in fluo-3 fluorescence reflects a change in Ca^{2+}, but without the same level of certainty. Instrumentation for detecting the dual emission wavelengths of indo-1 can be much less expensive and more reliable than implementing the excitation chopper required for fura-2. Furthermore, since detection of dual emission wavelengths requires no chopping, both wavelengths can be acquired simultaneously so that erratic changes in light scattering and lamp fluctuations cancel out. Finally, we typically find that, for a given cell type, indo-1/AM loads consistently better with more complete hydrolysis than is obtained with fura-2. In one T-cell clone loading, with 1 μM fura-2 consistently gave 92% hydrolysis, whereas loading with indo-1 under the same conditions gave 98% hydrolysis.

Although a single wavelength indicator, such as fluo-3, can be used to detect changes in Ca^{2+} using single cell or imaging approaches it is not really suitable for quantitative single cell studies. The ratiometric approach offers many advantages for single cell studies with the main disadvantage being that the current generation of indicators all require UV excitation. For imaging, the choice is almost always fura-2. There are no advantages in terms of chopping excitation vs emission wavelengths, but there are numerous disadvantages. When emission wavelengths are altered, the image shifts owing to the wavelength dependence of refraction. This causes severe problems in ratioing, because an image taken at one emission wavelength may be out of focus or shifted relative to the image at the other wavelength. Since fura-2 exhibits little shift in emission, these problems do not arise. The advantages of real-time ratio imaging using indo-1 may be realized using two video cameras with each one recording at one of the emission wavelengths. Here, the problems of focusing and image registration are still serious, but each camera could be focused and positioned independently. Indeed, instruments implementing this approach are already in use (Takamatsu and Wier, 1990).

6. Future Prospects

A useful fluorescence Ca^{2+} indicator is subject to many design constraints, and the indicators currently available represent both successful solutions and compromises dictated by those constraints. Given these constraints, what are the prospects that improvements can be attained? It seems likely that somewhat longer wavelength indicators exhibiting shifts in excitation or emission wavelengths will be attained. Even a modest increase in the excitation wavelength of an indicator, such as indo-1, would move the excitation away from the excitation peak of NADH and NADPH. It is not certain how far the wavelengths can be pushed, however, because, as pointed out by Tsien and Waggoner (1989), many of the characteristics of bright long-wavelength chromophores are at odds with the requirements for a good Ca^{2+} indicator. Furthermore, the factors involved in giving good shifts in excitation or emission shifts are not fully understood, so there is an element of trial and error associated with every new design. The process of synthesizing each new version of an indicator can take months, and there are few laboratories working on the problems. Even so, we anticipate that, with time, indicators with better working wavelengths will be forthcoming.

Improvements are also anticipated in the area of dye loading and compartmentation. Part of the problem with dye loading is that we know little about the esterases that hydrolyze the dye or their substrate specificity. A number of alternatives, such as straight-chain carbonate esters and glycolamide esters, have

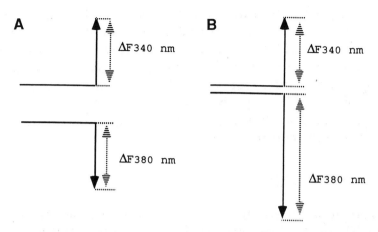

Fig. 7. A hypothetical illustration of how enhancement of 380 nm fura-2 fluorescence seen in a viscous environment would alter the change in fluorescence at 340 nm relative to that at 380 nm during a Ca^{2+} transient. Signals are recorded at both 340 and 380 nm, beginning with the cell at low resting levels of Ca^{2+} followed by a stimulus that raises Ca^{2+} suddenly. The elevation in Ca^{2+} causes a sudden deflection of both the 340- and 380-nm fluorescence in opposite directions (solid dark lines). **A** represents the signal that might be obtained if fura-2 acted as if in free solution. **B** represents the same change in Ca^{2+} for fura-2 in a viscous environment.

been synthesized, which give fura-2 good solubility properties, but fail to hydrolyze (A. Minta, R. Tsien, and M. Poenie, unpublished observations). This was surprising since these esters have been successfully incorporated into effective pro-drugs. Here there is still a large unexplored territory that has received little attention. The problem of compartmentation may be overcome by new indicator designs. The observation that rhod-2 does not compartmentalize in plant cells may provide a clue, since rhod-amine has a delocalized positive charge in addition to its carboxyl groups. Thus, we anticipate there will be improvements in these areas (*see* Fig. 7).

Another issue is the speed of the indicator. The rate of Ca^{2+} binding appears to be limited by the removal of coordinated water from the Ca^{2+} ion (Kao and Tsien, 1988). If this is true, it is unlikely that new designs based on bapta will give faster on rates. The off rate can be increased by reducing the affinity of the indicator, and this can be accomplished readily.

Some problems with differences between the intracellular environment and simple extracellular salt solutions are almost inevitable, and will probably not go away with new indicator designs. Fluorescence is influenced by the environment of the fluorophore, a feature that is central to some fluorescence applications. Perhaps these problems with the Ca^{2+} indicators may provide additional stimulus to study and understand the characteristics of the cytosol.

Acknowledgments

The preparation of this manuscript was supported by grants GM40605 from the National Institutes of Health and DCB 8858186 from the National Science Foundation.

References

Almers W. and Neher E. (1985) The Ca signal from fura-2 loaded mast cells depends strongly on the method of loading. *FEBS Lett.* **192**, 13–18.

Anderegg G. (1964) Komplexone XXXVI Reaktionsenthalpie und-entropie bei der Bildung der Metallkomplexe der hoheren EDTA-Homologen. *Helv. Chim. Acta.* **47**, 1801–1814.

Becker P. L. and Fay F. S. (1988) Photobleaching of fura-2 and its effect on the determination of calcium concentrations. *Am. J. Physiol.* **253**, c613–618.

Bers D. M. (1982) A simple method for the accurate determination of free $[Ca^{2+}]_i$ in Ca-EGTA solutions. *Am. J. Physiol.* **11**, C404-C408.

Bond M., Shuman H., Somlyo, A. P., and Somlyo A. V. (1984) Total cytoplasmic calcium in relaxed and maximally contracted rabbit portal vein smooth muscle. *J. Physiol.* **357**, 185–201.

Boyd S., Bryson A., and Nancollas G. H. (1965) Thermodynamics of ion association. Part XII. EGTA complexes with divalent metal ions. *J. Chem. Soc.* **5**, 7353–7358.

Cobbold P. H. and Rink T. J. (1987) Fluorescence and bioluminescence measurement of cytoplasmic free calcium. *Biochem. J.* **248**, 313–328.

Cohen L., Salzberg B. M., Davila H. V., Ross W. N., Landowne D., Waggoner A. S., and Wuang C. H. (1974) Changes in axon fluorescence during activity: Molecular probes of membrane potential. *J. Membr. Biol.* **19**, 1–36.

DiVirglio F., Steinberg T. H., and Silverstein S. C. (1990) Inhibition of fura-2 sequestration and secretion with organic anion transport blockers. *Cell Calcium* **11**, 57–62.

Drexhage K. H. (1973) Structure and properties of laser dyes, in *Dye Lasers. Topics in Applied Physics,* vol. 1 (Schafer F. P., ed.), Springer-Verlag, Berlin.

Fabiato A. and Fabiato F. (1979) Calculator programs for computing the composition of solutions containing multiple metals and ligands for experiments in skinned muscle cells. *J. Physiol. (Paris)* **75,** 463–505.

Garcia-Sancho J. (1984) Inhibition of glycolysis in the human erythrocyte by formaldehyde and Ca-chelator esters. *J. Physiol.* **357,** 60P.

Gerig J. T., Singh P., Levy L., and London R. E. (1987) Calcium complexation with a highly calcium selective chelator. Crystal structure of Ca(CaFBapta) 5H₂O. *J. Inorg. Chem.* **31,** 3111–3121.

Goligorsky M. S., Hruska K. A., Loftus D. J., and Elson E. L. (1986) Alpha₁-adrenergic stimulation and cytoplasmic free calcium concentration in cultured renal proximal tubular cells: Evidence for compartmentalization of quin-2 and fura-2. *J. Cell. Physiol.* **128,** 466–474.

Grynkiewicz G., Poenie M., and Tsien R. Y. (1985) A new generation of fluorescence Ca²⁺ indicators with greatly improved fluorescence properties. *J. Biol. Chem.* **260,** 3440–3450.

Gurney A. M., Tsien R. Y., and Lester H. A. (1987) Activation of a potassium current by rapid photochemically generated step increases of intracellular calcium in rat sympathetic neurons. *Proc. Natl. Acad. Sci. USA* **84,** 3496–3500.

Harafuji and Ogawa Y. (1980) Re-examination of the apparent affinity constant of ethylene glycol bis (*b*-aminoethyl ether) *N, N, N'N'* tetraacetic acid with calcium around neutral pH. *J. Biochem. (Tokyo)* **87,** 1305–1312.

Hellam D. C. and Podolsky R. J. (1969) Force measurements in skinned muscle fibres. *J. Physiol. (Lond.)* **200,** 807.

Hepler P. and Callahan D. A. (1987) Free calcium increases during anaphase in stamen hair cells of *Tradescantia. J. Cell. Biol.* **105,** 2137–2148.

Hesketh T. R., Smith G. A., Moore J. P., Taylor M. V., and Metcalfe J. C. (1983) Free cytoplasmic calcium concentration and the mitogenic stimulation of lymphocytes. *J. Biol. Chem.* **258,** 4876–4882.

Hollingworth S. and Baylor S. M. (1987) Fura2 signals from intact frog skeletal muscle fibers. *Biophys. J.* **51,** 549a.

Jackson A. P., Timmerman M. P., Bagshaw C. R., and Ashley C. C. (1987) The kinetics of calcium binding to fura-2 and indo-1. *FEBS Lett.* **216,** 35–39.

Jaffe F. L. (1986) in *Ionic Currents in Development in Progress.* Alan R. Liss, New York.

Johansson J. S. and Haynes D. (1988) Deliberate quin2 overload as a method for *in situ* characterization of active calcium extrusion systems and cytoplasmic calcium binding: Application to the human platelet. *J. Memb. Biol.* **104,** 147–163.

Kao J. P., Harootunian A. T., and Tsien R. Y. (1989) Photochemically generated cytosolic calcium pulses and their detection by fluo-3. *J. Biol. Chem.* **264,** 8179–8184.

Kao J. P. Y. and Tsien R. Y. (1988) Ca^{2+} binding kinetics of fura-2 and azo-1 from temperature-jump relaxation measurements. *Biophys. J.* 53, 635–639.

Kaplan J. H. and Elles-Davies G. C. R. (1988) Photolabile chelators for the rapid photorelease of divalent cations. *Proc. Natl. Acad. Sci. USA* 85, 6571–6575.

Keating S., Wensel T., and Stryer L. (1989) Reduction in rotational diffusion. *Biophys. J.* 55, 518a.

Klein M. G., Simon B. J., Szucs G., and Schneider M. F. (1988) Simultaneous recording of calcium transients in skeletal muscle using high and low affinity calcium indicators. *Biophys. J.* 53, 971–978.

Kohse K. P. and Heilmeyer L. M. G., Jr. (1981) The effect of Mg^{2+} on the Ca^{2+}-binding properties and Ca^{2+}-induced tyrosine fluorescence changes of calmodulin isolated from rabbit skeletal muscle. *Eur. J. Biochem.* 117, 507–513.

Konishi M., Olson A., Hollingworth S., and Baylor S. M. (1988) Myoplasmic binding of fura-2 investigated by steady state fluorescence and absorbance measurements. *Biophys. J.* 54, 1089–1104.

Korenbrot J. I., Ochs D. L., Williams J. A., Miller D. L., and Brown J. E. (1986) The use of tetracarboxylate fluorescent indicators in the measurement and control of intracellular free calcium ions, in *Optical Methods in Cell Physiology* (DeWeer P. and Salzberg B. M., eds.), Wiley, New York.

Lee H. C., Forte J. G., and Epel D. (1982) The use of fluorescent amines for the measurement of pHi: Applications in liposomes, gastric microsomes and sea urchin gametes, in *Intracellular pH: Its Measurement, Regulation and Utilization in Cellular Functions* (Nuccitelli, R., ed.), Alan R. Liss, New York.

Lev-Ram V. and Grinvald A. (1987) A citivity-dependent calcium transients in central nervous system myelinated axons revealed by the calcium indicator fura-2. *Biophys. J.* 52, 571–576.

Levy L. A., Murphy E., Raju B., and London R. E. (1988) Measurment of cytosolic free Mg^{2+} ion concentration by ^{19}F NMR. *Biochemistry* 27, 4041–4048.

Martell A. E. and Smith R. M. (1974) *Critical Stability Constants* (Plenum, New York).

Minta A., Kao J. P. Y., and Tsien R. Y. (1989) Fluorescent indicators for cytosolic calcium based on rhodamine and fluorescein chromophores. *J. Biol. Chem.* 264, 8171–8178.

Moisescu D. G. (1976) The kinetics of reaction in calcium activated skinned muscle fibres. *Nature* 262, 610–613.

Moisescu D. G. and Pusch H. (1975) A pH metric method for the determination of the relative concentration of calcium to EGTA. *Pfleugers Arch.* 355, 243.

Poenie M., Tsien R. Y., and Schmitt-Verhulst A. -M. (1987) Sequential activation and lethal hit measured by $[Ca^{2+}]_i$ in individual cytolytic T cells and targets. *EMBO J.* 6, 2223–2232.

Poenie M., Alderton J., Steinhardt R. A., and Tsien R. Y. (1986) Calcium rises abruptly and briefly throughout the cell at the onset of anaphase. *Science* **233,** 886–889.

Poenie M., Alderton J., Tsien R. Y., and Steinhardt R. A. (1985) Changes of free calcium with stages of the cell division cycle. *Nature* **315,** 147–149.

Poenie M. (1990) Alteration of intracellular fura-2 fluorescence by viscosity: A simple correction. *Cell Calcium* **11,** 85–91.

Popov E. G., Garrilov I. Y., Pozin E. Y., and Gabbasov Z. A. (1988) Multiwavelength method for measuring concentration of free cytosolic calcium using the fluorescent probe indo-1. *Arch. Biochem. Biophys.* **261,** 91–96.

Portzehl H. P., Caldwell C., and Ruegg J. C. (1964) The dependence of contraction and relaxation of muscle fibres from the crab *M. squinado* on the internal concentration of free calcium ion. *Biochim. Biophys. Acta* **79,** 581–591.

Raju B., Murphy E., Levy L. A., Hall R. D., and London R. E. (1989) A fluorescent indicator for measuring cytosolic free Mg²⁺. *Am J. Physiol.* **256,** C540–548.

Rink T. J. and Pozzan T. (1985) Using quin2 in cell suspensions. *Cell Calcium* **6,** 133–144.

Rink T. J., Sanchez A., and Hallam T. J. (1983) Dialcylglycerol and phorbol esters stimulate secretion without raising cytoplasmic free calcium in human platelets. *Nature* **305,** 317–319.

Roe M. W., Lemasters J. J., and Herman B. (1990) Assessment of fura-2 for measurement of cytosolic free calcium. *Cell Calcium* **11,** 63–73.

Scanlon M., Williams D. W., and Fay F. S. (1987) A calcium insensitive form of fura-2 associated with polymorphonuclear leukocytes. *J. Biol. Chem.* **262,** 6308–6312.

Schmidt T., Patton C., and Epel D. (1982) Is there a role for the Ca²⁺ influx during fertilization of the sea urchin egg? *Dev. Biol.* **90,** 284–290.

Schmidt R. W. and Reilly C. N. (1957) New complexon for titration of calcium in the presence of magnesium. *Anal. Chem.* **29,** 264–268.

Schwartzenbach G., Senn H., and Anderegg G. (1957) Komplexone XXIX. Ein grosser Chelateffekt besanderer. *Helv. Chim. Acta* **40,** 1886–1900.

Smith G. L. and Miller D. J. (1984) EGTA purity and the buffering of calcium ions in physiological solutions. *Am. J. Physiol.* C160-C166.

Smith G. L. and Miller D. J. (1985) Potentiometric measurements of stoichiometric and apparent affinity constants of EGTA for protons and divalent ions including calcium. *Biochim. Biophys. Acta* **839,** 287–299.

Speksnuder J. E., Miller A. L., Weisenseel M. H., Chen Tsung-Hsien, and Jaffe L. H. (1989) Calcium buffer injections block fucoid egg development by facilitating calcium diffusion. *Proc. Natl. Acad. Sci. USA* **86,** 6607–6611.

Takamatsu T. and Wier W. G. (1990) High temporal resolution video imaging of intracellular calcium. *Cell Calcium* **11,** 111–120.

Tsien R. Y. (1980) New calcium indicators and buffers with high selectivity against magnesium and protons: Design, synthesis and properties of prototype structures. *Biochemistry* **19,** 2396–2404.

Tsien R. Y. (1981) A non-disruptive technique for loading calcium buffers and indicators into cells. *Nature* **290**, 527,528.

Tsien R. Y. and Poenie M. (1986) Fluorescence ratio imaging: A new window into intracellular ionic signaling. *Trends Biochem. Sci.* **11**, 450–455.

Tsien R. Y. and Pozzan T. (1989) Measurement of cytosolic free Ca^{2+} with quin2. *Methods Enzymol.* **172**, 230–262.

Tsien R. Y., Pozzan T., and Rink T. J. (1982) Calcium homeostasis in intact lymphocytes: Cytoplasmic free calcium monitored with a new, intracellularly trapped fluorescent indicator. *J. Cell Biol.* **94**, 325–334.

Tsien R. Y. and Waggoner A. (1989) Fluorophores for confocal microscopy: Photophysics and photochemistry, in *Confocal Microscopy Handbook* (Pawley J., ed.), IMR, Madison, WI.

Von Tscharner V., Deranleu D. A., and Bagyiolini (1986) Calcium fluxes and calcium buffering in human neutrophils. *J. Biol. Chem.* **261**, 10163–10168.

White G. C., Levine S. N., and Steiner A. N. (1981) Platelet calcium-dependent proteins: Identification and localization of the calcium-dependent regulator, calmodulin, in platelets. *Am. J. Hematol.* **101**, 359–367.

Williams D. A. and Fay F. S. (1990) Intracellular calibration of the fluorescent calcium indicator fura-2. *Cell Calcium* **11**, 75–83.

Williams D. A., Fogarty K. E., Tsien R. Y., and Fay F. S. (1985) Calcium gradients in single smooth muscle cells revealed by the digital imaging microscope. *Nature* **318**, 558–561.

Winkler B. S. (1981) Glycolytic and oxydative metabolism in relation to retinal function. *J. Gen. Physiol.* **77**, 667–692.

Zeitz M., Lange K., Keller K., and Herken H. (1980) Distribution of acridine orange accumulating particles in neuroblastoma cells during differentiation and their characterisation by subcellular fractionation and electron microscopy. *J. Cell Biol.* **25**, 305.

Single-Cell Imaging Technology

Roger B. Moreton

1. Introduction

This chapter will describe and assess the technology by which optical indicators are used to produce images that show the distribution of a measured quantity, such as free ion concentration or electric potential, within individual cells. Common ions whose free concentrations, or activities, can be imaged presently include Ca^{2+}, Mg^{2+}, Na^+, K^+, Cl^-, and H^+. In addition, many specific subcellular structures can be imaged by attaching fluorescent tags or antibodies. For studying intracellular messengers and their effects, the measurement of free ion concentrations is of prime importance and I shall concentrate on this, but an imaging experiment can often be usefully extended by using a fluorescent dye to image a subcellular compartment or structure, such as endoplasmic reticulum, from which the ionic signal may be thought to emanate.

The primary requirement for imaging an intracellular ion-species is an indicator-dye that binds to the ions with a dissociation constant (K_d) comparable with the free ion concentration and whose optical properties are measurably changed in the presence of the ion. The indicators themselves are described in the chapter by Poenie. The technology and interpretation of imaging vary with the optical properties of particular indicators: for simplicity, I shall concentrate on the imaging of intracellular Ca^{2+}, which has so far attracted the most attention from cell biologists, and the fluorescent indicator fura-2, which is currently the

From: *Neuromethods, Vol. 20: Intracellular Messengers*
Eds: A. Boulton, G. Baker, and C. Taylor © 1992 The Humana Press Inc.

most popular dye in use for Ca^{2+} imaging. I will try to show how equipment and techniques need to be varied to take advantage of the special properties of some of the other indicators.

The stages in single-cell imaging are as follows:

1. Cells loaded with indicator are mounted, usually in a perfusion-chamber, on a microscope stage, and illuminated with light of selected wavelength, using a filter or a monochromator.
2. Light from the cells is collected by a sensitive videocamera, which scans the image, producing an electrical signal. Usually this is in standard video format, with 30 complete scans/s (25 in Europe): each complete scan is referred to as one video frame.
3. The camera output is digitized by a computer or specialized image-analyzer to produce a stream of numbers (pixels), each of which represents the image brightness at one point.
4. These numbers are processed arithmetically, to produce an output stream where the image brightness has been replaced by a new value, representing the measured ion activity or other parameter. Depending on the capability of the system, processing may be done either on-line, with each pixel being processed as it is acquired to produce a "real-time" output image, or off-line, by storing sequences of images that are later processed to give "snapshot" pictures of the experiment.
5. The output stream is then reconverted to analog form, producing a video output displayed on a monitor.
6. Output images are stored in analog or digital form, and can be further processed by the host computer to extract, for example, the average concentrations in one or more cells or parts of cells, or to provide graphical output for illustrations.

Optical indicators can respond to changes in the measured parameter by luminescence, by changes in optical absorption, or by changes in fluorescence when excited by blue or ultraviolet (UV) light. Fluorescence is the method of choice for almost all single-cell imaging. Luminescent indicators, such as aequorin, give sufficient light to provide a useful signal when averaged over a whole cell, for example, using a photomultiplier, but when divided into many pixels and scanned at video rate (25 or 30 times/s), the signal for each pixel becomes too weak to distinguish from background. Dyes, such as arsenazo-III, whose use is based on differential absorbance of infrared light, also produce very weak signals since they absorb at best only about 1% of the incident light.

Fluorescent indicators can be divided into three main groups according to their response:

1. Indicators showing a shift in excitation spectrum (i.e., changes in the wavelength at which high fluorescence is best excited);
2. A shift in emission spectrum; and
3. A change in fluorescence brightness, without any major spectral change.

The same basic techniques are applicable to imaging with any of these, the main difference being that with the first two groups, quantitative results can be obtained from the ratio of fluorescence at two different excitation or emission wavelengths. The ratio method of quantitation is very powerful, since it eliminates errors caused by unknown concentration or uneven distribution of the dye, and by variations in the optical pathlength through the cells. Fura-2 is suitable for use with dual-wavelength excitation; the Ca^{2+} dye indo-1 is used with dual-wavelength emission. Figure 1 summarizes the components of a complete system for imaging, using fura-2 and dual-wavelength excitation.

Quantitative interpretation of Ca^{2+} images is not simple. I shall devote some space to the steps needed to obtain the most reliable results, and I shall present them in the most intelligible way. The precise methods for quantitation and display in any imaging system have to be worked out and built into the procedures embodied in stage (4) above, before any serious experimentation can be done, but it is important to understand the principles and issues involved if images from different cell types are to be correctly interpreted.

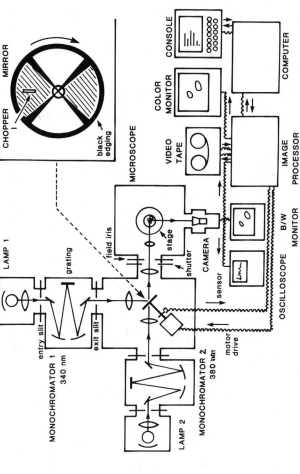

Fig. 1. Apparatus for real-time ratio-imaging, using monochromators to give dual-wavelength excitation. Light from each xenon arc lamp is passed through a grating monochromator, and focused by a quartz lens, which forms a real image (I: inset) of the monochromator exit slit at the front face of the chopper mirror. A black border, whose width is comparable to the width of this image, is attached to the edge of the mirror to provide clean switching between reflection of light from Lamp 1, or transmission of light from Lamp 2. Light of the selected wavelength passes through a shutter and focal-plane field iris, to illuminate the specimen (*see* Fig. 2 for details of optical pathway). Emitted fluorescence is collected by the camera to form a video signal, which is displayed on an oscilloscope and monochrome monitor for checking, and simultaneously digitized and processed by the *Imagine* image processor to form a new video signal representing the ratio image. This is displayed on

Finally, single-cell imaging is an elaborate and expensive technique, full of potential sources of error, so it is important to consider why one does it at all, instead of relying on the quicker, simpler, and more accurate measurements that can be made with a population of cells in a spectrofluorimeter. The reasons for using single-cell imaging are twofold. First, it shows the variability among cells: some agonists may produce the same response, at about the same time in almost every cell in a population, so that the population average seen in a fluorimeter is a faithful indicator of the response of any one cell; others may cause each cell to respond differently, or with a different latency, so that the population average will show only a feeble, smeared-out response. The faster and more transient the response of each cell, the more likely it is that different latencies may prevent it from showing properly in the population average. Second, video imaging can show how changes in the cytosolic free $[Ca^{2+}]$ are distributed throughout the cell. This is important, for example, if the Ca^{2+} signal is the trigger for exocytosis, which may be initiated best by a Ca^{2+} rise near to the plasma membrane (Cheek et al., 1989b). If intracellular Ca^{2+} is the common messenger for translating a range of hormonal and transmitter signals into different cellular events, the Ca^{2+} signals must be characterized by more than intensity to provide the necessary vocabulary. The spatial and temporal distribution of the Ca^{2+} signal may provide the key. Before setting out to look at single-cell images of intracellular free Ca^{2+}— or any other such parameter—one should consider how much one needs this spatial and temporal information. Given that the total amount of obtainable information is limited, one may also have to consider whether space or time is the more important feature in the behavior of the cells: Single-cell imaging systems are usually optimized for either temporal or (more often) spatial resolution. I shall try to show how such optimization is done, and what the limits are, so that the best choice of system can be made for each type of experiment.

2. Optics

Referring to Fig. 1, the optical equipment for single-cell imaging comprises a microscope, UV light source, and one or more videocameras.

2.1. Microscope

A standard epifluorescence microscope is suitable for single-cell imaging. Figure 2 (a) summarizes the optical components and lightpaths: Exciting radiation passes through a collimator lens and is reflected by a dichroic mirror, which is an interference-filter designed to reflect or transmit light with wavelength, respectively, below or above a given value. For UV light excitation of fura-2 or indo-1, a transition wavelength of 400 nm is suitable. The UV light then passes through a collecting lens and epifluorescence objective, which doubles as a condenser, to irradiate only the area of specimen under observation. Emitted (visible-wavelength) fluorescence is collected by the objective and transmitted through the dichroic mirror. A band-pass or long-pass interference filter cuts out nonspecific fluorescence and stray light, and the remaining light is distributed by a system of sliding prisms, either to the eyepieces, or to a side-port connected to the videocamera. Given that the spectra of most fluorescent indicators contain fairly broad peaks, a band-pass filter with a wavelength bandwidth of 10–20 nm between 50% transmission-points is about the best for exclusion of background, combined with good transmission of the wanted signal; for fura-2 the appropriate center-wavelength is 510 nm.

Alternatively, the band-pass filter may be omitted, and a pair of dichroic mirrors used to split the emitted light between two cameras (Fig. 2b). This arrangement can be used either for dual-emission work, with indo-1, for example, or one camera can be of an infrared-sensitive type. Illuminating the cells from above through an infrared filter will then allow simultaneous ratio imaging by dual-excitation, with detailed structural information from the infrared transmission image (Tsien and Harootunian, 1990).

The UV light used to excite many fluorescent indicators is poorly transmitted by most glasses. Thus, both the collimator and collector lenses need to be of fused quartz, and the objective lens, which doubles as a condenser, should be either of quartz or special, UV-transmitting glass. Also, given that it is desirable to minimize the dose of exciting radiation delivered to the cells, efficiency in collecting the emitted light is crucially important.

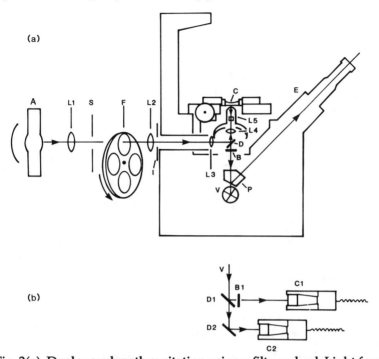

Fig. 2(a). Dual-wavelength excitation using a filter-wheel. Light from the arc (A) is focused by a quartz lens (L1) through a shutter (S) onto the rotating filter-wheel (F), which interposes different band-pass optical filters, in synchrony with the video signal. Two further quartz collimator lenses (L2, L3) and a field-iris (I) project a parallel beam onto a dichroic mirror (D), which reflects UV light through a collector lens (L4) into the objective (L5), which condenses it to illuminate the cells (C) uniformly over the field of view. Emitted fluorescence is transmitted by the dichroic mirror (D) through an optical filter (B), which may be of either long-pass ("barrier") or band-pass type, to a set of sliding prisms (P) that relay the image either to the eyepiece (E) or to the videocamera (not shown) on a side-port (V). In some systems, lenses L2, L3 are replaced by a fiberoptic bundle or liquid light-guide, which provides mechanical isolation of the microscope, and also facilitates uniform illumination of the field of view by "scrambling" the image of the light-source (Lindemann et al., 1990). (b). Twin-camera system for dual-emission measurements (after Takamatsu and Wier [1990]). Light from the side-port (V) is split by dichroic mirrors (D1, D2) and band-pass filter (B1) to feed a pair of ICCD cameras (C1, C2), arranged so that the two optical path lengths are identical. Camera C2 is mounted on a micromanipulator, allowing translation or rotation on all three axes, so that the two images can be made to coincide exactly. Alternatively, C2 may be an infrared-sensitive camera, allowing a structural image of the cells to be superimposed, for example, on a dual-excitation ratio image.

Objective lenses of high numerical aperture are therefore needed: the Nikon UV-F 100× glycerol- and 40× oil-immersion objectives (NA 1.3) provide a useful combination of efficiency and UV-light transmission, with working distances long enough to allow the use of a reasonably stout (No. 1.5) cover slip. Moore et al. (1990) give some figures for the UV-light transmission of these and other objective lenses, as well as some other glass microscope elements.

The inverted microscope allows cells to be imaged in an open chamber, with access from above for micropipets and perfusion outlets. The stage should combine accessibility with some immunity to stray light: Although it will normally be necessary to work in a darkened room, interference with experiments can easily be caused by operators wielding torches! The Nikon Diaphot TMD microscope satisfies these requirements quite well, and has the added advantage that focusing is achieved by moving the objective turret, rather than the stage. The stage thus remains at a fixed height, so that micropipets introduced from separately mounted manipulators will stay in focus.

Figure 3 shows the design of a simple, temperature-controlled stage for use with cultured cells. The cell chamber is made of stainless steel, for good thermal contact with the stage. Nevertheless, heat loss from the exposed face of the saline means that the temperature at the center of the cover slip may be 1–2°C lower than at the rim. For accurate work, use of a very small thermocouple is recommended to check this, and heating the objective lens with a small coil and thermostat may be useful (Tsien and Harootunian, 1990). An alternative is to enclose the whole stage in an environmental chamber, kept warm by a temperature controlled air-stream. This provides the most stable environment, but the restrictions on accessibility may prove crippling if manipulations, such as microinjection or patch-clamping, are contemplated.

2.2. UV Light Source

For most fluorescent indicators, excitation wavelengths lie in the range of 330–500 nm (an exception is fluo-3). Light in this range of wavelengths is conveniently provided by either mer-

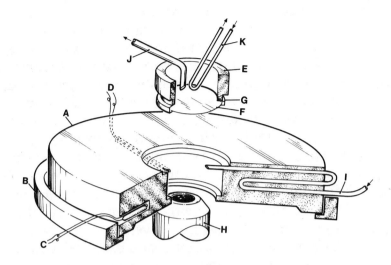

Fig. 3. Temperature-controlled stage and perfusion chamber. The stage is a block of aluminum alloy (A) mounted on a nylon ring (B) and heated by a pair of 1.2-Ω, 3-W resistors (C, only one shown) embedded in thermally conducting grease ("heat-sink compound"). The temperature sensor is a miniature head thermistor (D), similarly embedded, but protruding slightly to make direct contact with the outside of the perfusion chamber (E). The chamber is made of stainless steel, and its base is formed by a round cover slip (F), 22 mm in diameter, on which cells have been plated, and which is sealed in position by nonfluorescent silicon grease and a thin metal ring (G). This leaves a minimum of material below the cover slip, to avoid mechanical interference with the microscope objective (H). Perfusion medium enters the chamber through plastic tubing (I), embedded in a labyrinth to allow thermal equilibration, and is sucked away through an aspirator (J). Agonists are passed through the U-tube (K; Fenwick et al., 1982), which has a pin-hole positioned over the cell of interest, and is kept at a slight negative pressure by a peristaltic pump. Precisely timed application is achieved by cutting off the pump, using a solenoid valve; the control circuitry for this includes an event-marker, which records a continuous tone on one audio channel of the videotape recorder.

cury or high-pressure xenon arc lamps. For fura-2, the mercury arc has the disadvantage that much of its energy is concentrated into a spectral line at 365 nm, which is close to the isosbestic point (where fura-2 fluorescence is independent of [Ca^{2+}]). The xenon arc provides a smooth spectrum between 300 and 450 nm, and is to be preferred. (A xenon arc lamp in a quartz envelope will provide light at wavelengths down to about 200 nm, but this is not recommended, since production of atmospheric ozone becomes a hazard, and in any case, light below about 300 nm is not transmitted through most objectives. "Ozone-free" lamps, in glass envelopes, are more suitable.) A 150-W lamp provides enough light for most systems.

Excitation wavelengths can be controlled by grating mono-chromators, whose outputs are switched by a rotating mirror as in Fig. 1; alternatively, a single lamp can be used, with optical band-pass interference filters on a rotating filter-wheel (Fig. 2a), or on a solenoid-driven slider. In either case, the quantity of light output will be proportional to the wavelength bandwidth: As with the emission filter, for most indicators, a bandwidth of 10 nm between 50% transmission points represents a good com-promise between selectivity and brightness. Filters are simple and relatively cheap, and can be bought with a range of wave-lengths and bandwidths. Disadvantages are that, at best, they transmit only about 75% of the light over the designated wave-length-band, and changing wavelengths during the course of an experiment, for example, to make use of a fluorescent stain to localize cell components, is less easy than with monochromators. On the other hand, the use of monochromators restricts excita-tion to a maximum of two wavelengths in any one run, whereas with a filter-wheel, several different filters can be included, so that "double label" experiments with two dyes become possible—though not, so far, with real-time imaging.

Grating monochromators provide nearly 100% light transmission at the selected wavelength, and complete flexi-bility regarding wavelength, bandwidth (controlled by vary-ing the width of the output slit), and intensity (controlled by the input slit) of the exciting radiation. Disadvantages are high cost and difficulty in producing uniform illumination of a circular field from the image of a rectangular monochroma-tor slit, though the use of a light-guide or fiberoptic bundle can provide more uniform illumination, as well as allowing some mechanical isolation of the microscope (Lindeman et al., 1990).

For dual-excitation ratio-imaging, wavelengths need to be changed synchronously with the acquisition of images by the image processor. Where only a modest time-resolution (1 s or more between changes) is adequate, and image-processing is done off-line, then a protocol can be based on collecting and averaging several frames at one wavelength, changing wavelengths, optionally allowing a pause for the camera to adjust, and then repeating the average at the second wavelength. Wavelength changes are relatively infrequent, and the chopper or filter-wheel can be driven by direct command from the image- processor. Where time-resolution is important—as, for example, when attempting to follow changes in Ca^{2+} in neurons or muscle cells during excitation—and where the videocamera is capable of responding to rapid changes in image intensity, the wavelength needs to be changed synchronously with the video signal, possibly once every frame, and the chopper or filter-wheel must rotate continuously. Manufacturers have, so far, not produced any very satisfactory standard components for synchronizing a chopper mirror with the video signal, so that apart from microscope stage-design, this is the first major "do-it-yourself" input into the mechanics of imaging systems. Basically, the shaft carrying the chopper also carries an optosensor, whose position-signal can be used to drive a phase-locked loop that attempts to synchronize it with the video start-of-frame signal, by controlling the voltage supplied to the drive motor. Such a system is simple to install and works reasonably well, except that any disturbance, for example, owing to uneven friction in a drive-belt, is liable to produce slow oscillations in chopper speed, which look rather like oscillations in intracellular Ca^{2+}! An alternative that we now use is to drive the chopper with a stepper-motor, using pulses derived from the video line signal, and some simple logic to ensure that the motor will run a bit faster than this until the optosensor output is found to be synchronous with the video frame, at which time the drive is switched over to video-rate by a digital "latch."

For dual emission work only one UV-light wavelength is needed, so that a single monochromator or filter will suffice. The camera (Fig. 1) is replaced by a matched pair of cameras, with dichroic mirrors to separate the wavelengths (Fig. 2b). For quantitative ratio-imaging, it is clearly important that the images seen by the two cameras coincide exactly in position and magnification. Takamatsu and Wier (1990) have achieved this by mounting one camera on a micromanipulator.

The choice of excitation wavelength obviously depends on the spectrum of the dye in use. Excitation spectra are generally quite broad, however, and some adjustment of wavelengths is possible to suit particular instrumentation—for example, some workers prefer to use 345 or 350 nm as the shorter wavelength for exciting fura-2, because light at these wavelengths may be better transmitted by the microscope optics than at the more usual 340 nm.

2.3. Video Cameras

Even with the most highly fluorescent indicator dyes, the amount of light available for detection is not large. Considerations of photobleaching and cell damage resulting from irradiation dictate that the excitation intensity should be kept as low as possible, and concentrations of dye in the cells should also be kept low, to avoid toxic side-effects and buffering of intracellular Ca^{2+}. As a rough guide, if the fluorescence image from fura-2 can be seen through the microscope by the unaided, dark-adapted eye, then the loading and excitation are about as high as can be tolerated. A sensitive videocamera (or cameras, for dual-emission) is thus a vital component, and its choice will govern, to a large extent, the type of image that can be obtained. (A few authors, e.g., Timmerman et al. [1990], have used an image-intensifier with a film camera to image intracellular Ca^{2+} distribution. This technique is not suitable for quantitative imaging, however, not only because making measurements from film is tedious, but also because of the inherent nonlinearity of the photographic recording process.)

Two main types of videocamera are in use for single-cell imaging: Silicon intensified target (SIT), and intensified charge-coupled device (ICCD) cameras. For very low light-level work, an SIT camera may also be fitted with an image-intensifier (ISIT), and where time-resolution is not important, a CCD camera may be used without an intensifier, but with the sensor cooled to –50°C to reduce dark-current, and collecting each image over a period of 1 s or more, instead of the normal video frame-time. The output from such a camera is not generally in standard video format, and arrangements for processing it are relatively simple, given the ample time available. Also, intracellular events occurring over such long time-scales seem unlikely to be of primary interest to neurobiologists, so I do not propose to consider cooled CCD systems further. For some applications, *see* Connor et al. (1990) and Lindeman et al. (1990).

SIT cameras have, typically, high spatial resolution (better than 512 lines per frame), low noise, and good sensitivity. They suffer from minor problems of geometric distortion, "shading," whereby the brightness of the image varies slightly across the field, nonlinearity of response, and blemishes in the target, which can produce dead spots in the picture. Shading becomes a less serious problem when ratio-imaging is used, but should still be considered when the ratio method does not involve a straight division of one image by another (*see* Section 4.3. *below*), and where background subtraction is needed. Nonlinearity remains a problem with ratio-imaging, because if the ratio of intensities at the two wavelengths differs significantly from unity, then the ratio recorded by the camera will vary, for example, with differences in the amount of dye or the optical pathlength at different points in the cell, irrespective of any actual variations in $[Ca^{2+}]$. The major drawback of SIT cameras, however, is that the target does not respond instantly to changes in image brightness, but lags with a time-constant of about 70 ms (Moore et al., 1990). This means that after the excitation wavelength has been changed, a period of several video frames must be allowed for the camera to respond, before acquiring the next image. SIT cameras are thus not well suited to real-time ratio-imaging, and many experimenters are now turning to the newer technology offered by the ICCD.

A good ICCD camera has a microchannel plate image intensifier, coupled directly by a coherent fiberoptic bundle to the charge-coupled device, which is a photocathode overlaying a block of silicon on which an array of "potential wells" is created to trap photoelectrons emitted by the cathode. At intervals, the wells are emptied in sequence into a sensitive amplifier, whose output is converted into the video signal. Early ICCD cameras had only moderate spatial resolution, but more recent models can resolve up to 490 TV lines (576 in Europe). Sensitivity is similar to that of an SIT camera, and CCDs are virtually free from geometric distortion, blooming (overflow from bright regions into neighboring pixels), and shading. The major advantage, however, is that the temporal response of the CCD is limited only by that of the intensifier photocathode, which typically shows a lag-time of about 3 ms. It is thus possible to switch excitation wavelengths between video frames, and obtain successive images that are almost independent, making ICCD cameras the best choice for real-time ratio-imaging. The price to be paid for this is that the image intensifier and, to a lesser extent the CCD, are noisy devices. The image from an ICCD camera is thus punctuated by bright or dark "flashes," usually only one pixel wide, which can make quantitative interpretation difficult, and it is necessary to do some digital filtering or time-averaging before forming the ratio image. We find that a response time-constant of 200 ms is a realistic minimum for real-time ratio-imaging with fura-2. At this stage, we note that for dual-excitation recording, the best possible time-resolution (apart from considerations of noise-reduction) is two frame-times. For dual-emission work with two cameras acquiring data simultaneously, the minimum response-time of the system is, of course, halved.

Whichever type of camera is used, it is vital that any automatic gain or "black-level" controls be disabled. These may be fine for taking pictures, but they make nonsense of any quantitative imaging!

3. Image Processing

The processing required for single-cell imaging falls into three stages: digitization; initial processing, which includes background subtraction, shading corrections, scaling of data according to calibration curves, and for ratio-imaging, the combination of pictures taken at each of the two wavelengths (excitation or emission) to yield a single output picture; and further processing to extract numerical data and graphical output from the results. The further processing stage will almost always be done off-line, after the experiment is finished. Digitizing the video signal must, of course, be done at video rate, and is normally done on-line. The extent to which initial processing can also be done on-line, while the experiment is in progress, varies with the capability of the hardware: Most systems will be able to do some processing, such as background subtraction; a few can perform the whole calculation involving several arithmetic operations on each pixel as it is read from the camera, and producing an output ratio image, which is a "live" representation of intracellular Ca^{2+} (for example). The whole "feel" of the experiment will depend on how the processing is done: With off-line processing a rigid protocol is needed, in which cells are set up, stimuli are applied at chosen times, and images are acquired in a carefully designed sequence to display the response. Real-time processing, on the other hand, allows the experimenter to apply stimuli or agonists in a more relaxed manner, observe the response of the cells, and record a live sequence of Ca^{2+} images from an entire 60-min experiment on one videotape. The price of this freedom is expensive hardware and the need to set up the processing algorithms in advance, when some of the experimental parameters may not be accurately known. I shall consider the processing stages in order.

3.1. Digitization

The first stage in quantitative imaging is to use the framestore to convert video from the camera into a stream of numbers (pixels), of which there will normally be 480 lines of 640 pixels (for RS-170 or RS-330 standard video), or 512 lines of 768 pixels (for European CCIR standard video), representing an oblong TV picture. Each pixel is normally digitized with 8-bit resolution, giving 256 possible values (gray levels). Although this is not a severe limitation for straightforward images, if a ratio is to be formed from two images of widely differing brightnesses, then limited accuracy becomes a problem. For example, with fura-2 at high levels of Ca^{2+}, the fluorescence excited at 340 nm may be more than eight times as bright as that excited at 380 nm, so that to prevent overflow at 340 nm, intensities must be adjusted so that the 380-nm image is no brighter than 32 U on the 0–255 scale, and will not be digitized accurately. UV light intensities must then be chosen very carefully to make best use of the limited dynamic range: an oscilloscope, or the small monochrome monitor (Fig. 1), should be used to check that the video signal from the camera is comfortably within range. One imaging system, the Hamamatsu DVS-3000, offers freedom from this constraint, by providing 16-bit digitization (65,536 possible values). Alternatively, it is possible to control the gain of a CCD camera directly from the image analyzer (Brooker et al., 1990), and one could conceive of a system in which the camera gain for each frame is automatically optimized, based on the mean brightness of the last frame to be recorded at that wavelength. Provided that the image analyzer keeps track of the gain settings while computing the ratio image, the result could be quite sensible, but no attempt at this ambitious enterprise has yet been reported.

3.2. Initial Processing

The initial processing steps needed to form a ratio image are as follows:

1. Images are collected separately at each of two excitation (or emission) wavelengths, usually with filtering or time-averaging to reduce noise, and with correction where necessary to allow for zero offset or nonlinearity of the camera output.
2. From each image, a prerecorded background image may be subtracted. This may consist entirely of instrumental background, or it may be an image recorded from the cells before loading with dye, to include autofluorescence.
3. Each image may be multiplied (or divided) by a prerecorded shading correction picture, which is usually obtained by irradiating a chamber containing a thin layer of fura-2 solution, in order to represent any nonuniformity in the UV light distribution.
4. Some further improvement in signal/noise ratio can be gained at this stage by spatial filtering (combining neigh-

boring pixels to form a local mean), at the expense of spatial resolution.

5. The two images are combined to implement the appropriate ratio-imaging formula, to produce a resultant image representing [Ca^{2+}]. Usually, this will include a proviso that any pixel whose intensity at either or both wavelengths falls below an arbitrary threshold shall be set to zero in the output image, in order to give a clean background outside the cells.

Frame-store devices for initial image-processing range from a simple frame-store card mounted in a PC, and capable of storing two or four frames of video at a time, to larger, free-standing frame-store devices containing several megabytes of video memory, with varying degrees of processing capability, and linked to a host computer by a fast, parallel data-link, so that the computer can rapidly access the video memory to make further calculations. Processing by the frame-store will be either "hard-wired," giving a set of prearranged tasks that the host computer merely has to select or, in the newer frame-store devices, fully programmable. A fully programmable frame-store offers maximum flexibility, given that a tame and responsible programmer is available. On the other hand, most users will want a system with plenty of prewritten programs that can be run with a minimum of distraction from the experiment, and will control the acquisition and disposal of images, selection of wavelengths, and so on, automatically. Tsien and Harootunian (1990) put forward strong, if inevitably personal, views regarding how a frame-store and image analyzer should be set up and programmed.

The minimum initial processing requirement is that the frame-store should be able to average or summate a number of frames, typically 16–64, lasting about 0.5–2.1 s (stage 1 above). For dual-excitation ratio-imaging the wavelength is then changed, time is allowed (if necessary) for camera-lag, and a further sequence of images is averaged. For experiments in which time is not critical, it may be sufficient to store just one pair of such filtered images, which can then be collected and processed or stored on disk, by the host computer: typically, this will take a few seconds, after which a second image-pair can be acquired. Alternatively, if the frame-store has enough video memory, then a whole sequence of image-pairs can simply be stored, and the entire experiment collected by the computer afterward. The number of storable images can be increased by saving only a portion of each: at most, it is normal to store only the central 512 pixels of each line, occupying 0.25 Mbyte per image, and by storing only 128 × 128 pixels, Jacob (1990), for example, was able to fit 900 image pairs into 32 Mbytes, representing 15 min of continuous data at 1 image pair/s. Some authors (Marks and Maxfield, 1990; Tsien and Harootunian, 1990) have used high-speed optical disks to store data at this stage, in which case whole experiments can be recorded as digitized "raw" data: A 1 Gbyte disk will hold 2000 image-pairs of full 512 × 512 size, representing just over an hour's data at 1 image-pair/2 s, but disks are not reusable, and media costs may become considerable.

Other processing options commonly available for stage (1) include passing the video input through a "look-up table," for example, to implement a nonlinear calibration curve for the camera, by replacing each input pixel value by a precalculated value stored in the table. Some frame-stores can also allow for slightly nonstandard video input by adjusting the gain and offset of the digitizer. Some, for example, the Gould FD5000 (Tsien and Harootunian, 1990), can implement stages (2) and (3) by combining two images so that the input image may be corrected by subtracting a previously stored background picture, or multiplying by a shading-correction picture, to correct for uneven illumination of the specimen. Images can also be filtered spatially (stage [4]). Although they are "real-time," in the sense that any one of these operations can be completed within one frame-time, such corrections cannot be combined with capturing and filtering a new image, so they are not truly "real-time" in the context of a live experiment. The Joyce-Loebl Magiscan is able to perform real-time temporal averaging of video images, and combines this with an attractive package of purpose-written software for analyzing intracellular Ca^{2+} by off-line computation (*see below*).

Stage (5) requires the two images to be combined according to a formula derived from the spectral changes occurring when the dye is bound to the indicated ion, stoichiometry of the binding reaction, and the dissociation constant K_d. For fura-2 (and for any other dye with 1:1 stoichiometry and a spectral shift allowing dual wavelength ratio-imaging), the formula is

$$[Ca^{2+}] = K_d \, F \left(\frac{R - R_{min}}{R_{max} - R} \right) \tag{1}$$

where K_d is the dissociation constant for Ca-Fura; F is the ratio of fluorescences excited at the longer of the two wavelengths in use (usually 380 nm), respectively, in the presence of zero and saturating (1 mM) Ca^{2+}; R is the ratio of fluorescences from the cell at the two wavelengths; and R_{min} and R_{max} are values of R recorded with zero and saturating Ca^{2+} (Grynkiewicz et al., 1985; Fig. 4).

For a genuine "real-time" image of intracellular Ca^{2+}, this formula needs to be implemented at video rate by a high-speed image-processor. Given the specialized nature of the calculations, this means inevitably that the image-processor has to be fully programmable. The system in our laboratory (Fig. 1; Cheek et al., 1989a,b; Corps et al., 1989; O'Sullivan et al., 1989) is based on the *Imagine* frame-store [Synoptics Ltd., Cambridge, UK], which can perform several arithmetic operations on each pixel, as it passes through the processor. The ratio-imaging program runs synchronously with the video signal and lasts for two frame-times. During the first frame, the camera delivers an image obtained with illumination at 380 nm. *Imagine* combines this recursively with the previously stored 380 nm image (*see* Section 3.4.1.) to form a filtered 380 nm image, which is stored with 16-bit precision. During the second frame, the video image excited at 340 nm is combined with the stored, 380 nm image through a look-up table that is constructed in such a way that by using the two image-intensities as row and column indices into the table, an output value is obtained that implements Eq. (1) to represent log [Ca^{2+}]. This output value is combined, again by recursive filtering, with the ratio image that is stored in a different area of video memory, and is continuously output by *Imagine* as a monochrome video image to the output monitor, and also to an "Umatic" videotape recorder that can record up to 60 min of data, for later reprocessing. Because the calibration values are all incorporated into the ratio look-up table, which is calculated in advance, the function of the host computer at this

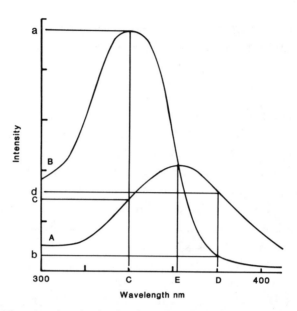

Fig. 4. Calibration for dual-wavelength ratio-imaging with fura-2. Fluorescence excitation spectra are plotted schematically for conditions of (A) zero and (B) saturating $[Ca^{2+}]$. C and D are the two wavelengths used for ratio-imaging (normally 340 and 380 nm). Wavelength E is the isosbestic point, where fluorescence is independent of $[Ca^{2+}]$. Parameters for Eq. (1) *(see text)* are $R_{max} = a/b$, $R_{min} = c/d$, $F = c/b$.

calculated in advance, the function of the host computer at this stage is limited to loading *Imagine's* look-up tables with the precalculated data, initializing and labeling the video display, including a gray-scale "wedge" along the bottom to give the Ca^{2+} scale, and starting the *Imagine* programs.

The disadvantages of real-time ratio-imaging with *Imagine* are, first, that precalculation of the ratio algorithm does not allow corrections, such as for autofluorescence or the effects of intracellular viscosity (Section 4.2.3.), which can be made after the experiment if live ratio-imaging is not used. It is, however, possible to incorporate a shading correction, to allow for non-uniformity in the distribution of UV light throughout the image area (Section 3.2.). Another possible source of difficulty is the inevitable loss of quality in playing back the videotape, as compared with the absolute fidelity with which digitally stored images can be retrieved. Care is needed to ensure that *Imagine* redigitizes the recorded material onto the correct scale, and there can be some loss of horizontal resolution caused by the limited bandwidth of the recorder, but we consider this a small price to pay for having the results of our experiments live on the screen, albeit only in monochrome. Clearly, any loss of accuracy in video-recording could be avoided for short experimental sequences, by recording instead on high-speed optical disks or in mass-storage video memory.

3.3. Image Display

Display of single-cell images, either digitally recorded, or redigitized from videotape, can in principle be done in monochrome: Only one parameter has to be displayed, namely, intracellular Ca^{2+} (for example), with 256 possible values for each pixel. In practice, this cheap and easy option is seldom satisfactory, because the eye can only readily distinguish about eight gray levels in a monochrome picture. Most users resort to a universal feature of image processors, whereby the output from an image can be split into three streams, controlling separately the red, green, and blue guns of the video monitor. The video signal for each stream is derived by using the pixel intensity as an index into one of three suitably programmed, 1-D look-up tables. The streams may be output on separate wires (RGB format), or combined into a composite video signal using one of the international standard formats called NTSC (America and Japan), PAL (Europe), or SECAM (France), but in any case, the experimenter can cause the value of each pixel to be encoded into a color, producing a pseudocolor image where colors graduate with increasing intensity, from black through blue, green, yellow, red, and white, to use the most common convention. Interpretation of the color code is aided by including in the picture a "wedge" of uniformly graded intensity (*see* Fig. 5), which can be labeled with appropriate concentrations.

As an additional refinement, having encoded the brightness of a ratio image into pseudocolor, one can encode a second parameter, for example, the average intensity of the two component images, into brightness in the pseudocolor image

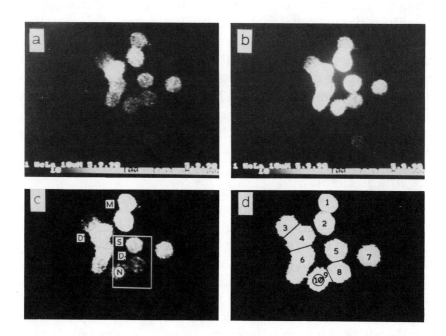

Fig. 5. Analysis of a field of HeLa cells, to give traces of mean intracellular [Ca²⁺] as a function of time. (Unpublished data kindly supplied by M. D. Bootman.) (a) Field of cells, showing basal [Ca²⁺], before application of agonist. Note the gray "wedge" below, giving the concentration scale. (b) The same field, showing initial transient response to 10 μM histamine. (c) Central portion from the same field, showing how cells are defined. The two cells at the top (M) have been defined manually by using the cursor to draw outlines, which the computer then fills in; the cell at the bottom (N) has been subdivided by the same method into central and peripheral regions; touching cells have been separated by drawing lines (D) between them; the region in the box (S) has been smoothed by replacing each pixel with the average of five nearest neighbors. (d) The computer then analyzes the field, using a simple threshold criterion to distinguish cells from background. For manually-defined cells the threshold is set at maximum brightness; otherwise, the threshold can be determined automatically by surveying representative "cell" and "background" areas. The result is a "segmented picture" showing the outline of each cell, together with its identification number. Note the more regular outlines of cells 1 and 2 (defined manually), and 5 and 8 (smoothed region). "Cells" 9 and 10 correspond to the separated peripheral and central regions of the lower cell. The tape can now be replayed, and identified cells scanned by the computer, to produce [Ca²⁺] traces.

splitting each 8-bit pixel, for example, into five bits for the ratio, plus three bits for the average fluorescence intensity. The ratio, and hence the output color, can thus have only $2^5 = 32$ possible values, and the average, controlling output brightness, can have $2^3 = 8$ values. The algorithm used to compute the ratio image, or the ratio look-up table, is adjusted to fit the ratio values into a scale of 0–31, and then add to each a multiple of 32, corresponding to the average intensity. For display, each video output look-up table is compressed and repeated eight times over the pixel range of 0–255. The advantage of such a dual-parameter display is that, as well as the intracellular Ca^{2+}, one also has some indication of the dye-distribution within each cell, so that checks can be made for dye leakage, or sequestration into subcellular compartments or organelles (*see* Section 4.4.1. *below*). If Mn^{2+}, which quenches fura-2 fluorescence, is used as a tracer for Ca^{2+} entry through plasma-membrane channels, the quenching can also be observed directly while still monitoring intracellular Ca^{2+} from the ratio image. The "average" field can be used to represent the quantity

$$I_{340} + I_{380} \cdot \left[\frac{(R_{max} - F \cdot R_{min})}{F - 1} \right] \tag{2}$$

which should be proportional to net fura-2 concentration (Moore et al., 1990; cf Eq. [1]). By comparison with fluorescence from saline droplets under the same conditions of illumination, this can be converted to an absolute measure of fura-2 loading in each cell.

The disadvantage is loss of potential accuracy in the computation and display of the ratio image, but in view of the other limitations on accuracy of the method, this is probably not serious. Clearly, one can choose greater accuracy in the ratio image, for example, by devoting six bits (64 values) to it, leaving only two bits (4 values) for the average.

Pseudocolored images make fine slides, and they can be translated into hard copy either by pointing a film camera at the video monitor, or more conveniently by using machines, such as the Polaroid Freeze-frame, which will convert a video signal directly into a film image, without any of the distortion or stray reflections that can creep into off-screen photographs. Publication of the prints is expensive, and monochrome or line-draw-

directly into a film image, without any of the distortion or stray reflections that can creep into off-screen photographs. Publication of the prints is expensive, and monochrome or line-drawing representations are often preferred. It is possible to make a contour-map of a cell, and some image-processors will do this, but a more common method is to get the computer to construct a perspective view of a 3-D "model" or "Y-mod plot," whose height above base represents the measured Ca^{2+} at each point. If the angle of view is chosen so that there is not too much dead ground, the Y-mod plot contains all the information that is in the pseudocolor picture. Given the necessary software, monochrome pictures and graphs can also conveniently be translated into a graphics output language, such as PostScript or HPGL, for immediate reproduction on a laser printer.

Most image-processors include in their video memory one or more "overlay-planes," each representing one bit per pixel, which can be set or reset independently of the main pixel value, to provide for line-drawing or annotation. Although such annotation will appear on a video-recording, it is not part of the image, and tends not to be stored along with it.

3.4. Further Processing

A 512 × 512 image contains 262 144 pixels, each with 256 possible values. Most of this information is not significant, and we find that at least as much time is spent in extracting data from experimental records as in doing the experiments. Most of the published literature on image analysis, as distinct from single-cell imaging as such, is devoted to objects such as electron micrographs or satellite images, in which noise-reduction and pattern-recognition are the name of the game. Noise-reduction is important in single-cell imaging, and will be dealt with briefly in this section. Automatic pattern-recognition is much less so, and I shall assume that apart from possibly using the computer to recognize the difference between cells and background, the experimenter will recognize cellular structures visually, perhaps from images of the cells taken after staining them at the end of the experiment, and will define them manually during analysis of the fluorescence images.

the image will have bright or dark specks continually flashing. In bad cases these fluctuations completely obscure the image, resulting in a "snowstorm" picture. Noise in single-cell imaging arises from three main sources: electronic and dark-current noise in the videocamera; quantization noise owing to the limited resolution in digitizing the video signal; and at very low light levels, "shot noise" owing to statistical variation in the number of photons contributing to each pixel. Camera noise arises from electronic components, with an important contribution from the image-intensifier, if present. Quantization noise is relatively unimportant in most single-cell imaging applications. Shot noise obeys Poisson statistics with a variance equal to the mean photon-count per pixel in each frame. In our system, using a Photonic Science Extended ISIS CCD camera, light levels can be used that result in <100 photons/pixel, giving standard deviations of 10% or more as a result of shot noise, which becomes the principal source of noise in the image. In ratio-imaging, the effect of camera noise is exacerbated by the use of two independent images to form each ratio. Also, the nonlinear formulas used, for example, for ratio-imaging with fura-2 (cf Eq. [3] *below*), can result in a ratio image in which the intensity distribution caused by noise is markedly skewed, so that simple calculations of mean and standard deviation fail to give reliable estimates of mean [Ca^{2+}]. A simple computer simulation of ratio image formation from two, equally noisy sources shows, for example, that if the standard deviation of either input image exceeds about 20% of the mean, then the mean value of the computed ratio begins to fall well below the value expected from the mean of each input image, even when large numbers of pixels are averaged. This is partly because, as the standard deviation of either input increases, a larger proportion of pixel values falls below threshold and produces "ratio" results of zero, but it is in part owing also to skewness in the output distribution, resulting from the nonlinearity of the ratio algorithm. Moore et al. (1990) have investigated the propagation of noise through their ratio-imaging system in some detail, giving a theoretical formula for the standard deviation (SD) of [Ca^{2+}] measurements, as a function of that of the measured fluorescence ratio R, as

tem in some detail, giving a theoretical formula for the standard deviation (SD) of $[Ca^{2+}]$ measurements, as a function of that of the measured fluorescence ratio R, as

$$SD[Ca^{2+}] = K_d \cdot F \cdot \left(\frac{R_{max} - R_{min}}{R_{max} - R^2} \right) \cdot SD_R \qquad (3)$$

Clearly, SD_R will be minimized if both components of the ratio are kept large, and hence easy to measure; also the error increases sharply at higher $[Ca^{2+}]$, when R approaches R_{max}. Choice of UV light intensities, such that experimental values of R are roughly evenly distributed around unity, is advantageous. For some measurements, it has been suggested that selection of excitation wavelengths may also be tailored to avoid very low fluorescence brightnesses, but Moore et al. (1990) show quite clearly that for the most accurate measurements over the widest concentration range, wavelength-pairs giving a wide range of ratios are most satisfactory, provided that the dynamic range of the camera is sufficient.

It is thus important to minimize noise in the input images, before the ratio is formed. The commonest method is to summate or average a set number of images at each wavelength: if n images are averaged, the signal/noise ratio is reduced by $\sqrt{n}$. This type of temporal filtering is referred to as "finite impulse response," because a step-change in the input will be transferred completely to the output during a time not greater than twice the duration of the averaging period. For systems in which time resolution is of secondary importance, or in which it is governed in part by the response-time of the camera, it is the simplest and probably the most suitable method of noise-reduction.

In systems where the ratio image is continuously computed, finite impulse response filtering may degrade the time-response unacceptably, and in this case, it can be shown (Tsien and Harootunian, 1990) that the best compromise between temporal resolution and noise-reduction is the "recursive filter," whereby a running average of each image is maintained, according to the formula

$$I_{n+1} = a\,I_{in} + b\,I_n \tag{4}$$

where I_{in} is the input video intensity, and I_n and I_{n+1} are existing and updated pixel intensities, respectively, in the stored image; a and b are constants, such that a + b = 1. If the time-interval between successive frames at the same wavelength is t, then it can be shown that the averaged image will respond exponentially to a step-change in input, with a time-constant equal to a/t. We generally use a time-constant of 200 ms for the ratio image.

Recursive filtering, with switching between wavelengths as frequently as the hardware will accommodate, has the added advantage that information acquired at the two wavelengths is as nearly as possible contemporaneous. Thus, averaging 32 frames, then changing wavelengths, allowing for possible camera lag, and then averaging another 32 frames may result in a ratio image derived from 340 nm excitation at time t, and 380 nm excitation at time $t + 1.5$ s. If Ca^{2+} is rising, for example, this will result in a consistent underestimate. By using recursive filtering, with wavelengths changed every frame, one continually compares fluorescent images separated in time by no more than 33 ms. Although the response of the output image may still be damped by the filtering, systematic errors owing to rapid changes are minimized.

3.4.2. Data Extraction

Manipulation of captured ratio images to extract local average concentrations or gradients, and to examine changes by subtracting background or baseline images, is a matter of standard image-processing: Most software packages provide facilities for such operations. Clearly, the output from imaging systems that provide a continuously varying ratio image can be reduced to a series of static pictures by "grabbing" frames at timed intervals; timing may be assisted by building event-markers into the record, for example, by applying a frequency-modulated tone to one audio channel of the videotape recorder. However, to make best use of the "live" ratio image, data, such as local averages, need to be extracted continuously from the image as rapidly as possible, and the results stored so that they can subsequently be analyzed and converted to graphical form. Our system operates as follows, using programs written under the SEMPER 6 image-processing package (Synoptics Ltd., Cambridge, UK).

The record from each experiment consists of a videotape, with recorded "live" ratio images. This is played back through *Imagine*, using a simple program that provides a pseudocolor display, with optional further recursive filtering. Figure 5 illustrates stages in the analysis: Cells (or areas within cells) are first defined by analyzing a representative image grabbed from the tape. This can be done automatically, using an intensity-threshold chosen by surveying sample "cell" and "background" areas, or manually by drawing round each cell with a pointing device, such as a "mouse" (this is sometimes best done on a transmitted-light image of the field of cells, taken before the start of the experiment, since the physical outlines of cells are not always distinguishable from the fluorescence alone). In either case, there is an option to separate overlapping cells by drawing lines between them with the mouse; this option can also be used to subdivide one cell into segments, in order to follow the progress of a Ca^{2+} "wave" across the cell, for example, or into central and peripheral zones (Fig. 5c). The program analyzes the image, assigning a number to each separate cell or region.

The tape is then replayed, as often as necessary, and up to five cells are scanned in a sequence, repeated at regular intervals. The interval between scans can be as small as 0.3 s, depending on the total number of pixels to be scanned and the capability of the host computer. (Dual-ported video memory, which gives the fastest possible access by the host computer by allowing data to be simultaneously written from video input and read by the host, is clearly an advantage here.) Each scan produces a set of mean (or median) values for the ratio image intensity, which are printed on the console screen and stored in a disk file, together with timing information, identification numbers and areas of cells scanned, and an event-marker trace derived from one audio channel on the tape. At a later stage, the image-processor converts the stored values for selected cells into graphs, for reproduction photographically or on a laser printer.

This method loses some time-resolution, compared with the 0.2-s time-constant used to construct the original ratio image, but this is rarely a serious limitation. It also presents difficulties if the cells move within the field of view during a scan. Some check can be kept on this by outlining the cells on the screen, using an overlay plane that is not part of the playback image; also, the scanning program can be set to disregard any pixels whose intensity falls below a given threshold—usually the same threshold that was used to differentiate cells from background during the initial analysis—so that although scanning is always restricted to within the predefined outline, if a cell moves, the survey will include only that part of the space within the outline that is still occupied by the cell. The included percentage is printed in the screen-display during scanning. We do not have, as yet, a system that will track the movements of contractile or motile cells.

4. Quantitative Interpretation of Single-Cell Images

Quantitative interpretation of single-cell images proceeds in three stages: calibration of the measuring equipment, to ensure accurate measurement of the fluorescence itself; correction for background, shading, and autofluorescence of the cells; and calibration of the indicator dye, to convert fluorescence measurements into $[Ca^{2+}]$, pH, and so on. The first two of these are straightforward physical measurements; the third is mostly physical chemistry, but because we are short of information about the detailed properties of the cytosol, and, in particular, about its effects on the fluorescence spectra of indicator dyes, a lot of guesswork and empirical measurement have to be used. In fact, such are the uncertainties that, in this author's view, absolute measurement, for example, of intracellular Ca^{2+} with fura-2, can be no more than semiquantitative at the single-cell level. Fortunately, this is not too severe a defect, because changes in cytosolic $[Ca^{2+}]$ resulting from physiological stimuli tend to be several times larger than the basal level, and are easily detected. Basal Ca^{2+} levels are probably correct within 50%, and the relative magnitudes of changes can be assessed with similar precision; absolute basal levels are still best determined on populations of cells in a fluorimeter, and it is the broad agreement between such measurements and the results of single-cell imaging that gives some confidence in the latter's findings.

4.1. Instrument Calibration

Quantitative imaging by epifluorescence is not generally done by making absolute measurements, either of the exciting UV light intensity, or of the intensity of the fluorescence itself. Rather, fluorescence intensities are compared between live cells and calibration solutions, or cells treated with ionophore to control cytosolic ion activities (see below), or in the ratio method, between outputs from live cells excited or monitored at different wavelengths. Absolute photometric measurements are thus not required, and calibration reduces to checking that the UV light source is adequately stable, and that the recording system is linear, so that ratios between fluorescence intensities are faithfully reported.

Instability of the UV light source is not usually a serious problem: Typically, the intensity from a high-pressure xenon arc lamp, driven from a stabilized DC power-pack, may fluctuate by a few percent over a period of minutes, but errors resulting from this are insignificant in comparison with those from other sources, particularly if the ratio method is used. Marks and Maxfield (1990) report stability of a mercury arc lamp as better than 1% over 5 min. More seriously, the arc may wander physically within the discharge tube. Arc wander is most often seen in aging lamps, and can cause larger fluctuations in UV light intensity, and uneven distribution of UV light across the field of view, a problem that is particularly noticeable with grating monochromators. Regular checks may be made using a standard fluorescent specimen, which can be a solution of fluorescent material, such as rhodamine B.

The videocamera is the experimenter's "eye," and linearity of response is clearly crucial, especially in ratio-imaging. Linearity is checked by imaging a standard fluorescent object, using a series of graded UV light intensities obtained with calibrated neutral-density filters (Goldman et al., 1990; MacGlashan, 1989), or by adjusting monochromator slits to give fluorescent outputs previously calibrated by recording from the same specimen with a photomultiplier. MacGlashan (1989) also used calibrated fluorescent beads, and separately compared the intensity profile obtained from a single, large fluorescent sphere with that which was theoretically expected. Generally, CCD cameras are approx linear over an intensity range of about 15:1, giving digitized outputs in the range of 10–180 (on a scale of 0–255). Using monochromator settings previously calibrated with a photomultiplier, we find that our ICCD camera is linear over the range of 10–220. Combined with the limitations of 8-bit digitization (Section 3.1.), this dynamic range severely limits the accuracy with which changes in intracellular Ca^{2+} can be measured. For example, using fura-2 with excitation at 340 and 380 nm, intensity-ratios may be expected to vary over a range of some 7:1, since the cytosolic $[Ca^{2+}]$ changes from 70 nM to 1 μM. For best results, camera gain is adjusted by reference to the video signal using an oscilloscope, and relative UV light intensities are set beforehand—for example, by adjusting monochromator slit-widths—to give actual ratios evenly spaced around unity. It is also possible to control the dynamic range of the measurements by using a different pair of excitation (or emission) wavelengths. SIT cameras (Jacob, 1990) have a similar dynamic range, but their linearity is inferior to that of a CCD.

4.2. Background, Shading, and Autofluorescence Corrections

Background radiation reaching the videocamera comes from stray light, from fluorescence by components in the microscope and experimental chamber, and possibly from fluorescent compounds in the extracellular medium. To the extent that all these sources are constant in intensity and distribution, it is sufficient to record a background image, or a pair of images for dual-wavelength work, using a chamber containing only extracellular medium, and the same UV light intensities as are to be used for measurement. The background image is then to be subtracted from each experimental image, before the ratio image is calculated. If this is not done, then the recorded intensity-ratios will be brought closer to unity. The effect of this on indicated Ca^{2+} levels depends on the relative UV light intensities in use (*see below*): Generally, the tendency will be toward a low-to-medium value—in our system about 70 nM—so that basal Ca^{2+} levels will be little affected, but stimulated increases will be underestimated. Figures for the absolute value of background fluorescence are seldom quoted in the literature, but it usually amounts to <5% of total light reaching the camera. Note, however, that at the extremes of the Ca^{2+} scale, fura-2 fluorescence at one of the two wavelengths may become very weak, so that background radiation becomes much more important. Thus in our system, the background from a saline-filled chamber amounts to about 1.3% of the fluorescence from 2 µM fura-2 in zero Ca^{2+} at 380 nM, or in saturating Ca^{2+} at 340 nM; but in zero Ca^{2+} at 340 nM the background accounts for 4.4% of the signal, and in saturating Ca^{2+} at 380 nM, fura-2 fluorescence is so weak that the background reaches 13% of total radiation.

Where background is spatially uniform, so that a constant value needs to be subtracted from each pixel intensity, this can be effected simply by adjusting the values of R_{min} and R_{max} in the ratio formula (Eq. [1]), or by using R_{min} and R_{max} recorded directly using an experimental chamber.

Shading corrections are needed where the intensity of the exciting radiation varies significantly across the field of view. Such variations will, of course, cancel out if the ratio method is used, except where there are differences in UV light distribution between the two wavelengths. In such cases, which are more likely when monochromators and a chopper mirror are used, than with band-pass filters on a filter-wheel, the procedure is to divide (or multiply) one component image, pixel by pixel, by a shading-correction picture that is the ratio of two images obtained by illuminating the same fluorescent standard from each UV light source, with both sources set to the same wavelength. The standard can be a film of fura-2 solution, rhodamine-B, or other fluorescent material, and need not be uniformly fluorescent, though a roughly even distribution of fluorescence is convenient for accurate calculation. In general, there will be regions where the shading correction should reduce the intensity of the corrected image, and regions where it should be increased. In order to prevent the corrected image from overrunning the 8-bit (0– 255) data range, the mean value of the shading-correction picture needs to be adjusted (Takamatsu and Wier, 1990), so that all pixel values in the input image are either reduced or unchanged. The other component image is then similarly scaled, or R_{max} and R_{min} are adjusted, before computing the ratio image.

The extra computational burden imposed by shading correction makes it difficult to incorporate into live ratio-imaging systems. The *Imagine* system is capable of carrying out a shading correction in real time, with limited precision (up to 32 gray levels in the correction picture). Without this correction, and with careful adjustment, UV light intensities from the two sources can be identically distributed over a central area 400 pixels in diameter, within better than 15%. The effect of such errors on the measured value of [Ca²⁺] is concentration-dependent, but inspection of calibration curves (Section 4.3.3.) shows that for [Ca²⁺] above 70 nM, the maximum error is nearly constant at 15–20%.

An additional source of background is autofluorescence from components within the cell itself. Autofluorescence is more difficult to correct, because its intensity and distribution will vary from cell to cell. Clearly, its significance is reduced by using heavier loadings of indicator, which will also improve accuracy by giving a more noise-free signal, but at the expense of possible toxic side effects and, in particular, of possible buffering of cytosolic [Ca²⁺] (Section 4.4.4.). Because the emission spectrum of autofluorescence is usually quite broad, it can also be reduced by using a narrow-band interference filter to limit the fluorescence reaching the camera: For fura-2, a 510 nm band-pass filter with 10 nm between 50% transmission points is appropriate.

For population measurements, autofluorescence can easily be assessed by examining the cells prior to loading with dye, to produce a combined figure for autofluorescence plus background. With single-cell imaging this is not usually possible, and one must rely on subtracting either an average value from unloaded cells, or a postexperiment value obtained by suppressing the dye fluorescence. With fura-2 this can be done by quenching with Mn^{2+}, or alternatively, cells can be permeabilized with digitonin so that the dye leaks out and is removed by the perfusion system (hopefully not accompanied by other fluorescent cell constituents). Clearly, this latter approach is not applicable to live ratio-imaging, in which all calibrations need to be done in advance, to establish the ratio algorithm. (In principle, a ratio image that also encodes average intensity (*see* Section 3.3. *above*) could be deconvoluted to restore the individual dual-wavelength fluorescences, these could be corrected for autofluorescence, shading, or anything else, and the ratio image recomputed; but we have not, so far, made any serious attempt at such an ambitious calculation!)

The importance of autofluorescence varies widely with cell type. When imaging with fura-2 in most mammalian cells, it is not a serious problem: Lindeman et al. (1990) quote values of 3–5% of total fluorescence from smooth muscle cells, and Takamatsu and Wier (1990) found background plus autofluorescence to be <8% of the total from cardiac myocytes. Correction for autofluorescence was not found necessary in imaging from 3T3 fibroblasts (Tucker et al., 1990), adrenal chromaffin cells (Cheek et al., 1989a), or sympathetic neurons (Lipscombe et al., 1988), and in smooth muscle cells (Goldman et al., 1990; Ozaki et al., 1990), only modest corrections were needed. With some invertebrate tissues, however, autofluorescence can present serious difficulties, and an attempt to image intracellular Ca^{2+} in intact blowfly salivary glands (Cheek et al., unpublished) was unsuccessful because autofluorescence largely swamped the fura-2 signal. Not all insect tissues suffer from this difficulty: Preliminary experiments in our laboratory (unpublished) have shown that useful images of intracellular Ca^{2+} can be obtained from Malpighian tubules of *Drosophila* and from isolated cockroach neurons in culture. An obvious precaution in all fluorescence measurements (Moore et al., 1990) is to check that fluorescent components are excluded from cell culture media, if necessary transferring cells to a nonfluorescent medium, and allowing enough time for fluorescent materials taken up by pinocytosis to clear from the cells, before starting the experiment.

4.3. Calibration of Dyes

Equation (1) (Section 3.2.) provides a general calibration formula for any dye with 1:1 stoichiometry and spectral properties suitable for dual-wavelength ratio-imaging. It requires four parameters: K_d, R_{min}, R_{max}, and F, of which all, except K_d, can be obtained from spectra at the two wavelengths, in the presence of zero and saturation activities of the measured ion. Figure 4 illustrates the necessary measurements for fura-2; for a dual-emission dye, such as indo-1, the same scheme can be used, but with emission wavelength as abscissa. Neher (1988) gives a convenient procedure for constructing fura-2 standards with zero and saturating Ca^{2+}, both at the same pH and ionic strength, for this purpose. To measure K_d, a series of carefully constructed Ca^{2+} buffers is needed, using EGTA of high purity; alternatively, published values can be used: For fura-2, these are 135 nM at 20°C, or 224 nM at 37°C (Grynkiewicz et al., 1985). Note that F is a property of the dye alone. R_{max} and R_{min}, being computed from measurements at two different wavelengths, include the ratio of UV light intensities reaching the dye, which depends on both the relative amounts of UV light reaching the microscope, and the transmission of the microscope optics at the two wavelengths. For example, the Nikon UV-F objectives transmit significantly less UV light at 340 nm than at 380 nm, which, coupled with the shape of the emission spectrum from a xenon arc, means that in our system only about half as much UV light reaches the specimen at 340 nm, as compared with 380 nm. The ratio R_{max}/R_{min}, on the other hand, should be instrumentation-independent, and may be compared with values given in the literature to test correctness of the measurements and purity of the fura.

4.3.1. In Vitro Calibration

The simplest method of calibration is to use buffer solutions of $CaCl_2$ and EGTA (e.g., Miller and Smith, 1984; Williams and Fay, 1990), placed in bulk in the experimental chamber or trapped as layers between cover slips (Tsien and Harootunian, 1990). Thickness of the layer should be well in excess of the depth of focus on the objective: A layer of solution about 1 mm thick is usually deep enough so that the fluorescence intensity recorded from it is almost independent of the plane of focus. For consistency, small glass beads may be used to focus at a known depth into the solution, the stage being moved so that no beads are in view, before taking measurements. We use a photomultiplier to read average intensities across the field for calibration, rather than the videocamera, to avoid inaccuracies resulting from the latter's limited dynamic range. Using 2 μM fura-2 in 150 mM KCl at pH 7.2 and 20°C, we measure R_{min} = 0.38, R_{max} = 15.3, F = 12.1. After correcting for background (measured in the same chamber, with no fura-2), we find R_{max}/R_{min} = 44.8; these data may be compared with F = 15.3, R_{max}/R_{min} = 45.7, reported by Grynkiewicz et al. (1985). Note that determination of R_{max} and R_{min} is independent of the fura-2 concentration, true to the principle of the ratio method. Determination of F, however, requires two solutions with equal fura-2 concentrations, or alternatively, that the relative concentrations of fura-2 be measured, for example, by comparing the fluorescence at 360 nm excitation, which should be independent of [Ca^{2+}] (*see* Poenie, this vol.). Note also that using the experimental chamber for calibration automatically incorporates the effects of instrumental background, and any effects owing to attenuation of UV light by the cover slip, into the measured values for R_{min} and R_{max}.

4.3.2. In Situ *Calibration*

The major drawback of such "in vitro" calibration methods is that the properties of intracellular indicator dyes are known to be affected by factors, such as ionic strength, pH, polarity, and viscosity of the medium, and also by the presence of proteins that bind the dyes, altering both their fluorescence spectrum and their dissociation constants. Thus, calibration data obtained with extracellular fura-2, for example, may be a poor model for its behavior in the cytosol, whose ionic strength and pH can be assessed, but whose polarity and viscosity are difficult to estimate, and where the status of proteins in solution is also a matter for conjecture. Moore et al. (1990) discuss some of the factors involved, and assess their likely contribution to errors in measured [Ca^{2+}]: One answer is to calibrate the fura-2 "*in situ*," by using a cocktail of ionophores to permeabilize the cells, so that cytosolic $[Ca^{2+}]$ can be controlled from outside to give empirical values for R_{max} and R_{min}. The ionophores most commonly used for this purpose are ionomycin or Br-A23187 (a nonfluorescent analog of the common fungal ionophore). Williams et al. (1985), for example, used 1–2 μM ionomycin to equilibrate intracellular Ca^{2+} in smooth muscle cells, with a range of extracellular Ca^{2+} buffers using EGTA. Intensity-ratios for fura-2, using excitation wavelengths of 340 and 380 nm, were reduced in the intracellular measurements, such that R_{min} was little changed, but R_{max} was reduced by about 5%. If these values are taken as representing the true intracellular behavior of fura-2, then using the extracellular values would cause $[Ca^{2+}]$ to be underestimated by about 30% over the range of 100 nM to 2 μM; at higher Ca^{2+} activities, the error increases sharply, as the curve relating fluorescence-ratio to Ca^{2+} activity flattens out at $[Ca^{2+}]$ in excess of K_d.

The success of such calibraton procedures obviously depends on how accurately the intracellular Ca^{2+} activity can be controlled. Williams and Fay (1990) describe a detailed protocol for equilibrating suspended smooth muscle cells, with extracellular Ca^{2+} buffers in a fluorimeter. To control intracellular Ca^{2+} activity, the ionophore must swamp all controlling active transport processes, including the ATPase-driven Ca^{2+} pump, and passive Na^+/Ca^{2+} exchange. Of these, the latter is driven by the Na^+ gradient across the plasma membrane, which can be reduced by promoting Na^+/H^+ exchange with the antibiotic monensin, and reducing the K^+ gradient by promoting K^+/H^+ exchange with nigericin, possibly with the addition of ouabain to further inhibit the Na^+/K^+ exchange pump. Thus, the cells are first treated with a mixture of monensin and nigericin, in a suspension medium whose ionic strength and viscosity are adjusted to resemble, as closely as possible, that of the cell cytosol. Fura-2 fluorescence should be unaffected by this treatment, confirming that the geometry of the cells has not changed. Ionomycin or Br-A23187 is then added progressively, taking care to ensure adequate dispersal of ionophore throughout the medium, until a constant fluorescence ratio is attained, at which point the intracellular Ca^{2+} activity is assumed to be clamped at the extra-cellular value. At this point, the fluorescence ratio was generally found to be within 5% of that recorded from fura-2 in the same calibration medium, without cells.

This procedure is not universally successful, however. Thus, Roe et al. (1990) report considerable difficulties in using iono-mycin to control intracellular Ca^{2+} in hepatocytes or in neonatal cardiac myocytes, since the necessary concentrations of iono-mycin caused partial permeabilization of the cells, resulting in loss of fura-2 into the bathing medium. Tucker et al. (1990) also report difficulty in equilibrating intracellular Ca^{2+} in 3T3 fibro-blasts or human FeSin mesenchymal cells using ionomycin: Intracellular Ca^{2+} activities, calculated using calibration values for fura-2 derived from measurements on Ca^{2+} buffers, reached peak levels of about 250 nM with ionomycin, but then declined rapidly to a plateau at about 70 nM. In our own experiments with chromaffin cells (unpublished), we have also found diffi-culty in achieving control over intracellular Ca^{2+}, without using concentrations of ionomycin high enough to cause loss of fura-2 from the cells.

4.3.3. Environmental Effects on Dye Performance

Given the uncertain success of *in situ* calibration and the lower inherent accuracy of quantitative single-cell imaging, as opposed to population determinations using a fluorimeter or photomultiplier, many authors judge it not worthwhile to attempt a separate calibration for every experiment, preferring instead to rely on in vitro calibrations performed in advance. To some degree, these can be made more realistic by adjusting the composition of the calibration media to resemble the cytosol, particularly regarding polarity and viscosity, which appear to affect the fluorescence of intracellular indicators by restricting the mobility of individual molecules. Direct measurement of these parameters in living cells is not easy, but their effect on fluorescence can be assessed by using exciting radiation that is polarized, and separately detecting fluorescence polarized par-allel and orthogonally to the excitation. This direct measure of the mobility of fluorescent molecules has been attempted by

Williams et al. (1985) and Williams and Fay (1990), who then constructed calibration media having similar equivalent properties. With fura-2, the general effect is to increase fluorescence excited at 380 nm, at the expense of the 340 nm signal, so that both R_{max} and R_{min} are reduced by amounts varying between 15–30% (Moore et al., 1990; Poenie, 1990; Roe et al., 1990).

The effect of errors in R_{max} and R_{min} on indicated values of intracellular Ca^{2+} can be assessed from Fig. 6, which shows indicated values of $[Ca^{2+}]$ as a function of the intensity-ratio R, for values of R_{max} and R_{min} close to those recorded in our system, and for values reduced by about 15%. The difference between the two curves may be taken as the extent to which $[Ca^{2+}]$ will be miscalculated if the values of R_{max} and R_{min}, used in the calculation, exceed the "true" values by this amount: The result is always an underestimate, with an error that is fairly constant at 25–30% for ratios between about 1 and 5, giving recorded $[Ca^{2+}]$ between 70 and 700 nM; outside this range the error increases progressively, exceeding 50% for $[Ca^{2+}]$, <20 nM or >2 µM. It is worth noting, however, that viscosity of the cytosol may also reduce the K_d of fura-2 (Roe et al., 1990), and that this will partly counteract the effect of reductions in R_{max} and R_{min}. Roe et al. (1990) consider the net effect on estimated $[Ca^{2+}]$ to be no greater than 20%. It is also found experimentally that errors resulting from viscosity can be reduced by using 340 and 365 nm as the two excitation wavelengths, instead of 340 and 380 nm. This procedure has the additional advantage of avoiding quantitation difficulties that arise at high $[Ca^{2+}]$, where the fluorescence excited at 380 nm is very weak, because the 365 nm fluorescence is almost independent of $[Ca^{2+}]$. However, this is at the expense of dynamic range in the measurements, because at high $[Ca^{2+}]$ the 340 nm fluorescence also becomes almost constant, whereas the residual fluorescence excited at 380 nm continues to decrease with increasing $[Ca^{2+}]$.

Roe et al. (1990) have also investigated the effects of environmental polarity on fura-2 fluorescence, using a range of organic solvents to replace water in the calibration media. Decreasing polarity causes the excitation spectrum to shift toward longer wavelengths—which would lead to a reduc-

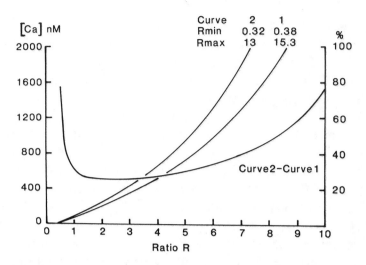

Fig. 6. Influence on measured [Ca²⁺], of a 20% reduction in R_{min} and R_{max} using the ratio-imaging formula of Eq. (1). The three curves show inferred [Ca²⁺] (left-hand ordinate) as a function of fluorescence intensity ratio, for the indicated values of R_{min} and R_{max}; and the percent error in [Ca²⁺] (right-hand ordinate), which would be caused by a shift in R_{min} and R_{max} between the two sets of values.

tion in 340/380 nm ratio—and the emitted fluorescence is shifted toward shorter wavelengths and a reduction in intensity. The polarity changes used to obtain these effects are quite extreme in biological terms, but nevertheless, the authors consider they may have some relevance, for example, to the behavior of fura-2 that has become bound to cytosolic proteins.

A minor source of possible error, concerned with the geometry rather than the physicochemical properties of the intracellular environment, is the so-called "inner filter" effect. Free fura-2 is more absorbent to light at 380 than at 340 nm; in high [Ca²⁺] the reverse is the case. Thus, depending on [Ca²⁺], one or other of the exciting wavelengths may be absorbed more strongly than the other, and hence have effectively a shorter optical path through the cell. Errors will result, because the ratio method assumes identical optical paths for the two wavelengths. This effect is discussed by Moore et al. (1990), who consider that the product of optical pathlength and fura-2 concentration should be kept below 12,000 $\mu M \cdot \mu m$ to keep differential attenuation of the UV light below 5%. For a 10 μm diameter cell, this corresponds to a fura-2 loading of 1.2 mM, which is well above the levels generally used in single-cell imaging.

4.4. Other Sources of Error in Quantitation

In the preceding section, I considered steps toward calibration of intracellular indicators, with the aim of obtaining absolute readings from single-cell images. Given careful calibration, it would appear that, at worst, estimates of intracellular Ca^{2+} activity using fura-2 and ratios calibrated by in vitro methods ought to be correct within about 30%. However, several further sources of possible error remain to be considered. First, dye loaded into the cells may not remain in the cytosol, and second, dye loaded by the ester method may not be completely hydrolyzed within the cells, so that the fluorescence signal is contaminated by emission from unhydrolyzed or partially hydrolyzed dye molecules, with different spectral properties. Neither of these problems is peculiar to single-cell imaging, but methods for handling them differ between single-cell and population experiments. A third factor is bleaching of dyes by prolonged exposure to UV light, together with damage to the cells caused either directly by the UV light, or by bleaching products. Finally, artifacts can arise if the cells are too heavily loaded with dye, so that enough dye is present to buffer changes, for example, in intracellular Ca^{2+}.

4.4.1. Intracellular Sequestration

Compartmentalization, or sequestration and binding of indicator dyes, is a problem that varies with cell type and loading conditions. In imaging applications, it is occasionally possible to observe directly as physical partition of the dye within the cell. Moore et al. (1990) give useful summaries of routes by which fura-2 can become sequestered or preferentially loaded into subcellular compartments. Membrane-permeable esters used for loading should be capable of penetrating all parts of the cell; the extent of accumulation in any one compartment depends on the presence of esterase activity, and under different conditions of temperature and incubation time, preferential loading has been recorded into nuclei and mitochondria (Steinberg et al., 1987), and into sarcoplasmic reticulum in muscle cells (Williams et al., 1985). Another important route for sequestration is the incorporation of dyes into pinocytotic vesicles (Malgaroli et al., 1987). Roe et al. (1990) developed a set of procedures to identify sites of sequestration, by labeling cells with fura-2 and also with fluorescent markers specific for known subcellular structures. Enzymic processes, and particularly pinocytosis, are temperature-dependent, and sequestration can often be avoided by adjusting the duration and temperature of loading: Roe et al. provide a table of best loading procedures for a variety of cell types.

Where macroscopic nonuniformities of dye distribution can be seen in the image, it is easy to discard badly loaded cells, or indeed to capitalize on the situation by aiming to load and study identified cell organelles specifically. Binding to diffuse cellular constituents is more difficult to detect: The best method is probably to permeabilize the cells at the end of the experiment, for example, with digitonin, and check that all the dye is released (Ryan et al., 1990). Clearly, sequestered dye is liable to be more or less inaccessible to changes in cytosolic [Ca^{2+}], so that the Ca^{2+} signal will be contaminated by a constant component, tending to damp the effect of any changes in [Ca^{2+}]. This effect will be indistinguishable from environmentally induced changes in spectral properties of the dye itself, and may well contribute to the observed discrepancies between in vitro behavior and *in situ* calibrations using ionophores. It is worth noting that more direct loading methods, such as electro- or hyposmotic permeabilization, microinjection, or infusion of dye from a patch-pipet in the whole-cell configuration, ought to be free, at least initially, from errors caused by sequestration. This advantage has to be balanced against the possible effects of disruption of the structure and composition of the cytosol, particularly in whole-cell patch-clamp, where major changes in ionic constitution, or even dialysis of the cytosol, may result.

A related problem to sequestration is the progressive loss of indicator dye from the cells during the course of the measurements. In imaging experiments, this is not a major source of error, since the dye is normally dispersed into a large volume of bathing medium as fast as it can leak from the cells. It can be a serious problem in population experiments, where the ratio of cells to medium is much higher, and in any case, it can result in progressive loss of signal, which can be ill afforded. Loss of fura-2 from cells results largely from nonspecific anion-transport processes, and some success has been obtained by using anion transport-inhibitors, such as probenecid (Di Virgilio et al., 1990). The effects of such treatment on other processes in cells remain to be studied in detail.

4.4.2. Incomplete Hydrolysis

Incomplete hydrolysis of esterified indicator dye within the cell may leave intact or partially hydrolyzed molecules whose fluorescence is independent of $[Ca^{2+}]$. To the extent that the excitation spectrum of fura-2/AM, for example, resembles that of fura-2 in the absence of Ca^{2+}, the contaminating signal from unhydrolyzed ester will have a generally depressing effect on all Ca^{2+} readings, particularly those at high $[Ca^{2+}]$. In population work, the dye can be released from the cells by controlled lysis after the experiment, and its spectrum recorded in a suspension-medium of known $[Ca^{2+}]$, for comparison with that of the free dye under identical conditions. The proportion of unhydrolyzed dye can then be calculated by subtraction of the two spectra (Roe et al., 1990; Oakes et al., 1988; Moore et al., 1990), and used to back-correct the spectra recorded from the live cells. This procedure is not available in single-cell experiments, because dye released from the cells by lysis is rapidly dispersed in the bathing medium, and the only resort is to record spectra from cells whose cytosolic $[Ca^{2+}]$ is controlled with ionophores. Again, there is no possibility of distinguishing between environmental degradation of dye response, and effects of sequestration or incomplete hydrolysis, so the outcome of such checks remains uncertain. Possibly the best method is to vary loading conditions until the recorded basal cytosolic $[Ca^{2+}]$ either reaches a constant value, or approaches that recorded from carefully controlled population measurements. Alternatively, basal $[Ca^{2+}]$ may be compared with that recorded with dye introduced directly by microinjection.

4.4.3. Photobleaching and Photodamage

Interaction of a UV light photon with a fluorescent dye molecule may result in irreversible damage rather than in emission of fluorescence. If the decay-product is not fluorescent, the result is simply a loss of signal, which will ultimately degrade the fluorescence image, but should not lead to quantitative errors if the ratio method is used. Unfortunately, photobleaching of fura-2 can produce a decay-product that is also fluorescent, but that responds to Ca^{2+} activity only in the millimolar range. At physiological $[Ca^{2+}]$, this results in a contamination of the fluorescence ratio, analogous to that introduced by incomplete hydrolysis of ester during loading. Becker and Fay (1987) have recorded excitation spectra from buffers containing partially bleached fura-2, and showed that bleaching of as little as 8% of the fura can cause $[Ca^{2+}]$ to be significantly underestimated, for example, by 20% at 200 nM.

Resistance to bleaching is an important factor in the design of intracellular fluorescent indicators, and generally, a fluorescent dye molecule can be expected on average to emit between 10^4 and 10^5 photons before it becomes damaged (but note much smaller values measured by Becker and Fay [1987] for fura-2). How long this will take depends on the intensity of illumination, which in turn depends, in practice, on the efficiency with which emitted photons can be collected. In cell population experiments, where fluorescence from a large bulk of cells is efficiently collected, UV light intensities can be kept very low and bleaching is not usually a problem. In single-cell imaging, however, photons from individual cells are collected, with an overall efficiency that is usually not much better than 1% (for example, a glycerol-immersion objective of 1.3 NA will collect about 40% of the emitted light, which, with the Extended ISIS CCD camera whose quantum-efficiency is about 10%, and some loss in the microscope optics, results in an overall quantum-efficiency of 3% at best), and divided among pixels that may be only 0.5 μm square, at 1000× magnification. Assuming for convenience a depth of focus of 4 μm, each pixel corresponds to an excited vol of 1 μm³, from which enough photons need to be collected every TV frame (33 ms) to provide a meaningful image. For a signal/

noise ratio of 10:1 (Section 3.4.1.), we need to collect at least 100 photons from each pixel, from which it can be calculated that in 1 min, molecules equivalent to a dye concentration of about 3 μM will be consumed. This is a purely hypothetical calculation; in practice, we find that with an ICCD camera, our system requires enough UV light to bleach 1.25 μM fura-2 in a test solution with a time-constant of 4 min, which is similar to that recorded by Marks and Maxfield (1990) from fura-2 loaded neutrophils. With the Extended ISIS camera (Photonic Science, Robertsbridge, UK), improved sensitivity allows the UV light intensity to be reduced about fourfold, leading to a bleaching time-constant on the order of 16 min. At this rate, and with careful avoidance of unnecessary exposure to the UV light, for example, by using an aperture-stop (Fig. 1) to limit exposure to the area of interest, it should be possible to observe for many minutes without serious loss of signal. Errors owing to fluorescent decay-products seem likely to be more serious, and although Becker and Fay (1987) have indicated a correction scheme based on additional measurements at a third wavelength close to the isosbestic point to monitor the extent of the bleaching, such measurements will inevitably introduce further noise, and will themselves increase the damaging UV light exposure. In practice, we find that within the limited accuracy of single-cell measurements, recorded values of basal cytosolic [Ca^{2+}] remain remarkably constant over periods of 30 min or more under continuous exposure to UV light.

Photobleaching may also result in toxic byproducts that damage the cell. Such damage, and of course direct damage caused by interaction of UV light with other cellular constituents, is more difficult to observe and control. Occasionally it may be obvious, as was found by Moore et al. (1990) with muscle cells, which contracted spontaneously after a certain exposure. Experimental protocols should be designed to make obvious any abnormalities in cell behavior that may appear progressively during continuous or repeated UV light exposure.

The control of bleaching artifacts requires the UV light intensity to be kept as low as possible, and this in turn implies that collection of fluorescent photons should be made as efficient as possible, by using high numerical aperture objectives and reducing the number of absorbing optical elements in the pathway for emitted light. Heavier loading of cells with fluorescent dye will increase the output for a given UV light intensity—though it will also increase the extent to which cytosolic [Ca^{2+}] is buffered by the dye (*see* Section 4.4.4.), and interference with cell behavior must always be guarded against. In addition, Becker and Fay (1987) have found that resistance to bleaching is dramatically enhanced by reducing the supply of oxygen. Though not many cells will function for long periods in the absence of oxygen, a useful degree of protection against bleaching may be obtainable by reducing oxygen tension in the bathing medium, for example, to 10 or 20% of normal.

4.4.4. Intracellular Buffering

It is a fundamental law that no measurement can be made without perturbing the measured process. For measurements with fluorescent indicators this constitutes binding to the indicator, which then acts as a buffer, exerting a damping effect on any changes recorded, particularly transient ones, and possibly upsetting the whole metabolism of the cell. Buffering artifacts are not specific to single-cell imaging measurements, except insofar as heavier loading with dyes may be required to give adequate images from single cells than is needed for population studies. Nor are they strictly artifacts associated with quantitation, but rather they concern the interpretation of what has been measured.

The seriousness of cytosolic buffering by an added indicator clearly depends on the loading, as compared with the intrinsic buffering capacity of the cell. In the case of Ca^{2+}, total content of cells is typically in the millimolar range, so that one might assume a net Ca^{2+} buffering capacity of this order (*see* chapter by Poenie in this vol.), but Ca^{2+} transients associated with cell signaling rarely exceed 1–2 μM in amplitude, suggesting that the immediate buffering capacity available to deal with them, for example, in mitochondria and endoplasmic reticulum, is also of this order (McBurney and Neering, 1987). The 100–200 μM loadings needed with earlier indicators like quin-2 will thus add considerably to cytosolic [Ca²⁺] buffering capacity, and have indeed been shown to affect cell behavior (DeFeo and Morgan, 1986; Nemeth and Scarpa, 1986; Tsien et al., 1984). Even the ten-times lighter loadings that can be used with fura-2 should be regarded as potentially invasive, and although most cells appear to behave "normally" after loading, it is still worth checking for interference, for example, by comparing results at two different loading levels. Regulation of cytosolic [Ca²⁺] is a complex and imperfectly understood process, and any treatments that may interfere with it ought to be considered with extreme caution.

4.5. Spatial and Temporal Resolution

Given that quantitative measurements of intracellular Ca²⁺, for example, can be made with modest confidence, it remains to consider how accurately we can localize these signals in space and time.

4.5.1. Imaging Geometry and Spatial Resolution

The output from a single-cell imaging system is a 2-D picture, derived from a 3-D cell. Ideally, one would like to consider the image as representing a horizontal section through the cell, along the plane of focus of the microscope, but in practice, such images are always contaminated by light from above and below. The contaminating light is not sharply focused, and in transmitted-light microscopy, it contributes a background that can be partly suppressed by adjusting viewing conditions to give the "best" image. In quantitative fluorescence microscopy, out-of-focus blur will also contaminate the image, because the exciting radiation normally penetrates through single cells with little attenuation, but the option of making arbitrary adjustments of brightness and contrast to reduce blur is not available, and in any case a Ca²⁺ image, for example, may not contain enough known structural detail to serve as a criterion for "in focus" information.

The effect on spatial resolution in single-cell imaging has been investigated in detail by Fay et al. (1989) (*see also* Fay et al., 1986). UV light is concentrated by the objective lens into a double cone of semivertical angle α determined by

$$n \sin \alpha = \text{NA} \qquad (5)$$

where NA = numerical aperture of the lens, and n = refractive index of the medium. For water, n = 1.33 so that with a lens of NA 1.3, α = 77°. At the apex, which is the plane of focus, light is concentrated into an area less than or equal to the field of view, so that the brightest fluorescence is to be expected from this plane. Nevertheless, fluorescence from some distance above and below this plane is also collected by the objective, so that each point in the fluorescence image contains light originating from the intersection between the specimen and a double cone of the same semivertical angle. Light from above and below the focal plane is defocused, and contributes to the background, though in practice, the efficiency of collection falls off toward the periphery of the cone, so that the effective semivertical angle is reduced. Fay et al. (1989) have demonstrated this empirically, using a series of images of a 0.2 μm diameter fluorescent bead, which is effectively a point-object, viewed through a 100× UVF planapo objective. At the plane of focus the image is a disk, with the expected diffraction-limited diameter of about 0.5 μm. Above and below this plane the image becomes progressively broader and fainter, doubling in diameter at distances of about 4 μm from focus, which is consistent with a semivertical angle of about 51° for the collection cone; similar measurements on our system using a Nikon UV-F 100× objective yielded an angle of 56°. A series of such images can be used to compute a "point spread function" for the objective; the image of a 3-D object is then given by the convolution of this function, with the true structure of the object. By recording a series of images at regularly spaced planes, in the manner of serial sections, Fay et al. (1989) were able to reverse this convolution by a computation relying on successive approximations, to get back to the 3-D structure of objects, such as muscle cells stained with fluorescent anti-α-actinin.

Although they can dramatically enhance the resolution and contrast in fluorescence images, such iterative procedures, requiring careful acquisition of serial "sections" and large amounts of computation, are clearly not feasible for monitoring the changing Ca^{2+} distributions in living cells. Probably the best that can be done is to assess the limitations on spatial resolution that are imposed by the large effective depth of focus in fluorescence imaging, and to be aware of any artifacts that may arise. Fig. 7 illustrates the situation when viewing a spherical cell and attempting, for example, to resolve differences in $[Ca^{2+}]$ between nucleus and cytoplasm (Goldman et al., 1990; Williams et al., 1985), or between central and peripheral regions of a cell attributable to sudden influx of Ca^{2+} through membrane channels (Cheek et al., 1989a). Each diagram represents a vertical section through the center of the cell, with a superimposed double cone in which brightness represents qualitatively the intensity of light gathered from each point in the vertical section, and contributing to the point in the image that corresponds to the axis of the cone. At a point corresponding to the midpoint of the cell (Fig. 7a and d), the image contains mainly light from the central region, but with contributions from the peripheral regions above and below; at points within the peripheral region (Fig. 7c and f), the principal contribution is from the periphery, but nearly half the collecting cone is now outside the cell, and receives only background radiation so that the overall intensity is reduced. The net effect is to "smear out" the distribution, leading to serious underestimates of such small-scale inhomogeneities. The precise effect is difficult to quantify, but in general, in a fura-2 ratio-imaging system, the signal from regions of low $[Ca^{2+}]$ will be more readily contaminated by neighboring high $[Ca^{2+}]$ compartments than vice versa, because of the stronger fluorescence of the dye at high $[Ca^{2+}]$. Fay et al. (1989) created a quantitative model system based on the simpler, cylindrical geometry of a smooth muscle cell, and calculated the extent to which nuclear and peripheral (sarcoplasmic reticulum) readings are likely to be contaminated by fluorescence from the intervening cytoplasm.

Although differences between regions are clearly distinguishable in practice in these 10 μm diameter cells (Williams et al., 1985), the model shows that the real differences are probably considerably higher than those seen on the smeared-out fluorescence images. Clearly, it is not realistic to quote a single figure for spatial resolution as such, because of the complex effects of cell geometry, but in general, one should assume that,

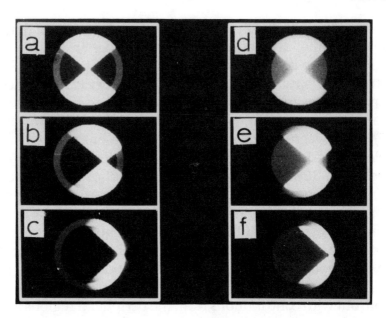

Fig. 7. Contribution to the epifluorescence image, from out-of-focus blur. Each panel shows a vertical cross-section through a spherical "cell" of diameter 10 μm, with a superimposed section through the double cone-shaped volume from which fluorescence is collected, to form one point in the image corresponding to the axis of the cone. The brightness at each point within the cone represents approximately the contribution to the image intensity, by light collected from that point. The figure shows contributions to three different points in the image, from cells with bright periphery (a–c) or bright center (d–f). The semivertical angle of the cone, and the distribution of intensity within it are derived from empirical measurements made with a Nikon UV-F 100× glycerol immersion objective (NA 1.35).

with a 1.3 NA objective, [Ca^{2+}] in regions of 1–2 μm will be partially resolved, whereas at 3–4 μm from a boundary, contamination is likely to be largely absent.

It has been assumed throughout the above discussion that the dye has been uniformly loaded into all parts of the cell. If this is not the case, as, for example, in plant cells where distribution of Ca^{2+} may need to be monitored in thin layers of cytoplasm around a large vacuole, then the degree of likely cross-contamination between differently loaded regions needs to be reassessed. Here it is useful to have an imaging system displaying total fluorescence, as well as [Ca^{2+}] deduced from the ratio (Section 3.3. above).

An important side effect of out-of-focus blur is that images from thick, translucent objects become very hard to interpret. Quantitative imaging using epifluorescence is thus restricted to cultured single cells or cell layers, or, at best, to very small tissue structures no more than one or two cells thick. The ultimate in 3-D spatial resolution clearly requires new optical techniques, such as confocal laser scanning, probably at the expense of temporal resolution, or of confining interest to limited areas or even single lines in the image (Hernandez-Cruz et al., 1990).

4.5.2. Temporal Resolution

The fastest rate of change that can be resolved in any fluorescent system is determined ultimately by the response-time of the fluorescent dye itself. Jackson et al. (1987) have determined the dissociation and association rate-constants for fura-2 and Ca^{2+} as $84.s^{-1}$, and of the order of $5 \times 10^8.M^{-1}.s$, respectively. This means that at $[Ca^{2+}] = 100$ nM, and with a fura-2 loading of 10 μM, Ca^{2+} transients ought to be resolvable on a time-scale of the order of 1 ms (for onset), and 12 ms (decay). Temporal resolution in single-cell imaging is not, therefore, limited by dye kinetics, but by the maximum rate at which fluorescence photons can be collected by the imaging system, and by the rate at which individual images are captured. The photon emission rate is limited by the maximum intensity of exciting radiation that can be tolerated, without too much bleaching, and by the amount of dye that can be loaded without seriously affecting the cells: In our system, this amounts to about 10^3 emitted photons/pixel in each video frame-time. The videocamera has an intensifier with quantum efficiency of about 10%, giving something like 100 photons/pixel, and a signal/noise ratio (discounting other sources of electronic noise) of 10:1 at best. In practice, this is not acceptable, particularly given that pairs of images are to be combined in the ratio method, and we use further temporal filtering of digitized images, increasing the final image time-constant to 200 ms (2.5 frame pairs).

Clearly, time-resolution in single-cell imaging can be improved either by increasing excitation intensity, for example, in short-duration experiments where the UV light exposure can be accurately timed by a shutter, or at the expense of spatial resolution, by averaging pixels in the individual images: For example, a twofold sacrifice of spatial resolution by averaging 2 × 2 blocks of pixels will gain a twofold improvement in signal/noise. Pixel averaging in real time is not possible with currently available programmable hardware, however, and can only be done by off-line processing, using short sequences of stored images: There is scope for development here.

Given that single video-frame images can be made acceptable, the response-time of a ratio image can be further improved by combining every frame with its predecessor, for example, frame 2/frame 1, then frame 2/frame 3, and so on (Marks and Maxfield, 1990). Alternatively, if one of the wavelengths is the isosbestic point, where fluorescence should change only slowly because of bleaching or dye loss, then a reference image at this wavelength need be taken only infrequently, so that the system can measure at full frame-rate for most of the time. Using a dual-emission dye, such as indo-1, produces a real doubling in temporal resolution, because two complete images are acquired simultaneously—provided that the problem of accurately aligning two videocameras on the same image can be overcome (Takamatsu and Wier, 1990).

5. Conclusion

Fluorescence digital imaging of single cells provides rapid and easy measurement of the distribution of a number of common ions, once the apparatus has been set up and calibrated. Quantitative measurements are possible, and by using the ratio method to compare fluorescence at two different excitation or emission wavelengths, the results can be made independent of dye distribution and optical pathway. However, uncertainties remain regarding how the fluorescence spectra of dyes are altered by viscosity, polarity, and ionic strength of the cytosol, so that absolute measurements are at present to be regarded as semiquantitative, with possible errors of up to 50%. Spatial resolution is limited because of blurring by out-of-focus fluorescence, to about 2–3 μm in 10 μm diameter cells, with better resolution in thin specimens, and very much poorer in cell aggregates or whole tissues. Temporal resolution is limited by the maximum rate of fluorescence emission that can be achieved without excessive bleaching or photodamage, and although faster rates can be achieved in principle by pooling pixels to give some spatial averaging, the video rate of 30 Hz (or 15 Hz/image-pair in dual-excitation ratio-imaging) is already too fast to give clean images in most systems.

Despite these limitations, there is still a wealth of useful information to be obtained from the technique, and attention should now be concentrated on more reliable quantitation through understanding of dye behavior, on techniques for interpreting the images, and on combining single-cell imaging with techniques, such as microinjection, timed intracellular release of caged agonists, and simultaneous patch-clamping, to integrate the imaging data with other dimensions of the behavior and metabolism of the cells. Confocal laser scanning microscopy will in due course provide much better spatial resolution, but is at present limited to dyes excitable by visible-wavelength light, which excludes all the Ca^{2+} indicators currently available for use with the ratio method. Once this problem can be resolved, much greater spatial detail will be available in three dimensions, though this will be at the expense of temporal resolution, because of the inferior quantum efficiency of the confocal microscope.

Acknowledgments

I am grateful to Martin Bootman for allowing me to use some of his, as yet, unpublished data in Fig. 5, and to Tim Cheek, through whose hard work our imaging system is kept running.

References

Becker P. L. and Fay F. S. (1987) Photobleaching of fura-2 and its effect on the determination of calcium concentrations. *Am. J. Physiol.* **253,** C613–C618.

Brooker G., Seki T., Croll D., and Wahlestedt C. (1990) Calcium wave evoked by activation of endogenous or exogenously expressed receptors in *Xenopus* oocytes. *Proc. Natl. Acad. Sci. USA* **87,** 2813–2817.

Cheek T. R., O'Sullivan A. J., Moreton R. B., Berridge M. J., and Burgoyne R. D. (1989a) Spatial localization of the stimulus-induced rise in cytosolic free Ca^{2+} in bovine adrenal chromaffin cells. *FEBS Lett.* **247,** 429–434.

Cheek T. R., Jackson T. R., O'Sullivan A. J., Moreton R. B., Berridge M. J., and Burgoyne R. D. (1989b) Simultaneous measurements of cytosolic calcium and secretion in single bovine adrenal chromaffin cells by fluorescent imaging of Fura-2 in cocultured cells. *J. Cell Biol.* **109,** 1219–1227.

Connor J. A., Kater S. B., Cohan C., and Fink L. (1990) Ca^{2+} dynamics in neuronal growth cones: regulation and changing patterns of Ca^{2+} entry. *Cell Calcium* **11,** 233–239.

Corps A. N., Cheek T. R., Moreton R. B., Berridge M. J., and Brown K. D. (1989) Single-cell analysis of the mitogen-induced calcium responses of normal and protein kinase C-depleted Swiss 3T3 cells. *Cell Regulation* **1,** 75–86.

DeFeo T. T. and Morgan K. G. (1986) A comparison of two different indicators: Quin 2 and Aequorin in isolated single cells and intact strips of ferret portal vein. *Pflügers Arch. Eur. J. Physiol.* **406,** 427–429.

Di Virgilio F., Steinberg T. H., and Silverstein S. C. (1990) Inhibition of fura-2 sequestration and secretion with organic anion transport blockers. *Cell Calcium* **11,** 57–62.

Fay F. S., Fogarty K. E., and Coggins J. M. (1986) Analysis of molecular distribution in single cells using a digital imaging microscope, in *Optical Methods in Cell Physiology (Society of General Physiologists Series,* vol. 40) (de Weer P. and Salzberg B. M., eds.), Wiley Interscience, New York, pp. 51–63.

Fay F. S., Carrington W., and Fogarty K. E. (1989) Three-dimensional molecular distribution in single cells analysed using the digital imaging microscope. *J. Microsc.* **153,** 133–149.

Fenwick E. M., Marty A., and Neher E. (1982) A patch-clamp study of bovine chromaffin cells and of their sensitivity to acetylcholine. *J. Physiol.* **331,** 577–597.

Goldman W. F., Bova S., and Blaustein M. P. (1990) Measurement of intracellular Ca^{2+} in cultured arterial smooth muscle cells using fura-2 and digital imaging microscopy. *Cell Calcium* **11,** 221–231.

Grynkiewicz G., Poenie M., and Tsien R. Y. (1985) A new generation of Ca^{2+} indicators with greatly improved fluorescence properties. *J. Biol. Chem.* **260,** 3440–3450.

Hernandez-Cruz A., Sala F., and Adams P. R. (1990) Subcellular calcium transients visualised by confocal microscopy in a voltage-clamped vertebrate neuron. *Science* **247,** 858–862.

Jackson A. P., Timmerman M. P., Bagshaw C. R., and Ashley C. C. (1987) The kinetics of calcium binding to fura-2 and indo-1. *FEBS Lett.* **216,** 35–39.

Jacob R. (1990) Imaging cytoplasmic free calcium in histamine stimulated endothelial cells and in fMet-Leu-Phe stimulated neutrophils. *Cell Calcium* **11,** 241–249.

Lindeman J. J., Harris L. J., Slakey L. L., and Gross D. J. (1990) Charge-coupled device imaging of rapid calcium transients in cultured arterial smooth muscle cells. *Cell Calcium* **11,** 131–144.

Lipscombe D., Madison D. V., Poenie M., Reuter H., Tsien R. W., and Tsien R. Y. (1988) Imaging of cytosolic Ca^{2+} transients arising from Ca^{2+} stores and Ca^{2+} channels in sympathetic neurons. *Neuron* 1, 355–365.

McBurney R. N. and Neering I. R. (1987) Neuronal calcium homeostasis. *Trends Neurosci.* 10, 164–169.

MacGlashan D., Jr. (1989) Single-cell analysis of Ca^{++} changes in human lung mast cells: graded vs. all-or-nothing elevations after IgE-mediated stimulation. *J. Cell Biol.* 109, 123–134.

Malgaroli A., Milani D., Meldolesi J., and Pozzan T. (1987) Fura-2 measurement of cytosolic free Ca^{2+} in monolayers and suspensions of various types of animal cells. *J. Cell Biol.* 105, 2145–2155.

Marks P. W. and Maxfield F. R. (1990) Local and global changes in cytosolic free calcium in neutrophils during chemotaxis and phagocytosis. *Cell Calcium* 11, 181–190.

Miller D. J. and Smith G. L. (1984) EGTA purity and the buffering of calcium ions in physiological solutions. *Am. J. Physiol.* 246, C160–C166.

Moore E. D. W., Becker P. L., Fogarty K. E., Wiliams D. A., and Kay F. S. (1990) Ca^{2+} imaging in single living cells: Theoretical and practical issues. *Cell Calcium* 11, 157–179.

Neher E. (1988) The influence of intracellular calcium concentration on degranulation of dialysed mast cells from rat peritoneum. *J. Physiol.* 395, 193–214.

Nemeth E. F. and Scarpa A. (1986) Cytosolic Ca^{2+} and the regulation of secretion in parathyroid cells. *FEBS Lett.* 203, 15–19.

Oakes S. G., Martin W. J., and Powis G. (1988) Incomplete hydrolysis of the calcium indicator fura-2 pentaacetoxymethyl ester (Fura-2 AM) by cells. *Anal. Biochem.* 169, 159–166.

O'Sullivan A. J., Cheek T. R., Moreton R. B., Berridge M. J., and Burgoyne R. D. (1989) Localization and heterogeneity of agonist-induced changes in cytosolic calcium concentration in single bovine adrenal chromaffin cells from video imaging of fura-2. *EMBO J.* 8, 401–411.

Ozaki H., Kwon S.-C., Tajimi M., and Karaki H. (1990) Changes in cytosolic Ca^{2+} and contraction induced by various stimulants and relaxants in canine tracheal smooth muscle. *Pflügers Arch.* 416, 351–359.

Paradiso A. M., Tsien R. Y., and Machen T. E. (1987) Digital image processing of intracellular pH in gastric oxyntic and chief cells. *Nature* 325, 447–450.

Poenie M. (1990) Alteration of intracellular fura-2 fluorescence by viscosity: a simple correction. *Cell Calcium* 11, 85–91.

Roe M. W., Lemasters J. J., and Herman B. (1990) Assessment of fura-2 for measurements of cytosolic free calcium. *Cell Calcium* 11, 63–73.

Ryan T. A., Millard P. J., and Webb W. W. (1990) Imaging $[Ca^{2+}]_i$ dynamics during signal transduction. *Cell Calcium* 11, 145–155.

Steinberg S. F., Bilezikian J. P., and Al-Awqati Q. (1987) Fura-2 fluorescence is localized to mitochondria in endothelial cells. *Am. J. Physiol.* 253, C744–C747.

Takamatsu T. and Wier W. G. (1990) High temporal resolution video imaging of intracellular calcium. *Cell Calcium* **11,** 111–120.

Timmerman M. P., Godber J. F., Walton A., and Ashley C. C. (1990) Imaging spatial distribution of release in single muscle fibres from *Balanus nubilus* using image intensification. *Cell Calcium* **11,** 211–220.

Tsien R. Y. and Harootunian A. T. (1990) Practical design criteria for a dynamic ratio imaging system. *Cell Calcium* **11,** 93–109.

Tsien R. Y., Pozzan T., and Rink T. J. (1984) Measuring and manipulating cytosolic Ca^{2+} with trapped indicators. *Trends Biochem. Sci.* **9,** 263–266.

Tucker R. W., Meade-Cobun K., and Ferris D. (1990) Cell shape and increased free cytosolic calcium $[Ca^{2+}]_i$ induced by growth factors. *Cell Calcium* **11,** 201–209.

Williams D. A. and Fay F. S. (1990) Intracellular calibration of the fluorescent calcium indicator fura-2. *Cell Calcium* **11,** 75–83.

Williams D. A., Fogarty K. E., Tsien R. Y., and Fay F. S. (1985) Calcium gradients in single smooth muscle cells revealed by the digital imaging microscope using Fura-2. *Nature* **318,** 558–561.

Analysis
of Protein Kinase C Function

Gillian M. Burgess

1. Introduction

Protein kinase C is a key element in signal transduction, cell regulation, and tumor promotion, influencing cellular responses to many stimuli at a variety of levels, including receptors, transduction systems, ion channels, and membrane pumps. Although it was once thought of as a single entity (Nishizuka, 1984), recent evidence from molecular cloning and enzymological analysis has revealed the existence of a family of subspecies (for a review, *see* Nishizuka, 1988). Members of this family are enzymes that require anionic phospholipids, in particular phosphatidylserine, the neutral lipid *sn*-1,2-diacylglycerol (DG), and in some cases, Ca^{2+} for activation (*see* Nishizuka, 1984,1986,1988 for reviews). In addition, protein kinase C has been identified as the major receptor by which the tumor-promoting phorbol esters, such as 4-β-phorbol-12,13-dibutyrate (PDBu) and phorbol 12-myristate-13-acetate (PMA), mediate their many effects on cells, and it has been demonstrated that these molecules can substitute for DG in activating the enzyme both in vitro and in vivo (Castagna et al., 1982). Physiologically, the enzyme is activated in response to transient receptor-induced generation of DG (normally present in cells at very low concentrations) from membrane phospholipids, in combination with increased cytostolic $[Ca^{2+}]$ ($[Ca^{2+}]_i$). Some subspecies of protein kinase C can also be activated in vitro by arachidonic acid and certain unsaturated

From: *Neuromethods, Vol. 20: Intracellular Messengers*
Eds: A. Boulton, G. Baker, and C. Taylor © 1992 The Humana Press Inc.

fatty acids (Sekiguchi et al., 1988; Shearman et al., 1989a), and it is possible that stimulus-evoked increases in the cellular levels of these fatty acids could provide an alternative pathway for activation of the enzyme. The enzymic reaction catalyzed by protein kinase C is the transfer of the γ phosphate of ATP to serine and threonine, but not tyrosine, residues of protein substrates (see Nishizuka, 1984).

In the brain, it has been possible to distinguish at least seven subspecies of protein kinase C, one of which (γ) appears to be expressed almost exclusively in nervous tissue. These subspecies appear to fall into two groups, and although there is quite a high degree of similarity within the groups, differences between them are marked. Initially, four cDNA clones that encode α, βI, βII, and γ were found in brain cDNA libraries (Parker et al., 1986; Coussens et al., 1986; Ono et al., 1986), and these comprise group one. They are composed of a single polypeptide chain (MW 77–85 kDa) with four conserved (C_1–C_4) and four variable (V_1–V_4) regions, made up of an amino-terminal regulatory domain and a carboxyl-terminal catalytic domain that interacts with ATP and the protein substrates of the enzyme. The conserved regions C_1 and C_2 of the regulatory domain contain the Ca^{2+}, DG, and phospholipid binding regions of the enzyme (Nishizuka, 1986), and in addition, the C_1 region contains a pseudosubstrate domain that is likely to be responsible for maintaining the enzyme in an inactive configuration in the absence of activator molecules (House and Kemp, 1987).

More recently, a further three cDNA clones have been found in a rat brain library (Ono et al., 1987,1988). These encode the subspecies δ, ε, and ζ, which make up the second group. They are also composed of a single polypeptide chain, but differ from the first group in that they lack the C_2 conserved region of the regulatory domain. Since the members of this group are insensitive to Ca^{2+} (Ono et al., 1988, Ohno et al., 1988; Schaap et al., 1989), it is thought likely that the C_2 region, present in the first group, contains the Ca^{2+}-binding site. In addition, the ζ subspecies lacks one of the two cysteine-rich sequences found in the C_1 region of all the other subtypes and which is thought to be associated with the phorbol ester -binding domain (Ono et al., 1989).

To date, three isozymes of protein kinase C have been purified from brain and various other tissues using hydroxyapatite chromatography. These have been designated I, II, and III (see, e.g., Huang et al., 1986; Kikkawa et al., 1987; Ono et al., 1987; Azhar et al., 1987; Kosaka, 1988; Shearman et al., 1987). The three fractions are immunologically distinct (Huang et al., 1986) and correspond to the γ, β, and α subspecies, respectively, on the basis of comparisons to expressed products in COS cells transfected with different cDNAs (Huang et al., 1987; Kikkawa et al., 1987). It seem likely that seven is a minimum estimate of the number of subspecies of protein kinase C, since several structurally undefined enzymes have have been reported in other tissues, e.g., heart, lung, and platelets (Kosaka, 1988; Tsukada et al., 1988).

Activation of protein kinase C is associated with translocation of the enzyme from the cytosol to the plasma membrane (Kraft and Anderson, 1983; Wolf et al., 1985a), and the ratio of cytosolic to membrane-bound enzymes is often used as a measure of protein kinase C activation (see Kraft and Anderson, 1983). This association of the activated enzyme with the membrane fraction suggests that many of its substrates are likely to be membrane-bound proteins, such as ion channels, pumps, and receptors.

Protein kinase C can be cleaved by limited proteolysis by the Ca^{2+}-dependent enzyme calpain (Kishimoto et al., 1989), yielding two fragments with mol wt of 51 and 26 kDa. The smaller hydrophobic fragment possesses the regulatory domain (Lee and Bell, 1986) and the larger, hydrophilic fragment the catalytic domain, which is active in the absence of Ca^{2+}, phospholipid, and DG (Kikkawa et al., 1982; Kishimoto et al., 1983), and which is rapidly removed from the cell (Kishimoto et al., 1983). It appears that the enzyme is more susceptible to proteolytic attack when it is activated (Kishimoto et al., 1983), and this may be related to the phenomenon of downregulation, in which persistent activation of protein kinase C by phorbol esters leads to translocation of the enzyme to the membrane and its subsequent degradation (Young et al., 1987). Some subspecies of protein kinase C may be more susceptible to downregulation than others (see Fournier et al., 1989; Isakov et al., 1990), and this may be related to the reported variation in their resistance to proteolytic attack by calpain (Kishimoto et al., 1983).

Most proteins kinase C are widely distributed in the tissues and organs of mammals and other organisms (Kuo et al., 1980; Minakuchi et al., 1981). Autoradiography using [³H]-PDBu (Worley et al., 1986), *in situ* hybridization histochemistry with oligodeoxynucleotide probes complementary to cDNAs from specific subspecies (Brandt et al., 1987), and immunocytochemical studies with antibodies directed against specific isozymes (Huang et al., 1987; Hidaka et al., 1988; Huang et al., 1989) suggest that the different subspecies of proteins kinase C have distinct cellular and subcellular localizations. The α subtype appears to be present almost ubiquitously, and the βI and βII subspecies are also widespread. Although it has been shown that some cells express only one subtype (McCaffrey et al., 1987; Shearman et al., 1988), this appears to be an exception, and most cell types contain several subspecies, often in different locations within the cell (*see* Kikkawa et al., 1989). The γ subspecies, which appears to be present almost exclusively in the brain and spinal cord, is particularly concentrated in the pyramidal cells of hippocampus, where it is associated with the membranes of dendrites (Huang, 1989), suggesting that it might play a role in synaptic transmission. The heterogeneity in the cellular expression of proteins kinase C suggests that different members of the family may have different substrates, and this, along with the fact that the isozymes isolated thus far have different enzymic properties (*see* Nishizuka, 1989; Huang et al., 1986; Asoaka et al., 1988), may imply that the subspecies have distinct roles in the modulation of cell growth and regulation.

Protein kinase C is particularly enriched in nervous tissue, where its activation has been related to a multitude of neural functions, including the sensitivity of neurotransmitter receptors (e.g., Vincenti et al., 1985), neurotransmitter release (Pozzan et al., 1984; Malenka et al., 1986,1987), the movement of ions across the plasma membrane via either ion channels (Madison et al., 1986; Strong et al., 1987; Baraban et al., 1985; Burgess et al. 1989) or exchange systems (Vincenti and Villereal, 1985; Dicker and Rosengurt, 1981), and growth and differentiation (e.g., Burgess et al., 1986; Spinelli and Ishii, 1983). In many cases, the exact mechanism by which protein kinase C exerts control of these functions is unknown.

Several major substrates that are phosphorylated by protein kinase C have been identified in the nervous system. These include P87, a functional role for which has yet to be determined (Albert et al., 1986); neuromodulin, an abundant neural-specific calmodulin-binding protein (Cimler et al., 1985) that is identical to GAP-43 and has been associated with nerve growth (Basi et al., 1987); and myelin basic protein (Wise et al., 1982). It is likely that at least some of the other ion channels and pumps whose activities appear to be regulated by protein kinase C are also substrates for the enzyme (for a review, *see* Shearman et al., 1989c).

Several methods have been used to elucidate the roles of protein kinase C in cellular function. Intracellular injection of purified enzyme (De Riemer et al., 1985) is an elegant way of providing direct evidence of its effects, but this technique is not feasible at most sites in the nervous system. Inhibitors of protein kinase C have, until recently, been of limited use because of their lack of specificity (Hidaka et al., 1984; Loomis and Bell, 1988). A more successful approach has been to monitor the effects of activators of the enzyme, such as phorbol esters and synthetic diacylglycerols, and to compare their effects to hormone, growth factor, or neurotransmitter-induced responses. This chapter will be devoted to a description of the methods that can be employed to investigate the possibility that protein kinase C might play a role in a particular cellular process or response, and will briefly describe techniques that could be used to determine the nature of the subspecies involved.

2. Is Protein Kinase C Activated?

To demonstrate that protein kinase C is implicated in a cellular response, it would be useful to show that the enzyme is activated under the conditions in which the response occurs. This section sets out some methods that might be used in attempting to answer the question "is protein kinase C activated?" Unfortunately, there is no direct way of measuring protein kinase C in vivo, although Alexander et al. (1990) have described a technique for measuring protein kinase C activity in permeabilized cells, using peptide substrates. The techniques to be described here provide only indirect evidence of the status of the enzyme.

2.1. Measurement of Diacylglycerol

Considerable evidence has accumulated to support the intracellular messenger function of DGs as physiological activators of protein kinase C. If the action of a hormone or neurotransmitter is thought to be mediated by activation of protein kinase C, then it is reasonable to expect that application of the stimulus will be followed by a transient increase in the level of this molecule. It would also be expected that such a rise in DG would precede the response. Elevated levels of DG can be measured in primary cultures of neurons, in neuronal cell lines, or in tissue slices by either of the following two methods. A rise in cellular arachidonic acid, which might also be directly responsible for activating the enzyme (see Sekiguchi et al., 1988), can also be detected using the first of the techniques described below.

2.1.1. Radiolabeling Studies

Neurons (10,000–30,300/sample) are incubated with [^{14}C]-arachidonic acid (~2 µC/mL) until quasi-equilibrium is achieved, usually between 60–120 min. During this period, [^{14}C]-arachidonic is incorporated into the cellular lipids, including phosphatidylinositol- 4,5-bisphosphate (PtdInsP_2). After equilibration, the cells are washed to remove unincorporated radioactive material, and exposed to hormone or neurotransmitter. After an appropriate period, the reaction is stopped by the addition of ice-cold chloroform:methanol solution (1:2). If the neurons are grown on glass cover slips, it is convenient to dip them into drug solution for the period of the experiment and then to plunge them into a test tube containing 1 mL of chloroform:methanol to terminate the reaction. The lipids are extracted for 15 min, and then 0.4 mL of chloroform and 0.4 mL of H$_2$O added to each tube. The tubes are vortexed, and after the extraction (a further 10 min), the tubes are centrifuged for 10 min at

2000g and the lower organic layers removed. The aqueous phase from each tube is reextracted with 0.8 mL of chloroform, and the second organic phase combined with the first and dried under a stream of nitrogen. The lipid samples can then be redissolved in 50 μL of 5% methanol in chloroform and spotted onto thin-layer chromatography plates (e.g., Whatman silica gel 60A-LK6D) along with authentic 1-stearyl-2-[^{14}C]-arachidonyl-glycerol. Lipids and free fatty acids are separated by developing the plates in one direction in benzene/diethyl ether/ethyl acetete/acetic acid (80:10:10:0.2) plus 0.01% butylated hydroxtoluene until the solvent is 16 cm from the origin *(see* Storry and Tuckley, 1967). ^{14}C-labeled lipids can be visualized by autoradiography following exposure of the thin-layer chromatography plates to film (e.g., Kodak X-OMAT) for between 5 and 14 days. Cellular [^{14}C]-DG should appear as a well separated doublet. The upper band, which has an approximate Rf value of 0.63, is [^{14}C]-1,3-DG, and the lower band, which has an Rf value of about 0.52, is [^{14}C]-1,2-DG. The DG should be clearly separated from [^{14}C]-arachidonic acid (Rf 0.27), which lies close to the origin. The radioactivity in individual spots can be quantified by scraping the relevant regions of the plate and counting in a liquid scintillation counter. The radioactivity in [^{14}C]-1,2,-DG should increase if there is a hormone or neurotransmitter-mediated rise in DG, whereas the number of counts in [^{14}C]-1,3,-DG should remain unaltered, making it a useful control. The amount of radioactivity in [^{14}C]-1,2,DG (or [^{14}C]-arachidonic acid) can be expressed as a percentage of the total radioactivity in the sample, by scraping and counting the remainder of the lane, or as suggested above, the change in [^{14}C]-1,2-DG can be expressed as a ratio of the radioactivity in [^{14}C]-1,3-DG. As an alternative to [^{14}C]-arachidonic acid, the increase in DG may be measured by labeling the cells with [^{3}H]-glycerol *(see* Polverino and Barritt, 1988). A similar protocol may be applied to tissues slices, although the extraction periods may have to be longer.

2.1.2. Mass Measurement of DG

Although labeling studies are technically simple and have proven to be useful for the study of DG metabolism, they do not define absolute levels of DG. In addition, it is possible that pools of lipid are preferentially labeled and that a relevant pool remains at a relatively low specific activity so that a rise in the cellullar concentration of DG, induced by a hormone, goes undetected. In order to avoid these problems, the mass of DG can be measured.

The mass assay most commonly used relies on the ability of DG kinase to quantitatively phosphorylate *sn*-1,2,-DG (but not 1,3,-DG) to [^{32}P]-phosphatidic acid (Preiss et al., 1986). In the protocol of Preiss et al. (1986), membranes, enriched in DG kinase, were obtained from *Eschericia coli* (strain N4830/pJW10). DG kinase from *E. coli* can be obtained from Lipidex, Middleton, WI, USA, or from Amersham, Aylesbury, UK, in the form of a 1,2-DG assay kit (code RPN200), which is based on the methods of Kennerly et al. (1979) and Preiss et al. (1986). In the assay described by Preiss et al. (1986) and in the Amersham DG assay kit, DG is extracted from the cells as described for the radiolabeling studies and dried down under a stream of nitrogen. Mixed micelles are then prepared by solubilizing each dried lipid extract in 20 μL of a detergent solution (7.5% octyl-β-D-glucoside [w/v] and 5 mM cardiolipin in 1 mM diethylenetriaminepentaacetic acid [DETAPAC]) using a bath-type sonicator for 15–20 s. This mixture should be kept in the dark and stored under nitrogen at –20°C.

The reaction mixture for the DG kinase phosphorylation assay contains:

1. 10 μL of DG kinase (approx 5 μg protein made up in 10 mM imidazole and 1 mM DETAPAC, pH 6.6);
2. 10 μL of 0.02M fresh dithiothreitol;
3. 50 μL of imidazole-HCl buffer (pH 6.6) containing: 100 mM NaCl, 25 mM MgCl$_2$, and 2 mM EGTA;
4. 20 μL of lipid sample made up in the detergent mixture; and
5. 10 μL of 10 mM [γ-^{32}P]-ATP made up in 100 mM imidazole and 1 mM DETAPAC (pH 6.6), about 1 μCi/tube.

The reaction is started by addition of the [^{32}P]-ATP to the tube. After mixing, the reaction is allowed to proceed at 25°C for

30 min, and then stopped by addition of 20 μL of 1% (v/v) perchloric acid and 450 μL of chloroform:methanol (1:2, v/v), vortexed thoroughly and left to extract for 10 min. After centrifugation at 2000g for 1 min, a further 150 μL of 1% perchloric acid and 150 μL chloroform are added to each tube, which is then vortexed and recentrifuged. The upper, aqueous phase is removed and discarded, and the lower phase washed twice with 1 mL of 1% perchloric acid. A sample (150 μL) from the lipid phase is then dried under nitrogen, redissolved in 50 μL of 5% methanol in chloroform, and spotted onto a thin-layer chromatography plate, which is developed in one direction with chloroform:methanol:acetic acid (65:15:5, v/v) and subjected to autoradiography as described above. The spot corresponding to [^{32}P]-phosphatidic acid (Rf 0.35), which can be identified by comparison with an authentic standard, is scraped and counted. The amount of 1,2-DG can then be calculated from the sample volume and the specific activity of the [^{32}P]-ATP in the assay. The conversion of DG to phosphatidic acid is normally linear in the range of 25 pmol to 25 nmol (Preiss et al., 1986). [^{3}H]-1,2-DG can be used as an internal standard to monitor recovery and conversion and, with a pure sample of *sn*-1,2-DG, this may be as great as 90%. If a range of standard DG concentrations (30–1000 pmol) is included in the assay, a standard curve can be constructed and used to calculate the amount of DG in the unknown samples. If chromatographic analysis consistently indicates that more than 90% of the radioactivity in the washed lipid sample is phosphatidic acid, it may be possible to eliminate the chromatography step and count the washed lipids directly. Other techniques for making mass measurements of DG include an assay based on HPLC separation of cellular lipids combined with quantification by on-line refractive index measurement (Bocckino et al., 1985).

It is now clear that phosphoinositides are not the only source of hormone-induced increases in DG. Phosphatidylcholine can also be degraded by what appears to be a phosphatidylcholine-specific phospholipase C (*see*, for example, Besterman et al., 1986; Irving and Exton, 1987; Martinson et al., 1989; Matozaki and Williams, 1989). If a mass measurement assay indicates that there is a net rise in DG, labeling studies would be necessary to determine the nature of the lipid responsible for this rise. This might include labeling the cells with [³H]-choline and [³H]-inositol to allow hormone-evoked increases in water-soluble [³H]-choline and/or [³H]-inositol containing compounds to be detected (e.g., Besterman et al., 1986; Grillone et al., 1988; Polverino and Barritt, 1988; Martinson et al., 1989; Matozaki and Williams, 1989).

In conclusion, mass measurement of DG allows the absolute levels of DG to be determined, but it is a relatively complicated procedure compared to labeling studies and does not allow the source of the DG to be determined. It measures net changes in DG and does not permit assessment of the relative effect of hormones on formation vs breakdown of the lipid. For a complete study of the effect of a hormone or neurotransmitter on cellular DG, it may be necessary to make both kinds of measurements. However, as an initial step, labeling studies may be more expedient.

2.2. Translocation of Protein Kinase C

The direct demonstration of protein kinase C activation in whole cells is not possible, but "translocation" of the enzyme provides an indirect index of activation. This movement of protein kinase C to the cell membrane has been shown to correlate with protein kinase C activation in a number of systems (e.g., Wise et al., 1982; Berridge and Irvine, 1984; Liles et al., 1986).

When cells are activated by DG or phorbol esters, protein kinase C is rapidly redistributed from the cytosolic compartment to the plasma membrane, where it binds with high affinity (Kraft et al., 1982; Ashendel et al., 1983; Wooten and Wrenn, 1984; Wolf et al., 1985a). Experiments with erythrocyte vesicles (Wolf et al., 1985a) showed that Ca^{2+} controls the binding of protein kinase C to membranes, and that phorbol esters and DG are involved in both the regulation of binding and activation. The membrane-bound form of the enzyme is able to interact with and phosphorylate endogenous and exogenous substrates (Wolf et al., 1985a). It has been suggested that protein kinase C, which remains membrane-bound in Ca^{2+}-free medium, is enzyme that has been activated (Wolf et al., 1985a,b; Gopalakrishna et al., 1986) and that translocation is an important, if not essential, event in signal transduction by protein kinase C (Nishizuka, 1986; Bell, 1986). Measurement of translocation normally involves a subcellular fractionation stage, followed by the determination of protein kinase C activity in the membrane and cytosolic fractions.

2.2.1. Preparation of Subcellular Fractions

After treatment of primary cultures of neurons or neural cell lines (grown, for example, on 100-mm dishes) with appropriate stimuli (it would be useful to include a culture treated with a phorbol ester, e.g., 0.1 µM PDBu as a positive control), the incubation is terminated by aspiration of the growth medium followed by several washes in ice-cold homogenization buffer, typically: HEPES, 20 mM, pH 7.5; EDTA, 1 mM; EGTA, 0.5 mM; phenylmethylsulfonyl fluoride, 2 mM; leupeptin, 10 µg/mL; dithiothreitol, 2 mM, as described by Kraft and Anderson (1983). Alternatively, the homogenization buffer described in Section 5.1. can be used. The cells are scraped off the plate and collected in 5 mL of ice-cold homogenization buffer. They are then disrupted either by homogenization with a tight-fitting Dounce homogenizer or by a brief sonication (usually two 5–10 s bursts). The homogenate is centrifuged at 1000g for 10 min to remove cell debris and nuclei. The supernatant from the low-speed spin is centrifuged at 100,000g for 1 h to generate the cytosolic and membrane fractions. The supernatant from the 100,000g spin is used as the source of cytosolic protein kinase C. The 100,000g pellet is resuspended in 2–3 mL of homogenization buffer by brief sonication or by further homogenization, and is used as the source of membrane-bound enzyme. If protein kinase C activity is to be measured by means of the phosphorylation assay described below, the enzyme in the membrane fraction should first be solubilized by the addition of the detergent Nonidet P-40 (0.2–1%). Nonidet P-40 at the same concentration is added to the cytosolic fraction to allow a comparison to be made between fractions.

At this stage, the fractions can be assayed for protein kinase C activity, but in some cases, investigators have first made a semipurified preparation of the enzyme by applying the solubilized fractions to DEAE-cellulose and eluting the protein kinase batchwise with 0.1M NaCl (*see*, for example, Liles et al., 1986). If this purification step is performed, it should be followed by a desalting step using Sephadex G25 columns, especially if protein kinase C is to be measured by means of the [³H]-PDBu binding assay (described below), which relies on binding of the enzyme to DEAE-cellulose in low salt to separate bound from free [³H]-PDBu. In most cases, however, translocation of protein kinase C has been followed successfully without further purification.

2.2.2. Phosphorylation Assays

2.2.2.1. "Standard" Phosphorylation Assay

The quantitative assessment of protein kinase C activity in each subcellular fraction can be followed by examination of phospholipid/Ca^{2+}-dependent incorporation of ^{32}P from [γ-^{32}P]-ATP into exogenous substrates (usually histone III-S).

All the stock solutions are made up in Tris-HCl, 20 mM, pH 7.5; magnesium acetate, 5 mM. The reaction mixture contains:

1. 25 µL of 5 mM $CaCl_2$;
2. 25 µL of 2 mg/mL histone III-S;
3. 25 µL of 50 µg/mL phosphatidylserine (Avanti polar lipids);
4. 25 µL of 8 µg/mL dioleoylglycerol (Avanti Polar lipids);
5. 25 µL of 100 µM [γ-^{32}P]-ATP (~1 µCi/tube);
6. 50 µL of enzyme, ~20–50 µg protein/mL; and
7. 75 µL of 20 mM Tris-HCl, pH 7.5, containing 5 mM magnesium acetate.

giving a total vol of 250 µL.

Background activity is determined in samples from which phosphatidylserine, dioleoylglycerol, and $CaCl_2$ have been omitted and to which 25 µL of 5 mM EGTA have been added. Phosphatidylserine and dioleoylglycerol should be kept in chloroform at –20°C in vials purged with N_2 and stocks in Tris-HCl, magnesium acetate made up just prior to each experiment by drying down the lipids under a stream of nitrogen and dispersing them into buffer by sonication. The phosphorylation reaction is started by addition of [^{32}P]-ATP and allowed to proceed for 4 min at 30°C. It is stopped by addition of 0.5 mL of ice-cold 10% trichloroacetic acid followed by 25 µL of bovine serum albumin (20 μg/mL) as a carrier. Blanks, consisting of samples to which trichloracetic acid is added before the [³²P]-ATP, should be included in the assay. The precipitates are collected on Whatman GF/F filters by vacuum filtration, and the filters are then washed twice with 10 mL of ice-cold 0.2 mM KCl, 1% trichloroacetic acid. Alternatively, the reaction can be terminated by spotting a 40-μL sample from each tube onto a 2 × 2 cm square of Whatman P-81 phosphocellulose paper and then washing the papers twice in beakers containing 500 mL of 30% acetic acid. The filters are counted for ³²P by liquid scintillation spectrophotometry or by Cerenkov counting.

Using this method to detect protein kinase C activity, Matthies et al. (1987) report almost complete transfer of the enzyme to the membrane fraction after a 30-min exposure of PC12 cells to 1 μM PMA. Similarly, Liles et al. (1986) saw a doubling of membrane-bound enzyme after a 5-min exposure of neuroblastoma cells to 1 mM carbachol.

2.2.2.2. MIXED MICELLE PHOSPHORYLATION ASSAY

An alternative to the phosphorylation assay described above entails presenting the lipids to protein kinase C in Triton X-100 mixed micelles (see Hannun et al., 1985). In the mixed micelle phosphorylation assay, phosphatidylserine (50 μL of a 20 mg/mL stock in chloroform:methanol, 95:5) is dried down in a glass test tube under a stream of nitrogen along with 1 μL of PMA (from a 0.01–1 mg/mL stock in methanol). The test tube should be rotated constantly while drying, so that the lipids form a white coat around the walls of the tube. The lipids are then redissolved in a 200-μL aliquot of a solution of 1% Triton X-100 in 20 mM Tris-HCl, pH 7.5, vortexed, incubated at 30°C for 10 min, and then revortexed. The mixed micelles (10 μL) are then added to a total vol of 40 μL of reaction mixture. The reaction mixture is made up as follows:

1. 10 μL of 5 mg/mL histone III-S in 0.2M HEPES, pH 7.5, containing 2 mM EGTA;
2. 5 μL of 0.1M MgCl₂;
3. 5 μL of 6 mM CaCl₂;
4. 10 μL mixed micelle solution;

5. 5 µL enzyme in 20 mM Tris-HCl, pH 7.5, containing 2 mM EDTA, 0.02% Triton X-100, and 2 mM fresh DTT; and
6. 5 µL of 1 mM [γ-^{32}P]-ATP (~500 cpm/pmol).

The reaction is started by addition of the [^{32}P]-ATP and stopped by spotting 25 µL of the mixture onto P-81 paper, or by the addition of 80 µL of 10% trichloroacetic acid, and filtering as decribed for the "standard" phophorylation assay. Basal activity is measured in the absence of lipid activators; blanks without enzyme, both in the presence and absence of lipid activators, should also be included.

Both the "standard" and the mixed micellar assay have been widely used to measure protein kinase C activity. It has been claimed, however (Hannun et al., 1985), that the mixed micelle assay permits a greater degree of reproducibility than the "standard" assay, since it controls the physical presentation of the activators to the enzyme. In our laboratory, we have found that the two assays give equally reliable results. The quality of the lipid activators is, however, important for obtaining reliable activation. As an alternative to histone, the peptide [Ser^{25}] PKC (19-36) (*see* House and Kemp, 1987) can be used as a substrate for protein kinase C, and to detect the kinase activity of expressed protein kinase C ε, it is necessary to use a synthetic peptide based on its pseudosubstrate domain (*see* Schaap et al., 1989) since histone is a very poor substrate. A kit that uses a peptide substrate to measure the activity of protein kinase C in cell or tissue extracts is available from Amersham.

2.2.3. [³H]-PDBu Binding

It is also possible to follow the movement of protein kinase C from the cytosol to the particulate fraction by measuring Ca^{2+}/phospholipid-dependent binding of [³H]-PDBu to the enzyme (e.g., Wolf et al., 1985a; *see also* Huang and Huang, 1990). For this assay, it is not necessary to solublize the enzyme present in the membrane fraction with detergent.

[³H]-PDBu binding is measured (in the presence and absence of phosphatidylserine) in a reaction mixture containing:

1. 25 µL of 5 mM CaCl$_2$;
2. 25 µL of 2.5 mM EDTA;

3. 25 μL of 5 mg/mL bovine serum albumin;
4. 50 μL of enzyme (~20–50 μg protein);
5. 25 μL of 120 nM [³H]-PDBu; and
6. 25 μL of 50 μg/mL phosphatidylserine.

All stock solutions are made up in 20 mM Tris-HCl, pH 7.5, containing 5 mM magnesium acetate.

Nonspecific binding is determined under the same conditions, but in the presence of 10 μM unlabeled PDBu. The assay is started by addition of [³H]-PDBu and proceeds at room temperature for 30 min.

For the cytosolic fraction, the reaction is stopped by addition of 0.5 mL of 30% DEAE cellulose in 20 mM Tris-HCl, pH 7.5, followed by incubation at 4°C for 30 min. The enzyme binds to the DEAE cellulose, and the bound and free ligands can be separated by filtering the samples through Whatman GF/C filters. The filters should be washed 3× with 2 mL ice-cold 20 mM Tris-HCl, pH 7.5. The radioactivity can then be determined by liquid scintillation counting after overnight soaking of the filters in scintillant.

For the membrane fraction, the samples can be diluted into 5 mL of ice-cold 20 mM Tris-HCl, pH 7.5, and filtered through Whatman GF/F filters to separate bound from free ligands. These filters are washed and counted as described for the cytosolic fraction.

2.3. Phosphorylation of Endogenous Substrates of Protein Kinase C

A third indirect possibility for investigating whether protein kinase C might be activated by a hormone or neurotransmitter is to determine whether a recognized substrate of the enzyme is phosphorylated after application of the stimulus to the tissue or cells. For example, in platelets, a 47-kDa protein is a recognized substrate for protein kinase C and is rapidly phosphorylated by agonists that are known to stimulate phosphoinositide breakdown and by phorbol esters (see, for example, Watson et al., 1988). The phosphorylation of this substrate in platelets can be detected quite easily after one-dimensional protein separation of a cellular extract.

There are a number of recognized neuronal substrates of protein kinase C. For example, an 87-kDa protein has been identified as a major substrate of protein kinase C in neuronal and other tissues (Wu et al., 1982; Walaas et al., 1983a,b), and neuromodulin (also named GAP43 or B50) is a nervous-tissue-specific protein substrate of protein kinase C (Oestreicher et al., 1981; Albert et al., 1986). Unfortunately, it is difficult to detect p87 and neuromodulin after in vivo phosphorylation by one-dimensional protein separation unless they are first immunoprecipated with specific antibodies (*see* Albert et al., 1986; Baudier et al., 1989; Wang et al., 1989; Oestreicher et al., 1986; De Graan et al., 1989). Unless such antibodies can be obtained or become commercially available, measurement of the phosphorylation of these endogenous substrates is not a staightforward option.

3. Direct Activation of Protein Kinase C

It is possible to circumvent receptor-induced activation of protein kinase C and activate the enzyme directly by application of phorbol esters (and synthetic diacylglycerols) to cells. These molecules have become the most widely used pharmacological tool for the elucidation of the cellular roles of protein kinase C.

In general, cellular responses elicited by phorbol esters are similar to those elicited by activators of phosphoinositide hydrolysis. However, one major difference is that agonist-stimulated increases in DG (and protein kinase C activation) are fairly short-lived, usually lasting <5 min, whereas activation by phorbol esters, which are metabolically relatively stable, is more prolonged.

Phorbol esters are useful, not only because they can be applied from the outside of cells, but also because the existence of a number of structural analogs, possessing varying potencies, allows the specificity of their actions to be studied. One of the most potent phorbol esters is PMA (12-*O*-tetradecanoylphorbol-13-acetate). The potency of other phorbol-12,13-diesters depends on the structure of the individual carboxylic acid moieties. For example, PMA is reported to have an IC_{50} of about 0.05 nM in displacing [^{3}H]-PDBu from purified enzyme, PDBu had an IC_{50} of about 10 nM, and phorbol-12,13-diacetate is much less active, with an IC_{50} of about 500 nM (Blumberg et al., 1984). Phorbol itself and 4α-phorbol-12,13-didecanoate do not activate protein kinase C, and can be used as negative controls. The potency of these agonists in activating a cellular function can be compared, therefore, to the known potency series in purified enzyme. In addition, it may be useful to determine whether the response to maximal concentrations of an active phorbol ester and the receptor agonist is additive. This can provide further evidence as to whether the hormone or neurotransmitter is acting via protein kinase C.

Synthetic diacylglycerols, such as *sn*-1,2-dioctanoylglycerol or *sn*-1,2-oleoyl-acetyl-glycerol, when applied to intact cells in the concentration range 1–30 μM can also activate protein kinase C (Kaibuchi et al., 1983; Davis et al., 1985; Lapetina et al., 1985; Osugi et al., 1986; Rane and Dunlap, 1986). The effects of the diacylglycerols are usually similar to those induced by phorbol esters, and the response to maximal concentrations of the two agents is not usually additive. The synthetic diacylglycerols are not usually as effective as phorbol esters. They may, however, be considered as a more physiological stimulus, since they undergo metabolism and do not have such a prolonged duration of action as the phorbol esters.

4. Inactivation of Protein Kinase C

The strongest evidence that can be obtained for implicating an enzyme as a mediator of a cellular event derives from experiments in which activation of the enzyme is prevented, and this, in turn, interferes with the process in question. Activation of protein kinase C can be prevented either by the use of enzyme inhibitors, or by prolonged incubation of cells or tissue with phorbol esters, which leads to downregulation of the enzyme and generates a preparation that is functionally deficient in protein kinase C. Both of these options will be discussed.

4.1. Downregulation of Protein Kinase C

Chronic incubation of a variety of cells with phorbol esters causes a marked loss of protein kinase C (e.g., Blackshear et al., 1988; Ballester and Rosen, 1985) by a mechanism thought to be associated with an increase in its rate of degradation (Young et al., 1987). This treatment has been employed to produce cells that are functionally deficient in protein kinase C and thus to obtain evidence that protein kinase C plays a role in the activation pathway of various cellular effectors (Blackshear et al., 1988; Pasti et al., 1986; Matthies et al., 1987; Mond et al., 1987; Young et al., 1987; Burgess et al., 1989). Recently, it has been suggested that some isozymes of protein kinase C are more sensitive to downregulation than others, both in vitro (Huang et al., 1989; Fournier et al., 1989) and in vivo (Cooper et al., 1989; Isakov et al., 1990), and this differential sensitivity may prove useful in matching specific subspecies to specific cellular tasks.

4.1.1. Chronic Treatment with Phorbol Esters

Cells are incubated with phorbol esters, normally either PDBu or PMA (0.2 –5 µM), for a period of between 18–36 h. This treatment has been shown to be effective in a variety of cell types, including rat glioma cells (Young et al., 1987) where incubation with 200 nM PMA for 18 h caused a severe loss of immunoreactive protein kinase C, in GH_3 cells (Ballester and Rosen, 1985) where PMA (0.4 µM) caused a 90% loss of [^{35}S]-methionine-labeled protein kinase C within a 24-h period, in PC12 cells (Matthies et al., 1987) where approx 90% of the cellular protein kinase C activity was lost after a 24-h treatment with 1 µM PDBu, and in cultured dorsal root ganglion neurons (Burgess et al., 1989) where an 18-h treatment with 300 nM PDBu caused a marked loss of PKC-related immunofluorescence. Since protein kinase is thought to control many aspects of growth and differentiation, it is important to ensure that the treatment has had no deleterious effects on cell viability measured, for example, by trypan blue exclusion.

Before making a comparison of the function or responses of the phorbol ester-treated cells with the control cells, the effectiveness of treatment should be determined. This can be done in several ways, e.g.:

1. The phosphorylation assays described in Section 2.2.2. can be used to measure protein kinase C activity. Total Ca^{2+} and phospholipid-dependent phosphorylation of exogenous histone by the cytosolic and membrane fractions should be measured in both control and phorbol ester-treated cells, and if the treatment has been successful, total activity in the downregulated cells should be significantly lower than in controls.

2. Total cellular specific [^{3}H]-PDBu binding can be measured as described in Section 2.2.3. Again, a comparison between control cells and phorbol ester-treated cells should be made, and if protein kinase C has been downregulated, there should be significantly less specific binding in the treated cells.

3. Immunoblot analysis of protein kinase C in cellular homogenates or immunocytochemical studies with antibodies directed against protein kinase C (*see* Fig. 1) can also be used to estimate the cellular level of protein kinase C. These two possibilities will be discussed in more detail below (Sections 4.1.2. and 4.1.3.).

4.1.2. Immunoblot Analysis of Protein Kinase C

A number of monoclonal antibodies directed against protein kinase C (some against specific isozymes) are commercially available. These include MC5, which recognizes both the α and β subspecies (available from Amersham), and three others (available from Seikagaku America Inc.) that recognize the α, β, and γ subspecies, respectively. In addition, there are numerous reports in the literature of successful attempts to raise antisera and monoclonal antibodies to purified proteins kinase C and to peptides derived from the known sequences of the subspecies (e.g., Ballester and Rosen, 1985; Huang et al., 1987; Hidaka et al., 1988; Makowske et al., 1987; Cazaubon et al., 1989). These antibodies can be used to detect protein kinase C in Western blot analysis of cellular homogenates.

Cells (both control and phorbol ester-treated), grown on 100-mm dishes, are washed with phosphate-buffered saline (PBS) and scraped into 1–3 mL of sample buffer (Laemmli, 1970) and boiled for 5 min. The extracts are then subjected to SDS (0.1–10%) polyacrylamide gel electrophoresis followed by electrophoretic

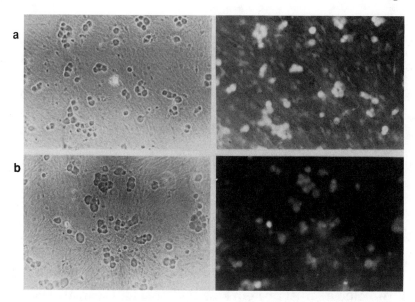

Fig. 1. Immunofluorescent staining of protein kinase C in isolated rat dorsal root ganglion neurons. The neurons were stained with MC5, an antibody that recognizes the α and β forms of protein kinase C. Figure 1a shows control cells and 1b shows cells that had been incubated with 300 n*M* phorbol dibutyrate for 18 h. The weak fluorescence seen in 1b was the same as the background level in cells that had not been exposed to MC5.

transfer of the proteins to nitrocellulose membrane. After blocking nonspecific binding sites by incubating the nitrocellulose with 3% bovine serum albumin in PBS containing 1 m*M* EDTA and 0.1% Triton X-100, strips of the nitrocellulose are incubated overnight (4°C) with appropriate antibody dilutions. Antibody binding is revealed by incubation with appropriate second antibodies (e.g., iodinated rabbit antimouse IgG for MC5). If downregulation has been successful, the band at 80 kDa, which corresponds to immunoreactive protein kinase C, will be fainter in the treated cells than in the controls.

4.1.3. Immunocytochemical Analysis of Protein Kinase C

An alternative to immunoblot analysis is to use suitable antibodies (e.g., MC5) for immunocytochemistry. Again, control cells are compared to cells that have been treated with phorbol esters. The cells, which are grown on glass cover slips, are washed with phosphate-buffered saline, and then fixed by placing a few drops of ice-cold paraformaldehyde (2–4%) on each cover slip for between 10 and 20 min. Subsequent steps are carried out at 4°C, unless otherwise noted. The cells are washed by dipping the cover slips into PBS and permeabilized by placing a few drops of a solution of 0.1% Triton X-100 and 10% horse serum in PBS on each cover slip for 5–10 min. Alternatively, fixation and permeabilization can be combined by a 5- to 15-min incubation in a solution of methanol:acetic acid (90:10), prechilled to –20°C. The cells are then washed in PBS with 10% horse serum, and incubated overnight in the presence or absence of an antibody against protein kinase C in PBS containing 0.1% Triton X-100 and 10% horse serum. After washing with PBS, the cells are exposed to an appropriate second antibody (e.g., a fluorescent-labeled or peroxidase-linked antimouse IgG), and diluted in PBS containing 0.1% Triton X-100 and 10% horse serum for 0.5–2 h to visualize the binding of the antibodies against protein kinase C. The cells that were not incubated with the first antibody should be used to determine the extent of nonspecific binding of the second antibody. If a peroxidase-linked second antibody is used, diaminobenzidine is employed as the substrate for the peroxidase, and the cells should be incubated with 1% H_2O_2 in methanol after permeabilization to block endogenous peroxidase.

Figure 1 shows immunofluorescent staining of protein kinase C in primary cultures of neonatal rat dorsal root ganglion neurons using MC5, an antibody that recognizes the α and β forms of protein kinase C. The cells were fixed with 3% paraformaldehyde and permeabilized with Triton X-100 as described above. MC5 binding was visualized using biotin-conjugated rabbit antimouse IgG, which was then exposed to fluorescein-labeled streptavidin. Figure 1a shows staining in control cells and Fig. 1b shows staining in cells that had been exposed to 300 nM PDBu for 18 h prior to fixation. MC5 has produced bright staining in about 60% of the neurons in the control culture (Fig. 1a,b). In the cells that were pretreated with PDBu (Fig. 1c,d), there is some weak staining at a level not significantly greater than the background in cells not exposed to the first antibody. This suggests that much of the PKC-associated immunoreactivity had been lost from the PDBu-treated cells.

For comparison, Fig. 2 shows an experiment in which $^{45}Ca^{2+}$ uptake was measured in control dorsal root ganglion neurons and in neurons that had undergone the same downregulation treatment described for the cells shown in Fig. 1b (*see also* Burgess et al., 1989). In the control cells, PDBu (300 nM) and bradykinin (100 nM) both induced an increase in the rate of $^{45}Ca^{2+}$ uptake into the neurons. In the downregulated cells, the response to both stimuli was abolished. It was noted that bradykinin was still able to activate polyphosphoinositidase C and increase the level of diacylglycerol in the downregulated cells, implying that the bradykinin receptor and its transduction mechanism were still functional. The conclusion drawn from these experiments was that activation of protein kinase C may be implicated in the $^{45}Ca^{2+}$-uptake response to both PDBu and bradykinin in dorsal root ganglion neurons.

4.2. Protein Kinase C Inhibitors

The use of inhibitors of protein kinase C activity is another approach to studying the role of protein kinase C in cellular function. A wide variety of compounds have been reported to inhibit protein kinase C, but until recently, these were either not very potent or not sufficiently selective for protein kinase C compared to other protein kinases (*see* Hidaka et al., 1984; Loomis and Bell, 1988). No isozyme-selective inhibitor of protein kinase C has yet been reported. To be of general use as a pharmacological tool, an inhibitor of protein kinase C should be both selective and cell permeable.

4.2.1. Catalytic Domain Inhibitors

Protein kinase C inhibitors can be classified into two groups, depending on their target zone on the enzyme. Compounds that inhibit by interfering with the interaction of ATP or the protein substrate are catalytic domain inhibitors.

The isoquinolinesulfonamides, such as H-7 and H-9, are commonly used catalytic domain inhibitors. They cause inhibition by competing with ATP (Hidaka et al., 1984), they are cell permeable and they act at micromolar concentrations. H-7, for example, has been shown to inhibit phorbol ester-induced phos-

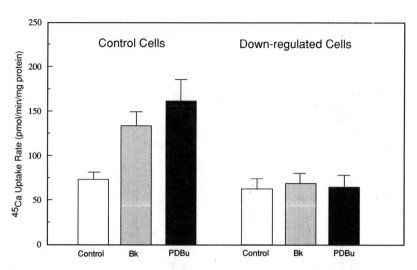

Fig. 2. Bradykinin and phorbol ester-induced $^{45}Ca^{2+}$ uptake into isolated rat dorsal root ganglion neurons in culture. The rate of uptake of $^{45}Ca^{2+}$, measured over 30 s, was determined in the presence or absence of either bradykinin (100 nM) or phorbol dibutyrate (300 nM) in control cells and in cells that had been pretreated with phorbol dibutyrate (300 nM) for 18 h to downregulate protein kinase C *(see* Fig. 1).

phorylation of the 47-kDa substrate in platelets (Kawamoto and Hidaka, 1984) and phorbol ester-induced differentiation of HL-60 cells (Matsui et al., 1986). The ATP-binding site for all protein kinases is, however, highly conserved, so although H-7 is often referred to as a protein kinase C inhibitor, it shows little selectivity for protein kinase C over other kinases *(see* Table 1) and, in the absence of other supporting information, it cannot be used to provide firm evidence for a role of protein kinase C in cellular activity.

In addition to the isoquinolinesulfonamides, sangivamycin, a nucleoside analog (Loomis and Bell, 1988), and a number of plant flavanoids, such as quercetin (Ferriola et al., 1989), also inhibit protein kinase C in the micromolar range by competing with ATP. The aminoacridines, such as acridine orange and acridine yellow G, are thought to have multiple interactions with the enzyme (Hannun and Bell, 1988), including the ATP-binding site. None of these compounds shows a high degree of selec-

Table 1
Inhibitors of Protein Kinase C

Inhibitor	Site of action	IC_{50} or K_i, μM			Selectivity factor	Cell permeable	Ref.
		PKC	MLCK	PKA			
H-7	ATP-binding site	6	97	3	0.5–16	Y	1
Staurosporine	ATP-binding site	0.003		0.015	5	Y	2
Partial staurosporine analog	ATP-binding site	0.01		1.5	150	?*	3
Pseudosubstrate peptide PKC (19–31)	Substrate-binding site	0.18	67	423	372–2350	N	4
Sphingosine	Regulatory domain	2–50	–	–	–	Y	5
Calphostin C	DG-binding site	0.05	>25	>50	>500	?*	6

[1]Hidaka et al., 1984.
[2]Meyer et al., 1989.
[3]Davis et al., 1990.
[4]House and Kemp, 1987.
[5]Hannun and Bell, 1986.
[6]Kobayashi et al., 1989a,b.
*Do show cellular activity, but at higher concentration than against enzyme.

tivity for protein kinase C, presumably because of the homology between the ATP-binding domains of protein kinases.

The microbial alkaloid staurosporine and its analogs K252a and UCN-01 are the most potent in vitro inhibitors of protein kinase C to date with K_i values of 2.7, 20, and 25 nM, respectively (Tamaoki et al., 1986; Kase et al., 1987). Staurosporine can be used in intact cells and tissues, and has been shown to inhibit phorbol ester- and bradykinin-induced activation of C-fibers (Dray et al., 1988), as well as some of the effects of phorbol esters and bradykinin in cultured sensory neurons (Burgess et al., 1989) at concentrations between 30 nM and 1μM. Staurosporine and its analogs do not interfere with the binding of phorbol esters, and they are able to inhibit the proteolytically generated catalytic domain of protein kinase C (Nakadate et al., 1988), which does not require lipids for activation. This is strong evidence that the site of action of these molecules is the catalytic domain, and it is thought that they act at or near to the ATP-binding site. In spite of their greatly increased potency, they are like the isoquinolinesulfonamides in that they show only limited selectivity among different protein kinases (*see* Rueegg and Burgess, 1989, for a review). Although it is claimed that UCN-01 exhibits some preference for protein kinase C over cAMP-dependent kinase (Takahashi et al., 1987), in general these compounds cannot be considered as selective protein kinase C inhibitors.

Recently, Davis et al. (1990) reported that partial structures of staurosporine were also active as protein kinase inhibitors and claimed that one of these (compound No. 3 in the above reference), which has an IC_{50} value against protein kinase C of 10 nM, was a factor of 150× less active against cAMP-dependent kinase and 1700× less active against Ca^{2+}/calmodulin-dependent protein kinase. This compound was also active, although less potent, in intact cells. It inhibited phorbol ester-induced phosphorylation of the 47-kDa substrate in platelets (IC_{50} value 0.7 μM), suggesting that it is able to penetrate cell membranes to some extent.

Synthetic peptides that resemble the pseudosubstrate sequences of protein kinase C are potent and selective inhibitors of the enzyme. House and Kemp (1987) have shown that a synthetic peptide corresponding to residues 19–31 of the first conserved region of the regulatory domain of PKCα, β, and γ is a selective inhibitor of protein kinase C with an IC_{50} value of 100 nM and a selectivity ratio of about 100 over cAMP-dependent kinase (see Table 1). In our hands, however (using the peptide [Ser^{25}] PKC [19–31] as a substrate for protein kinase C; see House and Kemp, 1987), it had an IC_{50} of between 3 μM and 10 μM. Although this 13 amino acid peptide is a valuable inhibitor of protein kinase C in vitro and in situations in which it is possible to introduce it inside cells (see, for example, Rane et al., 1989), it cannot penetrate intact cells and is not, therefore, of general use as a pharmacological tool.

4.2.2. Regulatory Domain Inhibitors

A second, relatively diverse group of compounds interferes with activation of protein kinase C by diacylglycerol, phosphatidylserine, or Ca^{2+}. These are the regulatory domain inhibitors.

Compounds such as trifluoperazine (Schatzman et al., 1981) and adriamycin (Katoh et al., 1981) are thought to inhibit protein kinase C by interfering with the binding of Ca^{2+} and phosphatidylserine, but are relatively nonselective since they also inhibit calmodulin-dependent kinases. Sphingosine (and some related compounds) interferes with lipid activation of protein kinase C in cell-free systems (Hannun et al., 1986; Bazzi and Nelsestuen, 1987), and it has been proposed that it may act as an endogenous regulator of protein kinase C activity (Hannun and Bell, 1986). Its exact mechanism of action is unclear, but it has been used, apparently successfully, in a number of studies in intact cells (e.g., Wilson et al., 1986; Hall et al., 1988). However, on the negative side, relatively high concentrations are required for inhibition (up to 100 μM), and solubility problems are often encountered. In addition, it has been claimed (Pittet et al., 1987) that sphinganine has nonspecific effects on intact cells, causing permeabilization of the plasma membrane of neutrophils at concentrations required to interfere with protein kinase C (although, see also Lambeth et al., 1988). More recently, it has been shown that sphingosine-1-phosphate (a metabolite of sphingosine) can release Ca^{2+} from intracellular organelles via a mechanism not thought to involve inhibition of protein kinase C (Ghosh et al., 1990), and that sphingosine has multiple nonspecific effects on adenylyl cyclase and cAMP accumulation (Johnson and Clark, 1990). It would seem, therefore, that the diverse actions of this class of compound would make studies in intact cells or tissues difficult to interpret.

In the last year, a family of compounds of microbial origin has been reported to inhibit protein kinase C by competing with diacylglycerol and phorbol esters. These molecules, the calphostins, appear to be highly potent and selective inhibitors of protein kinase C in vitro (Kobayashi et al., 1989a,b; Takahashi et al., 1989a). The most potent of these compounds is calphostin C (Fig. 3). It has been shown to inhibit the catalytic activity of semipurified protein kinase C and to displace the specific binding of [^{3}H]-PDBu with an IC_{50} value of 50 nM (Kobayashi et al., 1989a). The less potent analog, calphostin I (Fig. 3), is also equipotent (IC_{50} value ~200 nM) as an inhibitor of enzymic activity and as a displacer of [^{3}H]-PDBu. Calphostin C is reported to have no effect on the catalytic activity of the proteolytically generated catalytic domain of protein kinase C and, in assays on the whole enzyme, increasing the concentration of either Ca^{2+} or phosphatidylserine is apparently unable to relieve calphostin C-induced inhibition. Together these findings suggest that the calphostins inhibit the enzyme by interfering with the binding of diacylglycerol or phorbol ester to the regulatory domain.

The calphostins also appear to be quite selective. For example, calphostin C does not inhibit either cAMP-dependent kinase or tyrosine-specific protein kinase at up to 50 µM (*see* Table 1). Although these compounds have not yet been used extensively in biological experiments, the limited evidence that is available suggests that they may be of use in intact cells (Kobayashi et al., 1989a).

If futher studies indicate that calphostin C is cell permeable, then its selectivity may make it the inhibitor of choice for pharmacological studies of protein kinase C. Possible drawbacks might be its low solubility in aqueous solutions and its dark red color, which could interfere with colormetric assays.

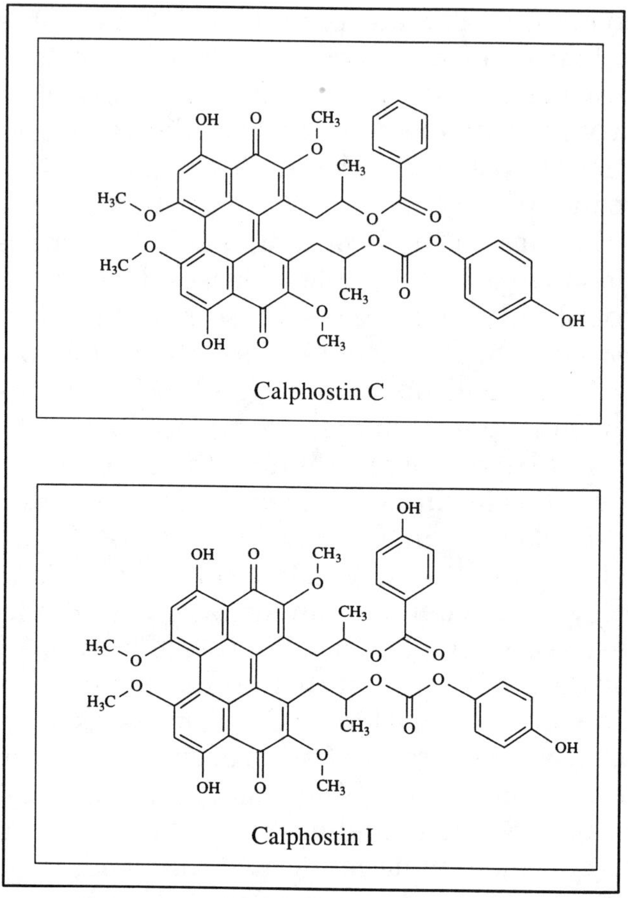

Fig. 3. Structures of calphostin-type microbial metabolites. The calphostins have been isolated from *Cladosporium cladosporoides*. They inhibit PKC and are thought to act at the regulatory domain by displacing diacylglycerol.

In summary, recent advances and discoveries in the field of protein kinase C inhibitors are quite impressive. The discovery of the calphostins and the development of the partial stauro-sporine analogs described above suggest that selective inhibition of protein kinase C with cell-permeable compounds may now be possible. At the time of writing, these compounds are not, however, commercially available. In the meantime, it is advised that the interpretation of results with the nonselective inhibitors of protein kinases that are currently available (e.g., H-7 and staurosporine) should not exclude the possibility that several protein kinases are being blocked simultaneously. Because of this, a combination of methods, including direct activation of protein kinase C with phorbol esters, downregulation to produce cells that are functionally deficient in the enzyme, and the cautious use of these nonselective compounds, is advised in attempting to define the role of protein kinase C in cellular activity.

5. Identification of Protein Kinase C Isozymes

The existence of multiple forms of protein kinase C (Nishizuka, 1988), their differential developmental expression (Yoshida et al., 1988), different enzymatic characteristics (Sekiguchi et al., 1987; Kosaka, 1988), and tissue or even cell-type-specific expression (e.g., Brandt et al., 1987; Hidaka et al., 1988) suggest specialized functions for the isozymic forms of protein kinase C (Ido et al., 1987).

It is possible to determine which subspecies are present in a tissue by means of protein purification (the chromatographic characterization of types I, II, and III, which correspond to γ, β, and α, respectively, are well-established; *see* Huang et al., 1986; Kosaka, 1988; Shearman et al., 1989b), immunological analysis, or Northern blot analysis of mRNA using cDNA probes specific to protein kinase C subspecies (*see*, for example, Brandt et al., 1987; Ono et al., 1988 for further details). This latter method could be used when antibodies are not available and in the case of the δ, ε (although *see* Schaap and Parker, 1990), and ζ subtypes where a method for protein purification has not been established.

5.1. Purification of Protein Kinase C Subspecies

5.1.1. Preparation of Tissue Fractions

All steps should be carried out at 4°C and, if possible, tissues freeze-clamped on removal. Tissues should be homogenized and cells lysed by sonication in about 10 vol of homogenization buffer containing: HEPES, 10 mM, pH 7.5; sucrose, 0.25M; EGTA, 10 mM; EDTA, 2 mM; dithiothreitol (freshly made up) 2 mM; phenylmethylenesulfonyl fluoride, 1 mM; leupeptin, 10 μg/mL; trypsin inhibitor, 4 μg/mL. The homogenate is given a low-speed spin (1000*g* for 10 min) to remove nuclei and cell debris, and the supernatant from this spin is then centrifuged at 100,000*g* for 1 h to generate cytosolic and membrane fractions. The supernatant contains the soluble enzyme. If the particulate enzyme is also to be analyzed, the pellet should be rehomogenized in homogenization buffer supplemented with 1% (v/v) Nonidet NP-40 and stirred at 4°C for 45 min followed by a further ultracentrifugation at 100,000*g* for 1 h. The supernatant from this last spin contains the particulate enzyme.

5.1.2. Chromatography

Both particulate and soluble fractions should then be applied to a DE-cellulose column (DE-52 from Whatman at approx 4 mL of gel for every gram wet wt of tissue), which has previously been equilibrated with Tris-HCl, 20 mM, pH 7.5, containing: EGTA, 0.5 mM; EDTA, 0.5 mM; and β-mercaptoethanol, 10 mM (buffer A). After application of the enzyme containing tissue fraction, the column is washed with 2 vol of buffer A, and the protein kinase C eluted with a linear gradient from 0–0.3M NaCl in buffer A (buffer B). Protein kinase C activity is assayed by means of the "standard" phosphorylation assay described in Section 2.2.2.1., and appears in the fractions collected between 45–75% of buffer B. Alternatively, the protein kinase C can be eluted batchwise (as described by Shearman et al., 1989b). The column is first washed with 15 vol 20 mM NaCl in buffer A, and then the protein kinase C is eluted by application of 3 vol of buffer A containing 120 mM NaCl. The first half-column vol is discarded. The combined fractions can be concentrated by dialysis against polyethyleneglycol.

The eluate from the DE-52 column is applied to a hydroxyapatite (Bio-Rad) column (1.5 × 5 cm), which has been preequilibrated with a solution of K$_3$PO$_4$, 20 mM; EGTA, 0.5 mM; EDTA, 0.5 mM; dithiothreitol, 1 mM; glycerol, 10% (v/v) (buffer C). The column is washed with a futher 50–60 mL of buffer C, and the protein kinase C isozymes are then eluted by applying a linear gradient composed of buffer C and buffer C containing 0.28 mM K$_3$PO$_4$ (buffer D) at a rate of 10 mL/h for 16 h. If, as described by Shearman et al. (1989b), the column is attached to a FPLC system (Pharmacia), buffer D contains 0.215 mM K_3PO_4, and the gradient is applied in a total of 84 mL at a rate of 0.4 mL/min. The peak activities elute at approx 70 mM, 90 mM, and 140 mM (depending on the column), corresponding to type I–γ, type II–β, and type III–α, respectively. The relative position of the three isozymes is constant, and the authenticity of the peaks can be verified by immunoblot analysis of the peak fractions with antibodies directed against specific isozymes as decribed in Section 4.1.2. Since the isozymes differ in their requirements for activators, it is also possible to characterize the fractions in terms of their enzymic activity. For example, the β isozyme does not have as great a requirement for Ca^{2+} as do the α and γ subtypes, and the γ isozyme can be activated by subnanomolar concentrations of arachidonic acid (*see* Nishizuka, 1988).

5.2. Immunological Analysis

The expression of protein kinase C in tissues or cells can be analyzed immunologically with antibodies directed against specific isotypes, either by immunocytochemical techniques or by immunoblot analysis as described in Sections 4.1.2. and 4.1.3. If a homogeneous population of cells, such as a cell line, is to be analyzed, then immunoblotting should provide answers about the subspecies present. This method does not provide information about the exact location of the enzyme within the cell other than whether it is in the cytosolic or membrane fraction (e.g., Lucas et al., 1990; Makowske et al., 1987). If, on the other hand, heterogeneous populations of cells, such as primary cultures of neurons containing fibroblasts and glial cells, or tissue slices are to be analyzed, then immunochemical staining would be more appropriate (*see* Huang et al., 1987; Saito et al., 1989).

Abbreviations

$[Ca^{2+}]_i$	intracellular free Ca^{2+} concentration
cAMP	adenosine 3',5' cyclic monophosphate
DG	1,2-diacylglycerol
DTT	dithiothreitol

PBS   phosphate-buffered saline
PDBu   4-β-phorbol-12,13-dibutyrate
PKC   protein kinase C
PMA   phorbol 12-myristate 13-acetate
SDS   sodium dodecyl sulfate

References

Albert K. A., Walaas S. I., Wang J. K. T., and Greengard P. (1986) Widespread occurrence of "87kDa," a major specific substrate for protein kinase C. *Proc. Natl. Acad. Sci. USA* **183,** 2822–2826.

Alexander D. R., Graves J. D., Lucas S. C., Cantrell, D. A., and Crumpton M. J. (1990) A method for measuring protein kinase C activity in permeabilized T lymphocytes by using protein substrates. *Biochem. J.* **268,** 303–308.

Ashendel C. L., Staller J. M., and Boutwell R. K. (1983) Identification of a calcium- and phospholipid-dependent phorbol ester binding activity in the soluble fraction of mouse tissues. *Biochem. Biophys. Res. Commun.* **111,** 340–345.

Asoaka Y., Kikkawa U., Sekiguchi K., Shearman M. S., and Kosaka Y. (1988) Activation of a brain-specific protein kinase C subspecies in the presence of phosphatidylethanol. *FEBS Lett.* **219,** 221–224.

Azhar S., Butte J., and Beaven E. (1987) Calcium-activated, phospholipid-dependent protein kinases from rat liver: Subcellular distribution, purification, and characterization of multiple forms. *Biochemistry* **26,** 7047–7057.

Ballester R. and Rosen O. M. (1985) Fate of immunoprecipitable protein kinase C in GH$_3$ cells treated with phorbol 12-myristae 13-acetate. *J. Biol. Chem.* **260,** 15,194–15,199.

Baraban J. M., Snyder S. H., and Alger B. E. (1985) Protein kinase C regulates ionic conductance in hippocampal pyramidal neurones: Electrophysiological effects of phorbol esters. *Proc. Natl. Acad. Sci. USA* **82,** 2538–2542.

Basi G. S., Jacobson R. D., Virag I., Schilling J., and Skene J. H. P. (1987) Primary structure and transcriptional regulation of GAP-43, a protein associated with nerve growth. *Cell* **49,** 785–791.

Baudier J., Bonner C., Kligman D., and Cole D. R. (1989) Protein kinase C substrates from bovine brain. *J. Biol. Chem.* **264,** 1824–1828.

Bazzi M. D. and Nelsestuan G. L. (1987) Mechanism of protein kinase C inhibition by sphingosine. *Biochem. Biophys. Res. Commun.* **146,** 203–207.

Bell R. M. (1986) Protein kinase C activation by diacylglycerol second messengers. *Cell* **45,** 631,632.

Berridge M. J. and Irvine R. F. (1984) Inositol phosphates and cell signalling. *Nature* **341,** 197–204.

Besterman J. M., Duronio V., and Cuatrecasas P. (1986) Rapid formation of diacylglycerol from phosphatidylcholine: A pathway for generation of a second messenger. *Proc. Natl. Acad. Sci. USA* **83**, 6785–6789.

Blackshear P. J., Nairn A. C., and Kuo J. F. (1988) Protein kinases 1988: A current perspective. *FASEB J.* **2**, 2957–2969.

Blumberg P. M., Jaken S., Konig B., Sharkey N. A., Leach K. L., Jeng A. Y., and Yeh E. (1984) Mechanism of action of the phorbol ester tumour promotors: Specific receptors for lipophilic ligands. *Biochem. Pharmacol.* **33**, 933–940.

Bocckino S. B., Blackmore P. F., and Exton J. H. (1985) Stimulation of 1,2-diacylglycerol accumulation in hepatocytes by vasopressin, epinephrine, and angiotensin II. *J. Biol. Chem.* **260**, 14,201–14,207.

Brandt S. J., Niedel J. E., Bell R. M., and Young W. S. III (1987) Distinct patterns of expressions of different protein kinase C mRNAs in rat tissues. *Cell* **49**, 57–63.

Burgess G. M., Mullaney I., McNeill M., Dunn P. M., and Rang H. P. (1989) Second messengers involved in the mechanism of action of bradykinin in sensory neurones in culture. *J. Neurosci.* **9**, 3314–3325.

Burgess S. K., Sahyoun N., Blanchard S. G., LeVine H., Chang K.-J., and Cuatrecasas P. (1986) Phorbol ester receptors and protein kinase C in primary neuronal cultures. *J. Cell Biol.* **102**, 312–319.

Casaubon S., Marais R., Parker P., and Strosberg A. D. (1989) Monoclonal antibodies to protein kinase C. Functional relationship between epitopes and cofactor binding sites. *Eur. J. Biochem.* **182**, 401–406.

Castagna M.,Takai Y., Kaibuchi K., Sano K., Kikkawa U., and Nishizuka Y. (1982) Direct activation of calcium-activated, phospholipid-dependent protein kinase by tumor-promoting phorbol ester. *J. Biol. Chem.* **257**, 7847–7851.

Cimler B. M., Andreasen T. J., Andreasen K. I., and Storm D. R. (1985) P-57 is a neural specific calmodulin binding protein. *J. Biol. Chem.* **260**, 10,784–10,788.

Cooper D. R., Watson J. E., Aceveda-Duncan M., Pollet R. J., Standaert M. I., and Farese R. V. (1989) Retention of specific protein kinase C isozymes following chronic phorbol ester treatment in RC3H-1 myocytes. *Biochem. Biophy. Res. Commun.* **161**, 327–334.

Coussens L., Parker P. J., Rhee L., Yang-Feng T. L., Chen E., Waterfield M. D., and Axel U. (1986) Multiple distinct forms of protein kinase C suggest diversity in cellular signalling pathways. *Science* **233**, 859–866.

Davis P. J., Hill C. H., Keech E., Lawson G., Nixon J. S., Sedgwick A. D., Wadsworth J., Westmacott D., and Wilkinson S. E. (1990) Potent selective inhibitors of protein kinase C. *FEBS Lett.* **259**, 61–63.

Davis R. J., Ganong B. R., Bell R. M., and Czech M. P. (1985) *sn* 1,2-Dioctanoylglycerol A cell permeable diacylglycerol that mimics phorbol diester action on the epidermal growth factor receptor and mitogenesis. *J. Biol. Chem.* **260**, 1562–1566.

De Graan P. N. E., Dekker L. V., Oestreicher A. B., Van Der Voorn L., and Gispen W. H. (1989) Determination of changes in the phosphorylation state of the neuron specific protein kinase C substrate B-50 (GAP-43). *J. Neurochem.* **52,** 17–23.

De Reimer S. A., Strong J. A., Albert K. A., Greengard, P., and Kaczmarek L. K. (1985) Enhancement of calcium current in Aplysia neurones by phorbol ester and protein kinase C. *Nature* **313,** 313–316.

Dray A. J., Bettaney J., Forster P., and Perkins M. N. (1988) Bradykinin-induced stimulation of afferent fibres is mediated through protein kinase C. *Neurosci. Lett.* **91,** 301–307.

Ferriola P. C. Cody V., and Middleton E., Jr. (1989) Protein kinase C inhibition by plant flavanoids. *Biochem. Pharmacol.* **38,** 1617–1624.

Fournier A., Hardy S. J., Clark K. J., and Murray A. W. (1989) Phorbol ester induces differential membrane-association of protein kinase C subspecies in human platelets. *Biochem. Biophys. Res. Commun.* **161,** 556–561.

Ghosh T. K., Bian J., and Gill D. L. (1990) Intracellular calcium release mediated by sphingosine derivatives generated cells. *Science* **248,** 1653–1656.

Gopalakrishna R., Barsky S. H., Thomas T. P., and Anderson W. B. (1986) Factors influencing chelator-stable, detergent-extractable, phorbol diester-induced membrane association of protein kinase C. *J. Biol. Chem.* **261,** 16,438–16,445.

Grillone L. R., Clark M. A., Godfrey R. W., Stassen F., and Crooke S. T. (1988) Vasopressin induces V_1 receptors to activate phosphatidylinositol- and phosphatidylcholine-specific phospholipase C and stimulates the release of arachidonic acid by at least two pathways in the smooth muscle cell line A-10 *J. Biol. Chem.* **263,** 2658–2663.

Hall F. L., Fernyhough P., Ishii D. N., and Vulliet R. P. (1988) Suppression of nerve growth factor-directed neurite outgrowth in PC12 cells by sphingosine, an inhibitor of protein kinase C. *J. Biol. Chem.* **263,** 4460–4466.

Hannun Y. A. and Bell R. M. (1986) Lysosphingolipids inhibit protein kinase C: Implications for the sphingolipidoses. *Science* **235,** 670–674.

Hannun Y. A. and Bell R. M. (1988) Aminoacridines, potent inhibitors of protein kinase C. *J. Biol. Chem.* **263,** 5124–5131.

Hannun Y. A., Loomis C. R., and Bell R. M. (1985) Activation of protein kinase C by Triton X-100 mixed micelles containing diacylglycerol and phosphatidylserine. *J. Biol. Chem.* **260,** 10,039–10,043.

Hidaka H., Inagaki M., Kawamoto S., and Sasaki Y. (1984) Isoquinolinesulfonamides: Novel and potent inhibitors of cyclic nucleotide dependent protein kinase and protein kinase C. *Biochemistry* **23,** 5036–5041.

Hidaka H., Tanaka T., Onoda K., Hagiwara M., Watanabe M., Ohta H., Ito Y., Tsurudone M., and Yoshida T. (1988) Cell type-specific expression of protein kinase C isozymes in the rabbit cerebellum. *J. Biol. Chem.* **263,** 4523–4526.

House C. and Kemp B. E. (1987) Protein kinase C contains a pseudosubstrate prototope in its regulatory domain. *Science* **238,** 1726–1728.

Huang F. L., Yoshida Y., Cunha-Melo J. R., and Beavan M. A. (1989) Differential downregulation of protein kinase C isozymes. *J. Biol. Chem.* 264, 4238–4243.

Huang F. L., Yoshida Y., Nakabayashi H., Knopf J. L., Young W. S. III, and Huang K.-P. (1987) Immunochemical identification of protein kinase C isozymes as products of discrete genes. *Biochem. Biophys. Res. Commun.* 149, 946–952.

Huang K.-P. (1989) The mechanism of protein kinase C activation. *Trends Neurosci.* 12, 425–432.

Huang K.-P. and Huang F. (1990) Differential sensitivity of protein kinase C isozymes to phospholipid-induced inactivation. *J. Biol. Chem.* 265, 738–744.

Huang K. -P., Nakabayashi H., and Huang F. L. (1986) Isozymic forms of rat brain Ca^{2+}-activated and phospholipid-dependent protein kinase. *Proc. Natl. Acad. Sci. USA* 83, 8535–8539.

Ido M., Sekiguchi K., Kikkawa U., and Nishizuka Y. (1987) Phosphorylation of the EGF receptor from A431 epidermoid carcinoma cells by three distinct types of protein kinase C. *FEBS Lett.* 219, 215–218.

Irving H. R. and Exton J. H. (1987) Phosphatidylcholine breakdown in rat liver plasma membranes. *J. Biol. Chem.* 262, 3440–3443.

Isakov N., McMahon P., and Altman A. (1990) Selective post-transcriptional down-regulation of protein kinase C isoenzymes in leukemic T cells chronically treated with phorbol esters. *J. Biol. Chem.* 265, 2091–2097.

Johnson J. A. and Clark R. B. (1990) Multiple non-specific effects of sphingosine on adenylate cyclase and cyclic AMP accumulation in S49 lymphoma cells preclude its use as a specific inhibitor of protein kinase C. *Biochem. J.* 268, 507–511.

Kaibuchi K., Takai Y., Sawamura M., Hoshijima M., Fujikura T., and Nishizuka Y. (1983) Synergistic functions of protein phosphorylation and calcium mobilization in platelet activation. *J. Biol. Chem.* 258, 6701–6704.

Kase H., Iwahashi K., Nakanishi S., Matsuda Y., Yamada K., Takahashi M., Murakata C., Sato A., and Kaneko M. (1987) K-252 compounds, novel and potent inhibitors of protein kinase C and cyclic nucleotide-dependent protein kinases. *Biochem. Biophys. Res. Commun.* 142, 436–440.

Katoh N., Wrenn R. W., Wise B. C., Shoji M., and Kou J. F. (1981) Substrate proteins for calmodulin-sensitive and phospholipid-sensitive calcium-dependent protein kinase in heart, and inhibition of their phosphorylation by palmitoylcarnitine. *Proc. Natl. Acad. Sci. USA* 78, 4813–4817.

Kawamoto S. and Hidaka H. (1984) 1-(5-Isoquinolinesulfonyl)-2-methylpiperazine (H7) is a selective inhibitor of protein kinase C in rabbit platelets. *Biochem. Biophys. Res. Commun.* 125, 258–264.

Kennerly D. A., Parker C. W., and Sullivan T. J. (1979) Use of diacylglycerol kinase to quantitate picomole levds of 1,2-diacylglycerol. *Anal. Biochem.* 98, 123–131.

Kikkawa U., Kishimoto A., and Nishizuka Y. (1989) The protein kinase C family and its implications. *Ann. Rev. Biochem.* **58,** 31–44.

Kikkawa U., Ono Y., and Ogita K. (1987) Identification of the structure of two types of protein kinase C: Alternative splicing from a single gene. *Science* **217,** 227–231.

Kikkawa U., Takai Y., Minakuchi R., Inohara S., and Nishizuka Y. (1982) Calcium-activated, phospholipid-dependent protein kinase from rat brain. Subcellular distribution, purification and properties. *J. Biol. Chem.* **257,** 13,341–13,348.

Kishimoto A., Mikawa M., Hashimoto K., Yasuda I., Tanaka S., Tominaga M., Kuroda T., and Nishizuka Y. (1989) Limited proteolysis of protein kinase C by calcium-dependent neutral protease (calpain) *J. Biol. Chem.* **264,** 4088–4092.

Kobayashi E., Nakano H., Morimoto M., and Tamaoki T. (1989a) Calphostin C (UCN-1028c), a novel microbial compound, is a highly potent and specific inhibitor of protein kinase C. *Biochem. Biophys. Res. Commun.* **159,** 548–543.

Kobayashi E., Ando K., Nakano H., Iida T., Ohno H., Morimoto M., and Tamaoki T. (1989b) Calphostins (UCN-1028), novel and specific inhibitors of protein kinase C. *J. Antibiot.* **41,** 1470–1475.

Kosaka Y. (1988) Multiplicity and tissue-specific expression of protein kinase C in rat tissues. *Kobe J. Med. Sci.* **34,** 271–285.

Kraft A. S. and Anderson W. B. (1983) Phorbol esters increase the amount of Ca^{2+}, phospholipid-dependent protein kinase C associated with plasma membrane. *Nature* **301,** 621–623.

Kraft A. S., Anderson W. B., Cooper H. L., and Sando J. J. (1982) Decrease in cytosolic calcium/phospholipid-dependent protein kinase activity following phorbol ester treatment of EL4 thymoma cells. *J. Biol. Chem.* **257,** 13,193–13,196.

Kuo J. F., Andersson R. G. G., Wise B. C., Mackerlova L., Salomonsson I., Bracket N. L., Katoh N., Shoji M., and Wrenn R. W. (1980) Calcium-dependent protein kinase: Widespread occurrence in various tissues and phyla of the animal kingdom and comparison of effects of phospholipid, calmodulin, and trifluoperazine. *Proc. Natl. Acad. Sci. USA* **77,** 7039–7043.

Laemmli U. K. (1970) Cleavage of structural proteins during the assembly of the head of bacteriophage T4. *Nature* **227,** 680–685.

Lambeth J. D., Burnham D. N., and Tyagi S. R. (1988) Sphinganine effects on chemoattractant-induced diacylglycerol generation, calcium fluxes, superoxide production, and on cell viability in the human neutrophil. *J. Biol. Chem.* **263,** 3818–3822.

Lapetina E. G., Reep B., Ganong B. R. and Bell R. M. (1985) Exogenous sn-1,2-diacylglycerols containing saturated fatty acids function as bioregulators of protein kinase C in human platelets. *J. Biol. Chem.* **260,** 1358–1361.

Lee M.-H. and Bell R. M. (1986) The lipid binding, regulatory domain of protein kinase C. *J. Biol. Chem.* **261,** 14,867–14,870.

Liles W. C., Hunter D. D., Meier K. E., and Nathanson N. M. (1986) Activation of protein kinase C induces rapid internalization and subsequent degradation of muscarinic acetylcholine receptors in neuroblastoma cells. *J. Biol. Chem.* **261,** 5307–5313.

Loomis C. R. and Bell R. M. (1988) Sangivamycin, a nucleoside analogue is a potent inhibitor of protein kinase C. *J. Biol. Chem.* **263,** 1682–1692.

Lucas S., Marais R., Graves J. D., Alexander D., Parker P., and Cantrell D. A. (1990) Heterogeneity of protein kinase C expression and regulation in T lymphocytes. *FEBS Lett.* **260,** 53–56.

Madison, D. V., Malenka R. C., and Nicoll R. A. (1986) Phorbol esters block a voltage-sensitive chloride current in hippocampal pyramidal cells. *Nature* **321,** 695–697.

Makowske M., Ballester R., Cayre Y., and Rosen O. M. (1987) Immunochemical evidence that three protein kinase C isozymes increase in abundance during HL-60 differentiation induced by dimethyl sulfoxide and tretinoic acid. *J. Biol. Chem.* **263,** 3402–3410.

Malenka R. C., Ayoub G. S., and Nicoll R. A. (1987) Phorbol esters enhance transmitter release in rat hippocampal slices. *Brain Res.* **403,** 198–203.

Malenka R. C., Madison D. V., Andrade R., and Nicoll R. A. (1986) Phorbol esters mimic some cholinergic actions in hippocampal pyramidal neurones. *J. Neurosci.* **6,** 475–480.

Martinson E. A., Goldstein D., and Brown J. H. (1989) Muscarinic receptor activation of phosphatidylcholine hydrolysis. *J. Biol. Chem.* **264,** 14,748–14,754.

Matthies H. J. G., Palfrey H. C., Hirning L. D., and Miller R. J. (1987) Downregulation of protein kinase C in neuronal cells: Effects on neurotransmitter release. *J. Neurosci.* **7,** 1198–1206.

Matozaki T. and Williams J. A. (1989) Multiple sources of 1,2-diacylglycerol in isolated rat pancreatic acini stimulated by cholecystokinin. *J. Biol. Chem.* **264,** 14,729–14,734.

Matsui T., Nakao Y., Koizumi T., Katakami Y., and Fujita T. (1986) Inhibition of phorbol ester-induced phenotypic differentiation of HL-60 cells by 1-(5-isoquinolinylsulfonyl)-2-methylpiperazine, a protein kinase inhibitor. *Cancer Res.* **46,** 583–587.

McCaffrey P. G., Rosner M. R., Kikkawa U., Sekiguchi K., Ogita K., Ase K., and Nishizuka Y. (1987) Characterisation of protein kinase C from normal and transformed cultured murine fibroblasts. *Biochem. Biophys. Res. Commun.* **146,** 140–146.

Meyer T., Regenass V., Fabbro D., Alter E., Rosel J., Müller M., Caravatti G., and Matter A. (1989) A derivative of staurosporine (CGP 41 251) shows selectivity for protein kinase C inhibition and *in vivo* antiproliferative as well as *in vivo* anti tumor activity. *Int. J. Cancer* **43,** 851–856.

Minakuchi R., Takai Y., Yu B., and Nishizuka Y. (1981) Widespread occurrence of calcium-activated, phospholipid-dependent protein kinase in mammalian tissues *J. Biochem. (Tokyo)* **89,** 1651–1654.

Mond J. J., Feurstein N., Finkelman F. D., Huang F., Huang K.-P., and Dennis G. (1987) B-lymphocyte activation mediated by anti-immunoglobulin antibody in the absence of protein kinase C. *Proc. Natl. Acad. Sci. USA* **84,** 8588–8592.

Nakadate T., Jeng A. Y., and Blumberg P. M. (1988) Comparison of protein kinase C functional assays to clarify mechanisms of inhibitor action. *Biochem. Pharmacol.* **37,** 1541–1545.

Nishizuka Y. (1984) The role of protein kinase C in cell surface signal transduction and tumour promotion. *Nature* **308,** 693–698.

Nishizuka Y. (1986) Studies and perspectives of protein kinase C. *Science* **233,** 305–312.

Nishizuka Y. (1988) The molecular heterogeneity of protein kinase C and its implication for cellular recognition. *Nature* **334,** 661–665.

Nishizuka Y. (1989) Studies and prospectives of the protein kinase C family for cellular recognition. *Cancer* **10,** 1892–1903.

Oestreicher A. B., Zwiers H., Schotman P., and Gispen W. H. (1981) Immunohistochemical localization of a phosphoprotein (B-50) isolated from rat brain synaptosomal membranes. *Brain Res. Bull.* **6,** 145–153.

Oestreicher A. B. and Gispen W. H. (1986) Comparison of the immunocytochemical distribution of the phosphoprotein B50 in the cerebellum and hippocampus of the immature and adult rat brain. *Brain Res.* **375,** 207–279.

Ohno S., Akita Y., Konno Y., Imajoh S., and Suzuki K. (1988) A novel phorbol ester receptor/protein kinase PKC, distantly related to the protein kinase C family. *Cell* **53,** 731–741.

Ono Y., Fujii T., Ogita K., Kikkawa U., Igarashi K., and Nishizuka Y. (1987) Identification of three additional members of rat protein kinase C family: δ-, ε-, and ζ subspecies. *FEBS Lett.* **226,** 125–128.

Ono Y., Fujii T., Ogita, K., Kikkawa U., Igarashi K., and Nishizuka Y. (1988) The structure, expression and properties of additional members of the protein kinase C family. *J. Biol. Chem.* **263,** 6927–6932.

Ono Y., Fujii T., Ogita, K., Igarashi K., Kuno T., Tanaka C., Kikkawa U., and Nishizuka Y. (1989) Phorbol ester binding to protein kinase C requires a cysteine-rich zinc-finger-like sequence. *Proc. Natl. Acad. Sci. USA* **86,** 4868–4871.

Ono Y., Kurokawa T., Kawhara K., Nishimura O., Marumoto R., Igarashi K., Sugino Y., Kikkawa U., Ogita K., and Nishizuka Y. (1986) Cloning of rat brain protein kinase C complementary DNA. *FEBS Lett.* **203,** 111–115.

Osugi, T., Imaizumi, T., Mizushima A., Uchida, S., and Yoshida, H. (1986) 1-Oleoyl-2-acetyl-glycerol and phorbol diester stimulate Ca^{2+} influx through Ca^{2+} channals in neuroblastoma × glioma hybrid NG108-15 cells. *Eur. J. Pharmacol.* **126,** 47–51.

Parker P. J., Coussens L., Totty N., Rhee L., Young S., Chen E., Stabel S., Waterfield M. D., and Ullrich A. (1986) The complete primary structure of protein kinase C—The major phorbol ester receptor. *Science* **233**, 853–859.

Pasti G., Lacal J.-C., Warren B. S., Aaronson S. A., and Blumberg P. M. (1986) Loss of mouse fibroblast cell response to phorbol esters restored by microinjection of protein kinase C. *Nature* **324**, 375–377.

Pittet D., Krause K.-H., Wolheim C. B., Bruzzone R., and Lew D. P. (1987) Non-selective inhibition of neutrophil functions by sphinganine. *J. Biol. Chem.* **262**, 10,072–10,076.

Polverino A. J. and Barritt G. B. (1988) On the source of the vasopressin-induced increases in diacylglycerol in hepatocytes. *Biochim. Biophys. Acta* **970**, 75–82.

Pozzan T., Gatti G., Vicentini L. M., and Meldolesi J. (1984) Ca^{2+}-dependent and -independent release of neurotransmitters from PC12 cells: A role for protein kinase C activation. *J. Cell Biol.* **99**, 628–638.

Preiss J., Loomis C. R., Bishop W. R., Stein R. Neidel J. E., and Bell R. M. (1986) Quantitative measurement of *sn*-1,2-diacylglycerol present in platelets, hepatocytes, and *ras*- and *sis*-transformed normal rat kidney cells. *J. Biol. Chem.* **261**, 8597–8600.

Rane S. G. and Dunlap K. (1986) Kinase C activator 1,2-oleoylacetylglycerol attentuates voltage-dependent calcium current in sensory neurons. *Proc. Natl. Acad. Sci. USA* **83**, 184–188.

Rane S. G., Walsh M. P., McDonald J. R., and Dunlap K. (1989) Specific inhibitors of protein kinase C block transmitter-induced modulation of sensory neuron calcium current. *Neuron* **3**, 239–245.

Rueegg U. T. and Burgess G. M. (1989) Staurosporine, K-252 and UCN-01: Potent but nonspecific inhibitors of protein kinases. *Trends Pharmacol. Sci.* **10**, 218–220.

Saito N., Kose A., Ito A., Hosoda K., Mori M., Hirata M., Ogita K., Kikkawa U., Ono Y., Igarashi K., Nishizuka Y., and Tanaka C. (1989) Immunocytochemical localisation of β_{II} subspecies of protein kinase C in rat brain. *Proc. Natl. Acad. Sci. USA* **86**, 3409–3413.

Schaap D. and Parker P. J. (1990) Expression, purification and characterization of protein kinase C-ε. *J. Biol. Chem.* **265**, 7301–7307.

Schaap D., Parker P. J., Bristol A., Kriz R., and Knopk J. (1989) Unique substrates specificity and regulatory properties of PKC-ε: A rationale for diversity. *FEBS Lett.* **243**, 351–357.

Schatzman R. C., Wise B. C., and Kuo J. F. (1981) Phospholipid-sensitive calcium-dependent protein kinase: Inhibition by antipsychotic drugs. *Biochem. Biophys. Res. Commun.* **98**, 669–676.

Sekiguchi K., Tsukada M., Ogita K., Kikkawa U., and Nishizuka Y. (1987) Three distinct forms of rat brain protein kinase C: Differential responses to unsaturated fatty acids. *Biochem. Biophys. Res. Commun.* **145**, 797–802.

Sekiguchi K., Tsukada M., Ase K., Kikkawa U., and Nishizuka Y. (1988) Mode of action and kinetic properties of three distinct forms of protein kinase C from rat brain. *J. Biochem. (Tokyo)* **103**, 759–765.

270   *Burgess*

Shearman M. S., Naor Z., Kikkawa U., and Nishizuka Y. (1987) Differential expression of multiple protein kinase C subspecies in rat central nervous tissue. *Biochem. Biophys. Res. Commun.* **147,** 911–919.

Shearman M. S., Berry N., Oda T., Ase K., Kikkawa U., and Nishizuka Y. (1988) Isolation of protein kinase C subspecies fom a preparation of human T lymphocytes. *FEBS Lett.* **234,** 387–391.

Shearman M. S., Naor Z., Sekiguchi K., Kishimoto A., and Nishizuka Y. (1989a) Selective activation of the γ subspecies of protein kinase C from bovine cerebellum by arachidonic acids and its lipoxygenase metabolites. *FEBS Lett.* **243,** 177–182.

Shearman M. S., Ogita K., Kikkawa U., and Nishizuka Y. (1989b) A rapid method for the resolution of protein kinase C subspecies from rat brain tissue. *Meth. Enzymol.* **168,** 347–351.

Shearman M. S., Sekiguchi K., and Nishizuka Y. (1989c) Modulation of ion channel activity: A key function of the protein kinase C enzyme family. *Pharmacol. Rev.* **41,** 211–237.

Spinelli W. and Ishii D. N. (1983) Tumor promoter receptors regulating neurite formation in cultured human neuroblastoma cells. *Cancer Res.* **43,** 4119–4125.

Storry J. E. and Tuckley B. (1967) Thin layer chromatography of plasma lipids by single development. *Lipids* **2,** 501,502.

Strong J. A., Fox A. P., Tsien R. W., and Kaczmarek L. K. (1987) Stimulation of protein kinase C recruits covert calcium channels. *Nature* **325,** 714–717.

Takahashi I., Asano K., Kawamoto I., Tamaoki T., and Kawamoto H. (1989a) UCN-01 and UCN-02, New selective inhibitors of protein kinase C. I. Screening producing organism and fermentation. *J. Antibiot.* **42,** 564–570.

Takahashi I., Saito Y., Yoshida M., Sano H., Nakano H., Morimoto M., and Tamaoki T. (1989b) UCN-01 and UCN-02, New selective inhibitors of protein kinase C. II. Purification, physico-chemical properties, structural determination and biological activities. *J. Antibiot.* **42,** 571–576.

Tamaoki T., Nomoto H., Takahashi I., Kato Y., Morimoto M., and Tomita F. (1986) Staurosporine, a potent inhibitor of phopholipid/Ca⁺⁺ dependent protein kinase. *Biochem. Biophys. Res. Commun.* **135,** 397–402.

Tsukada M., Asaoka Y., Sekiguchi K., Kikkawa U., and Nishizuka Y. (1988) Properties of protein kinase C subspecies in human platelets. *Biochem. Biophys. Res. Commun.* **155,** 1387–1395.

Vincenti L. M. and Villereal M. L. (1985) Activation of Na^+/H^+ exchange in cultured fibroblasts: Synergism and antagonism betweeen phorbol ester, Ca^{2+} ionophore, and growth factors. *Proc. Natl. Acad. Sci. USA* **82,** 8053–8056.

Vincenti L. M., Di Virgilio F., Ambrosini A., Pozzan A., and Meldolesi J. (1985) Tumor promotor phorbol 12-myristate, 13-acetate inhibits phosphoinositide hydrolysis and cytosolic Ca^{2+} rise induced by the activation of muscarinic receptors in PC 12 cells. *Biochem. Biophy. Res. Commun.* **127,** 310–317.

Walaas S. I., Nairn A. C., and Greengard, P. (1983a) Regional distribution of calcium and cyclic adenosine 3',5'-monophosphate protein phosphorylation systems in mammalian brain. I. Particulate systems. *J. Neurosci.* **3**, 291–301.

Walaas S. I., Nairn A. C., and Greengard P. (1983b) Regional distribution of calcium and cyclic adenosine 3',5'-monophosphate protein phosphorylation systems in mammalian brain. II. Soluble systems. *J. Neurosci.* **3**, 302–311.

Wang J. K. T., Walaas S. I., Shira T. S., Aderlem A., and Greengard P. (1989) Phosphorylation and associated translocation of the 87-KDa protein a major protein kinase C substrate, in isolated nerve terminals. *Proc. Natl. Acad. Sci. USA* **86**, 2253–2256.

Watson S. P., McNally J., Shipman L. J., and Godfrey P. J. (1988) The action of the protein kinase C inhibitor, staurosporine, on human platelets. *Biochem. J.* **249**, 345–350.

Wilson E., Olcott M. C., Bell R. M., Merrill, A. H. Jr., and Lambeth J. O. (1986) Inhibition of the oxidative burst in human neutrophils by sphingoid long-chain bases. *J. Biol. Chem.* **261**, 12,616–12,623.

Wise B. C., Glass D. B., Chou C.-H. J., Raynor R. L., Katoh N., Schatzman R. C., Turner R. S., Kibler R. F., and Kuo J. F. (1982) Phospholipid-sensitive Ca^{2+}-dependent protein kinase from heart. II. Substrate specificity and inhibition by various agents. *J. Biol. Chem.* **257**, 8489–8495.

Wolf M., LeVine H. III, May W. S. Jr., Cuatrecasas P., and Sahyoun N. (1985a) A model for intracellular translocation of protein kinase C involving synergism between Ca^{2+} and phorbol esters. *Nature* **317**, 346–349.

Wolf M., Cuatrecasas P., and Sayhoun N. (1985b) Interaction of protein kinase C with membranes is regulated by Ca^{2+}, phorbol esters, and ATP. *J. Biol. Chem.* **160**, 15,718–15,722.

Wooten M. W. and Wrenn R. W. (1984) Phorbol ester induces intracellular translocation of phospholipid/Ca^{2+}-dependent protein kinase and stimulates amylase secretion in isolated pancreatic acini. *FEBS Lett.* **171**, 183,184.

Worley P. F., Baraban J. M., and Snyder S. H. (1986) Heterogeneous localisation of protein kinase C in rat brain: Autoradiographic analysis of phorbol ester receptor binding. *J. Neurosci.* **6**, 199–207.

Wu W. C. S., Walaas S. I., Nairn A. C., and Greengard P. (1982) Calcium/phospholipid regulates phophorylation of M_r "87KDa" substrate protein in brain synaptosomes. *Proc. Natl. Acad. Sci. USA* **79**, 5249–5253.

Yoshida Y., Huang F. L., Nakabayashi, H., and Huang K.-P. (1988) Tissue distribution and developmental expression of protein kinase C isozymes. *J. Biol. Chem.* **263**, 9869–9873.

Young S., Parker P. J., Ullrich A., and Stabel S. (1987) Down-regulation of protein kinase C is due to an increased rate of degradation. *Biochem. J.* **244**, 775–779.

Synthetic Analogs
of Intracellular Messengers

Barry V. L. Potter

1. Introduction

Signaling pathways that convert extracellular stimuli into intracellular responses regulate many diverse physiological and biochemical processes, and share many characteristics. An external signal binds to a cell surface receptor and stimulates a transducer to activate an internal effector, which initiates production of an intracellular messenger. The intracellular messenger sets in motion events that culminate in the overall cellular response, often by means of actions on protein kinases (Cohen, 1985).

The two major transmembrane signaling pathways involving intracellular messengers known today have their origins in the 1950s. Adenosine 3',5'-cyclic monophosphate (cAMP; Fig. 1), a molecule well established as a ubiquitous intracellular messenger (for a review, *see* Schramm and Selinger, 1984), was discovered some 33 years ago (Sutherland and Rall, 1958). In 1953, already before the realization of the importance of cAMP, the phenomenon of agonist-stimulated, receptor-mediated phospholipid turnover had been demonstrated (Hokin, 1985). However, it was not until 1975 that the suggestion was made that this was linked to an increase in cytosolic $[Ca^{2+}]$ (Michell, 1975). The realization that phosphoinositidase C catalyzed cleavage phosphatidylinositol 4,5-bisphosphate [$PtdIns(4,5)P_2$] (Fig. 2[1]) to form two intracellular messengers, D-*myo*-inositol 1,4,5-trisphosphate (2) [$Ins(1,4,5)P_3$], which mobilizes intracellular Ca^{2+}

From: *Neuromethods, Vol. 20: Intracellular Messengers*
Eds: A. Boulton, G. Baker, and C. Taylor ©1992 The Humana Press Inc.

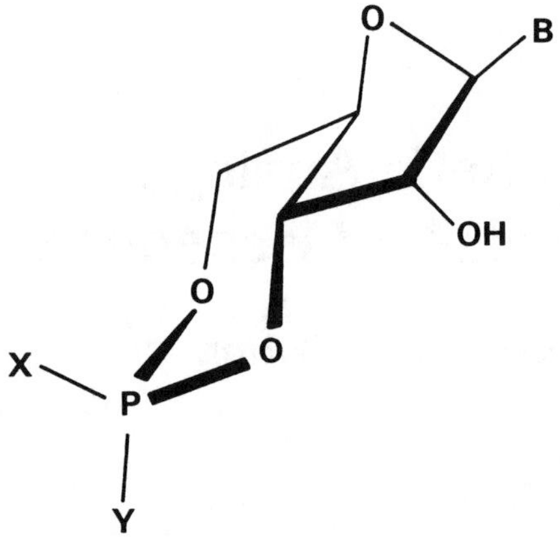

X = Y = O; B = A (cAMP); B = G (cGMP)

X = O; Y = S; B = A (Rp–cAMPS); B = G (Rp–cGMPS)

X = S; Y = O; B = A (Sp–cAMPS); B = G (Sp–cGMPS)

X = S; Y = S; B = A (cAMPS$_2$)

X = NMe$_2$; Y = O; B = A (Rp–cAMPNMe$_2$)

A =   G =

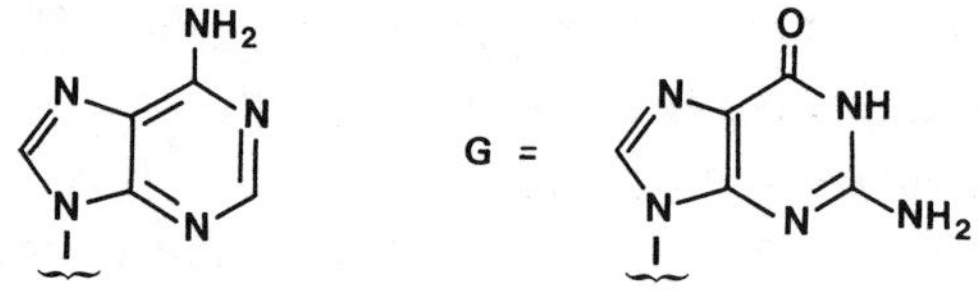

Fig. 1. Structure of adenosine- and guanosine-3',5'-cyclic phosphate and phosphorothioate analogs.

(Berridge and Irvine, 1984, 1989), and 1,2-diacylglycerol (3) (DG), which activates proteins kinase C (PKC) (Nishizuka, 1984, 1988), stimulated the recent upsurge of interest in the polyphosphoinositide signaling pathway (Berridge, 1987; Berridge and Irvine, 1989).

Advances in our understanding of the biology of both cAMP and polyphosphoinositide signaling pathways are now accompanied by chemical syntheses of analogs that might provide tools for the study of such pathways, or interfere with the actions or metabolism of intracellular messengers. Such pharmacological intervention may allow the receptors that bind intracellular messengers, and the cellular machinery responsible for generation

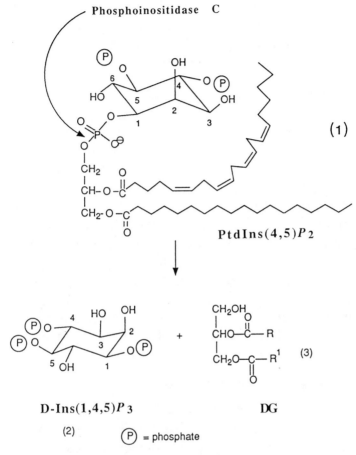

Fig. 2. Chemical structures of PtdIns(4,5)P_2, D-Ins(1,4,5)P_3, and DG.

and metabolism of messenger molecules, to become targets for rational drug design (Nahorski and Potter, 1989; Chilvers et al., 1989; Nahorski, 1990). For example, a study of 24 cAMP analogs showed that they could inhibit cancer cell growth at micromolar concentration (Katsaros et al., 1987). Indeed, since the discovery of cAMP, many hundreds of chemically modified analogs have been synthesized (Miller, 1978). Despite this, only two cAMP antagonists have been identified. Such efforts are only just beginning for Ins(1,4,5)P_3 and DG, and the chemistry, at least of the former molecule, is complex. This chapter will attempt to focus on the synthesis and biology of intracellular messenger analogs, with a bias toward the hydrophilic phosphorylated intracellular messengers, cAMP and Ins(1,4,5)P_3. The apparently successful and versatile phosphorothioate analogs of these messengers will be discussed in detail.

2. The cAMP Signaling System

cAMP is an intracellular messenger that activates cAMP-dependent protein kinase, and thus, mediates the action of various hormones and neurotransmitters. cAMP is synthesized from ATP by adenylyl cyclase, which, together with cAMP phosphodiesterase, regulates the steady-state and hormone-stimulated levels of cAMP. The activity of adenylyl cyclase is modulated by stimulatory and inhibitory G proteins linked to stimulatory and inhibitory receptors, respectively (Schramm and Selinger, 1984). For more detailed discussion of the biology of the cAMP signaling system, the reader is referred to reviews (Schramm and Selinger, 1984; Levitzki, 1988), a recent volume (Corbin and Johnson, 1988), and the chapters by Milligan and Murray in this volume.

2.1. Synthesis of cAMP Analogs

The nucleotide adenosine 3',5'-cyclic phosphate possesses fused five- and six-membered ring systems (Fig. 1). Chemical modification of this structure, with a view to determining structure–activity relationships and defining interactions between cAMP and cAMP-dependent protein kinase, has been focused to a great extent upon the purine ring, and many base-modified analogs have been prepared (Jastorff et al., 1979). Analogs prepared by modification of the environment around the single phosphorus atom, however, have also been of special interest and have led to the identification of the only known cAMP antagonists. The phosphorus atom in cAMP is a prochiral center within the chiral environment of a nucleotide. Replacement of one of the two peripheral oxygen atoms by another group produces diastereoisomers and, thus, for the phosphorothioates (e.g., the isomers of adenosine 3',5'-cyclic monophosphorothioate, cAMPS; Fig. 1), the introduction of a single sulfur atom at this center produces R_p and S_p epimers (Eckstein, 1985). Such diastereoisomers can possess significantly different properties (*see below*).

cAMPS was first synthesized in 1974 (Eckstein et al., 1974) by the phosphorylation of 2',3'-diacetyl adenosine with bis(*p*-nitrophenyl)thiophosphorochloridate and subsequent ring closure of the product. This method has now been modified and improved to provide a general preparation of nucleoside 3',5'-cyclic phosphorothioates, where only the base-protected nucleoside is phosphorylated and the product is then cyclized (Eckstein and Kutzke, 1986) (Fig. 3a). The unequal mixture of diastereoisomers resulting from this procedure is separated either by crystallization, ion-exchange chromatography, or by reverse-phase HPLC.

A simplified method for the cyclothiophosphorylation of unprotected nucleosides has now been reported (Genieser et al., 1988) (Fig. 3b), in which thiophosphorylation of the unprotected nucleoside and cyclization of the resulting thiophosphorodichloridate *in situ* give a good yield of the diastereoisomeric cyclic phosphorothioate mixture. This is particularly applicable to the synthesis of phosphorothioate analogs possessing a modified heterocyclic base.

A stereospecific approach, avoiding the sometimes tedious separation of diastereoisomeric mixtures, has also been devised (Baraniak and Stec, 1987), in which, for example for cAMPS, N^6-benzoyl-2'-O-benzoyl-adenosine-3',5'-cyclic phosphate is converted into separable diastereoisomeric anilidates, which, by reaction with potassium and CS_2, can be converted into diastereoisomerically pure isomers of cAMPS in a stereospecific fashion, with retention of configuration at phosphorus. This stereospecific method has been applied to other cyclic nucleotides (Stec, 1983), but has the disadvantage of requiring a multistep synthesis with low overall yields. Nevertheless, a vital reference point was, however, provided by material synthesized by this route, whereby the absolute configuration at phosphorus of nucleotides prepared via cyclization syntheses could be correlated. Additional reference has recently also been provided by a report of the X-ray structure of crystalline R_p-cUMPS (uridine

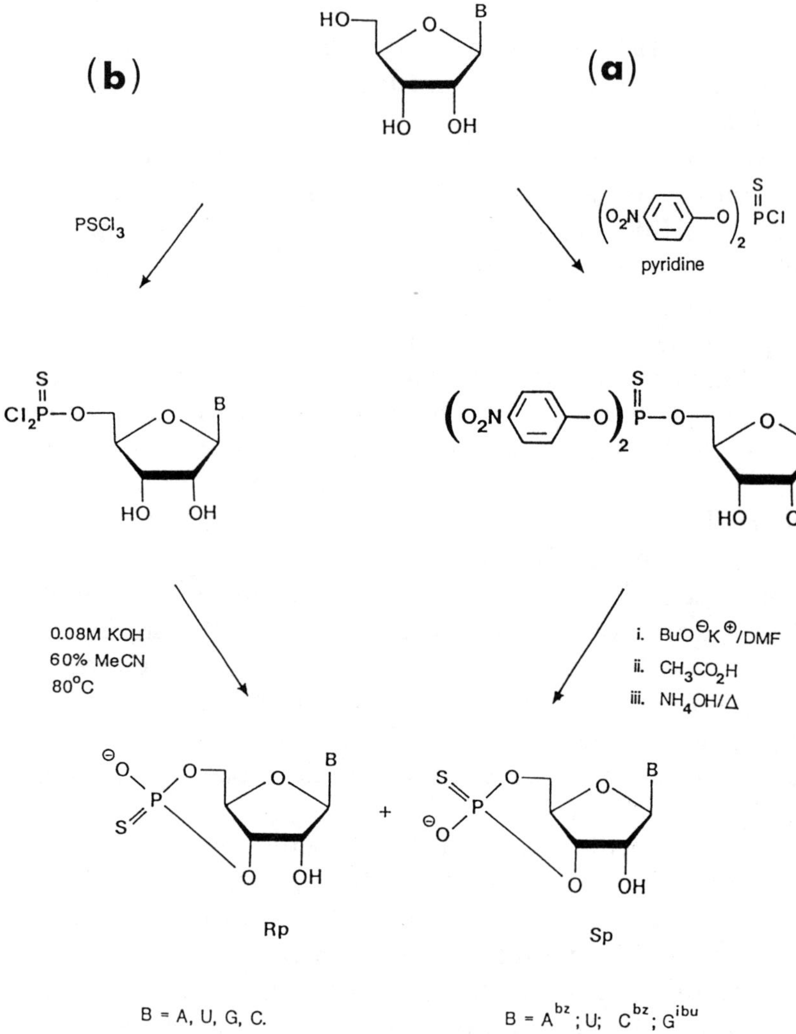

Fig. 3. Chemical synthesis of ribonucleoside 3',5'-cyclic phosphorothioates.

3',5'-cyclic monophosphate) (Hinrichs et al., 1987). The synthesis of the cAMP phosphorodithioate analog (cAMPS$_2$), in which both prochiral oxygens are replaced by sulfur (Fig. 1), has been accomplished by a similar route employing a cyclic thiophosphoranilidate (Baraniak and Stec, 1987).

A less commonly used route to cyclic nucleotide phosphorothioates draws upon the natural effector enzyme, which gen-

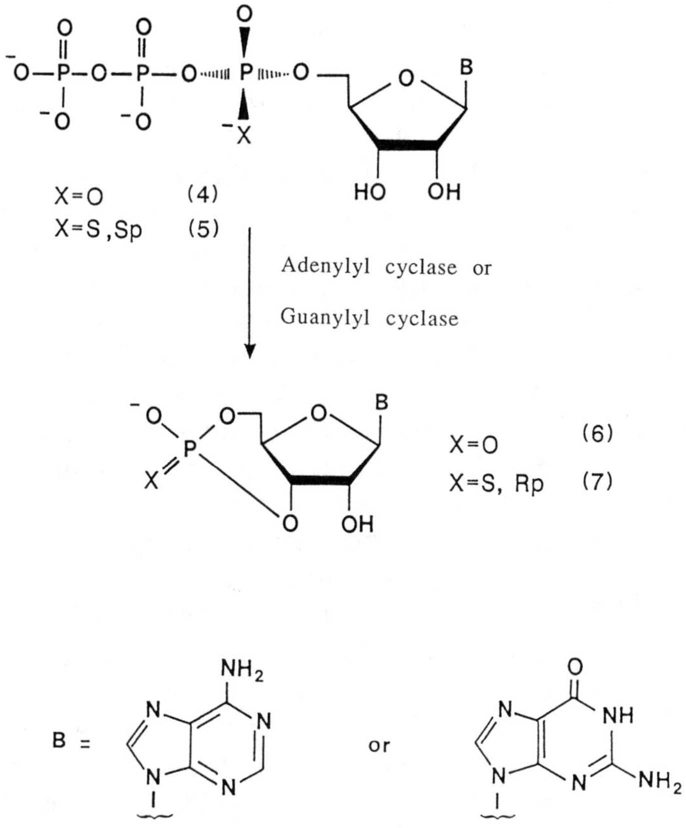

Fig. 4. Enzymic synthesis of cAMP, cGMP, R_p-cAMPS, and R_p-cGMPS.

erates the cyclic nucleotide (Fig. 4[6]) from a nucleoside triphosphate (4) and involves the stereoselective enzymatic cyclization of one diastereoisomer of the appropriate nucleoside 5'-O-(1-thiotriphosphate) (5) with inversion of configuration at phosphorus, to the nucleoside 3',5'-cyclic phosphorothioate (7) using the appropriate cyclase (Fig. 4). Both adenylyl cyclase and guanylyl cyclase have been used to prepare the R_p diastereoisomers of cAMPS (Eckstein et al., 1981) and cGMPS (Senter et al., 1983), respectively, from the appropriate S_p-α-thiotriphosphate. This method suffers from the drawback that the enzymes are not usually readily available, but also, more importantly, that the route only gives access to one diastereoisomer.

2.2. Biology of cAMP Analogs

Cyclic nucleotides are broken down by specific phosphodiesterases (*see* chapter by Murray). Considerable effort has been invested in the design of selective inhibitors of these enzymes (Weishaar and Bristol, 1990). Phosphorothioates are often turned over more slowly by enzymes than their phosphate counterparts, and cAMPS differs from cAMP in this respect by virtue of its enhanced metabolic stability, as shown by the slow hydrolysis of the S_p-cAMPS by beef heart and rabbit brain cAMP phosphodiesterases (Eckstein et al., 1974). (It was subsequently discovered that the experiments with these enzymes had been carried out on pure S_p-cAMPS and not on a mixture of diastereoisomers, as reported.) R_p-cAMPS was found not to be a substrate for beef heart cAMP phosphodiesterase, although both isomers are cleaved with inversion of configuration at phosphorus by baker's yeast phosphodiesterase (Jarvest et al., 1982).

S_p-cAMPS (axial sulfur substitution) activates both type I and type II cAMP-dependent protein kinases, but R_p-cAMPS (equatorial sulfur substitution) is an antagonist and does not cause release of the catalytic subunit of cAMP-dependent protein kinase, although it binds to the regulatory subunit (Parker Bothelo et al., 1988a). It has been proposed that the inability of R_p-cAMPS to stimulate a conformational change in the regulatory subunit is the result of interference of the equatorial sulfur with a hydrogen bond or salt bridge normally formed between the protein and the agonist (P-S has a much lower affinity for forming H-bonds than P-O [Hinrichs et al., 1987]).

Relative activation potencies for cAMP-dependent protein kinase of some 104 cAMP analogs, mostly modified in the adenine base, have been determined (Ogreid et al., 1985). Activation of cAMP-dependent protein kinase has been observed in hepatocytes, where the phosphorylase kinase–phosphorylase glycogenolytic cascade is affected by S_p-cAMPS. R_p-cAMPS, however, acts as an antagonist (Rothermel et al., 1983). In general, it appears that the ribose cyclophosphate, rather than the adenine moiety, makes the major contribution to cAMP binding. Alterations to the ribose cyclophosphate cause dramatic changes in activation of, and affinity for, protein kinase, whereas modifications in the adenine ring have much less drastic consequences and all such compounds appear to be full agonists. In a study of the interaction with mammalian protein kinase of a number of cAMP analogs modified in the adenine, ribose, and cyclophosphate moiety, R_p-cAMPS was found to be an antagonist, and the uncharged dimethylamino analogs R_p- and S_p-cAMPNMe$_2$ (Fig. 1) were apparent partial agonists (de Wit et al., 1984).

Glucagon-induced glycogenolysis is also inhibited by R_p-cAMPS (Rothermel et al., 1984). The phosphorodithioate analog of cAMP possessing double sulfur substitution at phosphorus, cAMPS$_2$ (Fig. 2), has also been found to be an antagonist of bovine heart type II cAMP-dependent protein kinase. This indicates that the most important structural requirement for the dissociation of the holoenzyme is an equatorial exocyclic oxygen (Parker Bothelo et al., 1988b). A recent study using purified enzyme has also been carried out (Rothermel and Parker Bothelo, 1988).

A number of reports testify to the ability of cAMPS to mimic the actions of cAMP. Thus, S_p-cAMPS can stimulate parotid cells to secrete amylase (Eckstein et al., 1976). It has also been used to study the cAMP-induced phosphorylation of microtubule-associated protein 2 in brain (Richter-Landsberg and Jastorff, 1985) as well as hormone-induced steroidogenesis (McMasters et al., 1986). cAMPS and other analogs have been employed in defining structure–activity relationships for the binding of cAMP to its cell surface receptor and chemotaxis in the slime mold *Dictyostelium discoideum* (Van Haastert and Kien, 1983), a system that shows no response with dibutyryl cAMP, another commonly used lipophilic analog (Hajjar, 1986). S_p-cAMPS binds to the *E. coli* cAMP receptor protein, as studied by ^{1}H NMR spectroscopy (Gronenborn and Clore, 1982). In investigations of stimulation of *lac* transcription by cAMP analogs, R_p-cAMPS was about 600-fold more potent at catabolite repression of *E. coli* β-galactosidase synthetase than S_p-cAMPS and about 50-fold more potent than cAMP (Scholuebbers et al., 1984). In hepatocytes, S_p-cAMPS inhibits triacylglycerol synthesis from glycerol, phosphatidylcholine from choline (Pelech et al., 1983), and in mast cells, the antigen-stimulated release of histamine (Eckstein and Foreman, 1978). cAMPS can penetrate cells and has an advantage over the widely used lipophilic analog, dibutyryl cAMP, which on hydrolysis releases butyric acid, which itself can exhibit a number of biological effects (Ito and Chou, 1984).

3. cGMP as an Intracellular Messenger

Guanosine 3',5'-cyclic monophosphate (cGMP) differs from cAMP in that the purine base guanine replaces adenine in the nucleotide (Fig. 1). cGMP appears to possess many of the characteristics of an intracellular messenger, but the precise details of its action are not yet clear, and a specific role for cGMP in the cell has yet to be established. It is in the retina, however, where light is tranformed into nerve impulses, that the most complete role for cGMP as an intracellular messenger in vision is currently available (Stryer, 1986, 1987; Chabre and Deterre, 1989; Matthews and Lamb, this vol.). cGMP is synthesized from GTP by guanylyl cyclase. Also, cGMP can activate a protein kinase, cGMP-dependent protein kinase (Edelman et al., 1987). Each subunit of this enzyme possesses two binding sites for cGMP. All reported direct activators of cGMP-dependent protein kinase retain a cyclic nucleotide structure.

Analogs of cGMP modified in the ribose cyclophosphate moiety produce ineffective cGMP-dependent protein kinase activators. A study of a large number of structurally modified analogs, especially of the purine ring, has addressed the factors responsible for binding of cGMP to the two sites of cGMP-dependent protein kinase (Corbin et al., 1986). Despite the gathering importance of cGMP, the synthetic diastereoisomers of cGMPS (Fig. 1) have not yet found wide application. However, they have been used as inhibitors of cyclic phosphodiesterase in rod outer segments (Zimmerman et al., 1985) and to determine the stereochemical course of the cGMP phosphodiesterase reaction as involving an inversion of configuration at phosphorus (Eckstein et al., 1988). R_p-cAMPS antagonizes the binding of cGMP to cGMP-dependent protein kinase, but S_p-cAMPS, which is a full agonist of cGMP-dependent protein kinase, is only a partial agonist for cGMP-dependent protein kinase (Hofmann et al., 1985).

4. *Myo*-Inositol 1,4,5-Trisphosphate as an Intracellular Messenger

The mechanisms underlying the generation and metabolism of Ins(1,4,5)P_3 will be outlined only briefly here. Sufficient information will be given for understanding the logic behind the design of certain analogs and their exploitation. For a fuller discussion of current developments in the polyphosphoinositide signaling pathway, the reader is referred to recent reviews (Berridge, 1987; Berridge and Irvine, 1989: Nahorski and Potter, 1989; Potter, 1990a,b; chapters by Godfrey and Taylor et al.) and books (Berridge and Michell, 1988; Michell et al., 1989).

Appreciation of the intracellular messenger role of Ins (1,4,5)P_3 in cells revealed the elusive link among the events of agonist-induced cell surface receptor stimulation, inositol phospholipid turnover, and the elevation of intracellular Ca^{2+}. Many cell surface receptors that are linked to the polyphos-phoinositide pathway are coupled via a G protein to the activation of phosphoinositidase C, which cleaves the minor lipid, phosphatidylinositol 4,5-*bis* phosphate (PtdIns[4,5]P_2; Fig. 2 [1]) into the intracellular messengers Ins(1,4,5)P_3 (2) and 1,2-diacyl-glycerol (DG) (3). DG is hydrophobic, remains in the membrane, and activates proteins kinase C. The intracellular messenger role of DG will not be discussed further here (*see* chapter by Burgess). However, synthetic DG analogs have been used to probe the interaction with protein kinase C (Ganong et al., 1986), and the structural requirements for PKC activation have been reviewed (Rando, 1988).

The hydrophilic Ins(1,4,5)P_3 diffuses into the cytosol and binds to a receptor on the endoplasmic reticulum, stimulating the release of Ca^{2+} from an internal store (*see* Taylor et al., this vol.). This receptor has now been cloned and sequenced (Furuchi et al., 1989), and shown to mediate Ca^{2+} release when reconstituted into lipid vesicles (Ferris et al., 1989).

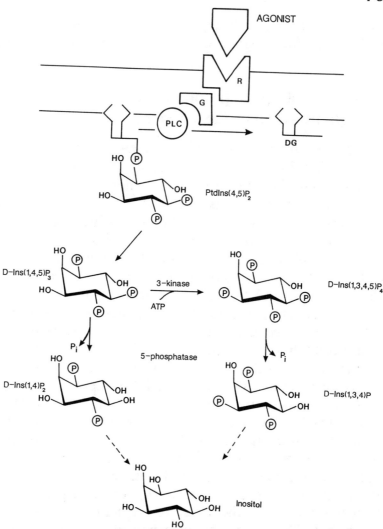

Fig. 5. Agonist-stimulated phosphodiesteratic (PIC) cleavage of PtdIns $(4,5)P_2$ via receptor (R) and G protein (G) produces DG and Ins(1,4,5)P_3. Ins(1,4,5)P_3 is metabolized by two metabolic pathways. Reproduced from Nahorski and Potter (1989) *Trends Pharmacol. Sci.* **10,** 139–144 by permission of Elsevier Publications.

The metabolism of Ins(1,4,5)P_3 is complex (Shears, 1989) two major pathways being operative (Fig. 5): Ins(1,4,5)P_3 is dephosphorylated by a specific 5-phosphatase, in what is almost certainly a deactivation pathway, to give Ins(1,4)P_2. Alternatively, phosphorylation of Ins(1,4,5)P_3 by a 3-kinase generates a higher poly-phosphate, inositol 1,3,4,5-tetrakisphosphate, $Ins(1,3,4,5)P_4$. The precise function of $Ins(1,3,4,5)P_4$ has not yet been defined, but it may play a role in gating the entry of extracellular Ca^{2+} or in controlling communication between $Ins(1,4,5)P_3$-sensitive and $Ins(1,4,5)P_3$-insensitive stores (Irvine et al., 1988; Irvine, 1990).

4.1. Synthesis of Inositol Phosphates and Analogs

Although rapid progress has been made, many details of the polyphosphoinositide signaling pathway are not yet clear (*see* Berridge, 1987; Berridge and Michell, 1988; Berridge and Irvine, 1989). Interest in the chemistry of this pathway has grown dramatically, and we can expect an increasing commitment by the medicinal chemist to the synthesis of novel inositol phosphate analogs to probe, and interfere with, the polyphosphoinositide system. Although most of the fundamental chemical problems relating to the synthesis of inositol phosphates have already been addressed, biological exploitation of inositol phosphate analogs is still undeveloped, and there is a clear need for specific chemically designed tools to rectify this. Recent developments in the chemistry of inositol phosphates have been reviewed (Billington, 1989; Potter, 1990a,b).

The chemistry of inositol phosphates had a long history before the biological developments of the 1980s gave it a new direction. In 1850, Scherer isolated a cyclohexanehexol from muscle, which he named "Inosit" (Scherer, 1850). The suffix -ol was subsequently added, and the existence of nine possible isomeric inositols was realized on stereochemical grounds, a situation that led to many difficulties with nomenclature. After the phosphoinositide lipids had been subsequently structurally characterized and were shown to contain *myo*-inositol (Figs. 2, 6 [8]), much chemical effort was invested in exploration of inositol chemistry. This original work has been thoroughly reviewed and is the subject of two comprehensive books (Posternak, 1965; Cosgrove, 1980).

We need, fortunately, within the context of recent developments, concern ourselves only with the *myo*-inositol isomer. *Myo*-inositol (Fig. 6 [8]) is a *meso*-molecule with five equatorial hydroxyl groups and a single axial hydroxyl group at C-2. It

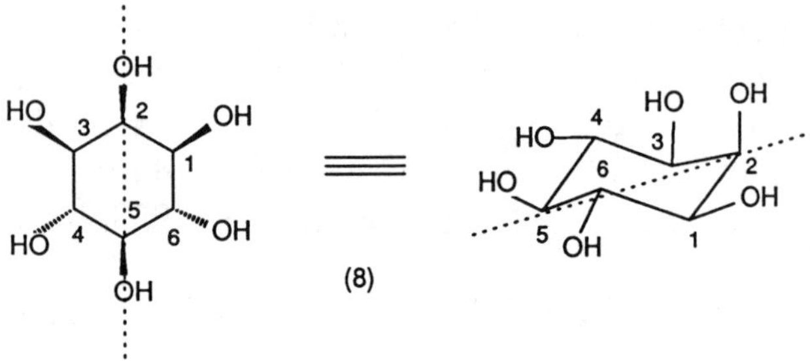

Fig. 6. Structures of *myo*-inositol, Ins(1,4,5)P_3, and analogs.

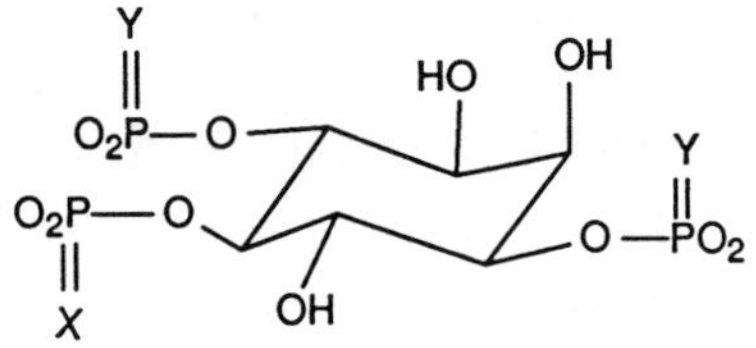

possesses a plane of symmetry through carbon atoms 2 and 5 (Fig. 6). Asymmetric substitution of the hydroxyl groups produces enantiomers, which are distinguished by D- or L-prefixes, e.g., the intracellular messenger D-*myo*-inositol-1,4,5-trisphosphate [Ins(1,4,5)P_3; Figs. 2, 6]. The nomenclature of inositols and their substitution products has caused many problems, and the reader is referred to an excellent account of inositol stereochemistry (Parthasarathy and Eisenberg, 1986) and some new rules (Numbering of atoms in *myo*-inositol [1989]).

Despite earlier work, a number of synthetic procedures in inositol phosphate chemistry were not fully optimized, particularly the chemistry of polyphosphate synthesis. Cyclitols, such as *myo*-inositol, possess vicinal diol systems, where phosphorylation, unless carefully controlled, can lead to the formation of a cyclic five-membered phosphate, rather than a vicinal bisphosphate. Four fundamental problems must principally be overcome in inositol phosphate synthesis: First, if starting with the readily available *myo*-inositol, the three hydroxyl groups of *myo*-inositol must be selectively protected; second, the remaining hydroxyl groups must be efficiently phosphorylated, without appreciable formation of cyclic phosphates; third, the fully blocked molecule must be smoothly deprotected using conditions that avoid possible phosphate migration between hydroxyl groups; finally, at some stage, in order to prepare an optically pure product, enantiomers resulting from the synthesis must be resolved.

Ins$(1,4,5)P_3$ was first synthesized in 1986 (Ozaki et al., 1986), although the phosphorylation reagent in the form of dianilido-phosphorochloridate was not ideal at that time. Subsequent syntheses proposed a phosphoramidite approach (Cooke et al., 1987a; Reese and Ward, 1987; Yu and Fraser-Reid, 1988), using more reactive P(III) phosphitylating reagents. This class of reagent had previously been shown to be applicable to the synthesis of a vicinal bisphosphate of *myo*-inositol (Hamblin et al., 1987a). Subsequently, together with the P(III) approaches, the adoption of an alternative reagent possessing the highly reactive phosphoanhydride linkage, tetrabenzylpyrophosphate (Billington and Baker, 1987; Watanabe et al., 1987), as an efficient alternative phosphorylating reagent has become widespread.

Thus, for example, in our own approach (Cooke et al., 1987a), the 2-, 3-, and 6-hydroxyl groups of *myo*-inositol were selectively protected using benzyl groups (Gigg et al., 1987) and the triol was phosphitylated using N,N-diisopropylamino (2-cyanoethyl) chlorophosphine. Further reaction to the trisphosphite and oxidation with t-BuOOH was followed by deprotection of all blocking groups using sodium in liquid ammonia to yield Ins $(1,4,5)P_3$ (Fig. 7; X = O). The mirror image enantiomer of the naturally occurring messenger L-Ins$(1,4,5)P_3$ (Fig. 8 [21]) was also synthesized in a similar fashion using resolved precursor.

A number of syntheses of Ins$(1,4,5)P_3$ have now been reported (Ozaki et al., 1986; Cooke et al., 1987a, 1989a; Reese and Ward, 1987; Meek et al., 1988; Yu and Fraser-Reid, 1988; Dreef et al., 1988; Watanabe et al., 1988; Ley and Sternfeld, 1988; Liu and Chen, 1989; Vacca et al., 1989; Tegge and Ballou, 1989; Stepanov et al., 1989; Falck and Yadagiri, 1989), including that for radiolabeled material (Maracek and Prestwich, 1989a), which will not be discussed in detail here. Rather, the emphasis will be upon unnatural inositol phosphates or analogs. Several groups have

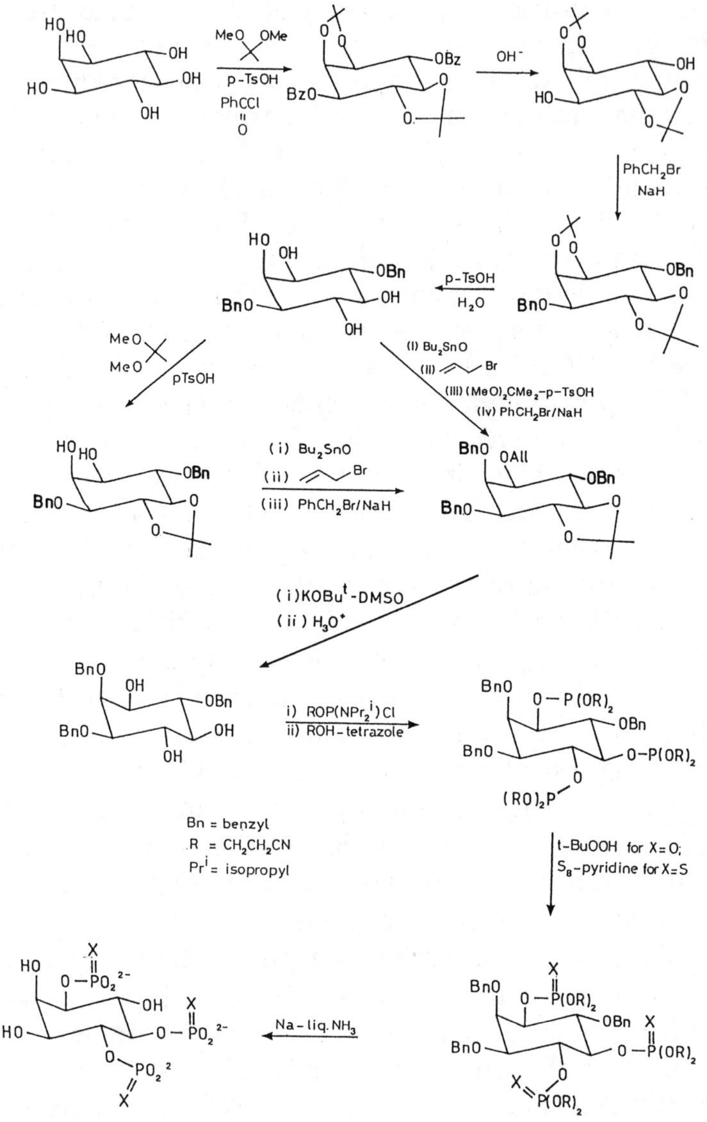

X = O (2)

X = S (9)

Fig. 7. Synthesis of Ins(1,4,5)P_3 and Ins(1,4,5)PS_3. Reprinted by permission of John Wiley and Sons, Ltd. from *Transmembrane Signalling, Intracellular Messengers and Implications for Drug Development* (Nahorski S.R., ed.), John Wiley and Sons, Ltd., pp. 207–239, copyright (1990).

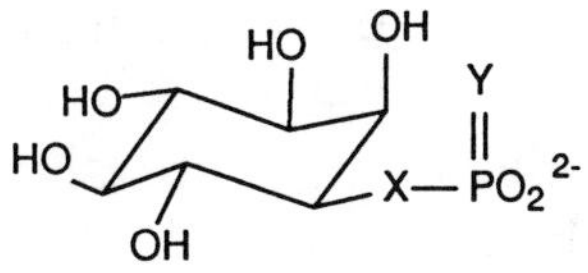

X=O, Y=S (11)

X=CH_2, Y=O (12)

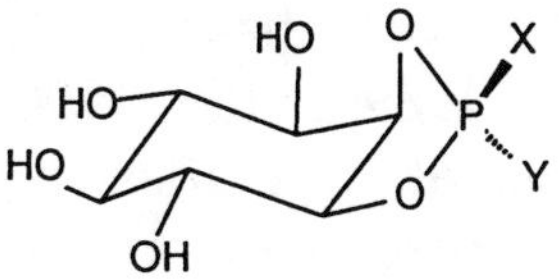

X=S, Y=O (13)

X=O, Y=S (14)

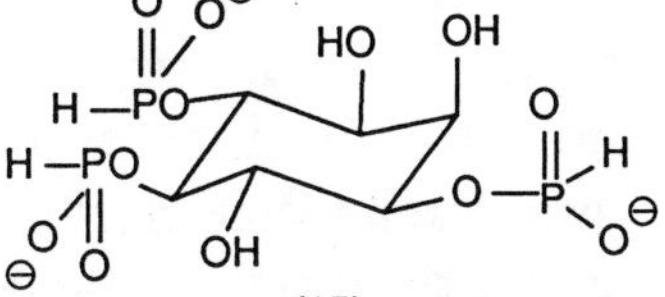

(15)

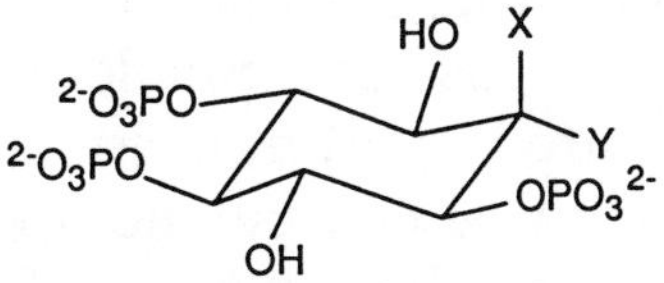

X=H, Y=F (16)

X=Y=F (17)

X=Y=H (18)

X= 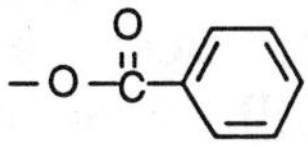 NH_2, Y=H (19)

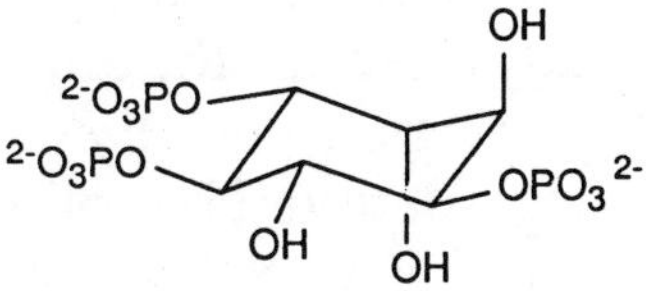

(20)

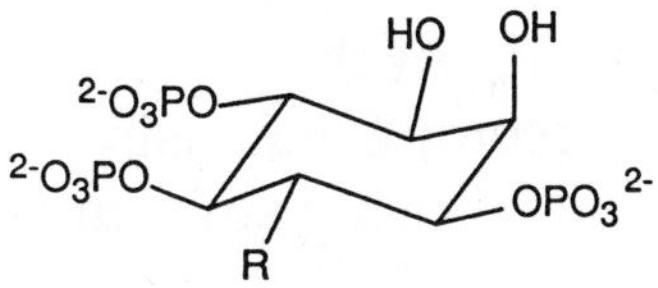

R=H (23)

R=F (24)

R=Me (25)

R=OMe (26)

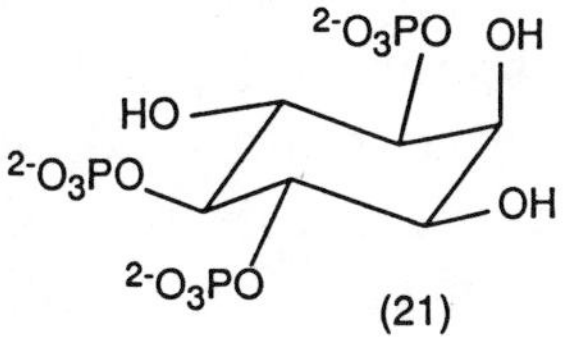

(21)

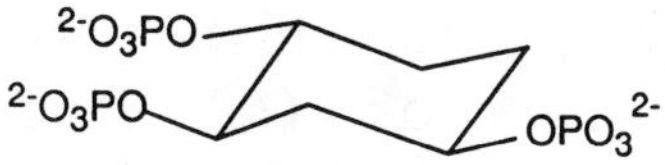

(27)

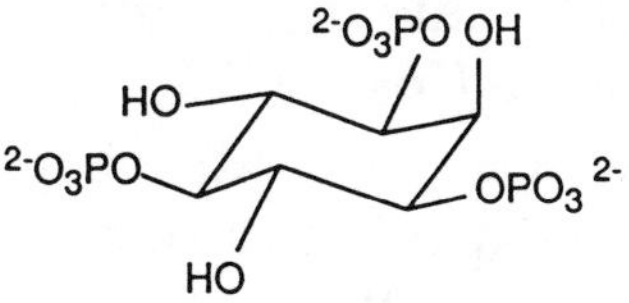

(22)

Fig. 8. Analogs of *myo*-inositol phosphates.

also addressed the synthesis of Ins(1,3,4,5)P_4, but since an intra-cellular messenger role for this molecule has not yet been def-initively proven, these syntheses will not be discussed here (*see* Billington, 1989; Potter, 1990a,b).

Now that difficulties with inositol phosphate synthesis have essentially been overcome, it is possible to envisage the synthe-sis of chemically designed Ins(1,4,5)P_3 analogs and, ultimately perhaps, a rationally designed small molecule Ins(1,4,5)P_3 antag-onist (at present, the only lead compound is the sulfated polysac-charide heparin [Ghosh et al., 1988]). Few structurally modified compounds possessing biological activity have, however, yet been prepared, apart from some inositol ring-modified analogs (Polokoff et al., 1988) and modifications at the 2-position (Hirata et al., 1989). The first example of such a compound was the non-hydrolyzable inositol 1,4,5-trisphosphorothioate (Ins[1,4,5] PS_3; Fig. 6 [9]) (Cooke et al., 1987b). It is already clear that the unique properties of inositol phosphorothioate analogs, especially their resistance to intracellular phosphatase degradation (Hamblin et al., 1987b), will make them valuable tools for investigation of the actions and metabolism of inositol phosphates. The synthe-sis of Ins(1,4,5)PS_3 was undertaken using a modification (X = S) of the route shown in Fig. 7 for Ins(1,4,5)P_3. At the stage of the protected trisphosphite, oxidation was accomplished using sul-fur in pyridine. Deprotection of the product using sodium and liquid ammonia gave Ins(1,4,5)PS_3.

Once the biological activity of Ins(1,4,5)PS_3 had been inves-tigated (*see below*) and the utility of the compound had been dem-onstrated, the potential importance of similar analogs possessing only one phosphate replaced by phosphorothioate was realized. In particular, the Ins(1,4,5)P_3 analog possessing only a 5-phos-phorothioate (Ins[1,4,5]P_3-5S, Fig. 6 [10]) seemed highly desir-able, since this would be nearer in structure to Ins(1,4,5)P_3, but would be expected to retain the advantages of phosphatase resistance. A synthetic route to inositol 1,4-bisphosphate-5-phos-phorothioate was developed (Cooke et al., 1989a), and incorpo-rated a novel combination of both P(III) and P(V) phosphorylation methods. Thus, phosphorylation of 2,3,6-tri-*O*-benzyl-4, 5-*O*-isopropylidene-*myo*-inositol with *bis*(2,2,2-trichloroethyl) phosphorochloridate was followed by removal of the isopropylidene group. Careful rephosphorylation of the product generated a mixture of 1,4- and 1,5-bisphosphates. The crystalline product that separated was analyzed by 2D-J resolved NMR spectroscopy and ^{1}H-COSY, and its structure was established to be the desired 1,4-isomer. After phosphitylation of this molecule, the intermediate was converted to the phosphorothioate triester and deprotected with sodium in liquid ammonia to give Ins $(1,4,5)P_3$-5S.

Syntheses of other inositol phosphate analogs have appeared (Fig. 8) since the initial preparation of these phosphorothioates. Inositol 1-phosphorothioate (IP_1-1S; Fig. 8 [11]) has been prepared by thiophosphorylation of 1,2,4,5,6-penta-O-acetyl-*myo*-inositol with thiophosphoryl chloride and quenching the intermediate thiophosphorodichloridate with potassium hydroxide, followed by removal of the acetyl groups using methanolic ammonia (Metschies et al., 1988). *Exo*- and *endo*-diastereoisomers of the five-membered (1,2-cyclic) phosphorothioates (Fig. 8 [13] and [14], respectively) have been synthesized by thiophosphorylation of 1,4,5,6-tetra-O-acetyl-*myo*-inositol using thiophosphoryl chloride. After quenching the intermediate with potassium hydroxide, the mixture of products was separated into *exo*- and *endo*-diastereoisomers by flash chromatography (Schulz et al., 1988). Another route used simultaneous monophosphitylation of 1,4,5,6-tetrabenzyl-*myo*-inositol at either the 1- or 2-positions, followed by cyclization of the resulting phosphoramidites to diastereoisomeric cyclic methoxyphosphites, oxidation to the diastereoisomeric cyclic phosphorothioates, separation of diastereoisomers, and deblocking (Lin and Tsai, 1989).

The preparation of phosphorothioate analogs of inositol 1,4,5-trisphosphate has also been achieved via a conceptually different approach by phosphorylation of the polyphosphoinositide lipids phosphatidylinositol (Fig. 9 [28]) and phosphatidylinositol 4-phosphate (Fig. 9 [29]) using human erythrocyte ghost kinases and ATPγS (Fig. 9). In this way, the highly desirable ^{35}S radioactively labeled material can be synthesized by employing ^{35}S-ATPγS. Incubation of ATPγ^{35}S with erythrocyte ghosts produced ^{35}S-labeled PtdIns(4,5)P_2 analogs, with ^{35}S label

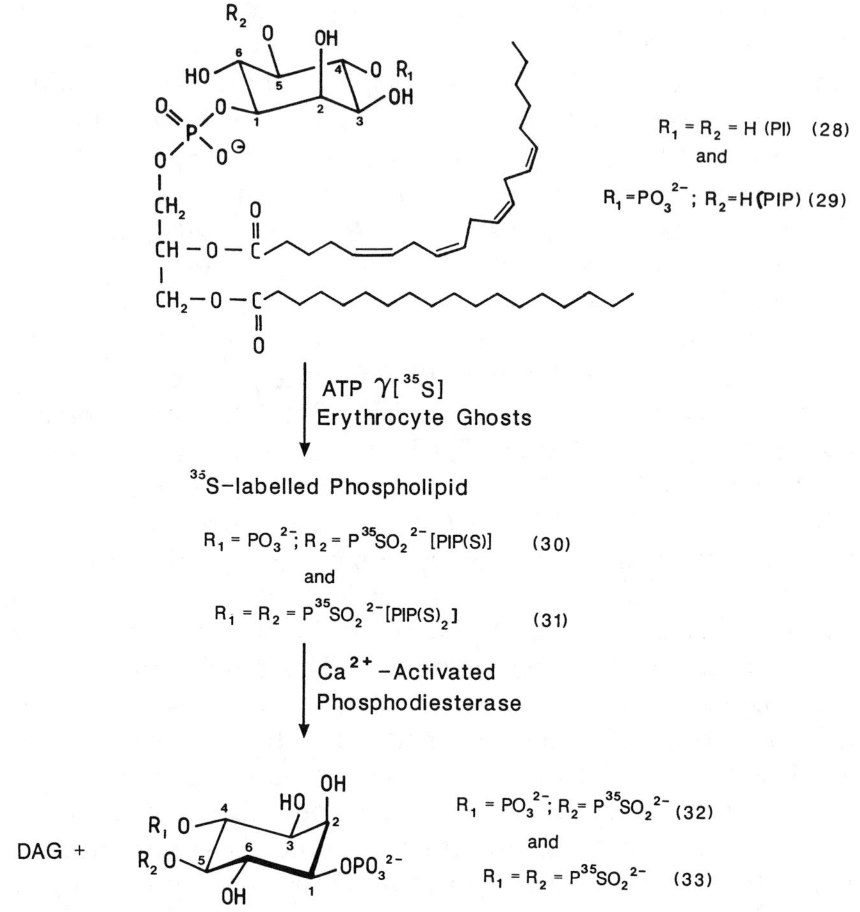

Fig. 9. Enzymatic synthesis of ^{35}S-labeled inositol phosphorothioates. Reprinted by permission of John Wiley and Sons, Ltd. from *Transmembrane Signalling, Intracellular Messengers and Implications for Drug Development* (Nahorski S.R., ed.), John Wiley and Sons, Ltd., pp. 207–239, copyright (1990). Adapted from Folk et al. (1988) *J. Labeled Comp. Radiopharm.* **25**, 793–803.

uniquely in the 5-position (Fig. 9 [30]) or in both the 4- and 5-positions (Fig. 9 [31]). This modified lipid was cleaved by the endogenous Ca^{2+}-activated phosphodiesterase to give a mixture of inositol 1,4-bisphosphate-5-[^{35}S]-phosphorothioate (Fig. 9 [32]) and inositol 1-phosphate-4,5-[^{35}S]-bisphosphorothioate (Fig. 9 [33]), which was demonstrated to be resistant to 5-phosphatase (Folk et al., 1988). Another lipid phosphorothioate analog has been synthesized in the form of 1,2-dipalmitoyl-*sn*-glycero-3-thiophospho-*myo*-inositol and has been used to determine the stereochemical course of the cleavage reaction catalyzed by phosphatidylinositol-specific phospholipase C (Lin and Tsai, 1989).

The inositol 1-phosphate analog inositol 1-phosphonate (Fig. 8 [12]) has been synthesized (Kulagowski, 1989). *Myo*-inositol-1,4,5-tris-H-phosphonate (Fig. 8 [15]) has been synthesized (Dreef et al., 1987), and fluorinated analogs of inositol 1,3,4-tris-phosphate have been prepared (Boehm and Prestwich, 1988) for $Ins(1,4,5)P_3$ (Fig. 8 [16] and [17]) (Maracek and Prestwich, 1989b). An analog with an axial hydroxyl group at the 3-position in place of the normal equatorial group, racemic *chiro*-inositol 2,3,5-trisphosphate (Fig. 8 [20]), has been synthesized from benzene via a photooxidation procedure (Carless and Busia, 1990). No biological activity for these analogs has, however, yet been reported.

A series of synthetic inositol phosphate analogs have been synthesized by the Merck group (Polokoff et al., 1988) (Fig. 8). As well as naturally occurring inositol phosphates and L-Ins $(1,4,5)P_3$ (21), $Ins(1,3,5)P_3$ (22), 6-O-methoxy-$Ins(1,4,5)P_3$ (26), and 1,2,4-cyclohexane trisphosphate (2,3,6-trideoxy-Ins[1,4,5]P_3) (27) have been prepared. An unusual route to D-6-O-methoxy-$Ins(1,4,5)P_3$ (26) has been developed using benzene as starting material and intermediates obtained by microbiological oxidation (Ley and Sternfeld, 1988). This route has also been adapted to synthesis of other 6-modified analogs, 6-deoxy-(23), 6-deoxy-6-fluoro-(24), and 6-deoxy-6-methyl-$Ins(1,4,5)P_3$ (25) (Ley et al., 1989). Inositol phosphate analogs modified at the 2-position on the inositol ring, including 2-deoxy-$Ins(1,4,5)P_3$ (18) and a number of 2-acylated analogs, for example, the 2-(4-aminobenzoyl) analog (19) (Fig. 8), have been chemically synthesized (Hirata et al., 1989).

$Ins(1,4,5)P_3$ analogs modified at the 1-phosphate position have been synthesized (Henne et al., 1988; Fig. 10 [35]–[38]). These analogs were prepared semisynthetically by chemical modification of the deacylated lipid, *sn*-glycero(3)-1-phospho-D-*myo*-inositol-4,5-bisphosphate (34). The phosphorothioate analog inositol 1-phosphorothioate-4,5-bisphosphate ($Ins[1,4,5]P_3$-1S [39]) has also been synthesized (Lampe and Potter, 1990).

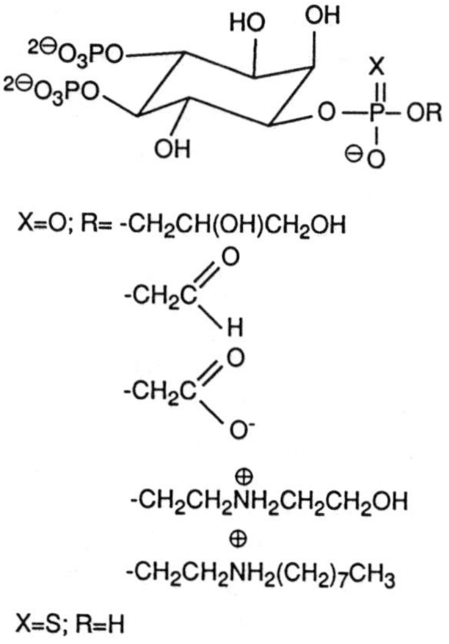

Fig. 10. Analogs of Ins(1,4,5)P_3 modified at the 1-phosphate position.

To assist in the location and purification of Ins(1,4,5)P_3 receptors, it is desirable to synthesize photoaffinity analogs of Ins(1,4,5)P_3. The first such compound was an aryl azide derivative of Ins(1,4,5)P_3 (Fig. 11 [40]), where p-azidobenzoic acid was coupled to Ins(1,4,5)P_3 using N,N'-carbonyldiimidazole (Hirata et al., 1985).

Fast kinetic measurements involving intracellular messengers can be made by photochemical release of an active molecule *in situ*, from an inactive or "caged" precursor. This is particularly important where diffusion of the active agent into tissues is a limitation (McCray and Trentham, 1989). Reaction of Ins(1,4,5)P_3 with 1-(2-nitrophenyl)diazoethane gave a mixture of singly and multiply caged Ins(1,4,5)P_3, which could be resolved by HPLC (Walker et al., 1989). These caged compounds (Fig. 11 [41] and [42]) released Ins(1,4,5)P_3 on irradiation (*see* chapter by Parker).

4.2. Biology of Inositol Phosphates and Analogs

Synthetic Ins(1,4,5)P_3 was active in binding to a stereospecific receptor site in cerebellum (Willcocks et al., 1987) and in

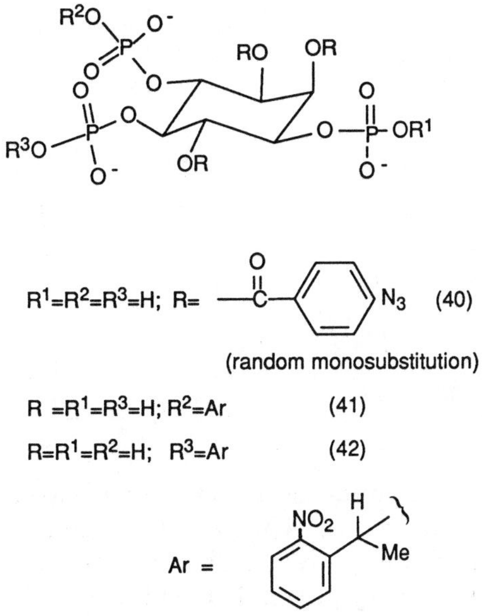

R¹=R²=R³=H; R= (structure with C=O, phenyl, N₃) (40)

(random monosubstitution)

R =R¹=R³=H; R²=Ar (41)

R=R¹=R²=H; R³=Ar (42)

Ar = (structure with NO₂, phenyl, CH(Me)(H))

Fig. 11. Photoaffinity and "caged" analogs of Ins(1,4,5)P_3.

hepatocytes (Nunn et al., 1990), as well as in mobilizing intracellular Ca^{2+} in several systems, including *Xenopus* oocytes (Taylor et al., 1988), permeabilized GH_3 cells (Strupish et al., 1988), Swiss 3T3 cells (Taylor et al., 1988; Strupish et al., 1988), and hepatocytes (Taylor et al., 1989). Unnatural L-Ins(1,4,5)P_3 (Fig. 8 [21]) was, however, very weak or essentially inactive in these systems, including in airway smooth muscle (Chilvers et al., 1990) and in *Limulus* photoreceptors (Payne and Potter, 1991), demonstrating the stereospecificity of diverse intracellular receptors controlling Ca^{2+} release.

The nonhydrolyzable phosphorothioate analog, inositol 1,4,5-trisphosphorothioate (Fig. 6 [9]), is active in binding to Ins(1,4,5)P_3 receptor sites in brain (Willcocks et al., 1988) and is only slightly less potent than Ins(1,4,5)P_3. Moreover, it is a potent agonist for the release of intracellular calcium in *Xenopus* oocytes (Taylor et al., 1988; DeLisle et al., 1990), permeabilized Swiss 3T3 cells (Taylor et al., 1988; Strupish et al., 1988), GH_3 cells (Strupish et al., 1988), hepatocytes (Taylor et al., 1989), pancreatic acinar cells (Thevenod et al., 1989; Wakui et al., 1989),

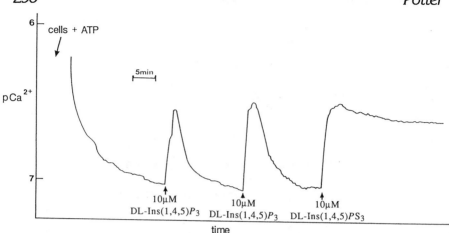

Fig. 12. Release of Ca²⁺ from electrically permeabilized SH-SY5Y neuroblastoma cells by synthetic Ins(1,4,5)P_3 and Ins(1,4,5)PS_3 using a Ca²⁺-specific electrode. Reproduced from Nahorski and Potter (1989) *Trends Pharmacol. Sci.* **10**, 139–144 by permission of Elsevier Publications.

and mouse lacrimal cells (Changya et al., 1989), being only some three- to fourfold less potent than Ins(1,4,5)P_3. As expected from the properties of phosphorothioates, it is resistant to 5-phosphatase-catalyzed dephosphorylation (Willcocks et al., 1988; Taylor et al., 1989) and can therefore produce a sustained calcium signal in cells (Nahorski and Potter, 1989; Taylor et al., 1989) (Fig. 12). Ins(1,4,5)PS_3 is a potent competitive inhibitor of 5-phosphatase (Cooke et al., 1989b), but is not bound by the 3-kinase and does not compete with Ins(1,4,5)P_3 for this enzyme (Taylor et al., 1989; Safrany et al., 1991). These properties are summarized in Fig. 13.

Inositol 1,4-bisphosphate-5-phosphorothioate (Ins[1,4,5]P_3-5S; Fig. 6 [10]) is nearer in structure to Ins(1,4,5)P_3, yet maintains the advantages of phosphatase resistance. Ins(1,4,5)P_3-5S behaves in a similar fashion to Ins(1,4,5)PS_3 in binding to the cerebellar receptor, mobilization of intracellular Ca²⁺, and inhibition of 5-phosphatase (Cooke et al., 1988), but is bound by the 3-kinase (Safrany et al., 1991), suggesting that the 4-phosphorothioate group of Ins(1,4,5)PS_3 may interfere sterically or electronically with the 3-hydroxyl group, thus inhibiting substrate binding and

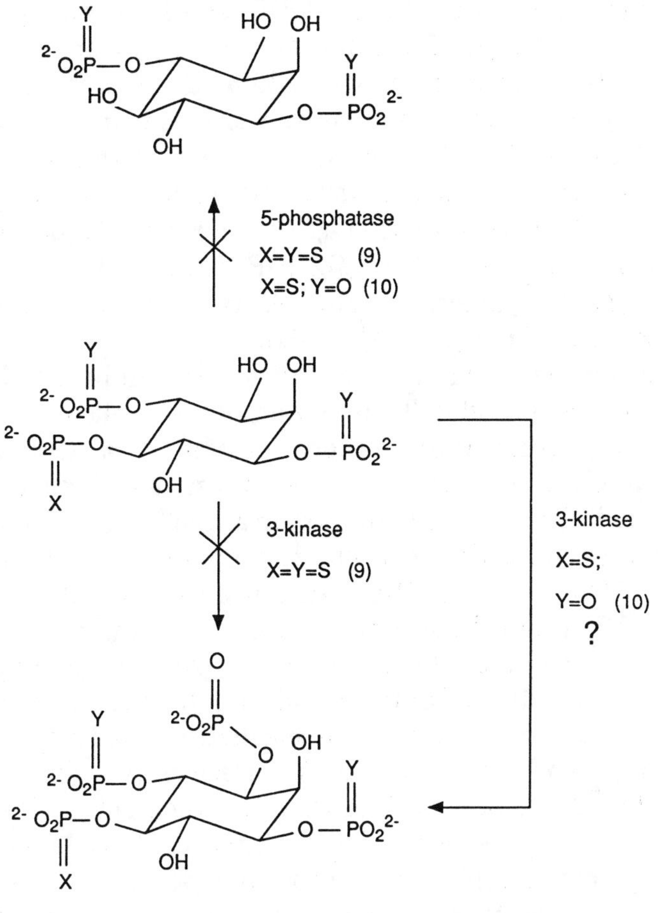

Fig. 13. Action of metabolic enzymes on Ins(1,4,5)PS$_3$ and Ins(1,4,5)P$_3$-5S.

kinase-mediated phosphorylation. It is not yet certain, although highly likely, that Ins(1,4,5)P$_3$-5S is a substrate for 3-kinase (summarized in Fig. 13).

Inositol phosphorothioates are finding considerable applications in studies of the polyphosphoinositide signaling pathway. Since the demonstration that Ins(1,4,5)P$_3$ and Ins(1,3,4,5)P$_4$ can operate synergistically in sea urchin eggs (Irvine et al., 1988), much effort has been expended in identifying a cellular role for Ins(1,3,4,5)P$_4$ and, in particular, examining its putative role in controlling extracellular Ca^{2+} entry. An apparent synergy has subsequently been identified only in one other system, in lacrimal acinar cells, where $Ins(1,3,4,5)P_4$ was found to be essential, together with $Ins(1,4,5)P_3$, in maintaining a Ca^{2+}-activated K^+ current, dependent upon extracellular Ca^{2+} (Morris et al., 1987). $Ins(1,4,5)PS_3$ alone gives rise to a single transient response, typical of $Ins(1,4,5)P_3$ in this system, but together with $Ins(1,3,4,5)P_4$, it evokes a sustained response, suggesting that the transient response to $Ins(1,4,5)P_3$ is not a consequence of rapid metabolism, and that $Ins(1,3,4,5)P_4$ is not acting by protecting Ins $(1,4,5)P_3$ against dephosphorylation by the common 5-phosphatase (Changya et al., 1989).

Rather than causing a continuous rise in intracellular Ca^{2+} levels in their target cells, many agonists evoke oscillations in internal Ca^{2+} concentration (Berridge et al., 1988; Cobbold et al., 1990). Much effort has been focused upon identifying the mechanisms underlying this phenomenon as well as its function in the cell, and mechanistic aspects have been reviewed (Petersen and Wakui, 1990). One mechanism invokes fluctuating levels of $Ins(1,4,5)P_3$ under the second-by-second control of the receptor, giving rise to a pulsatile release of Ca^{2+} (Meyer and Stryer, 1988; Berridge et al., 1988; Cobbold et al., 1990), and another is based upon the idea of a cytoplasmic oscillator, involving Ca^{2+} feedback loops and Ca^{2+}-induced Ca^{2+} release, possibly combined with interactions between communicating $Ins(1,4,5)P_3$-sensitive and $Ins(1,4,5)P_3$-insensitive intracellular Ca^{2+} pools (Berridge et al., 1988). $Ins(1,4,5)PS_3$ evokes repetitive pulses of Ca^{2+}-activated Cl^- current in perfused mouse pancreatic acinar cells (Wakui et al., 1989; Fig. 14), which are similar in amplitude and frequency to the response of such cells to acetylcholine. This demonstrates that pulsatile Ca^{2+} release is possible even at a constant level of this analog, and presumably $Ins(1,4,5)P_3$, and argues strongly against a receptor-controlled oscillator in the generation of Ca^{2+} spikes (Berridge et al., 1988; Meyer and Stryer, 1988; Cobbold et al., 1990) as well as any role for $Ins(1,4,5)P_3$ metabolism. $Ins(1,4,5)PS_3$ has also been used to generate repetitive oscillations in Ca^{2+}-dependent membrane potential in the *Limulus* photoreceptor, where the action of $Ins(1,4,5)P_3$ is rapidly terminated by metabolism (Payne and Potter, 1991).

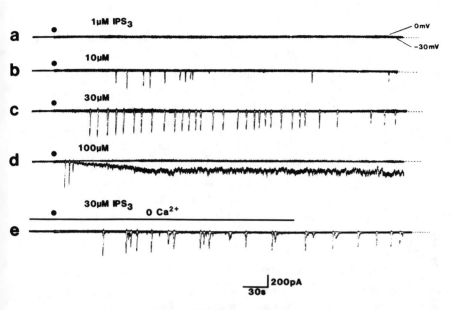

Fig. 14. Ins(1,4,5)PS_3-induced oscillations of Ca^{2+}-dependent Cl⁻ current in single perfused pancreatic acinar cells. Reprinted by permission from Wakui et al. (1989) *Nature* **339**, 317–320, copyright (1989), Macmillan Magazines.

Ins(1,4,5)PS_3 is also capable of generating oscillations of membrane potential (Taylor et al., 1988) and Ca^{2+}-dependent Cl⁻ current (DeLisle et al., 1990) in *Xenopus* oocytes. In the oocyte, however, such oscillations are not sustained, indicating that factors other than metabolism are important in terminating the response.

Ins(1,4,5)PS_3 has been used in rat pancreatic acinar cells to help distinguish functionally between Ins(1,4,5)P_3-sensitive and Ins(1,4,5)P_3-insensitive nonmitochondrial Ca^{2+} pools (Thevenod et al., 1989). By virtue of their properties as Ins(1,4,5)P_3 5-phosphatase inhibitors, Ins(1,4,5)PS_3 and Ins(1,4,5)P_3-5S have been employed to inhibit Ins(1,4,5)P_3 breakdown in electrically permeabilized neuroblastoma cells (Wojcikiewicz et al., 1990). Inhibition of 5-phosphatase-catalyzed metabolism of exogenously added 5[^{32}P]-Ins(1,4,5)P_3 was about 10 times greater than that of cell membrane-derived $[^3H]$-Ins$(1,4,5)P_3$, indicating the possibility of Ins$(1,4,5)P_3$ compartmentalization.

Another application of Ins$(1,4,5)PS_3$ has been in the investigation of apparent "quantal" release of intracellular Ca^{2+} by Ins$(1,4,5)P_3$ from permeabilized hepatocytes (Taylor and Potter, 1990; Taylor et al., this vol.), where it became apparent that the size of the Ins$(1,4,5)P_3$-sensitive Ca^{2+} pool was dependent upon the concentration of Ins$(1,4,5)P_3$. The metabolic stability of Ins$(1,4,5)PS_3$ allowed $^{45}Ca^{2+}$ efflux experiments to be performed at a high cell density where degradation of Ins$(1,4,5)P_3$ would have posed significant problems. The results suggested that the failure of submaximal concentrations of Ins$(1,4,5)P_3$ or Ins$(1,4,5)PS_3$ to empty the sensitive Ca^{2+} stores completely was not the result of either inactivation of the stimulus or receptor desensitization.

The metabolic stability of Ins$(1,4,5)PS_3$ was crucial for the observation that such a nonhydrolyzable analog of Ins$(1,4,5)P_3$ could activate a novel voltage-dependent K^+ conductance in CA1 pyramidal cells (McCarren et al., 1989). Ins$(1,4,5)P_3$ itself did not elicit this conductance, presumably because of its rapid breakdown in these cells. Thus, use of Ins$(1,4,5)PS_3$ may uncover activities of Ins$(1,4,5)P_3$ that may not be experimentally observed using the natural messenger.

Synthetic inositol phosphate analogs have been biologically evaluated by the Merck group (Polokoff et al., 1988). In addition to naturally occurring inositol phosphates and L-Ins$(1,4,5)P_3$ (Fig. 8 [21]), Ins$(1,3,5)P_3$ (22), racemic 6-O-methoxy-Ins$(1,4,5)P_3$ (26), and cyclohexane (1,2,4) trisphosphate (27) were examined (Fig. 8). In Ca^{2+} mobilization studies in permeabilized bovine aortic smooth muscle cells, all analogs were found to be full agonists, but were about 100–2000-fold less potent than the natural agonist. The analogs were not substrates of the aortic 5-phosphatase enzyme, but were quite potent inhibitors, even, interestingly, Ins$(1,3,5)P_3$, which does not possess a vicinal bisphosphate group. Although substrate properties were not measured, inhibition of $[^3H]$-Ins$(1,4,5)P_3$ turnover by the 3-kinase by these analogs was less apparent and was some 100–1000-fold less than Ins$(1,4,5)P_3$. It is clear, therefore, that in its recognition of inositol phosphates the 5-phosphatase is relatively nonspecific, whereas both the receptor and the 3-kinase are considerably more selective. Surprisingly, L-Ins(1,4,5)P_3 was only threefold less potent in binding to aortic 5-phosphatase than the natural substrate for this enzyme, D-Ins(1,4,5)P_3, and this is in excellent agreement with studies on erythrocyte 5-phosphatase using independently synthesized inositol L-Ins(1,4,5)P_3 (Cooke et al., 1989b).

Ins(1,4,5)P_3 analogs, semisynthetically modified at the 1-position, have been biologically evaluated (Henne et al., 1988; Fig. 10 [34]–[38]). The modified compounds were mostly relatively potent full agonists in Ca^{2+} release studies, indicating that modification of the 1-phosphate group of Ins(1,4,5)P_3 can be made without, or with only minor, loss of biological activity. We anticipate that Ins(1,4,5)P_3-1S (Fig. 10 [39]), by virtue of its nucleophilic sulfur, will also be of value in linking reporter groups to the molecule at the 1-position with minimal loss of biological activity.

An arylazide derivative of Ins(1,4,5)P_3 (Fig. 11 [40]) irreversibly inhibited Ins(1,4,5)P_3-induced Ca^{2+} release in saponin-permeabilized photoirradiated macrophages (Hirata et al., 1985). This could be prevented by a large excess of Ins(1,4,5)P_3. This photoaffinity label has also been shown to label three proteins in macrophages (Ishimatsu et al., 1988). Ins(1,4,5)P_3, photochemically caged at the 4- and 5-phosphate groups (Fig. 11 [41] and [42]), has been used to investigate the role of Ins(1,4,5)P_3 in excitation–contraction coupling in muscle (Walker et al., 1987) and has shown that this molecule may have an essential role in pharmacomechanical coupling in smooth, but not striated, muscle.

An important way to purify enzymes and receptors involved with the polyphosphoinositide signaling system is by the design and exploitation of an Ins(1,4,5)P_3 affinity chromatography column. Initial success in this area has been achieved with the employment of 2-acylated derivatives of Ins(1,4,5)P_3 (Hirata et al., 1989). Thus, the 2-(4-aminobenzoyl)-Ins(1,4,5)P_3 analog (Fig. 8 [19]) was diazotized and coupled with a derivative of Sepharose 4B to produce an immobilized Ins(1,4,5)P_3 analog (Fig. 15 [43]). This and another similar column were used to partially purify 3-kinase, 5-phosphatase, and Ins(1,4,5)P_3 cerebellar binding activity from rat brain.

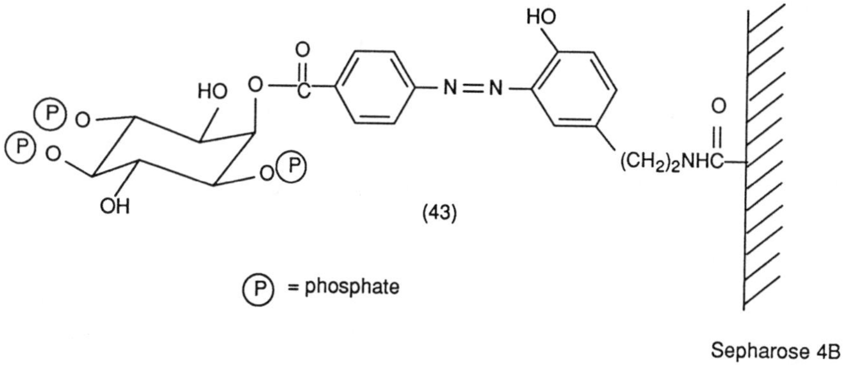

Fig. 15. Structure of Sepharose 4B column for Ins(1,4,5)P_3 affinity chromatography. Adapted from Hirata et al. (1990) *Biochem. Biophys. Res. Commun.* **168,** 379–386.

5. Outlook

The striking recent biological progress in unraveling the polyphosphoinositide signaling pathway has highlighted an ever-increasing dependence upon the chemist for the synthesis of novel intracellular messenger agonists, antagonists, enzyme inhibitors, cell-permeable analogs, and affinity and photoaffinity labels. There is an urgent need for a synthetic small molecule Ins(1,4,5)P_3 antagonist. We may hope that the lead compound heparin may provide important structural clues for the design of such a molecule. Phosphorothioate analogs have been extensively used in unraveling the complexities of many biological systems and are preeminent in the intracellular messenger field. Thus, GTPγS is a highly potent nonhydrolyzable stimulator of G proteins; S_p-cAMPS is a cAMP agonist sometimes more potent than cAMP itself; R_p-cAMPS and cAMPS$_2$ are the only known cAMP antagonists; R_p- and S_p-cGMPS have undoubtedly not yet been fully exploited; and most recently, Ins(1,4,5)PS_3 and Ins(1,4,5)P_3-5S are nonhydrolyzable Ins(1,4,5)P_3-like agonists and potent 5-phosphatase inhibitors. It is likely that such analogs, especially when radioactively labeled with ^{35}S, will, along with others, underpin many future biological developments in this flourishing area of research.

Acknowledgments

Work in the author's laboratory was supported by the Science and Engineering Research Council, The Research Corporation Trust, The Lister Institute of Preventive Medicine, and Merck Sharp & Dohme Research. B. V. L. P. is a Lister Institute Research Fellow.

Abbreviations

ATPγS	Adenosine 5'-[γ-thio]triphosphate
cAMP	Adenosine 3',5'-cyclic monophosphate
cAMPS	Adenosine 3',5'-cyclic monophosphorothioate (related analogs similarly abbreviated)
cGMP	Guanosine 3',5'-cyclic monophosphate
DG	1,2-diacylglycerol
GTP	Guanosine 5'-trisphosphate
GTPγS	Guanosine 5'-[γ-thio]trisphosphate
Ins(1,4,5)P_3	Inositol 1,4,5-trisphosphate
Ins(1,4,5)PS_3	Inositol 1,4,5-trisphosphorothioate
Ins (1,4,5)P_3-5S	Inositol 1,4-bisphosphate-5-phosphorothioate
PKC	Protein kinase C
PtdIns(4,5)P_2	Phosphatidylinositol 4,5-bisphosphate

References

Baraniak J. and Stec W. J. (1987) Ribonucleotide cyclic 3',5'-phosphoramidates: Synthesis, stereochemistry and conversion into ribonucleotide cyclic 3',5'-phosphorothioates and –[^{18}O]phosphates. *J. Chem. Soc. Perkin Trans.* I, 1645–1656.

Berridge M. J. (1987) Inositol trisphosphate and diacylglycerol: Two interacting second messengers. *Annu. Rev. Biochem.* 56, 159–193.

Berridge M. J. and Irvine R. F. (1984) Inositol trisphosphate, a novel second messenger in cellular signal transduction. *Nature* 312, 315–321.

Berridge M. J. and Irvine R. F. (1989) Inositol phosphates and cell signalling. *Nature* 341, 197–205.

Berridge M. J. and Michell R. H., eds. (1988) Inositol Lipids and Transmembrane Signalling, The Royal Society, London; *Philos. Trans. R. Soc. Ser. B* 320, 235–436.

Berridge M. J., Cobbold P. H., and Cuthbertson K. S. R. (1988) Spatial and temporal aspects of cell signalling. *Philos. Trans. R. Soc. Ser. B* 320, 325–343.

Billington D. C. (1989) Recent developments in the synthesis of myo-inositol phosphates. *Chem. Soc. Rev.* **18**, 83–122.

Billington D. C. and Baker R. (1987) Synthesis of myo-inositol 1,3,4,5-tetraphosphate and myo-inositol 1,3-bisphosphate. *J. Chem. Soc. Chem. Commun.* 1011–1013.

Boehm M. and Prestwich G. D. (1988) Fluorinated analogs and tritiated enantiomers of inositol 1,3,4-trisphosphate. *Tet. Lett.* **29**, 5217–5220.

Carless H. A. J. and Busia K. (1990) Total synthesis of chiro-inositol 2,3,5-trisphosphate: A myo-inositol 1,4,5-trisphosphate analogue from benzene via a photo-oxidation. *Tet. Lett.* **31**, 1617–1620.

Chabre M. and Deterre P. (1989) Molecular mechanism of visual transduction. *Eur. J. Biochem.* **179**, 255–266.

Changya L., Gallacher D. V., Irvine R. F., Potter B. V. L., and Petersen O. H. (1989) Inositol 1,3,4,5-tetrakisphosphate is essential for sustained activation of the Ca^{2+}-dependent K^+ current in singly perfused mouse lacrimal cells. *J. Membr. Biol.* **109**, 85–93.

Chilvers E. R., Kennedy E. D., and Potter B. V. L. (1989) Receptor-dependent formation and metabolism of inositol 1,4,5-trisphosphate: Sites for therapeutic intervention. *Drug News and Perspect.* **2**, 342–346.

Chilvers E. R., Challiss R. A. J., Willcocks A. L., Potter B. V. L., Barnes P. J., and Nahorski S. R. (1990) Characterisation of stereospecific binding sites for inositol 1,4,5-trisphosphate in airway smooth muscle. *Br. J. Pharmacol.* **99**, 297–302.

Cobbold P., Dixon J., Sanchez-Bueno A., Woods N., Daley M., and Cuthbertson R. (1990) Receptor control of calcium transients, in *Transmembrane Signalling, Intracellular Messengers and Implications for Drug Development* (Nahorski S. R., ed.), Wiley, Chichester, UK, pp. 185–286.

Cohen P. (1985) The role of protein phosphorylation in the hormonal control of enzyme activity. *Eur. J. Biochem.* **151**, 439–448.

Cooke A. M., Gigg R., and Potter B. V. L. (1987a) Synthesis of DL-myo-inositol 1,4,5-trisphosphate. *Tetrahedron Lett.* **28**, 2305–2308.

Cooke A. M., Gigg R., and Potter B. V. L. (1987b) myo-inositol 1,4,5-trisphosphorothioate: A novel analogue of a biological second messenger. *J. Chem. Soc. Chem. Commun.* 1525–1526.

Cooke A. M., Noble N. J., Gigg R., Willcocks A. L., Strupish J., Nahorski S., and Potter B. V. L. (1988) Myo-inositol 1,4-bisphosphate-5-phosphorothioate: Chemical synthesis and biological properties. *Biochem. Soc. Trans.* **16**, 992, 993.

Cooke A. M., Noble N. J., Payne S., Gigg R., and Potter B. V. L. (1989a) Synthesis of myo-inositol 1,4-bisphosphate–5-phosphorothioate. *J. Chem. Soc. Chem. Commun.* 269–271.

Cooke A. M., Nahorski S. R., and Potter B. V. L. (1989b) Myo-inositol 1,4,5-trisphosphorothioate is a potent competitive inhibitor of human erythrocyte 5-phosphatase. *FEBS Lett.* **242**, 373–377.

Corbin J. D. and Johnson R. A. (eds.) (1988) *Initiation and Termination of Cyclic Nucleotide Action*, in *Meth. Enzymol.*, vol. 159, Academic, London, pp. 1–850.

Corbin J. D., Ogreid D., Miller J. P., Suva R. H., Jastorff B., and Doskeland S. O. (1986) Studies of cGMP analog specificity and function of the two intrasubunit binding sites of cGMP-dependent protein kinase. *J. Biol. Chem.* 261, 1208–1214.

Cosgrove D. J. (1980) *Inositol Phosphates: Their Chemistry, Biochemistry and Physiology* (Elsevier, Amsterdam), pp. 1–191.

DeLisle S., Krause K.-H., Denning G., Potter B. V. L., and Welsh M. J. (1990) Effect of inositol trisphosphate and calcium on oscillating elevations of intracellular calcium in *Xenopus* oocytes. *J. Biol. Chem.* 265, 11,726–11,730.

de Wit R. J. W., Hekstra D., Jastorff B., Stec W. J., Baraniak J., van Driel R., and Van Haastert P. J. M. (1984) Inhibitory action of certain cyclophosphate derivatives of cAMP on cAMP-dependent protein kinase. *Eur. J. Biochem.* 142, 255–260.

Dreef C. E., van der Marel G. A., and van Boom J. H. (1987) Phosphorylation of benzyl-protected sugar derivatives via 1-H-phosphonate intermediates. Synthesis of DL-*myo*-inositol 1,4,5-tris–1-H-phosphonate. *Rec. Trav. Chim. Pays-Bas* 106, 512,513.

Dreef C. E., Tuinman R. J., Elie C. J. J., van der Marel G. A., and van Boom J. H. (1988) Total synthesis of optically active *myo*-inositol 1,4,5-trisphosphate and *myo*-inositol 1,3,4,5-tetrakisphosphate. *Rec. Trav. Chim. Pays-Bas* 107, 395–397.

Eckstein F. (1985) Nucleoside phosphorothioates. *Ann. Rev. Biochem.* 54, 367–402.

Eckstein F. and Foreman J. C. (1978) Action of adenosine and guanosine 3',5'-cyclic phosphorothioates on histamine secretion. *FEBS Lett.* 91, 182–185.

Eckstein F. and Kutzke U. (1986) Synthesis of nucleoside 3',5'-cyclic phosphorothioates. *Tetrahedron Lett.* 27, 1657–1660.

Eckstein F., Eimerl S., and Schramm M. (1976) Adenosine 3',5'-cyclic phosphorothioate: An efficient inducer of amylase secretion in rat parotid slices. *FEBS Lett.* 64, 92–94.

Eckstein F., Simonson L. P., and Baer H.-P. (1974) Adenosine 3',5'-cyclic phosphorothioate: Synthesis and biological properties. *Biochemistry* 13, 3806–3810.

Eckstein F., Karpen J. W., Critchfield J. M., and Stryer L. (1988) Stereochemical course of the reaction catalyzed by the cyclic GMP phosphodiesterase from retinal rod outer segments. *J. Biol. Chem.* 263, 14,080–14,085.

Eckstein F., Romaniuk P. J., Heideman W., and Storm D. R. (1981) Stereochemistry of the mammalian adenylate cyclase reaction. *J. Biol. Chem.* 256, 9118–9120.

306   Potter

Edelman A. M., Blumenthal D. K., and Krebs E. G. (1987) Protein serine/ threonine kinases. *Ann. Rev. Biochem.* **56**, 567–613.

Falck J. R. and Yadagiri P. (1989) Enantiospecific synthesis of D-*myo*-inositol 1,4,5-trisphosphate from (–)-quinic acid. *J. Org. Chem.* **54**, 5851, 5852.

Ferris C. D., Huganir R. L., Suppatapone S., and Snyder S. H. (1989) Purified inositol 1,4,5-trisphosphate receptor mediates calcium flux in reconstituted lipid vesicles. *Nature* **342**, 87–89.

Folk P., Kmonickova E., Krpejsova L., and Strunecka A. (1988) ^{35}S-Labelled thiophosphorylated derivative of inositol trisphosphate. *J. Label Comp. Radiopharm.* **25**, 793–803.

Furuchi T., Yoshikawa S., Miyawaki A., Wada K., Maeda N., and Mikoshiba K. (1989) Primary structure and functional expression of the inositol 1,4,5-trisphosphate binding protein P_{400}. *Nature* **342**, 32–38.

Ganong B. R., Loomis C. R., Hannun Y. A., and Bell R. M. (1986) Specificity and mechanism of protein kinase C activation by sn-1,2-diacylglycerols. *Proc. Natl. Acad. Sci. USA* **83**, 1184–1188.

Genieser H.-G., Dostmann W., Bottin U., Butt E., and Jastorff B. (1988) Synthesis of nucleoside 3',5'-cyclic phosphorothioates by cyclothiophosphorylation of unprotected nucleosides. *Tet. Lett.* **29**, 2803,2804.

Ghosh T. K., Eis P. S., Mullaney J. M., Ebert C. L., and Gill D. L. (1988) Competitive, reversible and potent antagonism of inositol 1,4,5-trisphosphate-activated calcium release by heparin. *J. Biol. Chem.* **263**, 11,075–11,079.

Gigg J., Gigg R., Payne S., and Conant R. (1987) The allyl group for protection in carbohydrate chemistry, part 18. Allyl and benzyl ethers of *myo*-inositol. Intermediates for the synthesis of *myo*-inositol trisphosphates. *J. Chem. Soc. Perkin Trans.* **I**, 423–429.

Gronenborn A. M. and Clore G. M. (1982) Proton nuclear magnetic resonance studies on cyclic nucleotides binding to the *Escherichia coli* adenosine cyclic 3',5' phosphate receptor protein. *Biochemistry* **21**, 4040–4048.

Hajjar D. P. (1986) Regulation of neutral cholesteryl esterase in arterial smooth muscle cells: Stimulation by agonists of adenylate cyclase and cyclic AMP-dependent protein kinase. *Arch. Biochem. Biophys.* **247**, 49–56.

Hamblin M. R., Gigg R., and Potter B. V. L. (1987a) Bisphosphorylation of a *vic*-diol using a phosphite approach: Synthesis of *myo*-inositol 4,5-bisphosphate. *J. Chem. Soc. Chem. Commun.* 626,627.

Hamblin M. R., Flora J. S., and Potter B. V. L. (1987b) *Myo*-inositol phosphorothioates, phosphatase-resistant analogues of *myo*-inositol phosphates. *Biochem. J.* **246**, 771–774.

Henne V., Mayr G. W., Grabowski B., Koppitz B., and Soeling H.-D. (1988) Semisynthetic derivatives of inositol 1,4,5-trisphosphate substituted at the 1-phosphate group. *Eur. J. Biochem.* **174**, 95–101.

Hinrichs W., Steifa M., Saenger W., and Eckstein, F (1987) Absolute configuration of R_p-uridine 3',5'-cyclic phosphorothioate. *Nucleic. Acids Res.* **15**, 4945–4955.

Hirata M., Sasaguri T., Hamachi T., Hashimoto T., Kukita M., and Koga T. (1985) Irreversible inhibition of Ca^{2+}-release in saponin-treated macrophages by the photoaffinity derivative of inositol 1,4,5-trisphosphate. *Nature* **317**, 723–725.

Hirata M., Watanabe Y., Ishimatsu T., Yanaga F., Koga T., and Ozaki S. (1990) Inositol 1,4,5-trisphosphate affinity chromatography. *Biochem. Biophys. Res. Commun.* **168**, 379–386.

Hirata M., Watanabe Y., Ishimatsu T., Ikebe T., Kimura Y., Yamaguchi K., Ozaki S., and Koga T. (1989) Synthetic inositol trisphosphate analogs and their effects upon phosphatase, kinase and the release of Ca^{2+}. *J. Biol. Chem.* **264**, 20,303–20,308.

Hofmann F., Gensheimer H.-P., Landgraf W., Hullin R., and Jastorff B. (1985) Diastereoisomers of adenosine 3',5'-monothionophosphate (cAMP[S]) antagonize the activation of cGMP-dependent protein kinase. *Eur. J. Biochem.* **150**, 85–88.

Hokin L. E. (1985) Receptors and phosphoinositide-generated second messengers. *Ann. Rev. Biochem.* **54**, 205–235.

Irvine R. F. (1990) Quantal Ca^{2+} release and the control of Ca^{2+} entry by inositol phosphates—a possible mechanism. *FEBS Lett.* **263**, 5–9.

Irvine R. F., Moor R. M., Pollock W. R., Smith P. M., and Wreggett K. (1988) Inositol phosphates: Proliferation, metabolism and function. *Philos. Trans. R. Soc. Ser. B* **320**, 281–298.

Ishimatsu T., Kimura Y., Ikebe T., Yamaguchi K., Koga T., and Hirata M. (1988) Possible binding sites for inositol 1,4,5-trisphosphate in macrophages. *Biochem. Biophys. Res. Commun.* **155**, 1173–1180.

Ito F. and Chou J. Y. (1984) Induction of placental alkaline phosphatase biosynthesis by sodium butyrate. *J. Biol. Chem.* **259**, 2526–2530.

Jarvest R. L., Lowe G., Baraniak J., and Stec W. J. (1982) A stereochemical investigation of the hydrolysis of cyclic AMP and the (S$_p$) and (R$_p$) diastereoisomers of adenosine cyclic 3',5' phosphorothioate by bovine heart and baker's-yeast cAMP phosphodiesterases. *Biochem. J.* **203**, 461–470.

Jastorff B., Hoppe J., and Morr M. (1979) A model for the chemical interactions of adenosine 3',5' monophosphate with the R subunit of protein kinase type I. *Eur. J. Biochem.* **101**, 555–561.

Katsaros D., Tortora, G ., Tagliaferri, P., Clair T., Ally S., Neckers L., Robins R. K., and Cho-Chung Y. S. (1987) Site-selective cyclic AMP analogs provide a new approach in the control of cancer cell growth. *FEBS Lett.* **223**, 97–103.

Kulagowski J. J. (1989) The synthesis of (+)-*myo*-inositol 1-phosphonate. *Tet. Lett.* **30**, 3869–3872.

Lampe D. and Potter B. V. L. (1990) Synthesis of *myo*-inositol 1-phosphorothioate 4,5,-bisphosphate: Preparation of a fluorescently labelled *myo*-inositol 1,4,5-trisphosphate analogue. *J. Chem. Soc. Chem. Commun.* 1500,1501.

Levitzki A. (1988) Transmembrane signalling to adenylate cyclase in mammalian cells and in *Saccharomyces cerevisiae*. *Trends Biochem. Sci.* **13**, 298–301.

Ley S. V. and Sternfeld F. (1988) Microbial oxidation in synthesis: Preparation from benzene of the cellular secondary messenger *myo*-inositol 1,4,5-trisphosphate (IP_3) and related derivatives. *Tetrahedron Lett.* **29**, 5305–5308.

Ley S. V., Parra M., Redgrave A. J., Sternfeld F., and Vidal A. (1989) Microbial oxidation in synthesis: Preparation of 6-deoxy cyclitol analogues of *myo*-inositol 1,4,5-trisphosphate from benzene. *Tet. Lett.* **30**, 3557–3560.

Lin G. and Tsai M.-D. (1989) Phospholipids chiral at phosphorus 18. Stereochemistry of the phosphatidylinositol-specific phospholipase C. *J. Am. Chem. Soc.* **111**, 3099–3101.

Liu Y.-C. and Chen C.-S. (1989) An efficient synthesis of optically active D-*myo*-inositol 1,4,5,-trisphosphate. *Tet. Lett.* **30**, 1617–1620.

Maracek J. F. and Prestwich G. D. (1989a) Synthesis of tritium-labelled enantiomers of *myo*-inositol 1,4,5-trisphosphate. *J. Labelled Comp. Radiopharm.* **27**, 917–925.

Maracek J. F. and Prestwich G. D. (1989b) Fluorinated analogues of $Ins(1,4,5)P_3$. *Tet. Lett.* **30**, 5401–5404.

McCarren M., Potter B. V. L., and Miller R. J. (1989) A metabolically stable analog of 1,4,5-inositol trisphosphate activates a novel K^+ conductance in pyramidal cells of the rat hippocampal slice. *Neuron* **3**, 461–471.

McCray J. A. and Trentham D. R. (1989) Properties and uses of photoreactive caged compounds. *Ann. Rev. Biophys. Chem.* **18**, 239–270.

McMasters K. M., Anderson D. M., Parker Bothelo L. H., McDonald G. J., and Moyle W. R. (1986) Use of a cAMP analogue which inhibits cyclic AMP-dependent protein kinase (A-kinase) to study the role of cyclic AMP and A-kinase as mediators of HCG and FSH-induced steroidogenesis. *Adv. Gene Technol. ICSU Short Rep.* **4**, 228, 229.

Meek J. L., Davidson F. L., and Hobbs F. W., Jr. (1988) Synthesis of inositol phosphates. *J. Am. Chem. Soc.* **110**, 2317–2319.

Metschies T., Schulz C., and Jastorff B. (1988) Synthesis of DL-*myo*-inositol 1-phosphate and its thiophosphate analogue. *Tet. Lett.* **29**, 3921, 3922.

Meyer T. and Stryer L. (1988) Molecular model for receptor-stimulated calcium spiking. *Proc. Natl. Acad. Sci. USA* **85**, 5051–5055.

Michell R. H. (1975) Inositol lipids and cell surface receptor function. *Biochim. Biophys. Acta* **415**, 81–147.

Michell R. H., Drummond A. H., and Downes C. P. (eds.) (1989) *Inositol Lipids in Cell Signalling* (Academic, London), pp. 1–534.

Miller J. P. (1978) Cyclic nucleotide analogs, in *Cyclic 3′,5′ Nucleotides: Mechanism of Action* (Cramer H. and Schulz J., eds.), Wiley, London, pp. 77–105.

Morris A. P., Gallacher D. V., Irvine R. F., and Petersen O. H. (1987) Synergism of inositol trisphosphate and tetrakisphosphate in activating Ca^{2+}-dependent K^+ channels. *Nature* **330**, 653–655.

Nahorski S. R. (ed.) (1990) *Transmembrane Signalling, Intracellular Messengers and Implications for Drug Development* (Wiley, Chichester, UK), pp. 1–248.

Nahorski S. R. and Potter B. V. L. (1989) Molecular recognition of inositol polyphosphates by intracellular receptors and metabolic enzymes. *Trends Pharmacol. Sci.* **10**, 139–144.

Nishizuka Y. (1984) The role of protein kinase C in cell surface signal transduction and tumour production. *Nature* **308**, 693–698.

Nishizuka Y. (1988) The molecular heterogeneity of protein kinase C and its implication for cellular regulation. *Nature* **334**, 661–665.

Numbering of atoms in *myo*-inositol. (1989) *Biochem. J.* **258**, 1, 2.

Nunn D. L., Potter B. V. L., and Taylor C. W. (1990) Molecular target sizes of inositol 1,4,5-trisphosphate receptors in liver and cerebellum. *Biochem. J.* **265**, 393–398.

Ogreid D., Ekanger R., Suva R. H., Miller. J. P., Sturm P., Corbin J. D., and Doskeland S. O. (1985) Activation of protein kinase isozymes by cyclic nucleotide analogues used singly or in combination. *Eur. J. Biochem.* **150**, 219–227.

Ozaki S., Watanabe Y., Ogasawara T., Kondo Y., Shiotani N., Nishii H., and Matsuki T. (1986) Total synthesis of optically active *myo*-inositol 1,4,5-tris(phosphate). *Tetrahedron Lett.* **27**, 3157–3160.

Parker Bothelo L. H., Rothermel J. D., Coombs R. V., and Jastorff B. (1988a) cAMP analog antagonists of cAMP action *Methods Enzymol.* **159**, 159–172.

Parker Bothelo L. H., Webster L. C., Rothermel J. D., Baraniak J., and Stec W. J. (1988b) Inhibition of cAMP-dependent protein kinase by adenosine cyclic 3′,5′-phosphorodithioate, a second cAMP antagonist. *J. Biol. Chem.* **263**, 5301–5305.

Parthasarathy R. and Eisenberg F., Jr. (1986) The inositol phospholipids: A stereochemical view of biological activity. *Biochem. J.* **235**, 313–322.

Payne R. and Potter B. V. L. (1991) Injection of inositol trisphosphorothioate into *Limulus* ventral photoreceptors causes oscillations of intracellular calcium ion concentration. *J. Gen. Physiol.* (in press).

Pelech S. L., Pritchard P. H., Brindley D. N., and Vance D. E. (1983) Fatty acids reverse the cyclic AMP inhibition of triacylglycerol and phosphatidylcholine synthesis in rat hepatocytes. *Biochem. J.* **216**, 129–136.

Petersen O. H. and Wakui M. (1990) Oscillating intracellular Ca^{2+} signals evoked by activation of receptors linked to inositol lipid hydrolysis: Mechanism of generation. *J. Membr. Biol.* **118**, 93–105.

Polokoff M. A, Bencen G. H., Vacca J. P., de Solms S. J., Young S. D., and Huff J. R. (1988) Metabolism of synthetic inositol trisphosphate analogues. *J. Biol. Chem.* **263**, 11,922–11,927.

Posternak T. (1965) *The Cyclitols* (Holden-Day), pp. 1–431.

Potter B. V. L. (1990a) Transmembrane signalling, second messenger analogues and inositol phosphates, in *Comprehensive Medicinal Chemistry*, vol. 3 (Hansch C., Sammes P. G., and Taylor J. B., eds.), Pergamon, Oxford, pp. 101–132.

Potter B. V. L. (1990b) Recent advances in the chemistry and biochemistry of inositol phosphates of biological interest. *Nat. Prod. Rep.* **7**, 1–24.

Rando R. (1988) Regulation of protein kinase C activity by lipids. *FASEB J.* **2**, 2348–2355.

Reese C. B. and Ward J. G. (1987) Synthesis of D-*myo*-inositol 1,4,5-trisphosphate. *Tetrahedron Lett.* **28**, 2309–2312.

Richter-Landsberg C. and Jastorff B. (1985) *In vitro* phosphorylation of microtubule-associated protein 2: Differential effects of cyclic AMP analogues. *J. Neurochem.* **45**, 1218–1222.

Rothermel J. D. and Parker Bothelo L. H. (1988) A mechanistic and kinetic analysis of the interactions of the diastereoisomers of adenosine 3',5'(cyclic) phosphorothioate with purified cyclic AMP-dependent protein kinase. *Biochem. J.* **251**, 757–762.

Rothermel J. D., Jastorff B., and Parker Bothelo L. H. (1984) Inhibition of glucagon-induced glycogenolysis in isolated rat hepatocytes by the R_p diastereoisomer of adenosine cyclic 3',5'-phosphorothioate. *J. Biol. Chem.* **259**, 8151–8155.

Rothermel J. D., Stec W. J., Baraniak J., Jastorff B., and Parker Bothelo L. H. (1983) Inhibition of glycogenolysis in isolated rat hepatocytes by the R_p diastereoisomer of adenosine cyclic 3',5'-phosphorothioate. *J. Biol. Chem.* **258**, 12,125–12,128.

Safrany S., Wojcikiewicz R. J. H., Stryish J., McBain J., Cooke A. M., Potter B. V. L., and Nahorski S. R. (1991) *Mol. Pharmacol.* Synthetic phosphorothioate-containing analogues of *myo*-inositol 1,4,5-trisphosphate mobilise intracellular Ca^{2+} stores and interact differently with inositol 1,4,5-trisphosphate 5-phosphatase and 3-kinase, in press.

Scherer J., (1850) Neue, aus dem Muskelfleische gewonnene Zuckerart. *Ann. Chem.* **73**, 322–330.

Scholuebbers H.-G., van Knippenberg P. H., Baraniak J., Stec W. J., Morr M., and Jastorff B. (1984) Investigations on stimulation of lac transcription *in vivo* in *Escherichia coli* by cAMP analogues. *Eur. J. Biochem.* **138**, 101–109.

Schramm M. and Selinger Z. (1984) Message transmission: Receptor controlled adenylate cyclase system. *Science*, **225**, 1350–1356.

Schulz C., Metschies T., and Jastorff B. (1988) Synthesis of thiophosphate analogues of DL-*myo*-inositol 1,2-cyclic phosphate. *Tet. Lett.* **29**, 3919, 3920.

Senter P. D., Eckstein F., Muelsch E., and Boehme E. (1983) The stereochemical course of the reaction catalysed by soluble bovine lung guanylate cyclase. *J. Biol. Chem.* **258**, 6741–6745.

Shears S. B. (1989) Metabolism of the inositol phosphates produced upon receptor activation. *Biochem. J.* **260**, 313–324.

Stec W. J. (1983) Wadsworth-Emmons reaction revisited. *Acc. Chem. Res.* **16**, 411–417.

Stepanov A. E., Runova O. B., Schlewer G., Spiess B., and Shvets V. I. (1989) Total synthesis of chiral sn-myo-inositol 1,4,5-trisphoshate and its enantiomer. *Tet. Lett.* 30, 5125–5128.

Strupish J., Cooke A. M., Potter B. V. L, Gigg R., and Nahorski S. R. (1988) Stereospecific mobilisation of intracellular Ca^{2+} by inositol 1,4,5-trisphosphate. *Biochem. J.* 253, 901–905.

Stryer L. (1986) cGMP cascade of vision. *Annu. Rev. Neurosci.* 9, 87–119.

Stryer L. (1987) The molecules of visual excitation. *Sci. Am.* 255, 42–50.

Sutherland E. W. and Rall T. W. (1958) Formation of a cyclic adenosine ribonucleotide by tissue particles. *J. Biol. Chem.* 232, 1065–1076; Fractionation and characterisation of a cyclic adenosine ribonucleotide formed by tissue particles. *Ibid.* 1077–1091.

Taylor C. W. and Potter B. V. L. (1990) The size of inositol 1,4,5-trisphosphate-sensitive Ca^{2+} stores depends on inositol 1,4,5-trisphosphate concentration. *Biochem. J.* 266, 189–194.

Taylor C. W., Berridge M. J., Cooke A. M., and Potter B. V. L. (1989) Inositol 1,4,5-trisphosphorothioate, a stable analogue of inositol trisphosphate which mobilizes intracellular calcium. *Biochem. J.* 259, 645–650.

Taylor C. W., Berridge M. J., Brown K. D., Cooke A. M., and Potter B. V. L. (1988) DL-myo-inositol 1,4,5-trisphosphorothioate mobilizes intracellular calcium in Swiss 3T3 cells and *Xenopus* oocytes. *Biochem. Biophys. Res. Commun.* 150, 626–632.

Tegge W. and Ballou C. E. (1989) Chiral syntheses of D- and L-myo-inositol 1,4,5-trisphosphate. *Proc. Natl. Acad. Sci. USA* 86, 94–98.

Thevenod F., Dehlinger-Kremer M., Kemmer T. P., Christian A.-L., Potter B. V. L., and Schulz I. (1989) Characterisation of inositol 1,4,5-trisphosphate-sensitive (IsCaP) and -insensitive (IisCaP) non-mitochondrial Ca^{2+} pools in rat pancreatic acinar cells. *J. Membr. Biol.* 109, 173–186.

Vacca J. P., deSolms S. J., Huff J. R., Billington D. C., Baker R., Kulagowski J. J., and Mawer I. M. (1989) The total synthesis of myo-inositol polyphosphates. *Tetrahedron* 45, 5679–5700.

Van Haastert P. J. M. and Kien E. (1983) Binding of cAMP derivatives to *Dictyostelium discoideum* cells. *J. Biol. Chem.* 258, 9636–9642.

Wakui M., Potter B. V. L., and Petersen O. H. (1989) Pulsatile intracellular calcium release does not depend on fluctuations in inositol trisphosphate concentration. *Nature* 339, 317–320.

Walker J. W., Feeney J., and Trentham D. R. (1989) Photolabile precursors of inositol phosphates. Preparation and properties of 1-(2-nitrophenyl) ethyl esters of myo-inositol 1,4,5-trisphosphate. *Biochemistry* 28, 3272–3280.

Walker J. W., Somlyo A. V., Goldman Y. E., Somlyo A. P., and Trentham D. R. (1987) Kinetics of smooth and skeletal muscle activation by laser pulse photolysis of caged inositol 1,4,5-trisphosphate. *Nature* 327, 249–252.

Watanabe Y., Nakahira H., Buyna M., and Ozaki, S (1987) An efficient method for the phosphorylation of inositol derivatives. *Tet. Lett.* **28**, 4179, 4180.

Watanabe Y., Ogasawara T., Nakahira H., Matsuki T., and Ozaki S. (1988) A verstile intermediate, D-4,5-bis(dibenzylphosphoryl)-*myo*-inositol derivative for synthesis of inositol phosphates. Synthesis of 1,2-cyclic-4,5-, 1,4,5- and 2,4,5-trisphosphate *Tet. Lett.* **29**, 5259–5262.

Weishaar R. E. and Bristol J. A. (1990) Selective inhibitors of phosphodiesterases, in *Comprehensive Medicinal Chemistry*, vol. 2, (Hansch C., Sammes P. G., and Taylor J. B., eds.), Pergamon, Oxford, pp. 501–530.

Willcocks A. L., Potter B. V. L., Cooke A. M., and Nahorski S. R. (1987) Stereospecific recognition sites for ^{3}H inositol (1,4,5) trisphosphate in particulate preparations of rat cerebellum. *Biochem. Biophys. Res. Commun.* **146**, 1071–1078.

Willcocks A. L., Potter B. V. L., Cooke A. M., and Nahorski S. R. (1988) *Myo*-inositol(1,4,5)trisphosphorothioate binds to specific [3H]inositol(1,4,5) trisphosphate sites in rat cerebellum and is resistant to 5-phosphatase. *Eur. J. Pharmacol.* **155**, 181–183.

Wojcikiewicz R. J. H., Cooke A. M., Potter B. V. L., and Nahorski S. R. (1990) Inhibition of D-*myo*-inositol(1,4,5)trisphosphate metabolism in permeabilized human SH-SY5Y neuroblastoma cells by phosphorothioate-containing *myo*-inositol (1,4,5) trisphosphate analogues. *Eur J. Biochem.* **192**, 459–467.

Yu K.-L. and Fraser-Reid B. (1988) A novel reagent for the synthesis of *myo*-inositol phosphates: N,N-diisopropyl dibenzyl phosphoramidite *Tet. Lett.* **29**, 979–982.

Zimmerman A. L., Yamanaka G., Eckstein F., Baylor D. A., and Stryer L. (1985) Interaction of hydrolysis-resistant analogues of cyclic GMP with the phosphodiesterase and light sensitive channel of retinal rod outer segments. *Proc. Natl. Acad. Sci. USA* **82**, 8813–8817.

Methods
in Cyclic Nucleotide Research

Kenneth J. Murray

1. Introduction

Cyclic nucleotides were discovered about 30 years ago,
becoming the prototype intracellular messengers (Robison et al.,
1971). Intracellular messenger signaling systems may be split
into the following divisions:

1. Generation of the signal;
2. Removal of the signal (generally this occurs by metabolism,
   although Ca^{2+} is an obvious exception);
3. Direct mediators of the signal; and
4. Ultimate effects of the signal.

Figure 1 shows a simplified scheme for the two cyclic nucleotides, adenosine 3',5'cyclic monophosphate (cAMP) and guanosine 3',5'cyclic monophosphate (cGMP). (Other cyclic nucleotides
are known, but are not discussed in this review.) The cyclic
nucleotide-generating systems are adenylyl cyclase and
guanylyl cyclase; these enzymes are discussed briefly in
Section 2., and the regulation of adenylyl cyclase is discussed in
more detail in the chapter by Milligan. The cyclic nucleotides
are hydrolyzed to their corresponding 5'-monophosphates and
thereby inactivated by a family of isoenzymes collectively known
as cyclic nucleotide phosphodiesterases (PDEs, discussed in
Section 5.). The direct mediators of cAMP in mammalian tissues
are the cAMP-dependent protein kinases (cAMP-PrK, Section 4.1.).
The homologous enzyme cGMP-dependent protein kinase

From: *Neuromethods, Vol. 20: Intracellular Messengers*
Eds: A. Boulton, G. Baker, and C. Taylor © 1992 The Humana Press Inc.

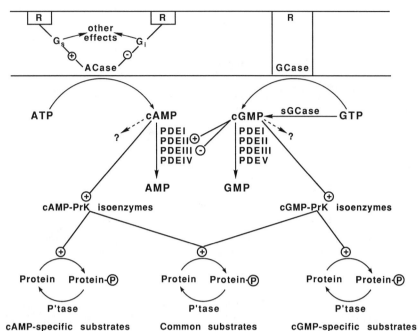

Fig. 1. Cyclic nucleotide metabolism and action in mammalian cells. The figure shows some of the potential ways in which cAMP and cGMP may act and be metabolized in mammalian cells. It is not intended to imply that all the enzymes and routes shown are present in every cell. On the contrary, as discussed in the text, its purpose is to point out the diversity that can occur between tissues. R, receptor; ACase, adenylyl cyclase; (s)GCase, (soluble) guanylyl cyclase; P'tase, phosphoprotein phosphatase isoenzymes (Cohen et al., 1990).

(cGMP-PrK) is one mediator of the effects of increased cGMP, others being some isoforms of PDE (*see* Section 5.) and ion channels (Light et al., 1990; Matthews and Lamb, this vol.). With the exception of direct effects on ion channels, the ultimate effects of increased cyclic nucleotide (cNMP) levels are changes in protein phosphorylation and concurrent changes in their activities. One way of mimicking the effects of receptor-stimulated increases in cyclic nucleotides is by using their cell-penetrant analogs; these are discussed in Section 3. and the chapter by Potter. Cyclic nucleotides in the central nervous system have previously been reviewed (Drummond, 1983,1984), and detailed reviews of methods are included in three volumes of *Methods in Enzymology* (Academic Press): vol. 38 (Hardman and O'Malley, 1974), vol. 99 (Corbin and Hardman, 1983), and vol. 159 (Corbin and Johnson, 1988).

1.1. Criteria for Establishing a Cyclic Nucleotide-Mediated Response

The original criteria proposed by Sutherland et al. (1968) to establish that cAMP is involved in mediating a particular physiological response have been previously discussed with respect to cardiac and smooth muscle contraction (Murray et al., 1989; Murray, 1990). Advances in the understanding of cyclic nucleotide action, especially the discovery of isoenzymes of both PDE and cNMP-PrK, and the realization that a receptor may be coupled to more than one signaling pathway, make it very difficult to satisfy rigorously these criteria. Nevertheless, the original criteria remain useful and are presented here in an amended form.

1. The physiological response and any biochemical assay should, whenever possible, be measured in the same sample or at least under identical conditions.
2. Receptor agonists, or direct activators of the cyclases, should cause a rise in intracellular levels of cNMP, as should inhibitors of PDE. This increase should also be observable in appropriate broken cell preparations (e.g., isolated enzymes or membrane preparations).
3. The effects of PDE inhibitors on agonist-induced responses should be investigated.
4. The effects of cell-penetrant cNMP analogs should also be investigated.
5. The effects on protein phosphorylation and activation of cNMP-PrK should be studied.

The first principle is clear. The second may be subject to technical limitations because some agonists, but especially PDE inhibitors, may cause only very small changes in cNMP levels, and in these instances, other assays may be considered. Since cAMP activates cAMP-PrK cooperatively, a small increase in cAMP can therefore cause a large activation of cAMP-PrK, and these assays may therefore give a better signal-to-noise ratio. Broken cell preparations may pose practical problems, because it is not always easy to demonstrate receptor-stimulated adenylyl cyclase activity in purified membranes.

The original suggestion that phosphodiesterase inhibitors should mimic or potentiate the actions of the first messenger was made when methylxanthines were the only available PDE inhibitors and before our current understanding of PDE isoenzymes. Two commonly used xanthines, theophylline and caffeine, are by current standards very poor PDE inhibitors (Beavo and Reifsnyder, 1990), and 3-isobutyl–1-methylxanthine (IBMX), although a reasonably potent nonselective PDE inhibitor, also has effects on adenosine receptors (Fredholm, 1980). Each cell has its own complement of PDE isoenzymes (Fig. 1, Section 5.). Knowing which are present therefore greatly aids the choice of inhibitor(s). Even so, the use of selective PDE inhibitors has potential pitfalls. Failure to potentiate the responses (physiological or biochemical) of a receptor agonist could indicate that either the agonist does not influence the cNMP pathway, or that the particular PDE inhibited is not the major route of cNMP metabolism in that tissue under those conditions. Any effect of a selective PDE inhibitor alone will similarly depend on the other PDEs present and on the rate of formation of cNMP. In both cases, a comparison of the effects of selective and nonselective inhibitors would be useful.

The utility of cell-penetrant cNMP analogs is comprised by the observation that some receptors, by interaction with G proteins, stimulate both adenylyl cyclase and certain ion channels (Birnbaumer et al., 1990). If the physiological response depends on activation of both adenylyl cyclase and the ion channel, then the response will clearly not be mimicked by a cAMP analog. Differences between agonists and analogs may also be explained by the physical distribution of cAMP-PrK (compartmentation) within a cell or by the isoenzyme content of the tissue, because some agonists cause selective isoenzyme activation (Livesey and Martin, 1988), whereas single cNMP analogs are nonselective activators (Section 3.2.).

Activation of cAMP-PrK (*see* Section 4.4. for assay) and consequent increases in protein phosphorylation (Morton, this vol.) are of interest because, being towards the end of the pathway, they would be expected to correlate in a concentration- and/or time-dependent manner with the physiological response. However, the signaling cascade has enormous potential for amplification of the initial signal (Swillens et al., 1988). Early points in the pathway (i.e., cAMP levels) may not necessarily be simply related to the end point (i.e., physiological response).

The cGMP pathway can operate through a number of mechanisms (Fig. 1): Activation of cGMP-PrK (*see* Section 4.2.), activation (*see* Section 5.2.), or inhibition (*see* Section 5.3.) of particular PDE isoenzymes and regulation of ion channels (Light et al., 1990). This complexity makes the cGMP pathway more difficult to study and makes the knowledge of the enzyme (cGMP-PrK, PDE) content of the tissue more important.

In addition to the problems described above, another reason for the dissociation of measured biochemical and physiological events may be the presence of compartments of cNMP and/or associated enzymes in the tissue. This has been discussed previously for cardiac and smooth muscle (Murray et al., 1989; Murray 1990), and for example, the subcellular distribution of cAMP-PrK (*see* Section 4.1.3.) is suggestive of the existence of compartments in brain. The development of intracellular indicators for cNMP and, possibly, cNMP-PrK will undoubtedly improve the understanding of this area.

2. Cyclases

Adenylyl and guanylyl cyclases catalyze formation of cAMP and cGMP from their respective nucleotide triphosphates. Receptor regulation of adenylyl cyclase, mediated by a family of G proteins, is described in the chapter by Milligan. Recent work on guanylyl cyclase has highlighted a number of differences from adenylyl cyclase. Whereas the latter is exclusively membrane bound, guanylyl cyclase is found as both soluble and particulate forms, neither of which appears to be regulated by G proteins. Molecular cloning of the membrane form of guanylyl cyclase has shown that the same protein acts as a cell surface receptor (Schulz et al., 1989). For other reviews of guanylyl cyclase, *see* Garbers (1989) and Tremblay et al. (1988).

The method for assay of adenylyl cyclase (Salomon et al. 1974) has been described in a previous volume of this series (Sulakhe et al., 1986). In this procedure, $[\alpha\text{-}^{32}P]$-ATP is converted to $[^{32}P]$-cAMP, which is then isolated by sequential chromatography on cation-exchange and alumina columns. An ATP-regenerating system is used to maintain ATP levels, and the effects of PDEs are minimized by inclusion of an excess of nonradioactive cAMP. Single-column methods have also been used for this assay (Alvarez and Daniels, 1990; Schultz and Jacobs, 1984). Guanylyl cyclase is assayed by an analogous method (Schultz and Bohme, 1984). A number of direct pharmacological modulators (as opposed to receptor agonists/antagonists) of cyclase are available. The best known is the adenylyl cyclase activator forskolin, although its effects are not necessarily specific (Laurenza et al., 1989); agents used as adenylyl cyclase inhibitors include 2',5'-dideoxyadenosine, SQ 22536 (Haslam et al., 1978), and MDL 12330A (Segal and Ingbar, 1989). LY83583 (Mulsch et al., 1988) and methylene blue (Gruetter et al., 1981) have been used as guanylyl cyclase inhibitors.

3. Cyclic Nucleotides

cAMP and cGMP are ubiquitous intracellular messengers, mediating the intracellular response to a wide variety of hormones and other stimuli that interact with receptors at the cell membrane. Measurements of cNMP levels can be used to identify receptors and their coupling to intracellular messenger systems. For example, 5-HT-induced increases of cAMP in brain (Dumuis et al., 1989) and human heart (Kaumann et al., 1990) played a large role in defining the 5-HT_4 receptor. However, changes in cNMP levels do not alone give the complete picture, because cAMP and cGMP interact with target proteins to produce a physiological response. At present, it is believed that the vast majority of the effects of cNMPs are through activation of their corresponding protein kinases (Section 4.), although there is increasing evidence that some ion channels may be directly gated by increases in cNMP (Light et al., 1990; Kolesnikov et al., 1990) and the effects of cGMP on ion channels in photoreceptors are now well established (Matthews and Lamb, this vol.). Electrophysiological techniques provide a method for determining the involvement of cNMP-PrK in the response, since the appropriate protein kinase can be injected into the cell and compared with the effects of raising cNMP. There is evidence in a number of tissues, including cardiac and smooth muscle, that there are compartments of cAMP and/or associated enzymes (Murray et al., 1989; Murray, 1990) This is obviously difficult, if not impossible, to study by any assay, such as those described below, that involves homogenization of the tissue. The importance of cellular structure is emphasized by the results of a recent study in which cAMP and cGMP were determined immunocytologically in microwave fixed cells (Barsony and Marx, 1990). The spatial pattern of cAMP or cGMP accumulation in these cells was agonist-specific, indicating that measurements of total cellular cNMP levels do not necessarily provide the full picture. However, determination of the cNMP levels in response to an agent provides a convenient first step for indicating the potential role of the cNMP.

3.1. Assays of Cyclic Nucleotides

There are two major methods for the quantification of cAMP in tissues or cells. The first involves the use of a specific "binding protein" or an antibody (radioimmunoassay). The second method requires incubation of the tissue with radiolabeled adenine to achieve intracellular labeling of ATP and, subsequently, cAMP. cGMP can be assayed similarly by either of these methods, although radioimmunoassay (RIA) is the most frequently used.

Commercial kits employing the "binding protein" or RIA are available, and the choice between them depends on the sensitivity required and cost. The "binding protein" assay (Brown et al., 1971; Gilman and Murad, 1974), available as a kit from Amersham (Bucks, UK) is cheaper, but less sensitive (range ~0.5–16 pmol/assay tube) than RIA. Homemade kits require the partial purification of the "binding protein," which is now known to be a regulatory subunit of cAMP-PrK (*see* Section 4.7.). After incubation, free cAMP is removed by adsorption onto charcoal followed by centrifugation. The timing of this step is critical, since it alters the equilibrium and cAMP will rapidly dissociate from cAMP-PrK. The use of ammonium sulfate instead of charcoal has been suggested (Santa-Coloma et al., 1987). The sensitivity and range of the assay are increased considerably by RIA (25–1600 fmol/tube). The major drawback of commercial kits (available from Amersham and NEN, Stevenage, Herts, UK, and Boston, MA, USA) is the increased cost, and of homemade kits, the need to raise antibodies (for details of antibodies, *see* Steiner et al., 1972). Recently, Amersham has introduced a scintillation proximity assay for cAMP, the main advantage being that separation of free and bound ligand is not required, which reduces the "hands on" time for the assay.

A number of methods can be used to prepare samples for the assays. We have found the method of Sharps and McCarl (1982) convenient for cells and tissues (e.g., Murray et al., 1990a). Following homogenization in 0.3M perchloric acid, the protein pellet is removed by centrifugation. The decanted supernatant is neutralized by extraction with a 1:1 (v/v) mixture of tri-n-octylamine and freon (1,1,2-trichlorotrifluoroethane). Following centrifugation and appropriate dilution, the upper aqueous layer is used in the cAMP assay. Other methods use trichloroacetic acid followed by ether extraction, or homogenization in a buffer containing EDTA and other PDE inhibitors followed by boiling the resultant extract or supernatant. The latter method is useful with small tissue samples when another assay (e.g., cAMP-PrK activity ratio; Section 4.4.) on the same sample is required. Generally homogenization in acid is the preferable method. It should be noted that, when cAMP is expressed relative to cell protein, different protein assays give different absolute values, and this should be remembered when comparing results of different workers.

The second major method, which involves the incubation of tissue with radiolabeled adenine, is well established in brain slices (Shimizu et al., 1969; Donaldson et al., 1988). The method is based on the continuous uptake of radiolabeled adenine by the tissue and its rapid conversion to ATP. The rate of labeling of cAMP is dependent on its rate of turnover (Barber and Butcher, 1988), and in most tissues and cells this is sufficiently rapid to allow its specific radioactivity to parallel that of the ATP. Under these circumstances, the measure of radioactivity will be the same as the chemical concentration of cAMP. The appropriate control experiment is that described by Maurice and Haslam (1990), who demonstrated that, in rabbit platelets, the prelabeling and RIA methods gave similar results. Following the tissue labeling and subsequent treatment, the incubations are terminated by addition of acid, and the labeled cAMP is isolated by sequential Dowex-alumina chromatography (*see* Donaldson et al., 1988; Haslam and Vanderwel, 1989). If the extracellular label has been removed, a measure of the total radioactivity in the sample will be proportional to the amount of tissue, and the results may be expressed as the amount of radioactivity in cAMP as a percentage of the total. An alternative method is to express cAMP as a percentage of the ATP and ADP counts eluted from the Dowex column. The method has several advantages over RIA, including time and cost. Moreover, unlike RIA, the assay is not limited to a range of [cAMP]. However, as already mentioned, it is important to ascertain that the prelabeling technique is measuring the chemical concentration of cAMP. Prelabeling is obviously restricted to cells and tissue slices, and therefore, lacks the more general applicability of RIA. A variant of this technique can be used to measure the turnover of cAMP in cells (Barber and Butcher, 1988). cGMP is measured by prelabeling with radiolabeled guanine (Maurice and Haslam, 1990). For any method, it may be necessary to establish that the measured cyclic nucleotide is intracellular, since cyclic nucleotides are released from some tissues (Brunton and Heasley, 1988).

3.2. Cyclic Nucleotide Analogs

Numerous derivatives of cAMP and cGMP are available. Although the cAMP-analogs were initially used mainly to "mimic" cAMP in cells, they have been used extensively to study cAMP-PrK (e.g., Rannels and Corbin, 1983; Ogreid et al., 1989) and PDEs (e.g., Erneux and Miot, 1988). An important consideration for this kind of study is the removal of any contaminant cAMP from the analog (Corbin et al., 1988). cAMP analogs are sometimes described as PDE inhibitors. However, most commonly used cAMP analogs have a similar potency to cAMP itself for activation of cAMP-PrK and are far less potent as PDE inhibitors. An exception is the commonly used analog N^6,O'-dibutyryl cAMP (=dibutyryl cAMP), which is a very poor activator of cAMP-PrK because it lacks the 2'-OH group required for activation (Ogried et al., 1985). Intracellular hydrolysis is therefore required for it to be effective. It has also been reported that some of the actions of dibutyryl cAMP are mediated by butyrate (Yusta et al., 1988).

Dibutyryl cAMP is commonly used to mimic cAMP in tissues and cells, since cAMP itself is not sufficiently lipophilic to be cell-penetrant (Braumann and Jastorff, 1985). In a number of tissues, the most potent cAMP analog is 8-(4-chlorophenylthio) cAMP (CPT-cAMP) (Beebe et al.. 1988; Murray, 1990), so that CPT-cAMP is a suitable analog for a first experiment. Many cAMP analogs penetrate tissues poorly, so that the physiological response occurs slowly with a poorly defined end point. Therefore, the potency of the analog is highly dependent on the time of incubation, and dose–response curves are difficult to construct. Another potential problem is that the analogs may be rapidly metabolized, and apparent effects of the analog may then be the result of the metabolites (Braumann et al., 1986a). It should also be noted that measurement of cAMP in analog-treated cells requires the separation of cAMP and the analog (Corbin et al., 1988), since analogs bind both to the R-subunit ("binding protein") and to cAMP antibodies. This is not a problem with the adenine prelabeling technique.

cAMP analogs may not exactly mimic the effects of a receptor agonist that increases intracellular cAMP. As previously mentioned, reasons could include effects of agonists on multiple signaling pathways, or specific activation of either the type I or type II isoenzyme of cAMP-PrK by an agonist (Livesey and Martin, 1988; *see* Section 4.1.1.). Although when used alone, cAMP analogs have insufficient selectivity to cause activation of a single isoenzyme, appropriate pairs of analogs can be used to achieve this. The method is based on the selectivity of certain analogs for the two cAMP-binding sites (A and B; Section 4.1.) and is described in detail by Beebe et al. (1988). The development of a new range of cAMP analogs increases the prospect that an isoenzyme-selective analog will be achieved (Dostmann et al., 1990). In addition to the cAMP analogs that activate cAMP-PrK, there are analogs that antagonize the activation by cAMP and its analogs. The best known antagonist is R_p-adenosine 3',5'-cyclic phosphorothioate (R_p-cAMPS; commercially available from BioLog, Bremen, Germany) (Parker Botelho et al., 1988; Potter, this vol.). R_p-cAMPS antagonizes the effects of glucagon in hepatocytes (Parker Botelho et al., 1988) and has been used in a number of other cells. However, it is not a very potent antagonist, and its "success" may be dependent on its rapid uptake by the tissue and the use of submaximal agonist concentrations or low-affinity cAMP analogs. Similar cGMP analogs (including Rp-cGMPS [Butt et al., 1990]) are available. Photoactivatable ("caged") derivatives of cAMP and cGMP have been described (Nerbonne et al., 1984; Parker, this vol.) and are commercially available (Calbiochem; Molecular Probes).

4. Cyclic Nucleotide-Dependent Protein Kinases

4.1. cAMP-Dependent Protein Kinase

The cAMP-dependent protein kinase (cAMP-PrK) is the only known eukaryotic cAMP-binding protein and, therefore, is thought to be responsible for mediating all the effects of changes in intracellular cAMP levels in mammalian cells. The cAMP-PrK isolated from striated muscle has been the most studied and has recently been reviewed (Beebe and Corbin, 1986; Taylor et al., 1990; Krebs, 1989; Taylor, 1989). cAMP-PrK is composed of two regulatory (R–) subunits and two catalytic (C–) subunits, which together form the tetrameric, inactive holoenzyme (R_2C_2). Binding of cAMP to the R-subunits results in dissociation of the R and C subunits (*see* Eq. 1); the free C subunits are now capable of phosphorylating substrate proteins.

$$R_2C_2 + 4cAMP \longleftrightarrow R_2cAMP_4 + 2C \qquad (1)$$
$$\text{inactive} \qquad\qquad\qquad \text{active}$$

There are various forms of both R- and C-subunits, but the following features are found in all R-subunits. The first 45 residues at the amino-terminus are required for dimer formation (Reimann, 1986). This is followed by the hinge region (so-called because of its susceptibility to proteolysis), which is the binding site for the catalytic subunit (Taylor et al., 1990). The carboxyl-terminal two-thirds of the protein contains the two contiguous highly conserved cAMP-binding sites located in domains (or sites) A and B (Weber et al., 1987). Despite the homology between the two binding sites, they interact with cAMP in a different manner. Site A selectively binds N^6-substituted cAMP derivatives, and exchange of cAMP is faster than that from site B, which selectively binds C^8-substituted derivatives (Rannels and Corbin, 1983; *see* Section 4.5.). The R-subunit is quantified by [^{3}H]-cAMP binding or may be covalently labeled by the photoaffinity probe 8-N_3-cAMP (*see* Section 4.5.1.).

The catalytic domains or, in the case of cAMP-PrK, the catalytic subunit of protein kinases is highly conserved and contains the consensus sequence Gly-X-Gly-X-X-Gly. This sequence is found in other nucleotide-binding proteins and represents the ATP-binding site (Hanks et al., 1988; Taylor et al., 1990). The cAMP-PrK phosphorylates Ser or Thr residues in substrate proteins, which are preceded by basic amino acids, usually in the sequence Arg-Arg-X-Ser/Thr or Arg-Arg-X-Y-Ser/Thr (Zetterqvist et al., 1990). Recent evidence suggests that Glu^{170} (of the catalytic subunit) is involved in recognition of the basic residues, and Asp^{328} and/or Glu^{332} may also play a role in substrate recognition (Buechler and Taylor, 1990). Cells contain a heat-stable protein (protein kinase inhibitor [PKI], "Walsh protein") that is a specific inhibitor of the cAMP-PrK catalytic subunit. Peptides derived from this protein have been used to demonstrate that basic residues are required for inhibition (Walsh et al., 1990). These peptides provide specific and potent inhibitors that are useful for studying cAMP-PrK in tissue extracts (Murray et al., 1990b). The C-subunit is myristylated at the amino terminus; however, this does not affect the properties of the protein in any obvious manner since recombinant C-subunit, with a free amino-terminus, behaves identically (Slice and Taylor, 1989; Clegg et al., 1989). The C-subunit is assayed by its phosphotransferase activity, ideally using a peptide as substrate (*see* Section 4.3.). Protein phosphorylation and the substrates of cAMP-PrK found in the brain are reviewed in Morton (this vol.), Hemmings et al. (1989), Nestler and Greengard (1989), and Nairn et al. (1985).

4.1.1. Isoenzymes and Isoforms*

It has long been established that cAMP-PrK exists as two isoenzymes and that these differ in their R-subunits; the isoenzymes were named type I and type II after the order of elution from DEAE-cellulose (Corbin et al., 1975). Other major differences are that only type I holoenzyme binds MgATP with high affinity and only the R^{II}-subunit is autophosphorylated by its own C-subunit. There are also differences between the cAMP-binding sites of the two R-subunits, in that cAMP derivatives, but not cAMP itself, show selectivity in their binding. The tissue distribution of the isoenzymes is poorly understood. Whereas adipose and brain from most species contain predominantly type II, many tissues show a large range of isoenzyme distribution, and it is not possible to predict which isoenzymes a particular tissue will contain. For a review on the differences between type I and II, *see* Beebe and Corbin (1986).

In addition to the well-established differences between the isoenzymes (I and II), it is now clear there are at least two isoforms (α and β) of each subunit, i.e., R_α^I, R_β^I, R_α^{II}, R_β^{II}, C_α, and C_β (*see* Table 1; reviewed in McKnight et al., 1988a,b). In general, the α isoform of each subunit is expressed in all tissues, whereas the β isoform is found in selected tissues, including brain, adrenal, and adipose. The differences between the α and β forms of the R-subunits are discussed below. Although there is current interest in the role of the C-subunit isoforms, C_α and C_β show

---

*In this chapter, the term isoenzyme is used to distinguish between families of enzymes that catalyze the same reaction (e.g., type I vs type II cAMP-PrK, or PDE I vs PDE III), whereas the term isoform is used to differentiate between members of the same family (e.g., R_α^I, R_β^{II}).

Table 1
Isoforms of Regulatory Subunits

R_α^I	Amino acid sequence from bovine skeletal muscle (Titani et al., 1984)
	Rat brain cDNA cloned, expressed, and studied by SDM* (Kuno et al., 1987, 1988, 1989)
	cDNA from bovine testis cloned, expressed, and studied by SDM* (e.g., Durgerian and Taylor, 1989)
	Human testis cDNA cloned (Sandberg et al., 1987)
R_β^I	Mouse brain cDNA cloned and expressed (Clegg et al., 1988)
R_α^{II}	Amino acid sequence from bovine heart (Takio et al., 1984)
	cDNA cloned from rat skeletal muscle and mouse brain (Scott et al., 1987)
	cDNA from human testis cloned (Oyen et al., 1989)
	Previously termed:
	Nonneural (e.g., Weldon et al., 1985)
	RII-H (e.g., Stein et al., 1987)
	$R\text{-}II_{54}$ (e.g., Jahnsen et al., 1986)
R_β^{II}	cDNA cloned from rat granulosa cells and human testis (Sandberg et al., 1988; Levy et al., 1988)
	Previously termed:
	Neural (e.g., Weldon et al., 1985)
	RII-β (e.g., Stein et al., 1987)
	$R\text{-}II_{51/52}$ (e.g., Jahnsen et al., 1986)

*SDM, site-directed mutagenesis.

>90% sequence identity (Uhler et al., 1986; Showers and Maurer, 1986), and to date, no physiological difference between them has been observed (Kinzel et al., 1987; Maurer, 1989; Olsen and Uhler, 1989). The C-subunit isoforms are not discussed further, except to note that brain is the major site of C_β mRNA expression (Uhler et al., 1986).

4.1.2. Brain cAMP-Dependent Protein Kinase

Brain from a number of species contains predominantly the type II holoenzyme (Rubin et al., 1979), and prior to the discovery of two genes for R^{II}, it was known that neural R^{II} differed from its nonneural counterpart (reviewed in Nairn et al., 1985). These proteins will now be termed R_α^{II} and R_β^{II} (*see* Table 1 for previous nomenclature), since it is clear that nonneural tissues can also express R_β^{II} and that the brain may also contain R_α^{II}. The differences in amino acid sequence and mol wt between R_α^{II} and R_β^{II} (Hartl and Roskoski, 1983; Stein and Rubin, 1985) have been confirmed and extended by molecular cloning (*see* Table 1), and there are also immunochemical differences between them because antibodies show limited crossreactivity (Erlichman et al., 1983; Weldon et al., 1985). It has been reported that brain type II cAMP-PrK is less sensitive to activation by cAMP than its muscle counterpart in rat (Bhatnagar et al., 1988) and cattle (Hartl and Roskoski, 1983). In contrast to most tissues where the majority of cAMP-PrK is soluble or cytosolic (Beebe and Corbin, 1986), most brain cAMP-PrK is associated with the membrane, cytoskeletal, or particulate fraction, and is particularly enriched in synaptic structures (Nairn et al., 1985). The subcellular location of cAMP-PrK is determined by the R-subunit, because treatment with cAMP releases C-subunits, but not R-subunits. The interaction among R^{II} and its various binding proteins is discussed below, and it is apparent that the subcellular location of R^{II} may be determined more by the binding proteins than by differences between R_α^{II} and R_β^{II}.

4.1.3. R^{II}-Binding Proteins

In general, two methods have been used to study the interaction of R^{II} with other proteins. The first method relies on the copurification of R^{II} and any putative binding protein. The tissue extract is passed through a cAMP-affinity column, and the R^{II}-binding protein complex is specifically eluted with cAMP and analyzed by SDS-PAGE. In the second technique, proteins are first separated by SDS-PAGE and then electrophoretically transferred to nitrocellulose ("Western blotting"). The R^{II}-binding proteins are located by incubating the nitrocellulose with R^{II}, and subsequently with antiR^{II} antibody and then with [^{125}I]-protein A. This method may be simplified by incubating the nitrocellulose directly with [^{32}P]-R^{II}, which is obtained by phosphorylating R^{II} with C-subunit (reviewed in Lohmann et al., 1988).

The relative merits of each of these methods have been discussed (Lohmann et al., 1988), and the major pitfall may be a failure to detect R^{II}-binding proteins rather than their false identification. The use of a cAMP-affinity column requires that the interaction between R^{II} and the binding protein is cAMP-independent, whereas the Western blotting procedure relies on the efficiency of transfer to the nitrocellulose and the renaturation of the R^{II}-binding site. The lack of detection of R^{II}-binding proteins is exemplified by the case of the C-subunit, which is not detected by either method. However, two major classes of R^{II}-binding proteins have been identified in brain. The first group is the microtubular-associated proteins-2 (MAP-2), and the second group binds both R^{II} and calmodulin.

MAP-2 is an integral, neuronal cytoskeletal protein, and is a substrate of cAMP-PrK and other protein kinases. The phosphorylation of MAP-2 results in the inhibition of tubulin assembly and may also affect its interaction with other cytoskeletal proteins (see Rubino et al., 1989). About one-third of the soluble cAMP-PrK in calf brain is associated with highly purified MAP-2 (Theurkaut and Vallee, 1982); interaction among the proteins can be seen by the overlay technique (Lohmann et al., 1984) and in tissue sections (DeCamilli et al., 1986). Rubino et al. (1989) have recently studied the interaction of R^{II} with cloned, human MAP-2. Fusion proteins were expressed, transferred to nitrocellulose, and examined by the overlay technique. The results showed that the R^{II}-binding site was located near the amino-terminus of MAP-2 and contained in a 31-residue amino acid sequence that was totally conserved between mouse and humans. Independently, Obar et al. (1989) reported that R^{II} could bind to all three products (MAP2A, MAP2B, MAP2C) of the MAP-2 gene and have apparently located the same R^{II}-binding site as Rubino et al. (1989). MAP2C appears to correspond to the 68-kDa R^{II}-binding protein of bovine brain (Lohmann et al., 1984). Although MAP-2 does not bind R^I, it will bind both R^{II}_α and R^{II}_β, and perhaps surprisingly, more R^{II}_α than R^{II}_β can be bound; there is no effect on binding of autophosphorylation of either isoform (Leiser et al., 1986; Rubino et al., 1989).

The interaction of R^{II} with a calmodulin-binding protein was first noted by Hathaway et al. (1981). A fraction containing cAMP-PrK was partially purified from bovine brain using ammonium sulfate fractionation followed by ion-exchange and size-exclusion chromatography. When this fraction was applied to a calmodulin-affinity column, 42% of the cAMP-PrK remained bound and could be eluted with EGTA. When the cAMP-PrK that did not bind to the column was reapplied, it still did not bind, suggesting that cAMP-PrK was bound to the column via a calmodulin- binding protein, which was identified as calcineurin. Sarkar et al. (1984) reported that preparations of bovine brain cytosolic R^{II} contained a tightly associated protein of 75-kDa (named P75). P75-bound calmodulin was a substrate for cAMP-PrK and protein kinase C, and was found predominantly in neural tissues. In contrast to MAP2, P75 binds more R_α^{II} than R_β^{II}, but similarly does not bind R^I and is not affected by the phosphorylation state of R^{II} (Leiser et al., 1986). Recently, the rat brain homolog of P75 has been cloned; this protein has a mol wt of 150 kDa and has been termed P150. Deletion analysis has shown that the 15 amino acids located at the carboxyl terminus of P150 are essential for R^{II} binding, but there are no homologies with the sequence found in MAP2 (Bregman et al., 1989).

4.1.4. Localization in Areas of the Brain

The localization of cAMP-PrK in various areas of the brain has been studied by two methods. The first involves cellular or subcellular fractionation of the brain followed by assays of cAMP-PrK activity (Sections 4.3. and 4.5.). Distribution may also be investigated immunocytochemically or by autoradiography using light or electron microscopy.

Stein et al. (1987) frationated 30 day-old rat cerebral cortex into purified populations of astrocytes, neurons, and oligodendrocytes, and used [^{3}H]-cAMP binding and immunochemical techniques to quantify the R-subunits. The three cell types contained similar levels of R-subunits and, in all cases, 70% was membrane bound. However, neurons and astrocytes contained >80% as R^{II}, whereas oligodendrocytes had relatively equal amounts of R^I and R^{II}. R^{II} in neurons and astrocytes was predominantly the β-isoform, whereas oligodendrocytes expressed mainly the α-isoform.

Binding of [^{3}H]-cAMP to rat and guinea-pig brain sections was specific to the R-subunits, and was assessed, semi-quantitatively, by visual inspection of autoradiograms (Gundlach and Urosevic, 1989). This method allows only particulate R-subunit to be studied, and [^{3}H]-cAMP will bind equally to all isoforms. The highest concentrations of R-subunits were found in neuronal cell layers (e.g., hippocampal pyramidal cells and cerebellar granule cells), less in cortical and limbic structures, and lower levels in hypothalmic, thalamic, and midbrain areas. Little species variation was noted. Excitotoxic legions were used to confirm the localization in the neuronal cells. Immunofluorescent techniques have also located R-subunits in granule and Purkinje cells (Cumming et al., 1981). De Camilli et al. (1986) studied the distribution of R^{II} in rat brain sections by immunocytochemistry; immunoreactivity was found in neurons, glia, neuroepithelial cells, and cells of mesenchymal origin, and the cytoplasmic R^{II} was associated with the golgi apparatus. However, since the antibody was against rat heart R_α^{II} and did not crossreact with rat brain R_β^{II}, it seems probable that this study specifically showed the distribution of R_α^{II} A recent study using antibodies to R_β^{II} has confirmed some of these observations and provided further evidence for the involvement of type II cAMP-PrK in postsynaptic function (Ludvig et al., 1990). Perhaps because of its lower abundance, little is known about the localization of R^I in the brain, although two reports indicate a nonneuronal distribution (Walter et al., 1978; Bradbury and Thompson, 1984).

4.1.5. Effects of Drug Treatment

Nestler et al. (1989) studied the effect of chronic antidepressant treatment on rat frontal cortex cAMP-PrK. They reported that imiramine, tranylcypromine, or electroconvulsive seizures decreased cAMP-PrK in the soluble fraction and increased it in a particulate, probably nuclear, fraction. This effect was not seen with other classes of psychotropic agents, suggesting that cAMP-PrK translocation could play a role in the long-term changes in the brain observed during antidepressant treatment. Moyer et al. (1986) had previously reported that desipramine, but not other antidepressants, caused a decrease in the soluble cAMP-PrK activity of rat pineal gland and that chronic desipramine treatment reduced the ability of isoproterenol to activate cAMP-PrK. Translocation of cAMP-PrK to the nucleus has also been reported for β-agonist-treated glioma cells (Schwartz and Costa, 1980) and in various nonneural tissues (reviewed in Lohmann and Walter, 1984). Morphine increased cAMP-PrK in rat locus cerulus, but not in several other areas of the brain (Nestler and Tallman, 1988). Ethanol appeared to decrease both the total amount and degree of phosphorylation of R^{II} in embryonic chick brain (Beeker et al., 1988). However, the above reports used histone as the substrate in the cAMP-PrK assay; even when the specific inhibitor protein (PKI) is included, this does not always provide the most satisfactory assay (*see* Section 4.4.).

4.2. cGMP-Dependent Protein Kinase

Although cAMP-PrK and cGMP-PrK are homologous enzymes, there are a number of significant differences. The most obvious difference is that of cyclic nucleotide specificity; this is proposed to be owing to a requirement for threonine in the cGMP-binding site as opposed to the alanine found in the cAMP-binding sites (Weber et al., 1989). However, autophosphorylation of cGMP-PrK increases its affinity for cAMP (Landgraf et al., 1986), and it has been proposed that cGMP-PrK is responsible for mediating some of the effects of increased cAMP in vascular smooth muscle (Lincoln et al., 1990). The other major difference from cAMP-PrK is that cGMP-PrK is a homodimer of 77-kDa subunits, each of which contains a regulatory and catalytic domain, and does not dissociate on cGMP binding (for reviews, *see* Lincoln and Corbin, 1983; Beebe and Corbin, 1986). cGMP-PrK is found in the brain. The highest levels are found in the cerebellum, where the enzyme may be concentrated in the Purkinje cells (*see* Nairn et al., 1985 for review). Recently, two isoforms of cGMP-PrK have been discovered, both by molecular cloning and protein isolation (Lincoln et al., 1988; Wolfe et al., 1989a; Wernet et al., 1989).

4.3. Phosphotransferase Assay

Equation 2 shows the general reaction for the phosphotransferase or catalytic activity of all protein kinases.

$$\text{Substrate} + \text{ATP} \longleftrightarrow \text{phosphosubstrate} + \text{ADP} \qquad (2)$$

The equilibrium lies heavily to the right, and special conditions are required for a protein kinase to catalyze the removal of phosphate from a protein (Flockhart, 1983). The forward reaction can, therefore, be followed by monitoring the appearance of either product. Details of the assay linked to the production of ADP are given in Roskoski (1983); this assay has advantages in that it is nonradioactive and can be monitored continuously, but has the major disadvantage that it is not suitable for tissue homogenates or preparations that contain significant ATPase activity. Essentially, this confines its use to highly purified preparations of cNMP-PrK and to situations where information about the cNMP-PrK, rather than its substrate, is required.

The assay using [γ-^{32}P]-ATP and quantifying the formation of ^{32}P-substrate is far more generally employed and can be used for all applications, e.g., assaying cNMP-PrK activity in tissue extracts (see Section 4.4.), studying the purified cNMP-PrK, and investigating the substrates of the cNMP-PrK. The general assay conditions for cAMP-PrK and cGMP-PrK have been described (Roskoski, 1983; Hardie, 1983). A wide range of buffers and pH values can be used, but it should be remembered that the buffer conditions may also affect the substrates of the cNMP-PrK. For example, bovine serum albumin is only phosphorylated by cAMP-PrK at alkaline pH in the presence of a reducing agent (Martin and Ekman, 1986), and other substrate-directed effects have also been reported (El-Maghrabi et al., 1983). Phosphate is a useful buffer because it will inhibit phosphoprotein phosphatase activity. Fluoride (20 mM) can also be used for this purpose, but not with phosphate, since the Mg^{2+}-fluoride-phosphate complex precipitates. Mg^{2+} (1–5 mM) is optimal for assaying the cAMP-PrK. Higher concentrations reduce the V_{max} (Cook et al., 1982); 50–100 mM Mg^{2+} is required for the cGMP-PrK (Gill et al., 1976). The K_m for ATP of both cNMP-PrKs is about 10 μM;

therefore, 10–100 µM [γ-^{32}P]-ATP is suitable. Care should be taken that this is not depleted by ATPase activities. There is a choice of protein and peptide substrates for both cNMP-PrKs (Zetterqvist et al., 1990; Glass, 1990). When compared with the most commonly used protein substrate, histone, synthetic peptides have the advantages of specificity and superior kinetic properties. Also, they do not allosterically interact with the kinases or cause substrate-directed effects (Murray et al., 1990b). The only disadvantage is their cost. The pyruvate kinase derived peptide, Leu-Arg-Arg-Ala-Ser-Leu-Gly (commonly referred to as kemptide), is a good substrate for both cNMP-PrKs; malantide shows higher selectivity for cAMP-PrK (Murray et al., 1990b), whereas Arg-Lys-Arg-Ser-Arg-Ala-Glu is more selective for cGMP-PrK (Glass and Krebs, 1982). To date, there is no peptide substrate with suitable kinetic properties that adequately discriminates between the two cNMP-PrKs so that the use of a specific inhibitor is required. The best inhibitor is a peptide derived from the naturally occurring protein inhibitor of cAMP-PrK (usually referred to as protein kinase inhibitor, [PKI] and previously as Walsh protein) (Kemp et al., 1988); in general, the peptide gives clearer results than the protein. Isoquinolinesulfonamides with selectivity for cAMP-PrK have recently been reported (H-89; Chijiwa et al., 1990). These compounds are competitive with ATP, and have been used more for intact cell or tissue experiments than for enzyme assays. Staurosporine and K-252a are also ATP-competitive inhibitors of cNMP-PrKs. However, these compounds show no selectivity with respect to other protein kinases (Murray and Warrington, 1990; Burgess, this vol.).

The protein kinase assay has been described previously (Corbin and Reimann, 1974; Roskoski, 1983). Briefly, the reaction is initiated, at timed intervals, by the addition of [γ-^{32}P]-ATP or kinase, and terminated by spotting an aliquot of the reaction mix onto numbered paper pieces, which are then placed in a beaker of acid. For protein substrates, the paper used is Whatman 3MM or 31ET, and the washing procedure is as described by Corbin and Reimann (1974) with the following exceptions: All washes may be carried out at room temperature, and 5% (w/v) sodium pyrophosphate is added to the first wash; this results in lower blank values. Whatman P81 paper is used for peptide substrates using the procedure described by Roskoski (1983). We have noted that certain batches of P81 produce high blanks, and this is apparently owing to binding of a contaminant in the $[\gamma\text{-}^{32}P]$-ATP. This may be overcome by including tetraphosphoric acid in the washing procedure (Murray et al., 1990b). However, under these conditions, kemptide does not bind quantitatively to the P81 paper, although the more basic substrate malantide does. This makes the general point that, when investigating new peptides or washing procedures, quantitative binding of the peptide should be demonstrated. Adaptations of the assay to permit a more rapid throughput include: termination of the assay with HCl to allow the use of repeating pipets (Murray et al., 1990b), a filtration microtiter plate-based assay (Buxbaum and Dudai. 1988), and the use of flat sheets, rather than individual squares of paper (Sahal and Fujita-Yamaguchi, 1987).

4.4. Activity Ratio Assay

This assay attempts to measure the activation of cAMP-PrK in intact tissues or cells. As shown previously in Eq. 1, binding of cAMP to the holoenzyme (R_2C_2) results in dissociation of the regulatory (R) and catalytic (C) subunits. A tissue homogenate is assayed in the absence of added cAMP to measure the activity of free catalytic subunits, and in the presence of added cAMP to ascertain the total kinase activity ($R_2C_2 + C$). This result is then expressed as an "activity ratio" or "fractional activity." This assay was first introduced by Corbin et al. (1973). Subsequently, problems and pitfalls of the procedure have been identified and discussed (Palmer et al., 1980; Corbin, 1983). Recently, the activity ratio assay has been reevaluated in respiratory (Giembycz and Diamond, 1990; Langlands and Rodger, 1990) and a number of other tissues (Murray et al., 1990b), the general conclusion being that the activity ratio assay is vastly improved when a peptide (malantide or kemptide) is used instead of histone as the phosphate acceptor. Specific disadvantages of histone are its low rate of phosphorylation, its lack of specificity for cAMP-PrK, and its ability to cause cAMP-PrK to dissociate in the test tube. These problems are not encountered with the peptide substrates.

The use of NaCl in the homogenization buffer to prevent trapping of free C-subunit has also been emphasized (Murray et al., 1990b). Although the brain has not been studied much with this assay, there is no inherent reason why interpretable results cannot be obtained provided suitable control experiments are performed (*see* Murray et al., 1990b). The use of a peptide substrate is essential because, when histone is the substrate, the majority of the phosphorylation catalyzed by guinea-pig brain extracts is owing to kinases other than cAMP-PrK (Murray et al., 1990b).

The activation of cGMP-PrK may be studied by an analogous assay (Fiscus and Murad, 1988), but because cGMP-PrK does not dissociate, the activation state is less easily preserved than that for the cAMP-PrK. There are other inherent problems in that most tissues, including brain, contain significantly more cAMP-PrK than cGMP-PrK, making a specific assay difficult. This is best achieved by the use of a selective peptide substrate (Glass and Krebs, 1982) and the cAMP-PrK inhibitor peptide (*see* Wolfe et al., 1989b), rather than histone and inhibitor protein, as described by Fiscus and Murad (1988).

4.5. Cyclic Nucleotide Binding Assays

Binding of cyclic nucleotides to cNMP-PrK may be conveniently assayed using [^{3}H]-cNMP and membrane filtration. Nonradioactive cNMP analogs are studied by their displacement of [^{3}H]-cNMP from cNMP-PrK (Rannels and Corbin, 1983). There are two classes of cyclic nucleotide binding sites on cAMP-PrK and cGMP-PrK (Beebe and Corbin, 1986; Taylor et al., 1990). Specific assay conditions are required to ensure quantitative binding to both sites. For the assay of [^{3}H]-cAMP binding to cAMP-PrK, the procedure originally described by Sugden and Corbin (1976) is widely used. Sample (10–20 µL) is mixed with 50 µL of buffer containing 50 mM potassium phosphate (pH 6.8), 1 mM EDTA, 2M NaCl, 0.5 mg/mL histone (Sigma type II-A), and 1 µM [^{3}H]-cAMP. After a suitable equilibration period, 1 mL wash buffer (10 mM potassium phosphate; 1 mM EDTA) is added to the assay tubes, and then the entire contents are removed with a Pasteur pipet and passed through a filter using vacuum filtration. The assay tube is rinsed with 1 mL buffer, and the filter washed twice with 5 mL each of buffer. No equilibration period is required for samples that do not contain bound cAMP (i.e., holo-enzyme), but a period of 60 min (at room temperature) is required for R-subunits containing bound cAMP. The filters (Millipore, HAWP, 0.45-μm pore size) are presoaked in ice-cold wash buffer to minimize blank values. If histone is omitted from the assay buffer, the amount of [^{3}H]-cAMP binding is reduced by 50%; for unknown reasons, cAMP does not appear to bind to site A under these conditions (Kerlavage and Taylor, 1982). NaCl is not required for [^{3}H]-cAMP binding, although it is needed to distinguish sites A and B by their exchange rates with unlabeled cAMP (Ogried et al., 1988); a description of the exchange rate method is given by Rannels and Corbin (1983). It has been reported that bovine brain R^{II} will only bind two molecules of cAMP per dimer (Weldon et al., 1985). MgATP increases the affinity for cAMP of type I, but not type II cAMP-PrK (Ringheim and Taylor, 1990). A modification to this method is required to study [^{3}H]-cGMP binding to cGMP-PrK, in that ammonium sul-fate is added to the assay tube to "freeze" the binding prior to the vacuum filtration. This method is also suitable for studying the binding of [^{3}H]-cAMP to cAMP-PrK. Full details are given in Doskeland and Ogried (1988). The use of [^{3}H]-cAMP binding is virtually specific for the R-subunits. The only other reported mammalian cAMP-binding protein being S-adenosylhomocys-teinase (see Doskeland and Ogried, 1988). In contrast, a number of phosphodiesterases are known to bind [^{3}H]-cGMP: The cGMP-stimulated PDE (PDE II) (Stroop et al., 1989), the cGMP-binding PDE (PDE V) (Francis and Corbin, 1988), and retinal PDE (Yamazaki et al., 1988). cGMP binding to PDE V is stimulated by various PDE inhibitors (Hamet and Tremblay, 1988).

4.5.1. Covalent Cyclic Nucleotide Binding

Radiolabeled 8-azido (8-N_3) cyclic nucleotides are routinely used to label covalently the cNMP-PrKs and, in the case of 8-N_3 cGMP, phosphodiesterases. [^{3}H]-8-N_3 cAMP is available from NEN. [^{32}P]-8-N_3 cAMP and [^{32}P]-8-N_3 cGMP are available from ICN. Methods for their synthesis have been published (Haley, 1977). For autoradiography, the [^{32}P]-labeled compounds are more convenient and produce quicker results. Methods for the photoaffinity labeling of cAMP-PrK with 8-N_3 cAMP have been described in detail (Walter and Greengard, 1983; Taylor et al., 1983), and general aspects of photoaffinity labeling are discussed by Guillory (1989). Particularly relevant to the use of cyclic nucleotides are:

1. IBMX can be used to prevent hydrolysis of the label by PDEs;
2. The label and any endogenous cyclic nucleotide should be allowed to equilibrate;
3. Cyclic nucleotide derivatives may be used to modulate the binding of the label (Rannels and Corbin, 1983); and
4. R^I binds 8-N_3 cAMP at both sites, whereas label is only covalently bound at one site of R^{II} (Ringheim et al., 1988).

Recently, new cAMP affinity probes have been described, and one of them, 8-[(4-bromo-2,3,dioxobutyl)thio]-cAMP, irreversibly inhibits PDE III (Grant et al., 1990); these probes have not been investigated with cAMP-PrK. The binding of [^{32}P]-8-N_3 cGMP to cGMP-PrK and retinal phosphodiesterases has recently been demonstrated (Thompson and Khorana, 1990).

4.6. Isoenzyme and Isoform Distribution

A number of methods allow the isoenzyme and isoform distribution of tissues or cell lines to be determined. This may be of particular interest since changes in the ratio of isoenzymes occur in response to certain stimuli (reviewed in Lohmann and Walter, 1984). Types I and II cAMP-PrK can be separated by anion-exchange chromatography (e.g., DEAE cellulose; Livesey and Martin, 1988), and assayed using either the cAMP-binding (Section 4.5.) or the phosphotransferase assay (Section 4.3.). The latter assay will give three peaks of activity corresponding to free C_2 R^I_2 C_2, and R^{II}_2 C_2, the latter two being cAMP-dependent. If dissociation of cAMP-PrK has occurred during this process, a false isoenzyme ratio may be obtained unless types I and II have been activated equally. However, if the dissociation state is preserved faithfully, then this assay can be used to investigate selective isoenzyme activation (*see* Livesey and Martin, 1988). Two peaks of [^{3}H]-cAMP binding are usually observed, and these correspond to mixtures of R^I_2 and $R^I_2 C_2$ in the first peak and R^{II}_2 and R^{II}_2, C_2 in the second. Under certain chromatographic conditions, R^I_2 and $R^I_2 C_2$ may be resolved to give a third peak. As long as the peaks are correctly identified and the binding assay is allowed time to reach equilibrium, then this method will give the "correct" isoenzyme ratio irrespective of the dissociation state.

Anion exchange does not resolve the α and β isoforms; to achieve this, the sample must be subjected to photoaffinity labeling with $[^{32}P]$-8-N_3 cAMP and analyzed by SDS-PAGE (Section 4.5.1.). R^{II}_β has a lower apparent mol wt (~51 kDa) than that of R^{II}_α (~55 kDa) (Table 1); conversely, R^I_β migrates more slowly than R^I_α (Clegg et al., 1988). Since types I and II cAMP-PrK covalently bind 8-N_3 cAMP with different stoichiometries (Ringheim et al., 1988), this must be taken into account to obtain quantitative results. Another consideration is that cAMP bound to the R-subunit may not exchange fully with the 8-N_3 cAMP, thereby decreasing the incorporation of label. However, this can be used to advantage and, by lowering the concentration of 8-N_3 cAMP, the assay has been used to show cAMP binding to R-subunit in extracts from hormonally treated cells (Middleton et al., 1989). Immunological methods may also be used to study isoenzyme distribution and activation (Erlichman et al., 1983; Schwoch et al., 1985), and a mini-gel assay has recently been described (Hohmann and Greene, 1990).

4.7. Protein Purification

cAMP-PrK R-subunits (Rannels et al., 1983) and cGMP-PrK (Lincoln, 1983) can be purified by affinity chromatography using cAMP columns (at the time of writing, Sigma is the only commercial source of these). cAMP-PrK C-subunit is prepared by cAMP elution of holoenzyme bound to DEAE-cellulose (Reimann and Beham, 1983). Although PKI is relatively easy to purify (Schlender et al., 1983; Whitehouse and Walsh, 1983), the synthetic PKI peptide gives better results and is commercially available from a number of sources. Commercial sources of some of these proteins are not always of the desired quality.

5. Phosphodiesterases

Twenty years ago, it was recognized that many tissues, including brain (Thompson and Appleman, 1971), contain multiple isoenzymes of cyclic nucleotide phosphodiesterase (PDE) (for earlier reviews, *see* Wells and Hardman, 1977; Palmer, 1981). However, it is only in the past five years or so that the enzymology of the various PDEs has been characterized in sufficient detail to allow comparisons between various workers to be made. Much confusion has followed the use of different systems for naming the PDEs; most workers have tended to number the isoenzymes according to their elution from anion-exchange chromatography. This practice is not satisfactory, because the number of peaks of PDE activity obtained is dependent on both the tissue and chromatographic conditions, and a peak may include several isoenzymes. An earlier recommendation on nomenclature (Strada and Thompson, 1984) has not been widely adopted, and a more recent proposal (Beavo, 1988; Beavo and Reifsnyder, 1990) will be followed in this chapter. The classification is based on protein sequence information (both primary amino acid sequence and that deduced from cDNA), but also takes into account the kinetic properties and physiological and synthetic modulators. Although the nomenclature cannot be complete since, undoubtedly, new isoenzymes and isoforms will be discovered and some may disappear (e.g., if it is found that they are proteolytic products of each other), it is to be hoped that this nomenclature will be widely adopted. There are a number of recent reviews of phosphodiesterases (Beavo, 1988; Beavo and Reifsnyder, 1990; Beavo and Houslay 1990; Strada and Thompson, 1984).

5.1. Ca²⁺-Calmodulin-Stimulated Phosphodiesterases (CaM-PDE, PDE I)

At present, seven isoforms of the CaM-PDE isoenzyme family have been identified (Beavo and Reifsnyder, 1990), and a number of these are present in brain. It should be noted that PDE V (Section 5.5.) elutes from DEAE-cellulose at similar salt concen-

Table 2
Brain Calmodulin-Stimulated Phosphodiesterases

Mol wt, kDalton	K_m, µM		V_{max} ratio	References
	cAMP	cGMP	cAMP/cGMP	
60	32	3	1.8	Sharma and Wang (1988)
63	10	1	0.3	Sharma and Wang (1988)
74	30	2	0.06	Shenolikar et al. (1985)
59	43	4	3.3	Kincaid and Vaughan (1988)

trations to CaM-PDE and so, in the past, may have been incorrectly included in this family. Table 2 shows the CaM-PDEs reported in brain. Two isoforms have been purified to homogeneity and their properties characterized; these have been termed 60-kDa isoform and 63-kDa isoform (reviewed in Sharma and Wang, 1988). Both enzymes are homodimers, although there is evidence that a heterodimer of the two forms also exists (Sharma et al., 1984).

The 60-kDa isoform is closely related to the 59-kDa heart CaM-PDE (Beavo, 1988), and has similar properties to the enzyme purified from brain by Kincaid and Vaughan (1988). The enzyme is a substrate for cAMP-PrK. Two moles of phosphate are incorporated per subunit, and this results in a 20-fold lowering in the affinity of the PDE for calmodulin. Increases in Ca^{2+} will favor dephosphorylation of the PDE, since binding of calmodulin to the enzyme prevents the phosphorylation by cAMP-PrK, and the PDE is dephosphorylated by the calmodulin-activated phosphoprotein phosphatase, calcineurin (Sharma and Wang, 1985). The 63-kDa isoform, a separate gene product from the 60-kDa isoform (Beavo and Reifsnyder, 1990), is phosphorylated by calmodulin-dependent protein kinase II (Hashimoto et al., 1989), but not by cAMP-PrK (Sharma and Wang, 1985). As with the 60-kDa isoform, phosphorylation occurs at multiple sites, is reduced by calmodulin binding to the PDE, and results in a decreased affinity (~1/6th) for calmodulin.

Shenolikar et al. (1985) have also characterized a Ca^{2+}-calmodulin-stimulated PDE in bovine brain, and this has been classed as a separate isoform by Beavo and Reifsnyder (1990).

However, as shown in Table 2, the K_m values for this enzyme are very similar to those reported for the 60-kDa isoform, although the V_{max} figures are different. In addition, the preparation of Shenolikar et al. (1985) contained a 59-kDa protein that was phosphorylated by cAMP-PrK, although this did not alter the activation of the PDE by calmodulin. This suggests that the PDE purified by Shenolikar et al. (1985) may be a mixture of isoforms, including the 60-kDa enzyme, and that it may be prudent to wait for protein sequence data before classifying it as a separate isoform. This also raises the general point that, if a tissue contains multiple isoforms of CaM-PDE, the use of a calmodulin-affinity column by itself will not result in a pure PDE. Since the 60-kDa and 63-kDa isoforms are difficult to separate by anion-exchange chromatography, it seems likely that many preparations of brain CaM-PDE are a mixture of isoforms. A calmodulin-fragment affinity column is useful because it binds the 60-kDa and 63-kDa isoforms, but not a number of other brain calmodulin-binding proteins (Draetta and Klee, 1988).

CaM-PDEs are inhibited, nonselectively, by IBMX; selectivity is increased by use of 8-methoxymethyl IBMX (Wells and Miller, 1988). CaM-PDEs may also be inhibited selectively with respect to other PDEs (but not with respect to other calmodulin-dependent enzymes) by calmodulin antagonists (e.g., W-7, trifluoperazine; Hidaka et al., 1988). A number of anthraquinone compounds inhibit CaM-PDE at both the active site and at the calmodulin-binding site (Matsuda et al., 1990). The cellular location of the 60-kDa isoform has been studied in rat brain with a specific antibody. In contrast to other calmodulin-dependent enzymes, this PDE has a restricted distribution (Kincaid et al., 1987), and its expression is regulated *trans*-synaptically (Balaban et al., 1989; Billingsley et al., 1990).

5.2. cGMP-Stimulated Phosphodiesterases (cGS-PDE, PDE II)

This PDE has been most studied in heart, liver, and adrenal, and appears to be the same in all these tissues. The enzyme is a homodimer, with a subunit mol wt of 105 kDa. The K_m for cAMP (~40 µM) is higher than that for cGMP (~10 µM), although the V_{max} is similar for both cyclic nucleotides. However, the distinguishing feature of this PDE is that the hydrolysis of cAMP is substantially increased by micromolar concentrations of cGMP. This stimulation can occur with physiological concentrations of cyclic nucleotides and has been proposed to be of importance in intact cells (Beavo, 1988). If so, the cGS-PDE could, under certain circumstances, fulfill the "Yin-Yang" hypothesis since increases in cGMP will result in lowering of cAMP levels. To determine whether cGS-PDE or cGMP-PrK is mediating the effects of increased cGMP, 8-Br cGMP could be used; 8-Br cGMP is a potent activator of cGMP-PrK (Corbin et al., 1986), but is less potent at increasing the hydrolysis of cAMP by cGS-PDE (Erneux et al., 1984). This method has been used by Fischmeister and Hartzell (1987) in their study of Ca^{2+} currents in frog heart.

Limited proteolysis and photoaffinity-labeling showed that cGS-PDE contains two functional domains (Stroop et al., 1989); one retains catalytic activity whereas the other binds cGMP. Recently, it has been shown that the cGMP-binding domain is homologous between cGS-PDE and the photoreceptor phosphodiesterases (Charbonneau et al., 1990). In addition to heart, liver, and adrenal, cGS-PDE has recently been identified in endothelial cells (Lugnier and Schini, 1990), although it does not appear to be present in smooth muscle (Murray, 1990). An early report that cGS-PDE was present in a brain particulate fraction (Beavo et al., 1971) has recently been confirmed in both rabbit (Whalin et al., 1988) and bovine brain (Murashima et al., 1990). As previously reported for liver (Pyne et al., 1986), the particulate and soluble bovine brain cGS-PDEs appear to be distinct isoforms. These observations confirm the presence of particulate PDEs and highlight the danger of determining the PDE isoenzyme content of a tissue from the soluble fraction alone. No selective inhibitors of cGS-PDE have been described.

5.3. cGMP-Inhibited Phosphodiesterases (cGI-PDE, PDE III)

The development of a new generation of cardiotonic agents that act by inhibition of cGI-PDE has meant that this enzyme has recently received intensive study. The bovine cardiac cGI-

PDE is well characterized; the K_m for cAMP and cGMP is similar (~0.1 μM), although the V_{max} for cGMP is a tenth of that for cAMP (6 U/mg). The hydrolysis of cAMP is potently inhibited by cGMP (K_i ~ 0.1 μM), hence the name cGI-PDE. The enzyme appears to be a dimer of two identical 110-kDa subunits, although the susceptibility of the enzyme to proteolysis is a problem (Beavo, 1988). Phosphorylation and activation of cGI-PDE have been observed in intact platelets (MacPhee et al., 1988) and adipocytes (Degerman et al., 1990). Potent selective inhibitors of cGI-PDE increase cAMP levels in heart (Silver, 1989), smooth muscle (Murray, 1990), adipose tissue (Manganiello et al., 1988), and platelets (Murray et al., 1990a). It has recently been proposed that cGMP inhibition of cGI-PDE may be of physiological significance in platelets (Maurice and Haslam, 1990). cGI-PDE inhibitors had no effect on basal or isoprenaline-stimulated cAMP levels in rat cerebral cortical slices (Challiss and Nicholson, 1990), and had little or no effect on histamine-stimulated cAMP accumulation in guinea-pig hippocampal slices (Stanley et al., 1989). These results suggest that, if cGI-PDE is present in the brain, it does not play a major role in regulating cAMP levels. In agreement with this, chromatographic separation and selective inhibitors have failed to reveal cGI-PDE in the brain (Nicholson et al., 1989; Kariya and Dage, 1988). The purification of cGI-PDE from bovine heart has been described (Harrison et al., 1988). Among the more commonly used cGI-PDE inhibitors are cilostamide (OPC3689, Otsuka), milrinone (WIN47203, Stirling Winthrop), trequinsin (HL725, Hoechst), indolidan (LY195115, Lilly), and siguazodan (SK&F94836, SmithKline Beecham).

5.4. cAMP-Specific Phosphodiesterases (PDE IV)

It is now apparent that, in some tissues at least, the so-called "low K_m PDE" was a mixture of the cGI-PDE and the cAMP-specific PDE (Reeves et al., 1987). The K_m for cAMP of PDE IV (~1 μM) is slightly higher than that of the cGI-PDE (~0.1 μM). However, cGMP is a substrate of only the latter enzyme, and in many cases, it is not possible to detect hydrolysis of cGMP by PDE IV. This enzyme appears to be a monomer of ~66 kDa. Recent work using molecular genetic techniques has shown that subforms of PDE IV may be present in brain and that PDE IV is the mammalian homolog of the *Drosophila* dunce mutant. This work has been made easier by the development and recognition of selective inhibitors of PDE IV, the most commonly used being rolipram (ZK62711, Schering), Ro 20-1724 (Hoffman-La Roche), and denbuffylline (SmithKline Beecham).

Schwabe et al. (1976) demonstrated the presence of a roli-pram-inhibited PDE in soluble and particulate fractions of a rat brain homogenate. This has been confirmed by more recent work (Kariya and Dage, 1988; Nicholson et al., 1989). Using a rolipram affinity column (Fougier et al., 1986), Némoz et al. (1989) puri-fied PDE IV from rat brain and heart. The enzymes from both tissues were very similar with respect to mol wt (44 kDa), and both gave two peaks of activity on isoelectric focusing. PDE IV appears to have a widespread tissue distribution (Beavo, 1988; Kariya and Dage, 1988; Pang et al., 1988).

Recently, PDE IV has been studied using molecular genetic techniques. Previous work on the *Drosophila* dunce mutant had shown that the defective gene coded for a low k_m, cAMP-spe-cific PDE (Davis and Kauvar, 1984). Using a probe representing this gene, homologous PDEs were found in mammalian tis-sues including brain (Davis et al., 1989; Swinnen et al., 1989a). Using a different approach, Colicelli et al. (1989) expressed a protein from a rat brain cDNA library that was also a low k_m, cAMP-specific PDE, but distinct from that reported by Davis et al. (1989). At present, it seems that there may be as many as eight members of the PDE IV family (Beavo and Reifsnyder, 1990). In rat sertoli cells, cAMP increased the transcription of one of these PDE IVs, providing evidence for regulation of cAMP levels via a feedback mechanism (Swinnen et al., 1989b). The proof that these genes code for PDE IV comes from the demonstration that the expressed PDEs are inhibited by rolipram and Ro 20-1724; sur-prisingly, the dunce enzyme is not inhibited (Swinnen et al., 1989b; Henkel-Tigges and Davis, 1990).

There is increasing evidence that PDE IV is the major cAMP metabolizing enzyme in brain, although it should be noted that, since these studies were conducted with brain slices, it does not preclude a role for other PDE isoenzymes in specific areas of the brain. In cerebral cortical and hippocampal slices, PDE IV inhibitors caused an increase in basal cAMP levels and also potentiated agonist-stimulated increases (Schwabe et al., 1976; Stanley et al., 1989; Donaldson et al., 1988; Challiss and Nicholson, 1990). In these studies, the effects of rolipram on agonist-stimulated cAMP levels were very similar to those of the nonselective inhibitor IBMX, suggesting that inhibition of PDEs other than PDE IV does not lead to further increases in cAMP levels. Based on kinetic evidence of cAMP turnover, Donaldson et al. (1988) suggested that PDE IV was the major metabolizing enzyme at low cAMP concentrations and that the contribution of other PDEs (probably PDE I) increases as the cAMP levels rise. Challiss and Nicholson (1990) incubated rat cerebral cortical slices with isoprenaline until a new steady-state concentration of cAMP had been reached; subsequent addition of a β-antagonist caused a rapid decline in cAMP levels. PDE IV inhibitors selectively decreased the rate of cAMP hydrolysis observed after the addition of the β-antagonist, again indicating a major role for PDE IV in the regulation of brain cAMP levels.

5.5. cGMP-Specific Phosphodiesterases (PDE V)

This family of phosphodiesterases is termed cGMP-specific, since they do not hydrolyze physiological concentrations of cAMP (K_m, cGMP ~ 0.5 μM; K_m cAMP > 100 μM). Although clearly related, this family may be divided into retinal and nonretinal isoforms; the former have been reviewed (Beavo, 1988) and are not discussed further here. The nonretinal enzyme has been most extensively studied in lung and platelet, where it has been termed the cGMP-binding PDE (Francis and Corbin, 1988; Hamet and Tremblay, 1988).

Although the physiological relevance of the cGMP binding is unknown, it does provide a convenient assay for the PDE. This is particularly so because IBMX stimulates cGMP binding to the PDE, but not to cGMP-PrK. Coquil et al. (1985) have used this property to demonstrate the presence of the PDE in rat brain, although the chromatography did not fully separate the IBMX-sensitive and IBMX-insensitive cGMP binding. The cGMP-specific PDE elutes early on DEAE chromatography and usually coelutes with Ca^{2+}-calmodulin-stimulated PDEs, which also have a high affinity for cGMP, making it difficult to distinguish the enzyme by this method. The most commonly used PDE V inhibitor, zaprinast (M&B22948, Rhone Poulenc), will also inhibit some PDE I isoforms making discrimination between the isoenzymes difficult. Dipyridamole will inhibit PDE V selectively, but also has effects on adenosine uptake (Harker and Fuster, 1986).

5.6. Assays of Phosphodiesterases

There are a number of methods available for the assay of PDE activity. Both the disappearance of the substrate (cNMP) and the appearance of the product (5'-NMP) can be quantified directly, or the 5'-NMP can be converted to the corresponding nucleoside for assay. For all assays, the pH of the buffer should be between 7.0–8.0 to reflect the optimum of most PDEs, and a number of buffers (e.g., HEPES, Tris) are suitable. Mg^{2+} (1–10 mM) is an essential cofactor, and a low concentration of bovine serum albumin is usually included. The other conditions, Ca^{2+}, calmodulin, inhibitors, and so on, are dependent on the exact purpose of the assay. As with many enzymes, the properties of PDEs may be changed by proteolysis, so protease inhibitors should be included when preparing the tissue extracts. The PDE assay itself is conveniently terminated by placing the tube in a boiling water bath or by stopping with a solution containing EDTA and IBMX.

Phosphodiesterase assays are most conveniently performed using appropriate radiolabeled cNMP and by quantifying the labeled reaction product, 5' NMP. This is achieved directly using a boronate column according to the method of Davies and Daly (1979). In our experience, this method, as modified by Reeves et al. (1987), produces reliable and reproducible results, and has the advantage that no conversion of the NMP is required. The assay is more time-consuming than the method described below, and another disadvantage is the high initial cost of the resin (Affigel 601, Bio-Rad), although the columns may be kept for over 1 yr.

The most commonly used phosphodiesterase assay is that introduced by Thompson and Appleman (1971), in which the NMP formed is converted to the corresponding nucleoside by snake venom nucleotidase (Sigma). The uncharged nucleoside is then separated from the charged cNMP with an anion-exchange resin. A number of different resins and buffers may be used (*see* Thompson et al. [1974] and Kincaid and Manganiello [1988] for a full description of this method). The nucleotidase reaction may be stopped by the addition of the resin followed by centrifugation or by passage through a column, the latter being more time-consuming. Provided adequate checks are kept on the recovery, there are no reported major disadvantages to this method, which has the benefit of speed and cost over the boronate column. A potential disadvantage is that PDE inhibitors could inhibit the nucleotidase, although, to date, this has not been reported and the appropriate control is easily performed.

Nonradioactive assays are also available, but these have reduced sensitivity and are not generally as convenient. One method is to quantify the phosphate released by the nucleotidase; this is less sensitive and therefore best performed with high substrate concentrations, although it has the advantage that it can be used in microtiter plates (Gillespie and Beavo, 1989). Other methods include various coupled assays (Chock and Huang, 1984) and measurement of pH change (Liebman and Evancuk, 1982); both of these are continuous assays. HPLC may be used to separate cNMP and NMP (Watterson and Lukas, 1988), and this procedure is also applicable to studying the hydrolysis of cNMP-derivatives (Braumann et al., 1986b). cNMP-derivatives may also be assayed by the disappearance of cNMP using the activation of the appropriate cNMP-PrK (Corbin et al., 1988). There are also histochemical methods suitable for the localization of phosphodiesterase (Ueno and Ueck, 1988).

5.7. Phosphodiesterase Inhibitors

The problems associated with using xanthine-based PDE inhibitors have been discussed previously (Wells and Kramer, 1981) and to some extent have been overcome by the second generation, selective PDE inhibitors (Beavo, 1988; *see* Table 1 of Beavo and Reifsnyder, 1990). It is worth reiterating that, whereas most of these new inhibitors are known to be selective (i.e., they inhibit only one PDE isoform), it is not always known whether they are specific (i.e., they do not interact with other proteins or cellular components). Therefore, although these inhibitors are powerful tools for PDE purification or analysis of the PDE profile of a tissue, their effects on cNMP levels in cells or tissues need to be interpreted with caution. Commonly used inhibitors have been mentioned in the sections on individual isoenzymes and have been recently reviewed (Beavo, 1988; Beavo and Reifsnyder, 1990). Selective PDE inhibitors have been used successfully with brain slices (*see* Section 5.4.); only PDE IV inhibitors increased cAMP levels, suggesting that this is the major cAMP-metabolizing isoform. However, the lack of effect of a particular PDE inhibitor should be treated with caution, since it is possible that the inhibitor does not penetrate all tissues equally. Similarly, the intracellular concentration of inhibitor is rarely known making it difficult to determine the actual selectivity achieved. Comparisons of selective inhibitors with IBMX may also be dangerous in view of the well-known adenosine receptor-blocking capabilities of this compound. Finally, although the initiation or potentiation of a physiological response by a PDE inhibitor is indicative of a cNMP-mediated response, this should be corroborated by measurement of cNMP levels. Ideally, both cAMP and cGMP should be measured, since a number of PDEs will hydrolyze both cNMPs and this will also help ascertain the selectivity of the PDE inhibitor.

5.8. Separation and Characterization of Phosphodiesterase Isoenzymes

It is obviously necessary to know which PDE isoenzymes are present in a tissue to have a fuller understanding of the physiological effects of PDE inhibitors. Traditionally, this has been achieved by subjecting a tissue extract to anion-exchange chromatography, usually on DEAE-cellulose. It is now clear that the isoenzymes are poorly resolved by this resin and that it is better replaced by high-performance columns, e.g., Mono Q (Lugnier and Schini, 1990) or DEAE SPW (Robicsek et al., 1989). However, at present there are few tissues for which chromatographic conditions have been found that will fully resolve all the isoenzymes, so that use must be made of physiological effectors and pharmacological inhibitors of the PDEs. Assaying the fractions with cAMP and cGMP as substrates will identify specific PDEs and those able to hydrolyze both substrates. With 1 μM cAMP as a substrate, inclusion of 10 μM unlabeled cGMP will identify PDE II (cGMP-stimulated) and PDE III (cGMP-inhibited) provided the two activities have been resolved. Although this is not always the case when DEAE-cellulose is used (e.g., Harrison et al., 1988), in our experience, chromatography on Mono Q provides good separation. Ca^{2+} and calmodulin are routinely used to identify Ca^{2+}/CaM-stimulated PDEs. Obviously, fractions may be assayed in the presence of selective inhibitors to determine the identity of the PDE, and care should be taken to use inhibitor concentrations that are truly selective.

It is usually necessary to further characterize the peaks identified by anion-exchange chromatography. Kinetic analysis is not a reliable method. Simple Michaelis-Menten kinetics does not necessarily mean the peak contains one PDE isoenzyme; conversely, complex kinetics could be owing to a mixture of isoenzymes or be a genuine feature of a single isoenzyme. The peaks are best characterized by further protein purification, taking advantage of any available affinity columns. Calmodulin-columns may be used to determine whether a peak is composed entirely of Ca^{2+}/CaM-stimulated isoform. Of course, this method will not separate members of the Ca^{2+}/CaM-stimulated family; specific antibodies are required to achieve this (Sharma and Wang, 1988; Hansen et al., 1988). It is sometimes difficult to resolve PDE II from either PDE I or PDE IV by anion-exchange chromatography, and a cGMP-affinity column may then be useful to bind PDE II (Harrison et al., 1988). Affinity columns for PDE III (Degerman et al., 1987) and PDE IV (Fougier et al., 1986) have been described, but require synthesis of the ligand.

More specialized techniques are required to distinguish between the isoforms of a particular family. Molecular genetic techniques may be used to give information at the nucleic acid level (e.g., Swinnen et al., 1989a,b) and may be the only practicable method when the differences between isoforms are small. However, as mentioned above, PDE I isoforms have been distinguished by an antibody (Sharma and Wang, 1988) and those of PDE II by peptide mapping (Murashima et al., 1990).

Although the PDE profile of a tissue gives valuable information, care must be taken in ascribing it any physiological significance. For example, usually only the soluble activities are chromatographed, ignoring any contribution of particulate PDE. Also, the proportion of one PDE to another is influenced by recovery on the column and the concentration of substrate in the assay. The cellular and subcellular localizations of the PDE are not determined by this method, and for the brain, this is of considerable interest and may help in ascertaining the physiological function of the PDE.

Acknowledgments

I am very grateful for the assistance of Nicky Dollimore and Peter McIntosh in the preparation of this manuscript, and also to David Mills, Jim Lynham, and Paul England for their useful comments.

Abbreviations

cAMP	Adenosine 3',5' cyclic monophosphate
CPT-cAMP	8-(4-chlorophenlthio)cAMP
cGMP	Guanosine 3',5' cyclic monophosphate
IBMX	Isobutylmethylxanthine
PAGE	Polyacrylamide gel electrophoresis
PDE	Phosphodiesterase
PrK	Protein kinase
RIA	Radioimmunoassay
SDS	Sodium dodecyl sulfate

References

Alvarez R. and Daniels D. V. (1990) A single column method for the assay of adenylate cyclase. *Anal. Biochem.* **187,** 98–103.

Balaban C. D., Billingsley M. L., and Kincaid R. L. (1989) Evidence of transsynaptic regulation of calmodulin-dependent cyclic nucleotide phosphodiesterase in cerebellar Purkinje cells. *J. Neurosci.* **9,** 2374–2381.

Barber R. and Butcher R. W. (1988) cAMP turnover in intact cells, in *Methods in Enzymology* 159 (Corbin J. D. and Johnson R. A., eds.), (Academic, New York), pp. 50–60.

Barsony J. and Marx S. J. (1990) Immunocytology on microwave-fixed cells reveals rapid and agonist-specific changes in subcellular accumulation patterns for cAMP or cGMP. *Proc. Natl. Acad. Sci. USA* 87, 1188–1192.

Beavo J. A. (1988) Multiple isozymes of cyclic nucleotide phosphodiesterase, in *Advances in Second Messenger and Phosphoprotein Research*, vol. 22 (Greengard P. and Robison G. A., eds.), Raven, New York, pp. 1–38.

Beavo J. A. and Houslay M. D. (eds.) (1990) *Molecular Pharmacology of Cell Regulation*, vol. 2, John Wiley & Sons, London.

Beavo J. A. and Reifsnyder D. H. (1990) Primary sequence of cyclic nucleotide phosphodiesterase isozymes and the design of selective inhibitors. *Trends Pharmacol. Sci.* 11, 150–155.

Beavo J. A., Hardman J. G., and Sutherland E. W. (1971) Stimulation of adenosine 3',5'monophosphate hydrolysis by guanosine 3',5'-monophosphate. *J. Biol. Chem.* 246, 3841–3846.

Beebe S. J. and Corbin J. D. (1986) Cyclic nucleotide-dependent protein kinases, in *The Enzymes XVII* (Boyer P. D. and Krebs E. G., eds.), (Academic, New York), pp. 43–111.

Beebe S. J., Blackmore P. F., Chrisman T. D., and Corbin J. D. (1988) Use of synergistic pairs of site-selective cAMP analogs in intact cells, in *Methods in Enzymology* 159 (Corbin J. D. and Johnson R. A., eds.), (Academic, New York), pp. 118–139.

Beeker K., Deane D., Elton C., and Pennington S. (1988) Ethanol-induced growth inhibition in embryonic chick brain is associated with changes in cytoplasmic cyclic AMP-dependent protein kinase regulatory subunit. *Alcohol. Alcohol.* 23, 477–482.

Bhatnagar D., Burton A. A., and Roskoski R., Jr. (1988) Differential sensitivity of neural and nonneural protein kinase isozymes to cyclic AMP. *Biochem. Biophys. Res. Commun.* 156, 801–806.

Billingsley M. L., Polli J. W., Balaban C. D., and Kincaid R. L. (1990) Development expression of calmodulin-dependent cyclic nucleotide phosphodiesterase in rat brain. *Dev. Brain. Res.* 53, 253–263.

Birnbaumer L., Abramowitz J., and Brown A. M. (1990) Receptor-effector coupling by G proteins. *Biochem. Biophys. Acta* 1031, 163–224.

Bradbury J. M. and Thompson R. J. (1984) Photoaffinity labelling of central-nervous-system myelin. *Biochem J.* 221, 361–368.

Braumann T. and Jastorff B. (1985) Physico-chemical characterization of cyclic nucleotides by reversed-phase high-performance liquid chromatography. II. Quantitative determination of hydrophobicity. *J. Chromatogr.* 350, 105–118.

Braumann T., Jastorff B., and Richter-Landsberg C. (1986a) Fate of cyclic nucleotides in PC12 cell cultures: Uptake, metabolism, and effects of metabolites on nerve growth factor-induced neurite outgrowth. *J. Neurochem.* **47**, 912–919.

Braumann T., Erneux C., Petridis G., Stohrer W.-D., and Jastroff B. (1986b) Hydrolysis of cyclic nucleotides by a purified cGMP-stimulated phosphodiesterase: Structural requirements for hydrolysis. *Biochem. Biophys. Acta* **871**, 199–206.

Bregman D. B., Bhattacharyya N., and Rubin C. S. (1989) High affinity binding protein for the regulatory subunit of cAMP-dependent protein kinase II-B. *J. Biol. Chem.* **264**, 4648–4656.

Brown B. L., Albano J. D. M., Ekins R. P., and Scherzi A. M. (1971) A simple and sensitive method for the measurement of adenosine 3',5'-cyclic monophosphate. *Biochem. J.* **121**, 561–572.

Brunton L. L. and Heasley L. E. (1988) cAMP export and its regulation by prostaglandin A_1, in *Methods in Enzymology* 159 (Corbin J. D. and Johnson R. A., eds.), (Academic, New York), pp. 83–93.

Buechler J. A. and Taylor S. S. (1990) Differential labeling of the catalytic subunit of cAMP-dependent protein kinase with a water-soluble carbodiimide: Identification of carboxyl groups protected by MgATP and inhibitor peptides. *Biochemistry* **29**, 1937–1943.

Butt E., van Bemmelen M., Fischer L., Walter U., and Jastorff B. (1990) Inhibition of cGMP-dependent protein kinase by (Rp)-guanosine 3',5'-monophosphorothioates. *FEBS Lett.* **263**, 47–50.

Buxbaum J. D. and Dudai Y. (1988) A microtiter-base assay for protein kinase activity suitable for the analysis of large numbers of samples, and its application to the study of *drosophila* learning mutants. *Anal. Biochem.* **169**, 209–215.

Challiss R. A. J. and Nicholson C. D. (1990) Effects of selective phosphodiesterase inhibition on cyclic AMP hydrolysis in rat cerebral cortical slices. *Br. J. Pharmacol.* **99**, 47–52.

Charbonneau H., Prusti R. K., LeTrong H., Sonnenburg W. K., Mullaney P. J., Walsh K. A., and Beavo J. A. (1990) Identification of a noncatalytic cGMP-binding domain conserved in both the cGMP-stimulated and photoreceptor cyclic nucleotide phosphodiesterases. *Proc. Natl. Acad. Sci. USA* **87**, 288–292.

Chijiwa T., Mishima A., Hagiwara M., Sano M., Hayash T. K., Inoue T., Naito K., Toshioka T., and Hidaka H. (1990) Inhibition of forskolin-induced neurite outgrowth and protein phosphorylation by a newly synthesized selective inhibitor of cyclic AMP-dependent protein kinase, N-12-(*p*-Bromocinnamylamino)ethyl-5-isoquinolinesulfonamide (H-89), of PC12D pheochromocytoma cells. *J. Biol. Chem.* **265**, 5267–5272.

Chock S. P. and Huang C. Y. (1984) An optimized continuous assay for cAMP phosphodiesterase and calmodulin. *Anal. Biochem.* **138**, 34–43.

Clegg C. H., Cadd G. G., and McKnight S. (1988) Genetic characterization of a brain-specific form of the type I regulatory subunit of cAMP-dependent protein kinase. *Proc. Natl. Acad. Sci. USA* **85,** 3703–3707.

Clegg C. H., Ran W., Uhler M. D., and McKnight G. S. (1989) A mutation in the catalytic subunit of protein kinase A prevents myristylation but does not inhibit biological activity. *J. Biol. Chem.* **264,** 20,140–20,146.

Cohen P. T. W., Brewis N. D., Hughes V., and Mann D. J. (1990) Protein serine/threonine phosphatases; an expanding family. *FEBS Lett.* **268,** 355–359.

Colicelli J., Birchmeier C., Michaeli T., O'Neill K., Riggs M., and Wigler M. (1989) Isolation and characterization of a mammalian gene encoding a high-affinity cAMP phosphodiesterase. *Proc. Natl. Acad. Sci. USA* **86,** 3599–3603.

Cook P. F., Neville M. E., Vrana K. E., Hartl F. T., and Roskoski R., Jr. (1982) Adenosine cyclic 3',5'-monophosphate dependent protein kinase: Kinetic mechanism for the skeletal muscle catalytic subunit. *Biochemistry* **21,** 5794–5799.

Coquil J. F., Brunelle G., and Guedon J. (1985) Occurrence of the methylisobutylxanthine-stimulated cyclic GMP binding protein in various rat tissues. *Biochem. Biophys. Res. Commun.* **127,** 226–231.

Corbin J. D. (1983) Determination of the cAMP-dependent protein kinase activity ratio in intact tissues, in *Methods in Enzymology* 99 (Corbin J. D. and Hardman J. C., eds.), (Academic, New York), pp. 227–232.

Corbin J. D. and Hardman J. C. (eds.) (1983) *Methods in Enzymology* 99 (Academic, New York).

Corbin J. D. and Johnson R. A. (eds.) (1988) *Methods in Enzymology* 159 (Academic, New York).

Corbin J. D. and Reimann E. M. (1974) Assay of cyclic AMP-dependent protein kinases, in *Methods in Enzymology* 38 (Hardman J. G. and O'Malley B. W., eds.), (Academic, New York), pp. 287–290.

Corbin J. D., Keeley S. L., and Park C. R. (1975) The distribution and dissociation of cyclic adenosine 3',5'-monophosphate-dependent protein kinases in adipose, cardiac, and other tissues. *J. Biol. Chem.* **250,** 218–225.

Corbin J. D., Soderling T. R., and Park C. R. (1973) Regulation of adenosine 3',5'-monophosphate-dependent protein kinase. I. Preliminary characterisation of the adipose tissue enzyme in crude extracts. *J. Biol. Chem.* **248,** 1813–1821.

Corbin J. D., Ogreid D., Miller J. P., Suva R. H., Jastorff B., and Doskeland S. O. (1986) Studies of cGMP analog specificity and function of the two intrasubunit binding sites of cGMP-dependent protein kinase. *J. Biol. Chem.* **261,** 1208–1214.

Corbin J. D., Gettys T. W., Blackmore P. F., Beebe S. J., Francis S. H., Glass D. B., Redmon J. B., Sheorain V. S., and Landiss L. R. (1988) Purification and assay of cAMP, cGMP, and cyclic nucleotide analogs in cells treated with cyclic nucleotide analogs, in *Methods in Enzymology* **159,** (Corbin J. D. and Johnson R. A., eds.), (Academic, New York), pp. 74–82.

Cumming R., Koide Y., Krigman M. R., Beavo J. A., and Steiner A. L. (1981) The immunofluorescent localization of regulatory and catalytic subunits of cAMP-dependent protein kinase in neuronal and glial cell types of the central nervous system. *Neuroscience* **6**, 953–961.

Davies C. W. and Daly J. W. (1979) A simple assay of 3',5'-cyclic nucleotide phosphodiesterase activity based on the use of the polyacrylamide-boronate affinity gel chromatography. *J. Cyclic. Nucleotide Res.* **5**, 65–74.

Davis R. L. and Kauvar L. M. (1984) *Drosophila* cyclic nucleotide phosphodiesterases, in *Advances in Cyclic Nucleotide and Protein Phosphorylation Research*, vol. 16 (Strada S. J. and Thompson W. J., eds.), Raven, New York, pp. 393–402.

Davis R. L., Takayasu H., Eberwine M., and Myres J. (1989) Cloning and characterization of mammalian homologs of the *drosophila* dunce[+] gene. *Proc. Natl. Acad. Sci. USA* **86**, 3604–3608.

De Camilli P., Moretti M., Donni S. D., Walter U., and Lohmann S. M. (1986) Heterogenous distribution of the cAMP receptor protein RII in the nervous system: Evidence for its intracellular accumulation on microtubules, microtubule-organizing centers, and in the area of the golgi complex. *J. Cell. Biol.* **103**, 189–203.

Degerman E., Belfrage P., Hauck Newman A., Rice K. C., and Manganiello V. C. (1987) Purification of the putative hormone-sensitive cyclic AMP phosphodiesterase from adipose tissue using a derivative of cilostamide as a novel affinity ligand. *J. Biol. Chem.* **262**, 5797–5807.

Degerman E., Smith C. J., Tornqvist H., Vasta V., Belfrage P., and Manganiello V. C. (1990) Evidence that insulin and isoprenaline activate the cGMP-inhibited low-K_m cAMP phosphodiesterase in rat fat cells by phosphorylation. *Proc. Natl. Acad. Sci. USA* **87**, 533–537.

Donaldson J., Brown A. M., and Hill S. J. (1988) Influence of rolipram on the cyclic 3',5'adenosine monophosphate response to histamine and adenosine in slices of guinea-pig cerebral cortex. *Biochem. Pharmacol.* **37**, 715–723.

Doskeland S. O. and Ogreid D. (1988) Ammonium sulfate precipitation assay for the study of cyclic nucleotide binding to proteins, in *Methods in Enzymology* 159 (Corbin J. D. and Johnson R. A., eds.), (Academic, New York), pp. 147–150.

Dostmann W. R. G., Taylor S. S., Genieser H-.G., Jastorff B., Doskeland S. O., and Ogreid D. (1990) Probing the cyclic nucleotide binding sites of cAMP-dependent protein kinases I and II with analogs of adenosine 3',5'-cyclic phosphorothioates. *J. Biol. Chem.* **265**, 10,484–10,491.

Draetta G. and Klee C. B. (1988) Purification of calmodulin-stimulated phosphodiesterase by affinity chromatography on calmodulin fragment 1-77 linked to sepharose, in *Methods in Enzymology* 159 (Corbin J. D. and Johnson R. A., eds.), (Academic, New York), pp. 573–581.

Drummond G. I. (1983) Cyclic nucleotides in the nervous system, in *Advances in Cyclic Nucleotide Research*, vol. 15 (Greengard P. and Robison G. A., eds.), Raven, New York, pp.373–494.

Drummond G. I. (1984) *Cyclic Nucleotides in the Nervous System*. Raven, New York.

Dumuis A., Sebben M., and Bockaert J. (1989) The gastrointestinal prokinetic benzamide derivatives are agonists at the non-classical 5-HT receptor (5-HT$_4$) positively coupled to adenylate cyclase in neurons. *Naunyn-Schmiedeberg's Arch. Pharmacol.* **340,** 403–410.

Durgerian S. and Taylor S. S. (1989) The consequences of introducing an autophosphorylation site into the type I regulatory subunit of cAMP-dependent protein kinase. *J. Biol. Chem.* **264,** 9807–9813.

El-Maghrabi M. R., Claus T. H. and Pilkis S. J. (1983) Substrate-directed regulatlon of cAMP-dependent phosphorylation, in *Methods in Enzymology* 99 (Corbin J. D. and Hardman J. C., eds.), (Academic, New York), pp. 212–219.

Erlichman J., Bloomgarden D., Sarkar D., and Rubin C. S. (1983) Activation of cyclic AMP-dependent protein kinase isoenzymes: Studies using specific antisera. *Arch. Biochem. Biophys.* **227,** 136–146.

Erneux C. and Miot F. (1988) Cyclic nucleotide analogs used to study phosphodiesterase catalytic and allosteric sites, in *Methods in Enzymology* 159 (Corbin J. D. and Johnson R. A., eds.), (Academic, New York), pp. 520–530.

Erneux C., Couchie D., Dumont J. E., and Jastorff B. (1984) Cyclic nucleotide derivatives as probes of phosphodiesterase catalytic and regulatory sites, in *Advances in Cyclic Nucleotide and Protein Phosphorylation Research*, vol. 16 (Strada S. J. and Thompson W. J., eds.), Raven, New York, pp. 107–118.

Fischmeister R. and Hartzell H. C. (1987) Cyclic guanosine 3',5'-monophosphate regulates the calcium current in single cells from frog ventricle. *J. Physiol.* **387,** 453–472.

Fiscus R. R. and Murad F. (1988) cGMP-dependent protein kinase activation in intact tissues, in *Methods in Enzymology* 159 (Corbin J. D. and Johnson R. A., eds.), (Academic, New York), pp. 150–159.

Flockhart D. A. (1983) Removal of phosphate from proteins by the reverse reaction, in *Methods in Enzymology* 99 (Corbin J. D. and Hardman J. C., eds.), (Academic, New York), pp. 14–20.

Fougier S., Nemoz G., Perigent A. F., Marivet M., Bourguignon J. J., Wermuth C., and Pacheco H. (1986) Purification of cAMP-specific phosphodiesterase from rat heart by affinity chromatography on immobilised rolipram. *Biochem. Biophys. Res. Commun.* **138,** 205–214.

Francis S. H. and Corbin J. D. (1988) Purification of cGMP-binding protein phosphodiesterase from rat lung, in *Methods in Enzymology* 159 (Corbin J. D. and Johnson R. A., eds.), (Academic, New York), pp. 722–729.

Fredholm B. B. (1980) Are methylxanthine effects due to antagonism of endogenous adenosine? *Trends Pharmacol. Sci.* **1**, 129–132.

Garbers D. L. (1989) Guanylate cyclase, a cell surface receptor. *J. Biol. Chem.* **264**, 9103–9106.

Giembycz M. A. and Diamond J. (1990) Evaluation of kemptide, a synthetic serine-containing heptapeptide, as a phosphate acceptor for the estimation of cyclic AMP-dependent protein kinase activity in respiratory tissues. *Biochem. Pharmacol.* **39**, 271–283.

Gill G. N., Holdy K. E., Walton G. M., and Kanstein C. B. (1976) Purification and characterization of 3',5'-cyclic GMP-dependent protein kinase. *Proc. Natl. Acad. Sci. USA* **73**, 3918–3922.

Gillespie P. G. and Beavo J. A. (1989) Inhibition and stimulation of photoreceptor phosphodiesterases by dipyridamole and M&B 22,948. *Mol. Pharmacol.* **36**, 773–781.

Gilman A. G. and Murad F. (1974) Assay of cyclic nucleotides by receptor protein binding displacement, in *Methods in Enzymology* 38 (Hardman J. G. and O'Malley B. W., eds.), (Academic, New York), pp. 49–61.

Glass D. B. (1990) Substrate specificity of cGMP-dependent protein kinase, in *Peptides and Protein Phosphorylation* (Kemp B. E., ed.), CRC, Boca Raton, pp. 209–238.

Glass D. B. and Krebs E. G. (1982) Phosphorylation by guanosine 3',5'-monophosphate-dependent protein kinase of synthetic peptide analogs of a site phosphorylated in histone 2b. *J. Biol. Chem.* **257**, 1196–1200.

Grant P. G., DeCamp D. L., Bailey J. M., Colman R. W., and Colman R. F. (1990) Three new potential cAMP affinity labels. Inactivation of human platelet low K_m cAMP phosphodiesterase by 8-[(4-bromo-2,3-dioxobutyl)thio]adenosine 3',5'-cyclic monophosphate. *Biochemistry* **29**, 887–894.

Gruetter C. A., Kadowitz P. J., and Ignarro L. J. (1981) Methylene blue inhibits coronary arterial relaxation and guanylate cyclase activation by nitroglycerin, sodium nitrite, and amyl nitrite. *Can. J. Physiol. Pharmacol.* **59**, 150–156.

Guillory R. J. (1989) Design, implementation and pitfalls of photoaffinity labelling experiments in in vitro preparations. *General Principles. Pharmac. Ther.* **41**, 1–25.

Gundlach A. L. and Urosevic A. (1989) Autoradiographic localization of particulate cyclic AMP-dependent protein kinase in mammalian brain using [^{3}H]cyclic AMP: Implications for organization of second messenger systems. *Neuroscience* **29**, 695–714.

Haley B. E. (1977) Adenosine 3',5'-cyclic monophosphate binding sites, in *Methods in Enzymology* 46 (Jakoby W. B. and Wilchek M., eds.), (Academic, New York), pp. 339–346.

Hamet P. and Tremblay J. (1988) Platelet cGMP-binding phosphodiesterase, in *Methods in Enzymology* 159 (Corbin J. D. and Johnson R. A., eds.), (Academic, New York), pp. 710–722.

Hanks S. T., Quinn A. M., and Hunter T. (1988) The protein kinase family: Conserved features and deduced phylogeny of the catalytic domains. *Science* **241**, 42–51.

Hansen R. S., Charbonneau H., and Beavo J. A. (1988) Purification of calmodulin-stimulated cyclic nucleotide phosphodiesterase by monoclonal antibody affinity chromatography, in *Methods in Enzymology* 159 (Corbin J. D. and Johnson R. A., eds.), (Academic, New York), pp. 543–557.

Hardie D. G. (1983) Cyclic nucleotide-dependent protein kinase, in *Methods of Enzymatic Analysis*, vol. III (Bergmeyer H. U., ed.), Verlag Chemie, Weinheim, pp. 481–487.

Hardman J. G. and O'Malley B. W. (eds.) (1974) *Methods in Enzymology* 38 (Academic, New York).

Harker L. A. and Fuster V. (1986) Pharmacology of platelet inhibitors. *J. Am. Coll. Cardiol.* **8**, 21B–32B.

Harrison S. A., Beier N., Martins R. J., and Beavo J. A. (1988) Isolation and comparison of bovine heart cGMP-inhibited and cGMP-stimulated phosphodiesterases, in *Methods in Enzymology* 159 (Corbin J. D. and Johnson R. A., eds.), (Academic, New York), pp. 685–702.

Hartl F. T. and Roskoski R., Jr. (1983) Cyclic adenosine 3',5'-monophosphate-dependent protein kinase. *J. Biol. Chem.* **258**, 3950–3955.

Hashimoto Y., Sharma R. K., and Soderling T. R. (1989) Regulation of Ca^{2+}/calmodulin-dependent cyclic nucleotide phosphodiesterase by the autophosphorylated form of Ca^{2+}/calmodulin-dependent protein kinase II. *J. Biol. Chem.* **264**, 10,884–10,887.

Haslam R. J. and Vanderwel M. (1989) Measurement of changes in platelet cyclic AMP in vitro and in vivo by prelabeling techniques: Application to the detection and assay of circulating PGI_2, in *Methods in Enzymology* 169 (Hawiger J., ed.), (Academic, New York), pp. 457–471.

Haslam R. J., Davidson M. M. L., and Desjardins J. V. (1978) Inhibition of adenylate cyclase by adenosine analogs in preparations of broken and intact platelets. Evidence for the unidirectional control of platelet function by cyclic AMP. *Biochem. J.* **176**, 83–95.

Hathaway D. R., Adelstein R. S., and Klee C. B. (1981) Interaction of calmodulin with myosin light chain kinase and cAMP-dependent protein kinase in bovine brain. *J. Biol. Chem.* **256**, 8183–8189.

Hemmings H. C., Jr., Nairn A. C., McGuinness T. L., Huganir R. L., and Greengard P. (1989) Role of protein phosphorylation in neuronal signal transduction. *FASEB J.* **3**, 1583–1592.

Henkel-Tigges J. and Davis R. L. (1990) Rat homologs of the *drosophila* dunce gene code for cyclic AMP phosphodiesterases sensitive to rolipram and Ro 20-1724. *Mol. Pharmacol.* **37**, 7–10.

Hidaka H., Inagaki M., Nishikawa M., and Tanaka T. (1988) Selective inhibitors of calmodulin-dependent phosphodiesterase and other enzymes, in *Methods in Enzymology* 159 (Corbin J. D. and Johnson R. A., eds.), (Academic, New York), pp. 653–660.

Hohmann P. and Greene R. S. (1990) Retinoid induced changes in cAMP-dependent protein kinase activity detected by a new minigel assay. *FEBS Lett.* **261**, 81–84.

Jahnsen T., Hedin L., Kidd V. J., Beattie W. G., Lohmann S. M., Walter U., Durica J., Schulz T. Z., Schiltz E., Browner M., Lawrence C. B., Goldman D., Ratoosh S. L., and Richards J. S. (1986) Molecular cloning, cDNA structure, and regulation of regulatory subunit of type II cAMP-dependent protein kinase from rat ovarian granulosa cells. *J. Biol. Chem.* **261**, 12,352–12,361.

Kariya T. and Dage R. C. (1988) Tissue distribution and selective inhibition of subtypes of high affinity cAMP phosphodiesterase. *Biochem. Pharmacol.* **37**, 3267–3270.

Kaumann A. J., Sanders L., Brown A. M., Murray K. J., and Brown M. J. (1990) A 5-hydroxytryptamine receptor in human atrium. *Br. J. Pharmacol.* **100**, 879–885.

Kemp B. E., Cheng H.-C. and Walsh D. A. (1988) Peptide inhibitors of cAMP-dependent protein kinase, in *Methods in Enzymology* 159 (Corbin J. D. and Johnson R. A., eds.), (Academic, New York), pp. 173–183.

Kerlavage A. R. and Taylor S. S. (1982) Site-specific cyclic nucleotide binding and dissociation of the holoenzyme of cAMP-dependent protein kinase. *J. Biol. Chem.* **257**, 1749–1754.

Kincaid R. L. and Manganiello V. C. (1988) Assays of cyclic nucleotide phosphodiesterase using radiolabeled and fluorescent substrates, in *Methods in Enzymology* 159 (Corbin J. D. and Johnson R. A., eds.), (Academic, New York), pp. 457–470.

Kincaid R. L. and Vaughan M. (1988) Purification and properties of calmodulin-activated cyclic nucleotide phosphodiesterase from mammalian brain, in *Methods in Enzymology* 159 (Corbin J. D. and Johnson R. A., eds.), (Academic, New York), pp. 557–573.

Kincaid R. L., Balaban C. D., and Billingsley M. L. (1987) Differential localization of calmodulin-dependent enzymes in rat brain: Evidence for selective expression of cyclic nucleotide phosphodiesterase in specific neurons. *Proc. Natl. Acad. Sci. USA* **84**, 1118–1122.

Kinzel V., Hotz A., Konig N., Gagelmann M., Pyerin W., Reed J., Kubler D., Hofmann F., Obst C., Gensheimer H. P., Goldblatt D., and Shaltiel S. (1987) Chromotagraphic separation of two heterogeneous forms of the catalytic subunit of cyclic AMP-dependent protein kinase holoenzyme type I and type II from striated muscle of different mammalian species. *Arch. Biochem. Biophys.* **253**, 341–349.

Kolesnikov S. S., Zhainazorov A. B., and Kosolapov A. V. (1990) Cyclic nucleotide-activated channels in the frog olfactory receptor plasma membrane. *FEBS Lett.* **266**, 96–98.

Krebs E. G. (1989) Role of the cyclic AMP-dependent protein kinase in signal transduction. *JAMA* **262**, 1815–1818.

Kuno T., Ono Y., Hirai M., Hashimoto S., Shuntoh H., and Tanaka C. (1987) Molecular cloning and cDNA structure of the regulatory subunit of type I cAMP-dependent protein kinase from rat brain. *Biochem. Biophys. Res. Commun.* **146**, 878–883.

Kuno T., Shuntoh H., Sakaue M., Saijoh K., Takeda T., Fukuda K., and Tanaka C. (1988) Site-directed mutagenesis of the cAMP-binding sites of the recombinant type I regulatory subunit of cAMP-dependent protein kinase. *Biochem. Biophys. Res. Commun.* **153**, 1244–1250.

Kuno T., Shuntoh H., Takeda T., Ito A., Sakaue M., Hirai M., Ando H., and Tanaka C. (1989) Activation of type I cyclic AMP-dependent protein kinases is impaired by a point mutation in cyclic AMP binding sites. *Eur. J. Pharmacol.* **172**, 263–271.

Landgrat W., Hullin R., Gobel C., and Hofmann F. (1986) Phosphorylation of cGMP-dependent protein kinase increases the affinity for cyclic AMP. *Eur. J. Biochem.* **154**, 113–117.

Langlands J. M. and Rodger I. W. (1990) Determination of soluble cAMP-dependent protein kinase activity in guinea-pig tracheal smooth muscle. *Biochem. Pharmacol.* **39**, 1365–1374.

Laurenza A., Sutkowski E. M., and Seamon K. B. (1989) Forskolin: A specific stimulator of adenylyl cyclase or a diterpene with multiple sites of action? *Trends Pharmacol. Sci.* **10**, 442–447.

Leiser M., Rubin C. S., and Erlichman J. (1986) Differential binding of the regulatory subunits (RII) of cAMP-dependent protein kinase II from bovine brain and muscle to RII-binding proteins. *J. Biol. Chem.* **261**, 1904–1908.

Levy F. O., Øyen O., Sandberg M., Taskén K., Eskild W., Hansson V., and Jahnsen T. (1988) Molecular cloning, complementary deoxyribonucleic acid structure and predicted full-length amino acid sequence of the hormone-inducible regulatory subunit of 3',5'-cyclic adenosine monophosphate-dependent protein kinase from human testis. *Mol. Endocrinol.* **2**, 1364–1373.

Liebman P. A. and Evancuk A. T. (1982) Real time assay of rod disk membrane cGMP phosphodiesterase and its controller enzymes, in *Methods in Enzymology* 81 (Packer L., ed.), (Academic, New York), pp. 532–542.

Light D. B., Corbin J. D., and Stanton B. A. (1990) Dual ion-channel regulation by cyclic GMP and cyclic GMP-dependent protein kinase. *Nature (Lond.)* **344**, 336–339.

Lincoln T. M. (1983) cGMP-dependent protein kinase, in *Methods in Enzymology* 99 (Corbin J. D. and Hardman J. C., eds.), (Academic, New York), pp. 62–71.

Lincoln T. M. and Corbin J. D. (1983) Characterization and biological role of the cGMP-dependent protein kinase, in *Advances in Cyclic Nucleotide Research* 15 (Greengard P. and Robison G. A., eds.), Raven, New York, pp. 139–192.

Lincoln T. M., Cornwell T. L., and Taylor A. E. (1990) cGMP-dependent protein kinase mediates the reduction of Ca^{2+} by cAMP in vascular smooth muscle cells. *Am. J. Physiol.* **258**, C399-C407.

Lincoln T. M., Thompson M., and Cornwell T. L. (1988) Purification and characterization of two forms of cyclic GMP-dependent protein kinase from bovine aorta. *J. Biol. Chem.* **263**, 17,632–17,637.

Livesey S. A. and Martin T. J. (1988) Selective activation of the cAMP-dependent protein kinase isoenzymes, in *Methods in Enzymology* 159 (Corbin J. D. and Johnson R. A., eds.), (Academic, New York), pp. 105–118.

Lohmann S. M. and Walter U. (1984) Regulation of the cellular and subcellular concentrations and distribution of cyclic nucleotide-dependent protein kinases. *Advances in Cyclic Nucleotide and Protein Phosphorylation Research* **18**, 63–117.

Lohmann S. M., De Camilli P., and Walter U. (1988) Type II cAMP-dependent protein kinase regulatory subunit-binding proteins, in *Methods in Enzymology* 159 (Corbin J. D. and Johnson R. A., eds.), (Academic, New York), pp. 183–193.

Lohmann S. M., DeCamilli P., Einig I., and Walter U. (1984) High-affinity binding of the regulatory subunit (R$_{II}$) of cAMP-dependent protein kinase to microtubule-associated and other cellular proteins. *Proc. Natl. Acad. Sci. USA* **81**, 6723–6727.

Ludvig N., Ribak, C. E., Scott, J. D., and Rubin C. S. (1990) Immunocytochemical localisation of the neural-specific regulatory subunit of the type II cyclic AMP-dependent protein kinase to postsynaptic structures in the rat brain. *Brain Res.* **520**, 90–102.

Lugnier C. and Schini V. B. (1990) Characterization of cyclic nucleotide phosphodiesterases from cultured bovine aortic endothelial cells. *Biochem. Pharmacol.* **39**, 75–84.

MacPhee C. H., Reifsnyder D. H., Moore T. A., Lerea K. M., and Beavo J. A. (1988) Phosphorylation results in activation of a cAMP phosphodiesterase in human platelets. *J. Biol. Chem.* **263**, 10,353–10,358.

Manganiello V., Degerman E., and Elks M. (1988) Selective inhibitors of specific phosphodiesterases in intact adipocytes, in *Methods in Enzymology* 159 (Corbin J. D. and Johnson R. A., eds.), (Academic, New York), pp. 504–520.

Martin S. C. and Ekman P. (1986) In vitro phosphorylation of serum albumin by two protein kinases: A potential pitfall in protein phosphorylation reactions. *Anal. Biochem.* **154**, 395–399.

Matsuda Y., Nakanishi S., Nagasawa K., and Kase H. (1990) Inhibition by new anthraquinone compounds, K-259-2 and KS-619-1, of calmodulin-dependent cyclic nucleotide phosphodiesterase. *Biochem. Pharmacol.* **39**, 841–849.

Maurer R. A. (1989) Both isoforms of the cAMP-dependent protein kinase catalytic subunit can activate transcription of the prolactin gene. *J. Biol. Chem.* **264**, 6870–6873.

Maurice D. H. and Haslam R. J. (1990) Molecular basis of the synergistic inhibition of platelet function by nitrovasodilators and activators of adenylate cyclase: Inhibition of cyclic AMP breakdown by cyclic GMP. *Mol. Pharmacol.* **37**, 671–681.

McKnight G. S., Clegg C. H., Uhler M. D., Chrivia J. C., Cadd G. G., Correll L. A., and Otten A. D. (1988a) Analysis of the cAMP-dependent protein kinase system using molecular genetic approaches, in *Recent Progress in Hormone Research* 44 (Clark J. H., ed.), (Academic, New York), pp. 307–335

McKnight G. S., Uhler M. D., Clegg C. H., Correll L. A., and Cadd G. (1988b) Application of molecular genetic techniques to the cAMP-dependent protein kinase system, in *Methods in Enzymology* 159 (Corbin J. D. and Johnson R. A., eds.), (Academic, New York), pp. 299–311.

Middleton J. P., Dunham C. B., Onorato J. J., Sens D. A., and Dennis V. W. (1989) Protein kinase A, cytosolic calcium, and phosphate uptake in human proximal renal cells. *Am. J. Physiol.* **257**, F631-F638.

Moyer J. A., Sigg E. B., and Silver P. J. (1986) Antidepressants and protein kinases: Desipramine treatment affects pineal gland cAMP-dependent protein kinase activity. *Eur. J. Pharmacol.* **121**, 57–64.

Mulsch A., Busse R., Liebau S., and Forstermann U. (1988) LY 83583 interferes with the release of endothelium-derived relaxing factor and inhibits soluble guanylate cyclase. *J. Pharmacol. Exptl. Ther.* **247**, 283–288.

Murashima S., Tanaka T., Hockman S., and Manganiello V. (1990) Characterization of particulate cyclic nucleotide phosphodiesterases from bovine brain purification of a distinct cGMP-stimulated isoenzyme. *Biochemistry* **29**, 5285–5292.

Murray K. J. (1990) Cyclic AMP and mechanisms of vasodilation. *Pharmac. Ther.* **47**, 329–345.

Murray K. J. and Warrington B. H. (1990) Protein kinases, in *Comprehensive Medicinal Chemistry*, vol. 2 (Hansch C., Sammes P. G., and Taylor J. B., eds.), Pergamon, Oxford, pp. 531–552.

Murray K. J., Reeves M. L., and England P. J. (1989) Protein phosphorylation and compartments of cyclic AMP in the control of cardiac contraction. *Mol. Cell. Biochem.* **89**, 175–179.

Murray K. J., England P. J., Hallam T. J., Maguire J., Moores K., Reeves M. L., Simpson A. W. M., and Rink T. J. (1990a) The effects of siguazodan, a selective phosphodiesterase inhibitor, on human platelet function. *Br. J. Pharmacol.* **99**, 612–616.

Murray K. J., England P. J., Lynham J. A., Mills D., Schmitz-Peiffer C., and Reeves M. L. (1990b) Use of a synthetic dodecapeptide (malantide) to measure the cyclic AMP-dependent protein kinase activity ratio in a variety of tissues. *Biochem. J.* **267**, 703–708.

Nairn A. C., Hemmings H. C., Jr., and Greengard P. (1985) Protein kinases in the brain. *Annu. Rev. Biochem.* **54**, 931–976.

Némoz G., Moueqqit M., Prigent A-F., and Pacheco H. (1989) Isolation of similar rolipram-inhibitible cyclic-AMP-specific phosphodiesterases from rat brain and heart. *Eur. J. Biochem.* **184**, 511–520.

Nerbonne J. M., Richard S., Nargeot J., and Lester H. A. (1984) New photoactivatable cyclic nucleotides produce intracellular jumps in cyclic AMP and cyclic GMP concentrations. *Nature (Lond.)* **310**, 74–76.

Nestler E. J. and Greengard P. (1989) Protein phosphorylation and the regulation of neuronal function, in *Basic Neurochemistry: Molecular, Cellular, and Medical Aspects* (Siegel G. J., ed.), Raven, New York, pp. 373–398.

Nestler E. J. and Tallman J. F. (1988) Chronic morphine treatment increases cyclic AMP-dependent protein kinase activity in the rat locus coeruleus. *Mol. Pharmacol.* **33**, 127–132.

Nestler E. J., Terwilliger R. Z., and Duman R. S. (1989) Chronic antidepressant administration alters the subcellular distribution of cyclic AMP-dependent protein kinase in rat frontal cortex. *J. Neurochem.* **53**, 1644–1647.

Nicholson C. D., Jackman S. A., and Wilke R. (1989) The ability of denbuffylline to inhibit cyclic nucleotide phosphodiesterase and its affinity for adenosine receptors and the adenosine re-uptake site. *Br. J. Pharmacol.* **97**, 889–897.

Obar R. A., Dingus J., Bayley H., and Vallee R. B. (1989) The RII subunit of cAMP-dependent protein kinase binds to a common amino-terminal domain in microtubule-associated proteins 2A, 2B, and 2C. *Neuron* **3**, 639–645.

Ogreid D., Doskeland S. O., Gorman K. B., and Steinberg R. A. (1988) Mutations that prevent cyclic nucleotide binding to binding sites A or B of type I cyclic AMP-dependent protein kinase. *J. Biol. Chem.* **263**, 17,397–17,404.

Ogreid D., Ekanger R., Suva R. H., Miller J. P., and Doskeland S. O. (1989) Comparison of the two classes of binding sites (A and B) of type I and type II cyclic-AMP-dependent protein kinases by using cyclic nucleotide analogs. *Eur. J. Biochem.* **181**, 19–31.

Ogreid D., Ekanger R., Suva R. H., Miller J. P., Sturm P., Corbin J. D., and Doskeland S. O. (1985) Activation of protein kinase isozymes by cyclic nucleotide analogs used singly or in combination. *Eur. J. Biochem.* **150**, 219–227.

Olsen S. R. and Uhler M. D. (1989) Affinity purification of the $C\alpha$ and $C\beta$ isoforms of the catalytic subunit of cAMP-dependent protein kinase. *J. Biol. Chem.* **264**, 18,662–18,666.

Øyen O., Myklebust F., Scott J. D., Hansson V., and Jahnsen T. (1989) Human testis cDNA for the regulatory subunit RII_{α} of cAMP-dependent protein kinase encodes an alternate amino-terminal region. *FEBS Lett.* **246**, 57–64.

Palmer G. C. (1981) Significance of phosphodiesterase in the brain. *Life Sci.* **28**, 2785–2798.

Palmer W. K., McPherson J. M. and Walsh D. A. (1980) Critical controls in the evaluation of cAMP-dependent protein kinase activity ratios as indices of hormonal action. *J. Biol. Chem.* **255**, 2663–2666.

Pang D. C., Cantor E., Hagedorn A., Erhardt P. and Wiggins J. (1988) Tissue specificity of cAMP-phosphodiesterase inhibitors: Rolipram, Amrinone, Milrinone, Enoximone, Piroximone, and Imazodan. *Drug Dev. Res.* **14**, 141–149.

Parker Botelho L. H., Rothermel J. D.. Coombs R. V. and Jastorff B. (1988) cAMP analog antagonists of cAMP action, in *Methods in Enzymology* 159 (Corbin J. D. and Johnson R. A., eds.), (Academic, New York), pp. 159–172.

Pyne N. J., Cooper M. E. and Houslay M. D. (1986) Identification and characterization of both the cytosolic and particulate forms of cyclic GMP-stimulated cyclic AMP phosphodiesterase from rat liver. *Biochem. J.* **234**, 325–334.

Rannels S. R. and Corbin J. D. (1983) Using analogs to study selectivity and cooperativity of cyclic nucleotide binding sites, in *Methods in Enzymology* 99 (Corbin J. D. and Hardman J. C., eds.), (Academic, New York), pp. 168–175.

Rannels S. R., Beasley A., and Corbin J. D. (1983) Regulatory subunits of bovine heart and rabbit skeletal muscle cAMP-dependent protein kinase isozymes, in *Methods in Enzymology* 99 (Corbin J. D. and Hardman J. C., eds.), (Academic, New York), pp. 55–62.

Reeves M. L., Leigh B. K., and England P. J. (1987) The identification of a new cyclic nucleotide phosphodiesterase activity in human and guinea-pig cardiac ventricle. *Biochem. J.* **241**, 535–541.

Reimann E. M. (1986) Conversion of bovine cardiac adenosine cyclic 3',5'-phosphate dependent protein kinase to a heterodimer by removal of 45 residues at the N-terminus of the regulatory subunit. *Biochemistry* **25**, 119–125.

Reimann E. M. and Beham R. A. (1983) Catalytic subunit of cAMP-dependent protein kinase, in *Methods in Enzymology* 99 (Corbin J. D. and Hardman J. C., eds.), (Academic, New York), pp. 51–55.

Ringheim G. E. and Taylor S. S. (1990) Dissecting the domain structure of the regulatory subunit of cAMP-dependent protein kinase I and elucidating the role of MgATP. *J. Biol. Chem.* **265**, 4800–4808.

Ringheim G. E., Saraswat L. D., Bubis J., and Taylor S. S. (1988) Deletion of cAMP-binding site B in the regulatory subunit of cAMP-dependent protein kinase alters the photoaffinity labeling of site A. *J. Biol. Chem.* **263**, 18,247–18,252.

Robicsek S. A., Krzanowski J. J., Szentivanyi A., and Polson J. B. (1989) High pressure liquid chromatography of cyclic nucleotide phosphodiesterase from purified human T-lymphocytes. *Biochem. Biophys. Res. Commun.* **163**, 554–560.

Robison G. A., Butcher R. W., and Sutherland E. W. (1971) *Cyclic AMP* (Academic, New York).

Roskoski R., Jr. (1983) Assays of protein kinase, in *Methods in Enzymology* 99 (Corbin J. D. and Hardman J. C., eds.), (Academic, New York), pp. 3–6.

Rubin C. S., Rangel-Aldao R., Sarkar D., Erlichman J., and Fleischer N. (1979) Characterization and comparison of membrane-associated and cytosolic cAMP-dependent protein kinases. *J. Biol. Chem.* **254**, 3797–3805.

Rubino H. M., Dammerman M., Shafit-Zagardo B., and Erlichman J. (1989) Localization and characterization of the binding site for the regulatory subunit of type II cAMP-dependent protein kinase on MAP2. *Neuron* **3**, 631–638.

Sahal K. and Fujita-Yamaguchi Y. (1987) Protein kinase assay by paper-trichloroacetic acid method: High performance using phosphocellulose paper and washing an ensemble of samples on flat sheets. *Anal. Biochem.* **167**, 23–30.

Salomon Y., Londos C., and Rodbell, M. (1974) A highly sensitive adenylate cyclase assay. *Anal. Biochem.* **58**, 541–548

Sandberg M., Levy F. O., Øyen O., Hansson V., and Jahnsen T. (1988) Molecular cloning, cDNA structure and deduced amino acid sequence for the hormone-induced regulatory subunit (RII$_\beta$) of cAMP-dependent protein kinase from rat ovarian granulosa cells. *Biochem. Biophys. Res. Commun.* **154**, 705–711.

Sandberg M., Taskén K., Øyen O., Hansson V., and Jahnsen T (1987) Molecular cloning, cDNA structure and deduced amino acid sequence for a type I regulatory subunit of cAMP-dependent protein kinase from human testis. *Biochem. Biophys. Res. Commun.* **149**, 939–945.

Santa-Coloma T. A., Bley M. A., and Charreau E. H. (1987) Improvement on the competitive binding assay for the measurement of cyclic AMP by using ammonium sulphate precipitation. *Biochem. J.* **245**, 923,924.

Sarkar D., Erlichman J., and Rubin C. S. (1984) Identification of a calmodulin-binding protein that co-purifies with the regulatory subunit of brain protein kinase II. *J. Biol. Chem.* **259**, 9840–9846.

Schlender K. K., Tyma J. L., and Reimann E. M. (1983) Preparation of partially purified protein kinase inhibitor, in *Methods in Enzymology* 99 (Corbin J. D. and Hardman J. C., eds.), (Academic, New York), pp. 77–80.

Schultz G. and Bohme E. (1984) Guanylate cylase, in *Methods of Enzymatic Analysis,* vol. 3 (Bergmeyer H. U., ed.), Verlag Chemie, Weinheim, pp. 379–389.

Schultz G. and Jacobs K. H. (1984) Adenylate cyclase, in *Methods of Enzymatic Analysis,* vol. 3 (Bergmeyer H. U., ed.), Verlag Chemie, Weinheim, pp. 369–378.

Schulz S., Chinkers M., and Garbers D. L. (1989) The guanylate cyclase/receptor family of proteins. *FASEB J.* **3**, 2026–2035.

Schwabe U., Miyake M., Ohga Y., and Daly J. W. (1976) 4-(3-Cyclopentyloxy-4-methoxyphenyl)-2-pyrrolidone (ZK 62711): A potent inhibitor of adenosine cyclic 3′,5′monophosphate phosphodiesterases in homogenates and tissue slices from rat brain. *Mol. Pharmacol.* **12**, 900–910.

Schwartz J. P. and Costa E. (1980) Protein kinase translocation following β-adrenergic receptor activation in C6 glioma cells. *J. Biol. Chem.* **255,** 2943–2948.

Schwoch G., Lohmann S. M., Walter U., and Jung U. (1985) Determination of cyclic AMP-dependent protein kinase subunits by an immunoassay reveals a different subcellular distribution of the enzyme in rat parotid than does determination of the enzyme activity. *J. Cyclic Nucleotide and Protein Phosphorylation Research* **10,** 247–258.

Scott J. D., Glaccum M. B., Zoller M. J., Uhler M. D., Helfman D. M., McKnight G. S., and Krebs E. G. (1987) The molecular cloning of a type II regulatory subunit of the cAMP-dependent protein kinase from rat skeletal muscle and mouse brain. *Proc. Natl. Acad. Sci. USA* **84,** 5192–5196.

Segal J. and Ingbar S. H. (1989) 3, 5, 3'-Triiodothyronine increases cellular adenosine 3',5'-monophosphate concentration and sugar uptake in rat thymocytes by stimulating adenylate cyclase activity: Studies with the adenylate cyclase inhibitor MDL 12330A. *Endocrinology* **124,** 2166–2171.

Sharma R. K. and Wang J. H. (1985) Differential regulation of bovine brain calmodulin-dependent cyclic nucleotide phosphodiesterase isozymes by cyclic AMP-dependent protein kinase and calmodulin-dependent phosphatase. *Proc. Natl. Acad. Sci. USA* **82,** 2603–2607.

Sharma R. K. and Wang J. H. (1988) Isolation of bovine brain calmodulin-dependent cyclic nucleotide phosphodiesterase isozymes, in *Methods in Enzymology* 159 (Corbin J. D. and Johnson R. A., eds.), (Academic, New York), pp. 582–594.

Sharma R. K., Adachi A.-M., Adachi K., and Wang J. H. (1984) Demonstration of bovine brain calmodulin-dependent cyclic nucleotide phosphodiesterase isozymes by monoclonal antibodies. *J. Biol. Chem.* **259,** 9248–9254.

Sharps E. S. and McCarl R. L. (1982) A high-performance liquid chromatographic method to measure ^{32}P incorporation into phosphorylated metabolites in cultured cells. *Anal. Biochem.* **124,** 421–424.

Shenolikar S., Thompson W. J., and Strada S. J. (1985) Characterization of a Ca^{2+}-calmodulin-stimulated cyclic GMP phosphodiesterase from bovine brain. *Biochemistry* **24,** 672–678.

Shimizu H. J., Daly J. W., and Creveling C. R. (1969) A radio-isotopic method for measuring the formation of cAMP in incubated brain slices. *J. Neurochem.* **16,** 1609–1619.

Showers M. O. and Maurer R. A. (1986) A cloned bovine cDNA encodes an alternate form of the catalytic subunit of cAMP-dependent protein kinase. *J. Biol. Chem.* **261,** 16,288–16,291.

Silver P. J. (1989) Biochemical aspects of inhibition of cardiovascular low (K_m) cyclic adenosine monophosphate phosphodiesterase. *Am. J. Cardiol.* **63,** 2A–8A.

Slice L. W. and Taylor S. S. (1989) Expression of the catalytic subunit of cAMP-dependent protein kinase in *escherichia coli*. *J. Biol. Chem.* **264,** 20,940–20,946.

Stanley C., Brown A. M., and Hill S. J. (1989) Effect of isozyme-selective inhibitors of phosphodiesterase on histamine-stimulated cyclic AMP accumulation in guinea-pig hippocampus. *J. Neurochem.* **52,** 671–676.

Stein J. C. and Rubin C. S. (1985) Isolation and sequence of a tryptic peptide containing the autophosphorylation site of the regulatory subunit of bovine brain protein kinase II. *J. Biol. Chem.* **260,** 10,991–10,995.

Stein J. C., Farooq M., Norton W. T., and Rubin C. S. (1987) Differential expression of isoforms of the regulatory subunit of type II cAMP-dependent protein kinase in rat neurons, astrocytes, and oligodendrocytes. *J. Biol. Chem.* **262,** 3002–3006.

Steiner A. L., Parker C. W., and Kipnis D. M. (1972) Radioimmunoassay for cyclic nucleotides. I. Preparation of antibodies and iodinated cyclic nucleotides. *J. Biol. Chem.* **247,** 1106–1113.

Strada S. J. and Thompson W. J. (eds.) (1984) *Advances in Cyclic Nucleotide and Protein Phosphorylation Research,* vol. 16. Raven, New York.

Stroop S. D., Charbonneau H., and Beavo J. A. (1989) Direct photolabeling of the cGMP-stimulated cyclic nucleotide phosphodiesterase. *J. Biol. Chem.* **264,** 13,718–13,725.

Sugden P. H. and Corbin J. D. (1976) Adenosine 3',5'-cyclic monophosphate-binding proteins in bovine and rat tissues. *Biochem. J.* **159,** 423–437.

Sulakhe P. V., Gupta R. C., and Jagadeesh G. (1986) Brain adenylate cyclase, in *Neuromethods,* vol. 5, *Neurotransmitter Enzymes* (Boulton A. A., Baker G. B., and Yu P. H., eds.), Humana, Clifton, NJ, pp. 503–517.

Sutherland E. W., Robison G. A., and Butcher, R. W. (1968) Some aspects of the biological role of adenosine 3',5' monophosphate. *Circulation* **37,** 279–306.

Swillens S., Boeynaems J.-M., and Dumont J. E. (1988) Theoretical considerations of the regulatory steps in the cAMP cascade system, in *Methods in Enzymology* 159 (Corbin J. D. and Johnson R. A., eds.), (Academic, New York), pp. 19–27.

Swinnen J. V., Joseph D. R., and Conti M. (1989a) Molecular cloning of rat homologues of the *drosophila melanogaster* dunce cAMP phosphodiesterase: Evidence for a family of genes. *Proc. Natl. Acad. Sci. USA* **86,** 5325–5329.

Swinnen J. V., Joseph D. R., and Conti M. (1989b) The mRNA encoding a high-affinity cAMP phosphodiesterase is regulated by hormones and cAMP. *Proc. Natl. Acad. Sci. USA* **86,** 8197–8201.

Takio K., Smith S. B., Krebs E. G., Walsh K. A., and Titani K. (1984) Amino acid sequence of the regulatory subunit of bovine type II adenosine cyclic 3',5'-phosphate dependent protein kinase. *Biochemistry* **23,** 4200–4206.

Taylor S. S. (1989) cAMP-dependent protein kinase. *J. Biol. Chem.* **264,** 8443–8446.

Taylor S. S., Beuchler J. A., and Yonemoto W. (1990) cAMP-dependent protein kinase: Framework for a diverse family of regulatory enzymes. *Ann. Rev. Biochem.* **59**, 971–1005.

Taylor S. S., Kerlavage A. R., and Zoller M. J. (1983) Affinity labeling of cAMP-dependent protein kinases, in *Methods in Enzymology* 99 (Corbin J. D. and Hardman J. C., eds.), (Academic, New York), pp. 140–153.

Theurkauf W. E. and Vallee R. B. (1982) Molecular characterization of the cAMP-dependent protein kinase bound to microtubule-associated protein 2. *J. Biol. Chem.* **257**, 3284–3290.

Thompson D. A. and Khorana H. G. (1990) Guanosine 3',5'-cyclic nucleotide binding proteins of bovine retina identified by photoaffinity labeling. *Proc. Natl. Acad. Sci. USA* **87**, 2201–2205.

Thompson W. J. and Appleman M. M. (1971) Multiple cyclic nucleotide phosphodiesterase activities from rat brain. *Biochemistry* **10**, 311–316.

Thompson W. J., Brooker G., and Appleman M. M. (1974) Assay of cyclic nucleotide phosphodiesterases with radioactive substrates, in *Methods in Enzymology* 38 (Hardman J. G. and O'Malley B. W., eds.), (Academic, New York), pp. 205–212.

Titani K., Sasagawa T., Ericsson L. H., Kumar S., Smith S. B., Krebs E. G., and Walsh K. A. (1984) Amino acid sequence of the regulatory subunit of bovine type I adenosine cyclic 3',5'-phosphate dependent protein kinase. *Biochemistry* **23**, 4193–4199.

Tremblay J., Gerzer R., and Hamet P. (1988) Cyclic GMP in cell function, in *Advances in Second Messenger and Phosphoprotein Research* 22 (Greengard P. and Robison G. A., eds.), Raven, New York, pp. 319–383.

Ueno S. and Ueck M. (1988) Cyclic nucleotide phosphodiesterase activity: Histochemical and cytochemical methods, in *Methods in Enzymology* 159 (Corbin J. D. and Johnson R. A., eds.), (Academic, New York), pp. 477–489.

Uhler M. D., Chrivia J. C., and McKnight G. S. (1986) Evidence for a second isoform of the catalytic subunit of cAMP-dependent protein kinase. *J. Biol. Chem.* **261**, 15,360–15,363.

Walsh D. A., Angelos K. L., Van Patten S. M., Glass D. B., and Garetto L. P. (1990) The inhibitor protein of cAMP-dependent protein kinase, in *Peptides and Protein Phosphorylation* (Kemp B. E., ed.), CRC, Boca Raton, pp.43–84.

Walter U. and Greengard P. (1983) Photoaffinity labeling of the regulatory subunit of cAMP-dependent protein kinase, in *Methods in Enzymology* 99 (Corbin J. D. and Hardman J. C., eds.), (Academic, New York), pp. 154–162.

Walter U., Kanof P., Schulman H., and Greengard P. (1978) Adenosine 3',5'-monophosphate receptor proteins in mammalian brain. *J. Biol. Chem.* **253**, 6275–6280.

Watterson D. M. and Lukas T. J. (1988) Analysis of phosphodiesterase reaction mixtures by high-performance liquid chromatography, in *Methods in Enzymology* 159 (Corbin J. D. and Johnson R. A., eds.), (Academic, New York), pp. 471–477.

Weber I. T., Shabb J. B., and Corbin J. D. (1989) Predicted structures of the cGMP binding domains of the cGMP-dependent protein kinase: A key alanine/threonine difference in evolutionary divergence of cAMP and cGMP binding sites. *Biochemistry* **28**, 6122–6127.

Weber I. T., Steitz T. A., Bubis J., and Taylor S. S. (1987) Predicted structures of cAMP binding domains of type I and II regulatory subunits of cAMP-dependent protein kinase. *Biochemistry* **26**, 343–351.

Weldon S. L., Mumby M. C., and Taylor S. S. (1985) The regulatory subunit of neuronal cAMP-dependent protein kinase II represents a unique gene product. *J. Biol. Chem.* **260**, 6440–6448.

Wells J. N. and Hardman J. G. (1977) Cyclic nucleotide phosphodiesterases, in *Advances in Cyclic Nucleotides Research*, vol. 8 (Greengard P. and Robison G. A., eds.), Raven, New York, pp. 119–143.

Wells J. N and Kramer G. L. (1981) Phosphodiesterase inhibitors as tools in cyclic nucleotide research: A precautionary comment. *Mol. Cell. Endocrinol.* **23**, 1–9.

Wells J. N. and Miller J. R. (1988) Methylxanthine inhibitors of phosphodiesterases, in *Methods in Enzymology* 159 (Corbin J. D. and Johnson R. A., eds.), (Academic, New York), pp. 489–496.

Wernet W., Flockerzi V., and Hofmann F. (1989) The cDNA of the two isoforms of bovine cGMP-dependent protein kinase. *FEBS Lett.* **251**, 191–196.

Whalin M. E., Strada, S. J., and Thompson W. J. (1988) Purification and partial characterization of membrane-associated type II (cGMP-activatable) cyclic nucleotide phosphodiesterase from rabbit brain. *Biochem. Biophys. Acta* **972**, 79–94.

Whitehouse S. and Walsh D. A. (1983) Inhibitor protein of the cAMP-dependent protein kinase: Characteristics and purification, in *Methods in Enzymology* 99 (Corbin J. D. and Hardman J. C., eds.), (Academic, New York), pp. 80–93.

Wolfe L., Corbin J. D., and Francis S. H. (1989a) Characterization of a novel isozyme of cGMP-dependent protein kinase from bovine aorta. *J. Biol. Chem.* **264**, 7734–7741.

Wolfe L., Francis S. H., and Corbin J. D. (1989b) Properties of a cGMP-dependent monomeric protein kinase from bovine aorta. *J. Biol. Chem.* **264**, 4157–4162.

Yamazaki A., Bitensky M. W., and Casnellie J. E. (1988) Photoaffinity labeling of high-affinity cGMP-specific noncatalytic binding sites on cGMP phosphodiesterase of rod outer segments, in *Methods in Enzymology* 159 (Corbin J. D. and Johnson R. A., eds.), (Academic, New York), pp. 730–736.

Yusta B., Ortiz-Caro J., Pascual A., and Aranda A. (1988) Comparison of the effects of forskolin and dibutyryl cyclic AMP in neuroblastoma cells: Evidence that some of the actions of dibutyryl cyclic AMP are mediated by butyrate. *J. Neurochem.* **51**, 1808–1818.

Zetterqvist O., Ragnarsson U., and Engstrom L. (1990) Substrate specificity of cyclic AMP-dependent protein kinase, in *Peptides and Protein Phosphorylation* (Kemp B. E., ed.), CRC, Boca Raton, pp.171–188.

Caged Intracellular Messengers and the Inositol Phosphate Signaling Pathway

Ian Parker

1. Introduction

The development of caged intracellular messenger compounds offers an elegant new approach to quantitative, time-resolved studies of intracellular signaling pathways. Several reviews of caged compounds have recently appeared (Kaplan and Somlyo, 1989; McCray and Trentham, 1989; Homsher and Miller, 1990; Somlyo and Somlyo, 1990), which may be consulted for details of their synthesis, properties, and use. The present chapter concentrates on practical aspects of the use of caged compounds, derived from work in the author's laboratory using caged inositol trisphosphate to study signaling in *Xenopus* oocytes.

1.1. Utility of Caged Compounds for Studying Intracellular Messenger Systems

The receptor sites for inositol 1,4,5-trisphosphate (InsP_3) and other second messengers are intracellular. Thus, in order to study their actions, it is necessary to have a means of introducing these compounds into the cytoplasm. Ideally, this should permit a rapid, precise, and spatially homogeneous control of the intracellular concentration.

From: *Neuromethods, Vol. 20: Intracellular Messengers*
Eds: A. Boulton, G. Baker, and C. Taylor © 1992 The Humana Press Inc.

Several approaches are currently in use for intracellular introduction of second messengers. One favored for biochemical studies is to permeabilize the cell membrane by various procedures, so that low-molecular-weight compounds applied in the external solution may gain access to the cell interior (*see* chapter by Taylor et al.).When used with small cells, this method gives good control of the intracellular concentration of applied second messenger and, with appropriate mixing devices, allows time resolution of the order of tens of milliseconds (e.g., Meyer and Stryer, 1990). Significant disadvantages include the possible loss of important constituents from the cytoplasm, the inability to localize messengers to particular parts of the cells, and disruption of the electrical properties of the plasma membrane.

For studies on intact single cells, a simple approach is to inject messenger compounds through an intracellular micropipet, by means of iontophoresis or pressure ejection (e.g., Sumikawa et al., 1989). This allows rapid (millisecond) applications and, with pressure injection, the total amount of compound injected may be measured. However, the resulting intracellular concentration is difficult to estimate and is likely to be spatially inhomogeneous since the compound diffuses away from the pipet tip. A related method of introducing messenger compounds is to allow them to diffuse into the cell from a whole-cell, patch-clamp recording pipet (e.g., Changya et al., 1989), which serves both for electrical recording and intracellular perfusion. Disadvantages of this technique are that the slow rate of access by diffusion from the pipet limits measurements of the time-course of responses, and metabolism of the intracellular messenger may raise further doubts as to its concentration (and even chemical identity) at the functional intracellular receptor sites.

The recent development and commercial availability of caged intracellular messenger compounds provide a new approach that circumvents many of the limitations inherent in the techniques described above. Caged compounds are inactive precursors of intracellular messenger that can be introduced into cells and allowed, at leisure, to diffuse throughout the cytoplasm. Illumination with near UV light is then used to liberate the active messenger compounds. Major advantages of this technique include the following:

1. Rapid (millisecond or shorter) jumps in intracellular concentration can be achieved;
2. The resulting intracellular concentration of second messenger can be precisely regulated; and
3. Manipulation of the photolysis light allows messenger molecules to be liberated uniformly throughout the cell or at any desired location.

A clear example illustrating the improvements available by using caged compounds as compared to previously available techniques is provided by experiments on hepatocytes perfused by a whole-cell pipet (Ogden et al., 1990). Inclusion of 10 μM InsP$_3$ in the pipet solution produced no consistent responses, whereas photorelease of a few hundred nanomoles of InsP$_3$ from caged InsP$_3$ evoked clear signals.

1.1.1. Properties of Caged Compounds

The term "caged" was coined by Kaplan et al. (1978) to describe ATP that was rendered inaccessible to the Na$^+$:K$^+$ pump by covalently linking its terminal phosphate to a photolabile nitrobenzyl group. Although this nomenclature should strictly refer only to clathrates, its use in the sense introduced by Kaplan et al. is now widespread and gives a clear idea of the principle involved.

A wide variety of caged compounds are presently available from suppliers, including Calbiochem (LaJolla, CA) and Molecular Probes (Eugene, OR), and the list will undoubtedly grow. Caged intracellular messengers include caged Ca^{2+} (Kaplan and Ellis-Davies, 1988; Adams et al., 1988), caged cAMP and cGMP (Karpen et al., 1988), caged GTP and GTPγS (Dolphin et al., 1988), and caged InsP$_3$ (Walker et al., 1987). Other caged compounds include caged neurotransmitters (Walker et al., 1986), caged protons (Janko and Reichert, 1987), and caged Ca^{2+} chelators (Kaplan, 1990). With the exception of caged Ca^{2+} compounds, all other caged intracellular messengers mentioned above are formed by esterifying a phosphate group with a photolabile nitrobenzyl group. Photolysis liberates the active messenger molecule, together with a proton and a nitrosoketone leaving group. A technically simple synthesis and purification technique for the preparation of these caged compounds has been described

(Walker et al., 1988,1989), and their structure and photochemistry have been reviewed (Gurney and Lester, 1987; Kaplan and Somlyo, 1989; McCray and Trentham, 1989). The kinetics of photorelease are still not fully understood, and depend on factors that include temperature, pH, and ionic strength (McCray and Trentham, 1989). Under physiological conditions, half-times for photolysis are of the order of several milliseconds (Homsher and Miller, 1990); e.g., 2.5–4 milliseconds for caged $InsP_3$ (Walker et al., 1989).

Two different families of caged Ca^{2+} compounds now exist. Both are based on the 2-nitrobenzyl chromophores, but otherwise show important differences in their properties (Adams et al., 1988; Kaplan and Ellis-Davies, 1988). Commercially available representatives of each type are nitr-5 and DM-nitrophen. The properties of these, and their relative advantages and disadvantages, are discussed by Kaplan (1990). Perhaps the most important differences between them are that DM-nitrophen shows a much larger change in affinity for Ca^{2+} following photolysis (from $5 \times 10^{-9}M$ to $3 \times 10^{-3}M$, as compared to $1.5 \times 10^{-7}M$ to $6.3 \times 10^{-6}M$ for nitr-5), whereas nitr-5 has a higher selectivity for Ca^{2+} over Mg^{2+}. Both show a very rapid ($3000 \ s^{-1}$) rate of Ca^{2+} release on photolysis.

2. Methodology

2.1. Light Sources

Photolysis of caged $InsP_3$ and other 2-nitrobenzyl compounds requires high light intensities in the near UV region (300–400 nm). Suitable light sources, listed in order of both increasing energy and increasing cost and complexity, include:

1. Continuous arc lamp equipped with shutter;
2. Xenon flash-lamp; and
3. Lasers.

The characteristics and relative advantages of each are discussed below. McCray and Trentham (1989) have reviewed the use of flash-lamp and laser systems, so I cover these only briefly and, instead, concentrate on continuous arc systems.

2.1.1. Continuous Arc

A continuously burning arc lamp provides a simple and very stable source of UV light, with several advantages. The main drawbacks are that the available light energy is limited and, because of mechanical inertia in the shutter, it is not possible to obtain flash durations shorter than 1 or 2 milliseconds.

For use with caged InsP$_3$, the limited light energy is not a great problem, because of the extremely high sensitivity of cells to intracellular InsP$_3$. Thus, cells can be loaded with a fairly high (several micromolar) concentration of caged InsP$_3$, so that photolysis of only a small fraction will liberate sufficient InsP$_3$ to evoke large responses. However, the situation is very different when using caged Ca^{2+} compounds. Because these act as powerful Ca^{2+} chelators, it is desirable to load cells with the minimum amount required and to photolyze a large fraction of the caged Ca^{2+} with a single flash. In my experience, a continuous arc provides only just enough energy in flashes of 10-milliseconds duration to be able to work usefully with DM-nitrophen and is woefully ineffective with nitr-5.

Commercially available high-speed electric shutters allow reproducible flashes to be obtained with durations as short as 1–2 milliseconds. This time is not much longer than the flash duration obtained from a flash-lamp device (Rapp and Guth, 1988), but is very long compared to the pulses obtained from a laser. In practice, however, there seems to be little advantage in having flash durations much shorter than 1 milliseconds. The half-time for photogeneration of InsP$_3$ is several millseconds (Walker et al., 1989), and in several cell types, the latency to onset of InsP$_3$-mediated responses is tens or hundreds of millseconds (Parker and Miledi, 1989; Ogden et al., 1990; Meyer and Stryer, 1990).

An advantage of a continuous arc system not shared by other light sources is that the flash duration may be readily varied by using an electronic timing circuit to control the shutter. This provides a convenient and highly reproducible way of controlling the total energy in each flash, and a range of tenfold or more can be achieved between the briefest shutter opening and a flash duration, which becomes appreciable in comparison to the onset latency of InsP$_3$-mediated responses. Another advantage is that paired flashes can be produced at any desired interval (limited only by the operating time of the shutter). The duration of each flash can be independently controlled, and the light intensity is stable for each flash. Furthermore, indefinitely long light exposures may be used to cause a sustained photorelease of InsP$_3$ and, thus, mimic more closely the situation that is likely to pertain during agonist activation (Parker and Miledi, 1989). For this application, the light output from the arc lamp is usually too intense, but can be attenuated by neutral density filters.

Regarding the choice of arc lamp, a mercury arc is expected to be preferable, since this has an intense spectral line at about 360 nm. However, in practice there seems to be little difference between mercury and Xenon lamps, possibly because the Xenon arc forms a very small (<0.5 mm) intense plasma ball that can be more efficiently focused onto small preparations. There is little point in using an arc lamp of higher power than 50 or 75 W. Although higher power lamps provide a greater total output, the size of the arc is larger. Thus, if the light is to be focused onto a small target (as is the case in most applications of caged compounds), the irradiance is no greater, and may even be less, with a higher power lamp. Quartz-halogen lamps produce very little output in the UV and are of no use for photolysis experiments. This does mean, however, that a conventional fiberoptic illuminator can be used without further filtering to view the preparation without causing appreciable photolysis.

The stability of light output from an arc lamp can be very good, provided that it is operated from a well-regulated, constant-current power supply and that adequate time is allowed for warm-up. A flash-to-flash repeatability of better than 1% can be achieved, taking into account both variations in lamp output and shutter timing (see Section 3.2.). An important point for stability is that the lamp be mounted vertically. It is possible to operate Xenon arc lamps horizontally, but convection currents then cause the arc to "wander," producing large and erratic changes in intensity of the focused beam.

2.1.2. Flash Lamps

Xenon arc lamp systems give a high energy output in a flash duration of 1 ms or shorter. A system designed for use with caged compounds is commercially available (Rapp and Guth, 1988), and produces 100–200 mJ at wavelengths between 300 and 400 nm when focused onto an area of 10 mm^2. The output of this flash-lamp during a 1-ms flash appears to be comparable to the output of a continuous mercury arc for 1 s (McCray and Trentham, 1989). Thus, the flash-lamp is much to be preferred for rapid kinetics studies in which a high proportion of a caged messenger must be rapidly photolyzed. However, the flash-to-flash stability is unlikely to be as good as with a continuous arc system, and independent control of each flash in paired-pulse experiments is difficult.

2.1.3. Lasers

Several types of laser are available that produce high-energy pulses at near UV wavelengths. The main advantages of pulsed lasers are that the parallel output beam can be readily directed and focused onto the specimen, the output is monochromatic, and the pulse duration is very short (generally tens of nanoseconds). Disadvantages include high cost, large size, and in some cases, the need for frequent maintenance and alignment.

A frequency-doubled ruby laser has been used in several studies on muscle (Goldman, 1986). This gives an output pulse at 347 nm with energies up to 300 mJ. However, a major drawback for some applications is that the maximum repetition rate is 1–0.1 Hz, thus effectively precluding paired-pulse experiments. Another alternative is the frequency-tripled neodymium-YAG laser, which gives pulses with energies of over 100 mJ at 355 nm and repetition rates up to 50 Hz. Finally, nitrogen lasers giving an output at 337 nm are much cheaper than the others, but the available pulse energies are only a few milliJoules.

For most purposes, the higher cost of lasers over flash-lamp systems is probably not justified. The total energy outputs are comparable and, as discussed above, the shorter pulse duration of lasers is of little practical benefit. Monochromicity of the laser output also offers little advantage, except that it may simplify rejection of flash artifacts when simultaneous measurements are made in another region of the spectrum (e.g., use of fluorescent probes to monitor free $[Ca^{2+}]$).

2.1.4. Two-Photon Excitation

All of the above-described light sources (lamps and lasers) allow photorelease to be spatially restricted by means of appropriate apertures in the light path. However, although the illumination can be controlled in the horizontal plane, it is not possible to obtain a controlled release of messenger throughout a "slice" at some particular depth into the cell. A recent development (Denk et al., 1990) using a colliding-pulse, mode-locked dye laser now offers the possibility of obtaining focal release of messengers at any desired x, y, and z coordinates.

The method involves focusing extremely brief (femtosecond) pulses of light through an objective lens of high numerical aperture. The wavelength of the laser is about double that normally required to cause photolysis, so that photorelease requires the simultaneous absorption of two photons by a caged molecule to combine their energy in order to reach the excited state. Thus, the extent of photorelease varies as the square of the light intensity, rather than being linear, as is the usual case. Because of the highly convergent cone of light formed by a lens with large numerical aperture, appreciable photorelease will occur only in a very restricted region around the beam waist. This will give rise to an effective point source release of messenger but, by rapidly scanning the photolysis spot, it may be possible to obtain near-simultaneous release throughout a thin section of the cell.

2.2. Optics

Several different approaches may be taken to collect the UV radiation from the light source and direct this onto the target cell. Factors to consider when choosing between these include the desired energy density at the specimen, and whether the light needs to be tightly focused (for example, to illuminate only a single cell or part of a cell) or whether it should diffusely cover a broad (several millimeters) area.

With a flash-lamp source, the simplest approach is to collect the light using a condenser lens or ellipsoidal mirror, pass this through a UV-transmitting filter, and place the target at the focus (Ogden et al., 1990). Inclining the light beam to an angle of 38° to the horizontal minimizes energy loss resulting from reflections at the fluid surface above the specimen. Advantages of this method are simplicity, a long working distance, and a high efficiency in collection of available light. Probably the main disadvantage is that the focused spot is fairly large (a few millimeters in diameter), so that the energy density is relatively low and there is little possibility to restrict the illuminated area spatially. Use of an ellipsoidal reflector for focusing gives very high (up to 80%) efficiency of light collection from the flash-lamp, and since additional focusing optics are not needed, there is no problem with glass optics absorbing at short wavelengths. In contrast, a condenser system provides a smaller spot size, at the expense of less efficient light collection (20% at best) and, unless expensive quartz lenses are used, a cutoff at wavelengths shorter than about 340 nm. Rapp and Guth (1988) compared the use of mirror and condenser optics, and found that a mirror gave a slightly higher energy density over a larger focus size.

A similar condenser or mirror system could, in principle, be used also with a shuttered continuous arc system. A difficulty, however, is that fast-acting shutters necessarily have small (1 cm or less) open apertures. Thus, the shutter needs to be mounted inconveniently close to the preparation in order not to obscure a large part of the light beam. Better approaches are to place the shutter at the focus of the condenser system, and then couple the light to the preparation through a fiberoptic or epifluorescence system.

One problem in focusing a flash-lamp or continuous arc directly onto the target is that the light source must be mounted fairly close to the preparation. This may be inconvenient because of the physical size and heat output of the system, and because the high-voltage ignition pulse introduces electrical artifacts into an associated electrophysiological recording system. One way to obviate these problems is to mount the lamp remotely (outside the screened recording "cage"), and to couple it to the setup through a fiberoptic light guide. Both high-grade fused silica and liquid light guides have good transmission in the UV, but the liquid light guide is probably to be preferred as having the better transmission in the near UV and a larger acceptance angle. At a wavelength of 350 nm, the total transmittance of a 1-m length of liquid light guide is about 50%.

Unlike other light sources, lasers produce a narrow, highly parallel output beam. This may simply be directed onto the preparation without any intervening optics. On the other hand, the highly collimated beam is ideal for focusing down to small spot sizes, to give a higher energy density, or to give spatially restricted stimulation.

The best approach to obtaining very small illuminated spot sizes is probably to modify an upright or inverted fluorescence microscope. The regular fluorescence illuminator is replaced by a flash-lamp or laser, or in the case of a shuttered continuous arc, it is simply necessary to mount a shutter in front of the existing arc lamp. For safety, it should not be possible to view the photolysis light flash through the eyepieces! By using objectives of different powers, the photolysis light can be focused to cover an area as great as 1 mm^2 or as small as a few micrometers.

2.3. Introduction of Caged Compounds into Cells

The majority of caged compounds (including caged InsP_3) are charged molecules and are thus impermeant through the cell membrane. For work with single cells, they may be introduced into the cytoplasm by injection through a micropipet or by diffusion from a whole-cell patch pipet. In the latter case, the resulting intracellular concentration should approximate that in the pipet, whereas the concentration following injection can be estimated from the amount of fluid expelled and the cell volume.

Clearly, mechanical injection of caged compounds will be impractical for studies on populations of cells and may also be problematic with especially large cells (e.g., skeletal muscle fibers), in which long periods of time would be required for the compound to equilibrate throughout the cell. A recent improvement in the application of caged compounds in such situations has been their introduction into cells permeabilized with staphylococcal α-toxin or saponin ester β–escin (Ahnert-Hilger et al., 1989; *see* chapter by Taylor et al.). Cells exposed to these agents become permeable to solutes with mol wt of about <1000 dalton and 17,000 dalton, respectively. Thus, caged compounds (e.g., caged InsP_3; mol wt 635 daltons) can be introduced while minimizing the loss of higher molecular-weight cytoplasmic constituents.

2.4. Artifacts

Possible artifacts in the use of caged compounds might arise from several mechanisms, including the device used to generate the photolysis light, the effect of light on the preparation, and actions of the caged compound or byproducts of the photolysis reaction on cell metabolism. So far, there appear to be no indications that any of these present major problems.

Potential difficulties with flash-lamp and laser systems include the high voltage pulse associated with their triggering. They may need careful electrical screening to avoid interference with sensitive electrophysiological recordings. Also, the chlorided silver wires commonly used to make contact with recording electrodes are photosensitive and may need to be shielded from the light. Regarding effects of light on the preparation, artifacts may arise through heating (especially if infrared radiation is not well blocked from arc and flash-lamp systems; Parker, 1989a), or from actions of the UV light on proteins and other cell constituents. A simple control for all of these is to illuminate the preparation before loading the caged compound. Effects of the caged compounds themselves may be checked by looking for responses evoked by their introduction into the cell and by testing that agonist-evoked responses (mediated by the second messenger system under study) are not altered following intracellular loading.

The byproducts of the photolysis reaction are 2-nitroso-acetophenone and a proton. Nitrosoketones react with thiols, including cysteines in proteins, and might thus perturb the cell. During experiments using caged ATP in muscle fibers, Goldman et al. (1984) observed a desensitization in contraction that appeared to arise through the formation of 2-nitrosoacetophenone.

Inclusion of reduced glutathione (GSH) in the intracellular solution gave protection from this effect. A good control for actions of photolytic byproducts is to load cells with a caged inactive analog of the messenger being studied. For example, caged $InsP_2$ has been used as a control in experiments with caged $InsP_3$ (Ogden et al., 1990); its photolysis produced no response and did not prevent subsequent agonist activation.

2.5. Flash Photolysis and Calcium Monitoring

One approach to monitoring the rise in intracellular free Ca^{2+} evoked by photorelease of $InsP_3$ is to record some Ca^{2+}-mediated process in the cell, such as the opening of Ca^{2+}-dependent membrane channels. However, interpretation is complicated by factors including possible cooperativity of Ca^{2+} in opening the channels, desensitization, and spatial variations in channel density. Thus, it is desirable to have a more direct measure of intracellular free Ca^{2+}. Until recently this was problematic, because the commonly used fluorescent indicators quin-2, fura-2, and indo-1 all require excitation at wavelengths in the near UV (Grynkiewicz et al., 1985). Measurement will, therefore, necessarily cause photolysis of caged $InsP_3$, although this can be minimized by using low-intensity fluorescence excitation (Gray et al., 1988). The introduction of new fluorescent Ca^{2+} indicators with excitation wavelengths in the visible spectrum (Minta et al., 1989) completely obviates this difficulty, and they have sucessfully been used in conjunction with photochemical generation of intracellular messengers (Kao et al., 1989; Parker and Ivorra, 1990a,b; Ivorra and Parker, 1990a).

Two long-wavelength indicators are currently available, fluo-3 and rhod-2. Their respective excitation and emission maxima are 506 nm and 526 nm for fluo-3, and 553 nm and 576 nm for rhod-2. Unlike the short-wavelength indicators fura-2 and indo-1, both fluo-3 and rhod-2 show only a fluorescence increase on binding Ca^{2+}, with no shift in either excitation or emission spectra. Thus, it is not possible to ratio signals at two wavelengths in order to obtain a calibration of free Ca^{2+} levels that is independent of variations in dye loading and path length. However, for many purposes, an uncalibrated index of intracellular Ca^{2+} transients is likely to be sufficient, and with some difficulty and uncertainty, it may be possible to calibrate the fluorescence by lysing the cell and titrating the Ca^{2+} level in the bathing solution. Dissociation constants of fluo-3 and rhod-2 for Ca^{2+} are 0.4 μM and 1 μM, respectively (Minta et al., 1989), so that both are able to resolve relatively high $[Ca^{2+}]$ that would saturate fura-2.

The visible excitation maxima of fluo-3 and rhod-2 allows the use of a quartz-halogen lamp as an efficient excitation source. Also, the spectra of the dyes match well to the usual fluorescein and rhodamine filter sets available for fluorescence microscopes. A quartz-halogen lamp has the advantages of low cost and, provided that it is operated from a well-stabilized DC power supply, a highly stable light output. For measurements on small cells where the excitation light must be tightly focused, the smaller source size of an arc lamp allows a higher energy density at the specimen, and the mercury arc has an intense peak at about 540 nm that matches well to the excitation maximum of rhod-2. Other possible light sources for fluo-3 include the 488-nm line of an argon ion laser and, for rhod-2, the 543-nm emission of the inexpensive green He-Ne laser (Parker and Ivorra, 1990b).

The green emission from fluo-3 is close to optimal for detection by photomultiplier tubes. However, most photomultipliers have a lower quantum efficiency in the red and will thus show less sensitivity for rhod-2. Photodiodes are much less sensitive than photomultipliers, but their performance improves at longer wavelengths, so that they might be used as an inexpensive detector with cells brightly stained with rhod-2.

Because of the wide separation in wavelengths optimal for photolysis of caged compounds and excitation of fluo-3 or rhod-2, it is possible to obtain good rejection of the photolysis flash in the fluorescence record. If a nonmonochromatic light source (i.e., flash-lamp or continuous arc) is used for photolysis, its output must be filtered to attenuate visible wavelengths. A Schott UG11 filter gives high transmission in the near UV (70% at 350 nm), together with strong ($>10^5$) attenuation at wavelengths longer than about 400 nm. Alternatively, a UG5 filter offers higher transmission in the UV over a broader range of wavelengths, but is unsuitable for use with fluo-3 since a "shoulder" in the transmission spectrum extends beyond 500 nm. A point to remember with both filters is that they show a transmission at wavelengths longer than about 700 nm. This needs to be blocked by a separate filter or dichroic mirror if the barrier filter shows appreciable transmission at these wavelengths.

3. Application of Caged Compounds to Study InsP_3 Signaling in Oocytes

Oocytes of *Xenopus laevis* possess a phosphoinositide signaling system by which the activation of cell surface receptors leads to the generation of an oscillatory Cl⁻ membrane current. InsP_3, which is formed in response to receptor activation, acts as an intracellular messenger to release Ca^{2+} from intracellular stores (Oron et al., 1985; Parker and Miledi, 1986) and to activate an influx of extracellular Ca^{2+} (Parker and Miledi, 1987). The resulting rise in cytoplasmic free Ca^{2+}, in turn, activates Ca^{2+}-dependent Cl⁻ channels to give the final membrane current response. For the purposes of studying this messenger pathway, the oocyte is a convenient model cell system, since its large size (>1 mm diameter) greatly facilitates such procedures as intracellular recording and microinjection.

The following sections describe equipment and procedures developed in the author's laboratory, together with results that illustrate some of the applications of caged compounds. The methods should be generally applicable also to cells much smaller than the oocyte, with the exception that different techniques may be required for intracellular recording, and for loading of caged compounds and indicators.

3.1. Optical System

Figure 1 shows the optical system employed to allow photolysis of caged InsP_3 or caged Ca^{2+}, together with recording of Ca^{2+}-dependent changes in fluorescence of long-wavelength indicators. The system is based on a Zeiss upright microscope, using two epifluorescence illuminators stacked one above the

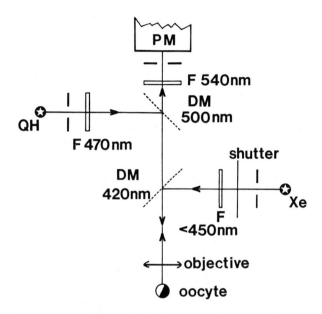

Fig. 1. Diagram of optical system for photolysis of caged compounds and fluorescence monitoring of intracellular Ca^{2+}. PM = photomultiplier, QH = quartz-halogen lamp (100 W), Xe = xenon arc lamp (75 W), DM = dichroic mirror, F = filter. *See text* for further details.

other. The lower illuminator provides precisely controlled flashes of UV light for photolysis, whereas the other provides excitation light for the Ca^{2+} indicator. To allow the two epifluorescence units to be physically mounted together and to improve the optical performance, the diverging and converging Telan lenses are removed from the upper and lower units, respectively. Further, the diverging Telan lens on the lower unit is replaced by a fused silica lens, since this element otherwise limits the short-wavelength cutoff for the entire optical path. A 6.3× Neofluar objective is usually fitted to the microscope. This has quite a good UV transmission and a long working distance to facilitate the insertion of micropipets. It also gives a field of view sufficient to visualize the entire oocyte.

The light source for photolysis is a continuous 75-W Xenon arc lamp mounted in a 100-W Zeiss lamp housing. The arc is operated from a constant-current DC power supply (Photon

Technology, NJ), and light is collected through a three-lens glass condenser and rear reflector. The outer arm of the epifluorescence unit (onto which the lamp housing mounts) is removed and replaced by a custom-built section holding a high-speed electric shutter (Newport Corp, CA) and a variable rectangular slit diaphragm. The shutter operates with a rise and fall time of about 1 millisecond, and its activation is controlled by a digital timer. The variable diaphragm allows the photolysis light to be focused as a square or rectangle of any desired size. Also, the diaphragm assembly is moveable, so that the illuminated square can be positioned anywhere on the oocyte surface. To assist in adjusting the photolysis light, the UV filter is temporarily replaced by a green filter, so that the focused light can be viewed on the oocyte without causing photolysis. A dichroic mirror reflects wavelengths shorter than about 420 nm from the photolysis light to the objective lens, and a Schott UG5 filter further restricts illumination to UV wavelengths. Because the dichroic mirror is transparent to visible wavelengths, it does not attenuate the fluorescence excitation light directed down from the upper epifluorescence unit or the emitted light collected by the objective. Neutral density filters are used to vary the intensity of the photolysis light, though, in most experiments, it is more convenient to keep the intensity constant and vary the extent of photolysis by altering the flash duration. Some control of light intensity is also possible by varying the arc-lamp current, but the available range is restricted, because at currents less than about one-half of the rated maximum, the arc becomes unstable. It is also possible to double the lamp output briefly (a few seconds) by overrunning it at 10 A. This is entirely in contradiction to the manufacturer's warnings, but so far, has not provoked any lamps to explode.

When operated at normal lamp current, the energy density available at the cell is about 10 μJ mm^{-2} from a 2-milliseconds flash (at wavelengths of about 340–400 nm). The illuminated area (using a 6× microscope objective) can be varied from 0 to a maximum of about 0.5 mm^2.

Fluorescence excitation for Ca^{2+} monitoring is provided by a quartz-halogen lamp mounted on the upper epifluorescence unit. Excitation and emission wavelengths are determined by standard Zeiss filter sets, which can therefore be readily interchanged to permit measurements with either fluo-3 or rhod-2. For convenience, the second position in the sliding filter holder is left empty, so that the filters can be slid out of the way for viewing the oocyte while positioning electrodes. The photolysis light is usually focused on the oocyte as a spot concentric with, and slightly smaller than, the photolysis light. However, because the diaphragms in the photolysis and fluorescence emission systems are independent, it is also possible to record from regions of the cell distant from that stimulated. Fluorescence emission is monitored by a photomultiplier (EMI 9524B) mounted on the microscope phototube.

3.2. Intracellular Loading of Caged InsP$_3$

Caged InsP$_3$ obtained from Calbiochem is dissolved at a concentration of 1 mM in an aqueous solution including 5 mM HEPES (at pH 7.0) and 50 μM EDTA (to chelate contaminating Ca^{2+} and prevent activation of Cl⁻ currents on injection). This solution is passed through a 0.22-μm Millipore filter to remove any particles that might block the injection pipet, and a few microliters are loaded into the back of a micropipet. Pipets for injection are pulled using a conventional microelectrode puller, from fiber-filled glass tubing, and the tip is broken to a diameter of a few micrometers. Judicious tapping of the pipet will usually (but not always) ensure that the shank of the pipet becomes filled without bubbles. After mounting in the micromanipulator, the pipet is connected by flexible tubing to a "Picopump" (World Precision Instruments, New Haven, CT), which supplies pneumatic pressure pulses (usually set to 20 psi for 50 milliseconds) to eject fluid. The volume of fluid expelled by each pulse is estimated by measuring the diameter of the fluid droplet expelled with the pipet tip raised in the air.

Insertion of the pipet into the oocyte is monitored by a transient inward current in the voltage-clamp record. After allowing a few minutes for the cell membrane to seal around the pipet, sufficient pressure pulses are then applied to load the oocyte with a total of 0.5–10 pmol of caged InsP$_3$. This corresponds to a fluid vol of 0.5–10 nL of solution, giving a final intracellular concentration of roughly 0.5–10 μM.

Caged InsP$_3$ appears to be stable under normal room lighting, and even illumination of the oocyte by a powerful fiberoptic lamp causes no detectable response, but to be safe, stock solutions should be kept frozen in the dark when not in use. Greater care is needed, however, with stray light from the arc lamp, which, even though it may appear dim, contains a high proportion of UV. Thus, remove the injection pipet from the oocyte after loading, and displace it several millimeters away from the objective lens. Stray light from the base of the lamp housing can also cause appreciable photolysis of caged InsP$_3$ in the pipet over a few hours and needs to be shielded.

Even after taking all the above precautions, injections of freshly prepared solutions of caged InsP$_3$ obtained from Calbiochem still evoked oscillatory chloride currents, like those generated by free InsP$_3$ (Parker and Miledi, 1989). Most probably, these currents arose because the samples of caged InsP$_3$ were slightly contaminated by free InsP$_3$, or by InsP$_3$ caged at the 1-phosphate position. Whatever the explanation, these responses do not pose any great problem for work in the oocyte, because the oscillations die away after several minutes, presumably as the physiologically active contaminant is metabolized. We normally wait for at least 30 min after loading caged InsP$_3$ to allow the compound to distribute throughout the cell and to allow the size of the response evoked by light flashes to stabilize. However, in other preparations, this trick of using the cell itself to "clean up" the caged InsP$_3$ may not work so effectively, and it would certainly be better to have more pure preparations available.

Figure 2 illustrates the changes in peak size of Ca^{2+}-dependent current responses evoked by identical light flashes at intervals after loading an oocyte with about 5 pmol of caged InsP$_3$. The injection of caged InsP$_3$ itself elicited an oscillatory current, which ceased after about 8 min. No responses were evident to the light flashes until 15 min following injection, and the currents then grew progressively before stabilizing at a maximal value after about 30 min. In this experiment, the part of the

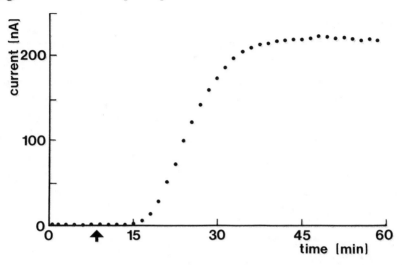

Fig. 2. Sizes of peak membrane currents evoked by constant light flashes applied at 90-s intervals following intracellular loading of an oocyte with caged InsP₃. The arrow indicates the time at which oscillatory currents ceased.

oocyte illuminated by the photolysis light was close (about 100 μm) to the injection site, so that the lack of responses during the first 15 min was probably not owing to a restricted spread of the caged InsP₃ to the illuminated area. Further, a similar delay to the onset of responses was seen even when the photolysis light was focused at the injection site. Instead, it may be that responses to photorelease of InsP₃ are inhibited because the intracellular free Ca^{2+} level is elevated for some time after loading (Parker and Ivorra, 1990), as a result of Ca^{2+} liberation from intracellular stores and from influx of extracellular Ca^{2+} (Parker and Miledi, 1986, 1987).

After allowing sufficient time for stabilization, the current responses evoked by successive stimuli are very stable. For example, the SD of the last 12 responses in Fig. 2 is only about 0.7% of the mean current amplitude.

A rough estimate of the amount of caged InsP₃ photolyzed can be obtained from the energy density of the photolysis light at the preparation. Ogden et al. (1990) estimate that a flash with an energy density of about 20 mJ/mm⁻² produces about 50% conversion of caged InsP₃ to free InsP₃. The photolysis system described here gives an energy density of about 0.2 mJ mm^{-2} in a typical flash of 20-milliseconds duration, and is thus expected to photolyze <1% of the caged InsP$_3$ exposed to the light. Furthermore, in most experiments, only a small part of the oocyte is illuminated and, because of the opacity and light scattering of the cytoplasm, the photolysis light penetrates to a depth of only a few tens of micrometers into the cell. Thus, each flash consumes only a negligible fraction of the total amount of caged InsP$_3$ loaded into the oocyte, allowing tens or hundreds of reproducible responses to be evoked without causing appreciable depletion of caged InsP$_3$.

3.3. Responses Evoked in Oocytes by Photoreleased InsP$_3$ and Ca^{2+}

Figure 3 shows superimposed records obtained from two oocytes, loaded with caged InsP$_3$ or caged Ca^{2+} (DM-nitrophen), and illustrates the time resolution made possible by the use of caged compounds. Both oocytes were stimulated identically (by a 10-milliseconds flash of near UV light at the arrow). Simultaneous measurements of the resulting elevations in intracellular free Ca^{2+} were obtained by recording the Ca^{2+}-activated Cl$^-$ membrane current under voltage clamp (lower traces), and by monitoring fluorescence of fluo-3 loaded into the oocyte together with the caged InsP$_3$.

Photolysis of caged InsP$_3$ within the oocyte evokes transient membrane current and fluorescence signals. Following brief light flashes like that in Fig. 3, the responses arise from liberation of Ca^{2+} from intracellular stores, since they are not altered by removal of extracellular Ca^{2+} (Parker and Miledi, 1989). However, release of larger amounts of InsP$_3$ by prolonged illumination also induces responses that arise through the activation of an influx of extracellular Ca^{2+} (Parker, I., unpublished observation).

3.3.1. Kinetics

Both the fluorescence calcium signal and the membrane current begin following a latency of a few hundred milliseconds after photolysis of caged InsP$_3$. Since photolysis of caged InsP$_3$ is

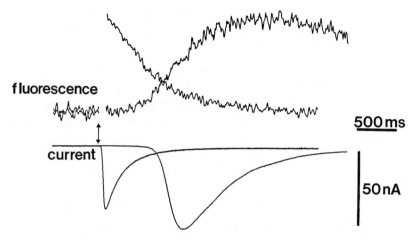

Fig. 3. Records of fluorescence and membrane current (clamp potential = –60 mV) obtained in two oocytes loaded intracellularly with caged Ca^{2+} (DM-nitrophen) or caged $InsP_3$. Traces from each oocyte are shown superimposed to facilitate comparison of responses evoked by photoreleased Ca^{2+} and $InsP_3$. Brief (10-milliseconds) light flashes were given at the arrow. Both oocytes were loaded with fluo-3, and an increase in fluorescence (upward deflection) corresponds to increasing free Ca^{2+}. For both the fluorescence and current traces, the more rapidly rising and falling records were obtained in the oocyte loaded with caged Ca^{2+}. The amplitudes of the fluorescence signals are not calibrated, but were scaled to give similar peak sizes for both oocytes. The fluorescence traces are blanked out during and shortly after the light flashes, because stray light saturated the photomultiplier. Records are from Ivorra and Parker (1990a).

expected to be complete within a few milliseconds, this long latency presumably arises from the processes within the cell responsible for $InsP_3$-induced Ca^{2+} release. In agreement, experiments using $InsP_3$ microinjected into the oocyte by pressure injection or ionophoresis showed a similar latency (Miledi and Parker, 1989). Furthermore, photorelease of Ca^{2+} from DM-nitrophen evoked fluorescence and current signals beginning with short latency, demonstrating that activation of the Ca^{2+}-dependent Cl^- conductance occurs rapidly.

A further difference between the responses to photoreleased Ca^{2+} and $InsP_3$ is the much longer durations of the signals evoked by $InsP_3$. The fluorescence signal evoked by Ca^{2+} declines with a time constant of a few hundred milliseconds, and presumably reflects the rate at which Ca^{2+} is buffered and sequestered by the cell. On the other hand, the fluorescence signal evoked by $InsP_3$ is maintained for several seconds, suggesting that $InsP_3$ causes a relatively prolonged liberation of Ca^{2+}. An interesting point remaining to be explained is the fact that the membrane current evoked by $InsP_3$ is much more transient than the fluorescence Ca^{2+} signal. This does not appear to arise through desensitization of the Cl^- channels, but may reflect spatial differences in free Ca^{2+} close to the cell membrane (Parker and Ivorra, 1990a).

3.3.2. Dose–Response Relationship and Facilitation

One advantage of the use of caged compounds is the ability to control precisely the amount of messenger released. Thus, by varying the duration or intensity of light flashes applied to oocytes loaded with caged $InsP_3$, we have been able to construct dose–response relationships for the membrane current and intracellular Ca^{2+} signals (Parker and Miledi, 1989; Parker, 1989b). For several reasons (including uncertainties as to the intracellular concentration of caged $InsP_3$ loaded and the light energy penetrating into the cell), it is difficult to estimate the absolute concentration of intracellular $InsP_3$ resulting from a given light flash. Nevertheless, the flash duration or intensity provides a relative indication of the resulting concentration of $InsP_3$ and, because even the strongest stimuli photolyze only a small percentage of the caged $InsP_3$, the extent of photolysis is expected to be linearly proportional to the energy of the flash.

An important finding is that the light flash must be greater than a certain threshold intensity or duration before any detectable calcium or membrane current signals are seen (Parker and Miledi, 1989; Ivorra and Parker, 1990b). Previous experiments (Parker and Miledi, 1987) measuring responses evoked by different doses of microinjected $InsP_3$ had indicated that the phosphoinositide signaling pathway in the oocyte operated in a highly nonlinear manner, but the use of caged $InsP_3$ provided a more precise approach to the study of this phenomenon, and facilitated the identification of the stage in the pathway responsible.

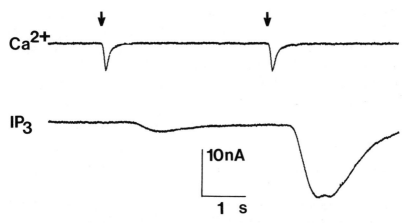

Fig. 4. Comparison of membrane current responses evoked by paired light flashes in two oocytes loaded with caged Ca^{2+} (DM-nitrophen) (upper trace) or caged $InsP_3$ (lower trace). In both cases, identical light flashes (10-milliseconds duration) were applied at an interval of 3 s, as indicated by the arrows.

Associated with the threshold phenomenon is a pronounced facilitation seen with paired light flashes that are each close to threshold. For example, the lower trace in Fig. 4 shows membrane currents evoked by two identical light flashes applied to an oocyte loaded with caged $InsP_3$. The first flash evoked only a small current, whereas the second gave a much larger response. This facilitation might arise if a preceding flash somehow "primed" the caged $InsP_3$ so that a subsequent flash caused greater photolysis. Such an effect (at intervals of several seconds) is, however, not expected from what is known of the photochemistry of caged $InsP_3$. Instead, it seems that the facilitation arises within the cell, a conclusion that is supported by the finding that intracellular injections of low doses of $InsP_3$ can also facilitate responses to subsequent light flashes. The question then is whether the threshold and facilitation arise because a certain level of $InsP_3$ is required before Ca^{2+} is released from intracellular stores, or because a threshold level of Ca^{2+} is needed to activate the Cl^- membrane conductance. One experiment to distinguish between these hypotheses is shown in Fig. 4. Paired light flashes delivered to an oocyte loaded with caged Ca^{2+} evoked membrane current responses of almost identical sizes, whereas an oocyte loaded with caged $InsP_3$ showed a clear facilitation when stimulated in the same way. Experiments using fluorescent probes to monitor intracellular Ca^{2+} also lead to the same conclusion: That a threshold level of $InsP_3$ is required before Ca^{2+} release is detected and Ca^{2+} signals are facilitated by paired flashes (Ivorra and Parker, 1990b; Parker and Ivorra, 1990b).

3.3.3. Interactions Between Intracellular Messengers

Unfortunately, all the caged compounds presently available are photolyzed by UV light of about the same wavelengths, so that it is not possible to load a cell with two or more caged compounds and photolyze each of these selectively. The development of caged compounds activated by light of well-separated wavelengths would allow some fascinating experiments, but for the moment, it is still feasible to study interactions between a photoreleased messenger compound and another compound that is microinjected into the cell. Using this approach we have, for example, shown that responses to photolysis of caged $InsP_3$ are facilitated by microinjection of $InsP_3$ (Parker and Miledi, 1989), but are depressed by microinjection of Ca^{2+} (Parker and Ivorra, 1990a).

3.3.4. Spatial Localization

A great advantage of using light as a stimulus is the ease with which it can be focused to any desired shape and location, thus permitting spatially localized release of intracellular messengers. One application is to "map" the spatial sensitivity of the oocyte to intracellular $InsP_3$. The photolysis light is focused on the oocyte surface as a small (ca. 50 μM) spot, and membrane currents are recorded in response to identical flashes delivered with the spot moved to various positions across the cell. (Pigmentation in the animal hemisphere of normal oocytes complicates this experiment, but this problem can be avoided by the use of oocytes from albino frogs.) Another example is the measurement of the localization of Ca^{2+} liberation induced by a spatially restricted photorelease of $InsP_3$. The optical system described above allows the light spots used for photolysis and fluorescence Ca^{2+} monitoring to be positioned independently. Thus, $InsP_3$ can be released at one area on the oocyte and

Ca^{2+} signals recorded from sites at various distances (Parker and Ivorra, 1990b).

Acknowledgments

I would like to thank Isabel Ivorra for excellent help with many of the experiments described here. Work in the author's laboratory is supported by grants GM39831 and NS23284 from the NIH.

Abbreviations

cAMP	Adenosine 3',5' cyclic monophosphate
cGMP	Guanosine 3',5' cyclic monophosphate
GTP	Guanosine 5'-triphosphate
GTPγS	Guanosine 5'-[γ-thio]triphosphate
InsP_2	Inositol bisphosphate (isomer specified in text)
InsP_3	Inositol trisphosphate (isomer specified in text)

References

Adams S. R., Kao J. P. Y., Grynkiewicz G., Minta A., and Tsien R. Y. (1988) Biologically useful chelators that release Ca^{2+} upon illumination. *J. Am. Chem. Soc.* **110**, 3212–3220.

Ahnert-Hilger G., Mach W., Fohr K., and Gratzl M. (1989) Poration by alpha-toxin and streptolysin O: An approach to analyze intracellular processes. *Methods in Cell Biol.* **31**, 63–90.

Changya L., Gallacher D. V., Irvine R. F., Potter B. L. V., and Petersen O. H. (1989) Inositol 1,3,4,5-tetrakisphosphate is essential for sustained activation of the Ca^{2+}-dependent K$^+$ current in single internally perfused mouse lacrymal cells. *J. Membr. Biol.* **109**, 85–93.

Denk W., Strickler J. H., and Webb W. W. (1990) Two-photon laser scanning fluorescence microscopy. *Science* **248**, 73–76.

Dolphin A. C., Wooton J. F., Scott R. H., and Trentham D. R. (1988) Photoactivation of intracellular guanosine trisphosphate analogues reduces the amplitude and slows the kinetics of voltage-activated calcium channel currents in sensory neurones. *Pflugers Arch.* **411**, 628–636.

Goldman Y. (1986) in *Optical Methods in Cell Physiology* (DeWeer P. and Salzburg, B. M., eds.), Wiley-Interscience, New York, pp. 397–415.

Goldman Y. E., Hibberd M. G., and Trentham D. R. (1984) Relaxation of rabbit psoas muscle fibres from rigor by photochemical generation of adenosine-5'-triphosphate. *J. Physiol. (Lond.)* **354,** 577–604.

Gray P. T. A., Ogden D. C., Trentham D. R., and Walker J. W. (1988) Elevation of cytosolic free calcium by rapid photolysis of 'caged' inositol trisphosphate in single rat parotid acinar cells. *J. Physiol. (Lond.)* **405,** 84P.

Grynkiewicz G., Poenie M. J., and Tsien R. Y. (1985) A new generation of Ca^{2+} indicators with greatly improved fluorescence properties. *J. Biol. Chem.* **260,** 3440–3450.

Gurney A. M. and Lester H. A. (1987) Light-flash physiology with synthetic photosensitive compounds. *Physiol. Rev.* **67,** 583–617.

Homsher E. and Miller N. C. (1990) Caged compounds and striated muscle contraction. *Ann. Rev. Physiol.* **52,** 875–896.

Ivorra I. and Parker I. (1990a) Simultaneous recording of membrane currents and intracellular free Ca^{2+} transients evoked by photolysis of caged compounds in *Xenopus* oocytes. *J. Physiol. (Lond.)* **424,** 31P.

Ivorra I. and Parker I. (1990b) Intracellular Ca^{2+} transients evoked by inositol trisphosphate in *Xenopus* oocytes show a threshold and facilitation. *J. Physiol. (Lond.)* **424,** 32P.

Janko K. and Reichert J. (1987) Proton concentration jumps and generation of transmembrane pH-gradients by photolysis of 4-formyl-6-methoxy-3-nitrophenoxyacetic acid. *Biochim. Biophys. Acta* **905,** 409–416.

Kao J. P. Y., Harootunian A. T., and Tsien R. Y. (1989) Photochemically generated cytosolic calcium pulses and their detection by fluo-3. *J. Biol. Chem.* **264,** 8179–8184.

Kaplan J. H. (1990) Photochemical manipulation of divalent cation levels. *Ann. Rev. Physiol.* **52,** 897–914.

Kaplan J. H. and Ellis-Davies G. C. R. (1988) Photolabile chelators for the rapid photorelease of divalent cations. *Proc. Natl. Acad. Sci USA* **85,** 6571–6575.

Kaplan J. H. and Somlyo A. P. (1989) Flash photolysis of caged compounds: New tools for cellular physiology. *Trends Neurosci.* **12,** 54–59.

Kaplan J. H., Forbrush B. III, and Hoffman J. F. (1978) Rapid photolytic release of adenosine 5'-trisphosphate from a protected analogue: Utilization by the Na:K pump of human red cell ghosts. *Biochemistry* **17,** 1929–1935.

Karpen J. W., Zimmerman A. L., Stryer L., and Baylor D. A. (1988) Gating kinetics of the cyclic-GMP-activated channel of retinal rods: Flash photolysis and voltage-jump studies. *Proc. Natl. Acad. Sci. USA* **85,** 1287–1291.

McCray J. A. and Trentham D. R. (1989) Properties and uses of photoreactive caged compounds. *Ann. Rev. Biophys. Chem.* **18,** 239–270.

Meyer T. and Stryer L. (1990) Transient calcium release induced by successive increments of inositol 1,4,5-trisphosphate. *Proc. Natl Acad. Sci. USA* **87,** 3841–3845.

Miledi R. and Parker I. (1989) Latencies of membrane currents evoked in *Xenopus* oocytes by receptor activation, inositol trisphosphate and calcium. *J. Physiol. (Lond.)* **415**, 189–210.

Minta A., Kao J. P. Y., and Tsien R. Y. (1989) Fluorescent indicators for cytosolic calcium based on rhodamine and fluoresceine chromophore *J. Biol. Chem.* **264**, 8171–8178.

Ogden D. C., Capiod T., Walker J. W., and Trentham D. R. (1990) Kinetics of the conductance evoked by noradrenaline inositol trisphosphate or Ca^{2+} in guinea-pig isolated hepatocytes. *J. Physiol. (Lond.)* **422**, 585–602.

Oron Y., Dascal N., Nadler E., and Lupu M. (1985) Inositol 1,4,5-trisphosphate mimics muscarinic response in *Xenopus* oocytes. *Nature (London)* **313**, 141–143.

Parker I. (1989a) Ionic and charge-displacement currents evoked by temperature jumps in *Xenopus* oocytes. *Proc. R. Soc. Lond. [Biol.]* **237**, 379–387.

Parker I. (1989b) Latency threshold and facilitation in phosphoinositide signalling, in *Membrane Technology. Serono Symposia* vol. 64 (Verna, R., ed.), Raven, New York, pp. 39–56.

Parker I. and Ivorra I. (1990a) Inhibition by Ca^{2+} of inositol trisphosphate-mediated Ca^{2+} liberation: A possible mechanism for oscillatory release of Ca^{2+}. *Proc. Natl. Acad. Sci. USA* **87**, 260–264.

Parker I. and Ivorra I. (1990b) Localized all-or-nothing Ca^{2+} liberation by inositol trisphosphate. *Science* **250**, 977–979.

Parker I. and Miledi R. (1986) Changes in intracellular calcium and in membrane currents evoked by injection of inositol trisphosphate into *Xenopus* oocytes. *Proc. R. Soc. Lond. [Biol.]* **228**, 307–315.

Parker I. and Miledi R. (1987) Inositol trisphosphate activates a voltage-dependent calcium influx in *Xenopus* oocytes. *Proc. R. Soc Lond. [Biol.]* **231**, 27–36.

Parker I. and Miledi R. (1989) Nonlinearity and facilitation in phosphoinositide signalling studied by the use of caged inositol trisphosphate in *Xenopus* oocytes. *J. Neurosci.* **9**, 4068–4077.

Rapp G. and Guth K. (1988) A low cost high intensity flash device for photolysis experiments. *Pflugers Arch.* **411**, 200–203.

Somlyo A. P. and Somlyo A. V. (1990) Flash photolysis studies of excitation-contraction coupling, regulation, and contraction in smooth muscle. *Ann. Rev. Physiol.* **52**, 857–874.

Sumikawa K., Parker I., and Miledi R. (1989) Expression of neurotransmitter receptors and voltage-operated channels from brain mRNA in *Xenopus* oocytes. *Methods Neurosci.* **1**, 30–45.

Walker J. W., Feeney J., and Trentham D. R. (1989) Photolabile precursors of inositol phosphates. Preparation and properties of 1(2-nitrophenyl) ethyl esters of *myo*-inositol 1,4,5-trisphosphate. *Biochemistry* **28**, 3272–3280.

Walker J. W., McCray J. A., and Hess G. P. (1986) Photolabile protecting groups for an acetylcholine receptor ligand. Synthesis and photochemistry of a new class of *o*-nitrobenzyl derivatives and their effects on receptor function. *Biochemistry* **25**, 1799–1805.

Walker J. W., Reid G. P., McCray J. A., and Trentham D. R. (1988) Photolabile 1-(2-nitrophenyl)ethyl phosphate esters of adenine nucleotide analogues. Synthesis and mechanism of photolysis. *J. Am. Chem. Soc.* **110**, 7170–7177.

Walker J. W., Somlyo A. V., Goldman Y. E., Somlyo A. P., and Trentham D. R. (1987) Kinetics of smooth and skeletal muscle activation by laser pulse photolysis of caged inositol 1,4,5-trisphosphate. *Nature (London)* **327**, 249–252.

Protein Phosphorylation

David B. Morton

1. Introduction

It is now well established that protein phosphorylation is one of the most common forms of posttranslational modification of proteins and that it is of major importance in regulating the activity of proteins (Krebs and Beavo, 1979; Wold, 1983). Furthermore, protein phosphorylation regulates many aspects of cell physiology (e.g., Cohen, 1982). In 1976, Paul Greengard first described a model for explaining the molecular basis by which various neurotransmitters act via the intracellular messenger, cAMP. (Greengard, 1976). He proposed that these effects were mediated by the action of cAMP-dependent protein kinase (A-kinase), which catalyzed the phosphorylation of various proteins in the postsynaptic membrane. Furthermore, he also described how presynaptic effects could be regulated by the action of cAMP and A-kinase (Greengard, 1976). Since then, many studies have revealed the types of proteins phosphorylated in the nervous system, not by only A-kinase, but by other kinases, and how protein phosphorylation may regulate and modulate various aspects of neuronal function (Browning et al., 1985; Nairn et al., 1985b; Greengard, 1987).

The aim of this chapter is to give a brief summary of the types of kinases believed to be important in the nervous system and how their activity is regulated. Details of the methods used to identify which kinases are involved in a particular system will be described and also how the proteins that are phosphorylated by those kinases can be identified.

From: *Neuromethods, Vol. 20: Intracellular Messengers*
Eds: A. Boulton, G. Baker, and C. Taylor © 1992 The Humana Press Inc.

2. Assays for Protein Kinases and Their Substrates

2.1. Cyclic Nucleotide-Dependent Protein Kinases

There are three identified protein kinases in nervous tissue, whose activity is dependent on the presence of either adenosine 3',5' cyclic monophosphate (cAMP) or guanosine 3',5' cyclic monophosphate (cGMP). Two forms of kinase are preferentially activated by cAMP, cAMP-dependent protein kinase (A-kinase) types I and II, and one type is preferentially activated by cGMP, cGMP-dependent protein kinase (G-kinase). All three forms have been sequenced and show considerable sequence homology (Taylor, 1989). Both types of A-kinase have four subunits: Two identical regulatory subunits (R), which bind cAMP, and two indentical catalytic subunits (C), which catalyze the phosphorylation of substrate phosphoproteins (Taylor, 1989). Type I and type II A-kinase share the same C subunits, but have different R subunits, RI and RII, respectively. When the R and C subunits are associated, the C subunit is inactive. Upon binding of cAMP to the R subunit, all the subunits dissociate, producing the active, monomeric C subunit. The reaction has the following stoichiometry:

$$\underset{\text{inactive}}{R_2C_2 + 4cAMP} \quad \longleftrightarrow \quad \underset{\text{active}}{R-cAMP_2 + 2C} \qquad (1)$$

G-kinase is composed of two identical subunits, each of which contains both the cGMP binding site and the catalytic site. The regulatory region of the subunit has considerable sequence homology with both RI and RII subunits of A-kinase, and the catalytic region has homology with the C subunit of A-kinase. Upon binding cGMP, a conformational change takes place that activates the catalytic site. The reaction has the same stoichiometry for cyclic nucleotide as does A-kinase:

$$\underset{\text{inactive}}{E_2 + 4cGMP} \quad \longleftrightarrow \quad \underset{\text{active}}{E_2 \cdot cGMP_4} \qquad (2)$$

Both A- and G-kinases have similar, but not identical, protein substrate specificities and use ATP as the phosphate donor. The consensus sequence for the phosphorylation site of the substrate protein is similar for both kinases and is usually on a serine residue, with basic residues two amino acids toward the amino terminal from the serine (Beavo and Mumby, 1982). Both enzymes require Mg^{2+} for activity, although the optimal concentration is different for the two enzymes. This can be used to separate their activities, since A-kinase is inhibited by high Mg^{2+} concentrations whereas G-kinase is not.

2.1.1. Assays for the Kinase

The simplest assay for kinase activity is to measure the incorporation of labeled phosphate from $[\gamma^{32}P]$-ATP to a suitable protein. The phosphoprotein is then separated from unreacted ATP by precipitation with trichloroacetic acid (TCA). The TCA-soluble and TCA-insoluble fractions can either be separated by spotting the solution onto filter paper disks, followed by extensive washing, or by collecting the precipitated protein by centri-fugation in 1.5-mL centrifuge tubes, again followed by extensive washing. The filter paper method is commonly used and is espe-cially useful when dealing with large numbers of samples. I have found, however, that it tends to give higher blanks (no kinase) than the centrifugation method.

A number of different proteins can be used as the substrates for phosphorylation, but the most common is histone, usually histone 2b (type VII, Sigma, St. Louis, MO). The enzyme source will depend on the experiment. Crude homogenates, partially purified preparations that separate A- and G-kinases, or pure enzymes can all be used. Purification of both A- and G-kinases is easily accomplished by published protocols (Reimann and Beham, 1983; Rannels et al., 1983; Lincoln, 1983).

The reaction is conveniently carried out in 1.5-mL centri-fuge tubes with the composition of the reaction mixture shown in Table 1. The tubes are held on ice, the reaction started by the addition of labeled ATP, and the tubes placed in a 30°C water bath. After 10 min, 1 mL of 10% TCA is added plus 100 µL bovine serum albumin (BSA; 1 mg/mL) as a carrier and the tube

Table 1
Reaction Mixture Composition for Assaying
the Activity of Cyclic Nucleotide-Dependent Protein Kinases

	μL	
	A-kinase	G-kinase
H_2O	20	10
Phosphate buffer[a]	5	5
0.1M $MgCl_2$	5	15
H_2O or cAMP	5	–
H_2O or cGMP	–	5
Kinase or enzyme source[b]	5	5
Histone (10 mg/mL)	5	5
[γ ^{32}P]-ATP, 5 mM, 0.1 mCi/mL	5	5

[a]The phosphate buffer consists of 0.3M sodium phosphate, pH 6.8, containing 20 mM dithiothreitol.
[b]The kinase or enzyme source should be 2–40 U/mL or 0.2–1 mg protein/mL.

left on ice for 15 min. After centrifugation at 10,000g for 5 min, the supernatant is removed, the pellet redissolved in 100 μL 1M NaOH, and the protein reprecipitated with 1 mL 10% TCA. These steps are repeated twice, the pellet dissolved in 100 μL 1M NaOH, and the radioactivity counted. A small aliquot of the [^{32}P]-ATP is saved to determine the exact specific activity of the ATP used. From this value, the specific activity of the kinase can be calculated and expressed as picomoles of phosphate transferred/min/mg protein. Typical results are shown in Table 2.

The A-kinase inhibitor is recommended to distinguish between A- and G-kinases (Whitehouse and Walsh, 1983). This heat- and acid-stable protein specifically inhibits A-kinase, but not G-kinase, and is commercially available (Sigma). A specific modulator of G-kinase activity has also been reported, which activates G-kinase when histone is used as a substrate; it is not commercially available, but can be easily purified (Kuo and Kuo, 1976). Contaminants of the enzyme can interfere with the assay, but ATPase activity can be inhibited with 40 mM fluoride; 0.25 mM EDTA will chelate Ca^{2+}, and 1 mM isobutyl methyl xanthine will inhibit phosphodiesterase.

Table 2
Example of the Results Obtained for Assaying A-Kinase and G-Kinase
Prepared from the CNS of the Tobacco Hornworm, *Manduca sexta*

	Activity	Fold stimulation
A-kinase[a]		
No addition	127 ± 5	–
$10^{-8} M$ cAMP	246 ± 7	1.9
$10^{-7} M$ cAMP	319 ± 31	2.5
$10^{-8} M$ cGMP	153 ± 5	1.2
$10^{-7} M$ cGMP	226 ± 8	1.8
G-kinase[a]		
No addition	26 ± 2	–
$10^{-8} M$ cGMP	37 ± 4	1.4
$10^{-7} M$ cGMP	44 ± 6	1.7
$10^{-8} M$ cAMP	22 ± 2	1.0
$10^{-7} M$ cAMP	32 ± 1	1.2

[a]All kinase activities are given as pmol phosphate transferred/min/mg protein.
The kinases were purified according to the methods of Kuo and Greengard (1970)
and Kuo et al. (1971).

An alternative method for assessing the levels of kinase is
to measure the binding of cAMP or cGMP to their respective
kinases. It must be remembered, however, that, since cAMP only
binds to the R subunit of A-kinase and the R and C subunits
exist separately or bound, cAMP binding need not reflect the
true level of A-kinase activity.

The reaction mixtures for such binding assays are given in
Table 3. The reactions are best carried out in 13 × 100 mm glass
tubes. The reaction is started by addition of the enzyme and
incubated at room temperature for 5 min. The reaction is stopped
by addition of 5 mL ice-cold 30 mM phosphate buffer (pH 6.8)
and 10 mM MgCl$_2$, filtered through Millipore GF/C filters. The
filters are washed twice with 5 mL of the buffer, and the amount
of bound nucleotide is determined by scintillation counting of
the filters. To prevent binding of [^{3}H]-cAMP to G-kinase, 0.1 μM
unlabeled cGMP can be added; conversely, unlabeled cAMP (0.1
μM) will prevent [^{3}H]-cGMP binding to A-kinase.

Photoaffinity labels can also be used to study cyclic nucleotide-binding sites of kinases directly (Walter and Greengard,
1983). [^{32}P]-labeled photoaffinity analogs for cGMP and cAMP

Table 3
Reaction Mixture Composition for Assaying the Cyclic Nucleotide
Binding Capacity of Cyclic Nucleotide-Dependent Protein Kinases

	μL	
	A-kinase	G-kinase
H_2O	20	20
Phosphate buffer[a]	5	5
$0.1M$ $MgCl_2$	5	5
[3H]-cAMP, 2 μM, 10–30 Ci/mmol	10	–
[3H]-cGMP, 2 μM, 10–30 Ci/mmol	–	10
BSA, 10 mg/mL	5	5
Kinase or enzyme source[b]	5	5

[a]The phosphate buffer consists of $0.3M$ sodium phosphate, pH 6.8, containing $0.02M$ dithiothreitol.
[b]The kinase or enzyme source should be 2–40 U/mL or 0.2–1 mg protein/mL.

are commercially available (ICN, Costa Mesa, CA) and can be used to label G-kinase and the R subunit of A-kinase covalently. Tissue homogenates are incubated with the ligand and photolyzed; the proteins are then separated by SDS-PAGE, and the binding proteins identified by autoradiography. Alternatively, [32P]-labeled photoaffinity analogs of ATP (ICN) can be used to label the ATP-binding sites of all the kinases (Taylor et al., 1983).

2.1.2. Identification of Substrate Phosphoproteins

There are two major approaches to examining potential substrate phosphoproteins. One examines protein phosphorylations in vitro in response to particular intracellular signals, such as cyclic nucleotides, Ca^{2+}, or diacylglycerol. The second approach is to examine protein phosphorylation in intact cells in response to extracellular signals, such as neurotransmitters, hormones, growth factors, or the physiological stimuli. A combined approach is usually necessary. This section will concentrate on the methods used for an extracellular signal that acts via cyclic nucleotides, but the principles will apply to other systems, and any special features will be dealt with in the appropriate section.

2.1.2.1. CELL HOMOGENATES

The methods given here can be applied to studies looking for an unknown phosphoprotein or for those of a previously identified protein to see if it is phosphorylated by a given intracellular messenger and/or kinase. When investigating unknown proteins, a tissue homogenate is made, endogenous protein kinase(s) is (are) activated, and proteins in the homogenate are phosphorylated with [γ-^{32}P]-ATP. When studying a particular protein, however, one might be dealing with partially purified proteins or proteins purified to homogeneity. In this case, an exogenous kinase is needed, and except for A-kinase, there is, at present, no commercial source of protein kinases. There are, however, published protocols for purifying or partially purifying most of the common protein kinases (Corbin and Hardman, 1983).

The following reaction mixture can be used for phosphorylating proteins in a tissue homogenate using endogenous cyclic nucleotide-dependent protein kinases. The tissue should be homogenized in 50 mM HEPES buffer (pH 7.0), 1 mM dithiothreitol (DTT), and 5 mM EDTA. The final protein concentration should be adjusted to 0.5–1 mg/mL. An aliquot (100 µL) is incubated with 40 µL 250 mM HEPES buffer (pH 7.0), 20 µL 100 mM MgCl$_2$, 20 µL 1 mM cAMP or cGMP, and 20 µL 10 µM ATP containing 10 µCi [γ-^{32}P]-ATP. The reaction is started by addition of ATP and is incubated for 5 min at 30°C. An equal vol of SDS sample buffer (0.125 M Tris-HCl, pH 6.8, 20% glycerol, 4.6% SDS, 10% 2-mercaptoethanol [2-ME], and 0.05% bromophenol blue) is added to stop the reaction, and the mixture heated at 95°C for 5 min. An aliquot of this mixture can be separated by SDS-PAGE, and the radioactive proteins identified by autoradiography or fluorography using standard protocols (Hames and Rickwood, 1981). If the amount of labeled proteins in the sample is low, the proteins should be concentrated by lyophilization or precipitation (with TCA or ethanol) before solubilization in SDS sample buffer.

Autoradiographs from incubations with and without cyclic nucleotides are compared to find proteins whose phosphorylation is stimulated by the addition of cyclic nucleotide. The bands can be cut from the gel and the radioactivity measured directly by scintillation counting. Alternatively, there are a number of densitometers on the market, all of which have extensive software for quantifying the density of images on autoradiographs. However, the relationship between image density and amount of radioactivity is not linear at all densities. A typical fluorogram made from an SDS gel is shown in Fig. 1. The samples were prepared as described above, using CNS homogenates from the tobacco hornworm, *Manduca sexta*.

The addition of exogenous kinase has been used in the author's laboratory to aid in the partial purification of the eclosion hormone- and cGMP-regulated phosphoproteins (EGPs; Morton and Truman, 1988). Although the EGPs are phosphorylated in vivo through the action of G-kinase, they can be phosphorylated in vitro by A-kinase (Morton and Truman, unpublished results). This has enabled us to use commercially available A-kinase to monitor purification of the EGPs. The reaction mixture consists of 100 µL protein fraction (0.5 mg/mL), usually in 10 mM phosphate buffer, pH 7.4, 20 µL catalytic subunit of A-kinase (500 U/mL; Sigma), 20 µL 100 mM $MgCl_2$, 40 µL 250 mM HEPES buffer, pH 7.0, and 20 µL 10 µM ATP containing 10 µCi [γ-^{32}P]-ATP. The reaction is initiated by addition of the ATP, incubated at 30°C for 5 min, and stopped by addition of 20 µL 100% TCA. After centrifugation, the proteins are dissolved in sample buffer and separated by SDS-PAGE.

2.1.2.2. Intact Cells

There are two approaches to identify the proteins phosphorylated in response to extracellular signals on intact cells. The direct approach is to preload the cells or tissue with [^{32}P]-H_3PO_4, which will be taken up by cells and converted to [γ-^{32}P]-ATP.

The degree to which cells or tissue can be preloaded with [^{32}P]-H_3PO_4 varies considerably with species and tissue. It is important to note, however, that the incubation medium should be free of protein, since this may bind the [^{32}P] and interfere with the electrophoresis step, and the media should be free of phosphate to prevent dilution of the [^{32}P]-specific activity. The following incubation conditions are a few examples from the literature: *Aplysia* bag cells, 0.25 mCi/mL, 22–24 h at 14°C (Jennings et al., 1982), lobster opener muscle, 0.25 mCi/mL, 4 h at 12°C (Goy et al., 1984), locust CNS, 0.08 mCi/mL, 3 h at 30°C (Rotondo et al., 1987), *Limulus* eyes and optic nerve, 2.3 mCi/

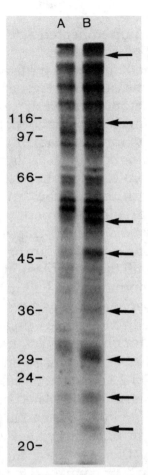

Fig. 1. Cyclic GMP-stimulated phosphorylation of proteins in the CNS of *Manduca*. Nervous tissue from *Manduca* was homogenized, and the proteins phosphorylated with $[\gamma\text{-}^{32}P]$-ATP as described in the text. The proteins were separated by SDS-PAGE, and the radioactive proteins visualized with fluorography. Lane A is from a sample phosphorylated in the absence of cGMP, and lane B is from an identical sample phosphorylated in the presence of 0.1 m*M* cGMP. The numbers to the left of the gel represent the positions of various mol wt markers (in kDa), and the arrows to the right show the positions of proteins whose phosphorylation is increased in the presence of cGMP.

mL, 1.5 h at 22°C (Edwards and Battelle, 1987), bovine chromaffin cells, 0.5 mCi/mL, 60 min at 37°C (Haycock et al., 1988). After the initial incubation, the hormone, neurotransmitter, or stimulating agent is applied and incubation continued for the appropriate amount of time. The cells or tissue can then be solubilized in SDS sample buffer, heated at 95°C for 5 min, and the proteins separated by SDS-PAGE.

An alternative methodology is the technique of "back phosphorylation," which measures the level of the dephospho form of the substrate phosphoprotein (Nestler and Greengard, 1982). Thus, if a treatment stimulates phosphorylation of a protein, there will be less dephospho form available for subsequent labeling with ^{32}P phosphorylating conditions. This method is particularly useful for looking at protein phosphorylation correlated with a physiological event, since it can be carried out on the intact animal, cells, or tissue in culture or in a physiological preparation.

The animal or tissue is stimulated appropriately—for example, by adding hormone to the medium or stimulating a particular nerve—and left until the physiological effect has just begun to take place or the hormone has had its effect. At this point, proteins that are phosphorylated as a result of the stimulation are phosphorylated with endogenous, unlabeled phosphate and, thus, are unavailable for phosphorylation with exogenous [^{32}P]-ATP. The tissue is then homogenized and phosphorylated as described in Section 2.1.2.1., and the proteins separated by SDS-PAGE. Control tissue is left unstimulated, and homogenized and phosphorylated with [γ-^{32}P]-ATP in parallel with the experimental tissue. Any proteins phosphorylated with unlabeled phosphate, as a result of the physiological stimulus, will then incorporate little, if any, ^{32}P compared to the controls. This procedure is shown schematically in Fig. 2. The method was used to identify the proteins phosphorylated by the action of the neuropeptide eclosion hormone on the CNS of *Manduca sexta* (Morton and Truman, 1986, 1988).

One potential problem with this method is that, after homogenization, any active phosphatases present will dephosphorylate any phosphoproteins present, and thus, no differences will be seen between experimentals and control (*see* Section 4. for specific inhibitors of phosphoprotein phosphatases). If this is a problem, the proteins can first be separated by SDS-PAGE and transferred to nitrocellulose. The phosphorylation can then be carried out directly on the nitrocellulose itself (Valtorta et al., 1986). Briefly, this is done as follows. Immediately after the in

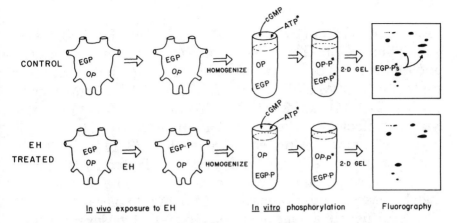

CONTROL

EH
TREATED

In <u>vivo</u> exposure to EH   In <u>vitro</u> phosphorylation   Fluorography

Fig. 2. Diagrammatical representation of a "back phosphorylation" experiment, showing that the neuropeptide, eclosion hormone (EH), stimulates
the phosphorylation of the EGPs in the CNS of *Manduca*. The nervous system of *Manduca* contains proteins whose phosphorylation is regulated by
cGMP and EH (the EGPs; Morton and Truman, 1988). In addition, there are
other proteins that are phosphorylated in the presence or absence of cGMP,
but are unaffected by EH (OPs). Back phosphorylation can be used to distinguish the EGPs from the OPs. Untreated nervous tissue (Control) is homogenized and phosphorylated in the presence of cGMP and [γ-^{32}P]-ATP (ATP*).
After the proteins have been separated by 2D SDS-PAGE, fluorography
reveals the EGPs and OPs as labeled proteins (EGP-P* and OP-P*). If the
nervous tissue is exposed to EH (EH-treated), however, the EGPs are phosphorylated with endogenous, unlabeled phosphate (EGP-P), whereas the
OPs remain dephosphorylated. After homogenization, the EGPs are unavailable for phosphorylation with [γ-^{32}P]-ATP and remain unlabeled, whereas
the OPs are phosphorylated with ^{32}P (OP-P*), in an identical manner to the
control sample. After 2D SDS-PAGE and fluorography, fluorograms from
the EH-treated sample will be identical to the one from the control sample,
with the exception that the EGPs will be unlabeled.

In some situations, the proteins stimulated by the extracellular signal will
not be completely phosphorylated. This will lead to only a reduction and
not a complete inhibition of incorporation of label, when compared to controls and OPs.

vivo stimulation is complete, the tissue sample is homogenized
in SDS sample buffer. This will immediately denature any phosphatases and "freeze" any proteins in their present phosphorylation state. The proteins are separated by SDS-PAGE and
transferred to nitrocellulose by standard protocols (Towbin et
al., 1979). The strips of nitrocellulose are then blocked for 1 h in

50 mM Tris-HCl, pH 7.4, 200 mM NaCl, 0.4% Ficoll 400, and 0.1% Triton X-100. The proteins on the nitrocellulose are then phosphorylated by incubating the strips for 1 h in a solution containing 50 mM HEPES, pH 7.4, 25 mM NaCl, 10 mM MgCl$_2$, 1 mM EGTA, 0.1 mM 2-ME, 0.1% Triton X-100, 0.1–1 μM catalytic subunit of A-kinase (Sigma), and 10 nM–10 μM [γ^{32}P]-ATP (5–600 μCi/nmol). After extensive washing (3–24 h) in 50 mM Tris-HCl, pH 7.4, 200 mM NaCl, and 0.1% Tween 20, the nitrocellulose is dried and exposed to X-ray film. The proteins can then be stained with Coomassie blue or amido black.

A further advantage of this method is that the level of endogenously phosphorylated substrate proteins (unlabeled) can be verified by treating the homogenate with exogenous protein phophatase (alkaline phosphatase works well) before separating by SDS-PAGE and transferring to nitrocellulose. After phosphorylation on the nitrocellulose, the amount of label incorporated will represent the maximum amount that can be incorporated into the protein and can be compared with incorporation after various treatments of the tissue. This method has been used successfully to show that stimulation of the efferent nerve to *Limulus* lateral eyes stimulates the phosphorylation of a 122-kDa protein (Edwards et al., 1990).

2.2. Ca^{2+}-Dependent Protein Kinases

Most of the methodology for studying the presence or activity of Ca^{2+}-dependent kinases is very similar to that used for cyclic nucleotide-dependent kinases. There are two major classes of Ca^{2+}-dependent protein kinases: those requiring phospholipid, 1,2-diacylglycerol (DG), and Ca^{2+}, and those requiring calmodulin and Ca^{2+}. The former are termed protein kinases C (C-kinases) and the latter Ca^{2+}/calmodulin-dependent protein kinases (CaM kinases).

2.2.1. Protein Kinase C

There are at least seven different subspecies of C-kinase that have been identified in mammalian tissues by probing cDNA libraries (Nishizuka, 1988). They show a high degree of sequence homology and can be divided into two main groups based on their structure: groups A and B (Nishizuka, 1988). All subspecies exist as a single polypeptide chain with a calculated mol wt of 70–80 kDa. All C-kinases require phospholipid, usually phosphatidylserine (PtdSer), and DG for activity, and most, but not all, also require Ca^{2+}. The various subspecies have slightly different affinities for these compounds, and their requirements in vitro vary according to the phosphate acceptor used.

Most studies investigating the mechanism of action of C-kinase have used the group A enzymes, all of which have an absolute requirement for PtdSer, and require DG and Ca^{2+} for maximal activation. Kinetic analysis of activation shows that DG dramatically increases the affinity of C-kinase for Ca^{2+}, such that maximal activity is obtained at physiological concentrations of Ca^{2+} (Kikkawa and Nishizuka, 1986). Therefore, full activation, in vivo, can be achieved without an increase in Ca^{2+}. This has led to the hypothesis that physiological activation of C-kinase follows the liberation of DG from phosphoinositides after agonist-activation of phosphoinositidase C (Kikkawa and Nishizuka, 1986). More recent evidence, however, has shown that the mechanism of activation is more complicated, and may involve translocation of the enzyme from the cytoplasm to the plasma membrane and additional sources of DG (*see* Huang, 1989).

The presence and activity of C-kinase in various tissues or protein preparations can be easily measured in an assay similar to that described for cyclic nucleotide-dependent kinases. The reaction components are shown in Table 4. Suitable substrate proteins are lysine-rich histone (type III, Sigma) or C-kinase itself, which is autophosphorylated. The reaction is started by addition of the label, incubated for 2 min at 30°C, and stopped by adding 20 µL 100% TCA. The labeled protein is then separated from unreacted label, as described above, with sequential TCA precipitation and washing.

The same reaction mixture can also be used successfully for studying the endogenous proteins that are phosphorylated, in homogenates, by the action of C-kinase. The reaction is carried out without the addition of exogenous protein. After the reaction is stopped, the labeled proteins are dissolved in SDS sample buffer and separated by SDS-PAGE using standard protocols (Hames and Rickwood, 1981).

Table 4
Reaction Mixture Composition for Assaying
the Activity of Calcium-Dependent Protein Kinases

	μL	
	C-kinase	Ca²⁺/CaM kinase
Kinase or enzyme source	50ᶜ	50
Tris buffer^a	30	30
EDTA or Ca²⁺/CaMᵇ	–	10
EDTA or Ca²⁺/PtdSer/DGᵇ	10	–
[γ ³²P]-ATP, 0.5 mM, 1 mCi/mL	10	10

^aThe tris buffer consists of 30 mM Tris-HCl, pH 7.5, 30 mM MgCl$_2$, and 3 mM dithiothreitol.

ᵇThe concentration of EDTA or Ca²⁺, when used, is 10 mM. Calmodulin (CaM) is used at 100 μg/mL, phosphatidylserine (PtdSer) at 500 μg/mL, and diacylglycerol (DG) at 50 μg/mL.

ᶜWhen the sample is not going to be separated by PAGE, the enzyme source can be substituted for 25 μL of enzyme plus 25 μL 1 mg/mL histone (type III, Sigma).

One of the principal reasons for the great interest in the study of C-kinase over recent years was the discovery that the tumor-promoting phorbol esters directly activate C-kinase in intact cells (Kikkawa et al., 1984). Thus, a phorbol ester, such as 12-*O*-tetradecanoyl phorbol-13-acetate, can be used as a powerful tool to stimulate the phosphorylation of C-kinase substrates in intact cells. It can also be used to stimulate physiological processes that are assumed to be mediated by C-kinase. For further details of methodology related to C-kinase, *see* the chapter by Burgess in this vol.

2.2.2. Ca²⁺/Calmodulim-Dependent Protein Kinases

The second group of Ca²⁺-dependent protein kinases is a ubiquitous and diverse group of enzymes, all of which require Ca²⁺ and the Ca²⁺-binding protein, calmodulin, for maximal activity. There are four major Ca²⁺/calmodulin-dependent protein kinases (CaM kinases) found in the mammalian brain: CaM kinases I and II, myosin light-chain kinase, and phosphorylase kinase (Nairn et al., 1985b). In addition, a number of other distinct kinases that also require Ca²⁺ and calmodulin for activity have been identified in the brain: CaM kinase III (Nairn et al., 1985a), hydroxymethylglutaryl CoA reductase kinase (Beg et al., 1987), and a CaM kinase enriched in cerebellar granule cells (Ohmstede et al., 1989). It is likely that yet more distinct enzymes will be discovered. Most of these enzymes appear to be unrelated, except in their requirement for Ca^{2+} and calmodulin for activity. Where known, sequence homologies tend to be only in the catalytic and calmodulin-binding domains. Often the only functional difference among the different CaM kinases is their substrate specificity. CaM kinase II has a very broad range of substrates, whereas types I and III have a much narrower range. Myosin light-chain kinases and phosphorylase kinase are very specific phosphorylating enzymes (Nairn et al., 1985b).

The most extensively studied enzyme in this group is CaM kinase II; because of its broad substrate specificity, it was identified independently in a number of different laboratories. It has been implicated in a variety of neuronal processes and is found in a wide variety of organisms (Colbran et al., 1989). Generally, CaM kinase II is a heteropolymer, with a native mol wt of 550–650 kDa. It consists of two or three similar subunits, with mol wt of 50–62 kDa; the numbers of each subunit vary with species, tissue, and developmental age (Colbran et al., 1989). The different subunit compositions presumably determine the slightly different subcellular distributions and physiological functions of the enzymes.

Although CaM kinase II requires Ca^{2+} and calmodulin for maximal activity, it is rapidly autophosphorylated, which renders it partially Ca^{2+}-independent (Colbran et al., 1989). Thus, the enzyme will remain active for some time after a transiently elevated Ca^{2+} concentration has returned to basal levels. This has been shown to occur in vivo (Fukunaga et al., 1989) and has been implicated as a mechanism for synaptic plasticity, such as long-term potentiation (Schwartz and Greenberg, 1987; Kennedy, 1988).

To assay CaM kinase activity or to study endogenous substrates in a particular tissue, a broadly similar procedure to those described above can be used. The principal difference is that the assay is carried out in the presence of Ca^{2+} and calmodulin, and that there is no one particular exogenous substrate that can be used. Many studies have employed synapsin I as a substrate to aid in the purification of CaM kinases or to measure activity levels (Kennedy and Greengard, 1981; DeRiemer et al., 1984; Ohmstede et al., 1989). Synapsin I, however, is not commercially available and must be purified in the lab (Ueda and Greengard, 1977).

A suitable reaction mixture for the study of CaM kinases is shown in Table 4. The reaction is started with the addition of [γ-^{32}P]-ATP and reacted for 5 min at 30°C. The reaction is stopped with the addition of 20 μL of 100% TCA, the proteins collected by centrifugation, and separated by SDS-PAGE. By comparing the phosphorylation patterns between samples incubated with Ca^{2+}/calmodulin and those incubated with EDTA, substrates for CaM kinase can be identified.

For studying phosphorylation in intact tissue, any method that has been shown to elevate intracellular Ca^{2+} levels is a suitable stimulus, e.g., depolarization, Ca^{2+} ionophores, or addition of neurotransmitters or hormones. As with other kinases, phosphorylation can be studied directly after incubation with [^{32}P]-H$_3$PO$_4$ or using "back phosphorylation" methods.

2.3. Tyrosine Kinases

All of the cyclic nucleotide- and Ca^{2+}-dependent protein kinases specifically phosphorylate proteins on serine or threonine residues. Indeed, in many cells, virtually all of the phosphoamino acid content of proteins is phosphoserine or phosphothreonine (Sefton et al., 1980). Just over 10 yr ago, however, it was first discovered that many animal cells also contained small levels of phosphotyrosine in proteins (Sefton et al., 1980). Over the last 10 yr, there has been a tremendous increase in our knowledge of the kinases that specifically phosphorylate proteins on tyrosine residues (tyrosine kinases) and of their substrate proteins.

The first tyrosine kinases identified were viral transforming proteins, especially those of the oncogenic retroviruses (Hunter and Cooper, 1985). Subsequently, homologs were found in untransformed cells that also showed tyrosine kinase activity (Hunter and Cooper, 1985), of which the best studied is the pp60$^{c\text{-}src}$ protein kinase. In almost all cases, there is no known function for the cellular homologs of the viral transforming proteins, although tyrosine kinase activity is clearly associated with cell transformation and growth (Sefton et al., 1980).

The tyrosine kinase pp60$^{c\text{-}src}$ is found in high levels in neural tissue, and the highest levels of activity are detected after cell proliferation, during the time of neuronal differentiation (Sorge et al., 1984; Brugge et al., 1987). Glycoproteins derived from neuronal growth cones are enriched in tyrosine kinase activity, compared to glycoproteins from mature synaptosomes (Cheng and Sahyoun, 1990), which further suggests that the tyrosine kinases are involved in neuronal differentiation. Mutation analysis of *Drosophila* also supports a role for tyrosine kinases in neuronal development. *Drosophila* mutants, deficient in fasciclin I protein, a neural cell-adhesion molecule (Zinn et al., 1988), show no gross defects of the embryonic CNS. Likewise, mutants deficient in the Abelson tyrosine kinase, a tyrosine kinase homologous to a viral oncogene (Hoffman et al., 1983), are also viable and show no gross defects in the embryonic CNS. Double mutations of these genes, however, show a major disruption in the patterns of axon guidance during CNS development, suggesting that an interaction between these two proteins is crucial for neural development (Elkins et al., 1990).

The other major group of tyrosine kinases are various growth factor receptors, which include the epidermal growth factor, platelet-derived growth factor, and insulin receptors (Hunter and Cooper, 1985). The binding of the growth factor to the receptor stimulates an integral tyrosine kinase activity, but the physiological function of kinase activity is generally unknown. Insulin-stimulated tyrosine phosphorylation of the insulin receptor, however, renders the kinase activity independent of insulin (Rosen, 1987). Usually the major protein phosphorylated by the growth factor receptor is the receptor itself, by autophosphorylation (Hunter and Cooper, 1985).

The two groups of tyrosine kinases share a number of characteristics. The known sequences of the tyrosine kinases all show considerable homology, and there are also homologies with the A-kinase catalytic region (Hunter and Cooper, 1985). All the tyrosine kinases show high specificity for tyrosine phosphorylation, will not phosphorylate serine, threonine, or hydroxyproline in synthetic peptide substrates, and prefer Mn^{2+} for full activity rather than Mg^{2+} (Hunter and Cooper, 1985).

To assay for tyrosine kinase activity, similar methods to those described above can be used, using a synthetic peptide that contains tyrosine as a phosphate acceptor (Swarup et al., 1983; Hirano et al., 1988); a commercially available substrate is poly (glutamate$_4$-tyrosine)(Sigma). The reaction is carried out in the following buffer: 20 mM Tris-HCl, pH 7.4, 5 mM $MgCl_2$, 2 mM $MnCl_2$, 0.1 mM $NaVO_3$, and 1 mM EDTA. Mn^{2+} is used either in place of, or in addition to, Mg^{2+} since it is a potent tyrosine kinase stimulator, and VO_3^- is a tyrosine phosphatase inhibitor and an activator of insulin receptor tyrosine kinase activity (Pang et al., 1988). Tubes are set up with tissue homogenate in the presence or absence of 200 μg poly (Glu$_4$-Tyr) (90 μL total vol). The reaction is started with the addition of 10 μL 0.1 mM [γ-^{32}P]-ATP. The mixture is incubated for 10 min at 30°C, and the reaction stopped by addition of 20 μL 100% TCA. After precipitation, the protein is separated as described above. An increase in the incorporation of ^{32}P into the tube containing poly (Glu$_4$-Tyr) is taken as a measure of tyrosine kinase activity. To unequivocally confirm this, the proteins have to be hydrolyzed and the identity of the phosphoamino acids determined (Section 3.1.). Typical results of a tyrosine kinase assay are shown in Table 5 for CNS homogenates from *Manduca sexta*.

To identify the endogenous substrates for tyrosine kinases, the proteins containing phosphotyrosine have to be identified, preferably by using an antibody directed against phosphotyrosine to identify proteins on Western blots. The proteins from intact nervous tissue or homogenates incubated as described below are separated by SDS-PAGE and transferred to Immobilon PVDF membrane (Millipore) or nitrocellulose using standard methodology (Towbin et al., 1979). Nonspecific binding to the membrane is blocked by incubation in 5% nonfat dry milk, BSA, or gelatin in phosphate-buffered saline (10 mM phosphate buffer, 0.9% NaCl, pH 7.4) (PBS) for 8 h (Hauri and Bucher, 1986). The membrane is then incubated with antiphosphotyrosine antibody (ICN) at a 1:1000 dilution in PBS containing 0.05% Tween-20

Table 5
Example of the Result Obtained for Assaying the Levels of Tyrosine
Kinase Activity in the CNS of the Tobacco Hornworm, *Manduca sexta*

Amount of tissue homogenate, μg	Absence of poly Glu/Tyr[a]	Presence of poly Glu/Tyr[a]	Difference[a]
0	79	78	0
10	2045	2412	367
20	1856	2990	1134
40	1700	3322	1622

[a] All values are in counts per minute.

(PBST) and 2% normal goat serum overnight at 4°C. The membrane is then washed in PBST (3 × 30 min washes), and incubated in alkaline phosphatase conjugated goat antimouse antibody (BioRad, 1:3000 dilution) for 2 h at room temperature, washed in PBST (3 × 30 min washes), and incubated in alkaline phosphatase substrate (BioRad, Richmond, CA) until the proteins containing phosphotyrosine are visible.

The other method of detecting tyrosine kinase substrates relies on the fact that phosphotyrosine is more stable to high pH than phosphoserine and phosphothreonine (Plimmer, 1941). Thus, after the proteins have been phosphorylated, either with $[\gamma\text{-}^{32}P]$-ATP in homogenates or with $[^{32}P]\text{-}H_3PO_4$ in intact cells, and separated by SDS-PAGE, the gel is treated with $1M$ KOH at 55°C for 2 h (Cooper and Hunter, 1981; Cooper et al., 1983). This treatment will remove the majority of $^{32}P]$-phosphate from serine and threonine residues, but will leave phosphotyrosine residues relatively intact. Unambiguous identification of tyrosine phosphorylation will require excising the phosphorylated protein band from a gel and subjecting it to phosphoamino acid analysis (*see* Section 3.1.).

2.4. Casein Kinases

Another group of serine/threonine protein kinases, present in the nervous system, are the casein kinases. Their activity is independent of any known intracellular messenger, and they preferentially phosphorylate acidic proteins, such as casein and phosvitin (Nairn et al., 1985b). They have been identified in many cell types, including brain, based on their ability to phosphorylate exogenously added casein. Two distinct types have so far been identified, casein kinase I and casein kinase II.

Very little is known about endogenous substrates for these kinases, their function, or their physiological stimulation. One identified protein in the brain that has been shown to be phosphorylated by casein kinase II is DARPP-32 (dopamine and cAMP-regulated phosphoprotein; Girault et al., 1989). DARPP-32 is a protein found in neurons with dopamine D1 receptors, is phosphorylated by dopamine stimulation of these receptors, via cAMP and A-kinase, and when phosphorylated, acts as an inhibitor of protein phosphatase type 1 (Hemmings et al., 1987). Casein kinase II phosphorylates DARPP-32 at a different site than does A-kinase, but will not phosphorylate other closely related protein phosphatase inhibitors (Girault et al., 1989). Phosphorylation of DARPP-32 by casein kinase II does not affect its ability to inhibit protein phosphatase type 1, but instead, facilitates phosphorylation of DARPP-32 by A-kinase. This interaction suggests that casein kinase II might have a role in modulating DARPP-32-containing neurons (Girault et al., 1989). Assays for casein kinases involve similar procedures to those described above, but require the use of acidic proteins, such as casein or phosvitin, as phosphate acceptors (Hathaway and Traugh, 1983).

3. Further Analysis of Phosphoproteins

After a particular protein has been identified as a substrate for a kinase, a number of simple analyses can be carried out to characterize the site(s) of phosphorylation. Each can use an isolated band from an SDS-PAGE gel as starting material. The first aim is to identify which amino acid(s) is (are) phosphorylated. This is especially important if the identity of the kinase is not certain, as is the case for phosphorylation of proteins in crude cell homogenates. Second, generation of phosphopeptide maps after limited proteolysis can be used to ascertain the number of different phosphorylation sites and also, if more than one kinase phosphorylates the protein, if they phosphorylate at the same or different sites.

3.1. Identification of the Phosphoamino Acid

The sample is first labeled with ^{32}P as described in Section 2. and the proteins separated by SDS-PAGE. The phosphoprotein of interest is then localized with autoradiography. The appropriate band is cut from the gel and incubated in 1 mL 5.7M HCl at 110°C for 1 h in a 13 × 100 mm screw cap glass tube. This procedure only results in the partial hydrolysis of the protein, which is desirable since the phosphate esters themselves are acid labile. Longer incubation times tend to reduce the recovery of phosphotyrosine and phosphoserine, but increase the recovery of phosphothreonine (Cooper et al., 1983). After hydrolysis, 4 mL H_2O is added to the tube, which is centrifuged and the supernatant lyophilized. The yellow residue is dissolved in 10 mL H_2O and 0.4 mL of a 50% (v/v) slurry of ion exchange resin added (Dowex AG1-X8, Cl⁻ form). The pH of the mixture should be adjusted to 7.5–8.5 with 10 mM NH_4OH. Internal standards of unlabeled phosphoserine, phosphotyrosine, and phosphothreonine are added (1 μL of 1 mg/mL) and the mixture gently shaken for 4 h at room temperature. The mixture is then poured into a small disposable column (BioRad) and washed with H_2O; the phosphoamino acids are eluted with 1 mL 0.1M HCl. After lyophilization, the identity and relative amounts of the phosphoamino acids can be determined with a number of different methods, depending on the equipment available.

The simplest method of separation is with TLC, using cellulose thin-layer plates and a buffer of isobutyric acid:0.5M NH_4OH (5:3). In some instances, this does not give good separation of all three phosphoamino acids, and 2D electrophoretic separation on cellulose thin-layer plates is recommended. The first dimension is carried out in formic acid:acetic acid:water (1:10:89) at 500 V and then the second dimension in acetic acid:pyridine: water (19:1:89) also at 500 V. The locations of the labeled phosphoamino acids are revealed by autoradiography and compared to the location of unlabeled phosphoamino acid markers, which are visualized with 1% ninhydrin in acetone. High-performance liquid chromatography (HPLC) allows separation of all three phosphoamino acids on anion exchange columns (Swarup et al., 1981; Yang et al., 1982) with more accurate quantification of radioactivity.

3.2. Limited Proteolysis
and Phosphopeptide Maps

A number of phosphoproteins have multiple phosphorylation sites that can be phosphorylated by different kinases. A good
example is synapsin, which is phosphorylated by both A-kinase
and CaM kinase on multiple sites. Furthermore, the degree of
phosphorylation at each site varies with the kinase (Huttner and
Greengard, 1979). To determine whether a phosphoprotein is
phosphorylated on multiple sites, the protein is first subjected
to limited proteolysis, the resulting peptides separated, and the
phosphopeptides identified. The pattern of peptides produced,
the phosphopeptide map, is a characteristic of the protein and
the digesting enzyme.

The proteolytic step can be very conveniently carried out
directly on proteins isolated from SDS gels since the proteolytic
enzymes used are still active in low concentrations of SDS (Cleveland et al., 1977). Gel pieces containing the phosphoprotein are
soaked in 125 mM Tris-HCl, pH 6.8, containing 0.1% SDS for 1 h
and placed directly in the slots of an SDS-PAGE (15% stacking
gel). The gel pieces are then overlaid with 100 µL of the same
buffer containing 15% glycerol, a trace of pyronin Y, and 10 µg
Staphylococcus aureus V8 protease, and electrophoresed at 60 V
until the dye reaches the bottom of the gel. The phosphopeptides
are then localized with autoradiography. Other proteases, such
as chymotrypsin, trypsin, or papain, can also be used and will
generate different peptide patterns.

More extensive characterization of the phosphopeptides can
be achieved by separating the peptides by 2D TLC or reverse-
phase HPLC. In both cases, analysis can be carried out from
excised gel bands. The gel bands are first lyophilized, and then
rehydrated with 50 mM NH_4HCO_3 containing 150 µg trypsin
and digested for 20 h at 37°C. The gel piece is then removed and
washed in the same buffer. The wash is combined with the incubation buffer and lyophilized.

For TLC separation, the peptides are resuspended in 1%
$(NH_4)_2CO_3$, pH 8.9, and spotted onto the corner of cellulose TLC
plates. The first dimension is run in the same buffer at 500 V,

and the second dimension is ascending chromatography in 1-butanol:acetic acid:water:pyridine (15:3:12:10). The phospho-peptides are again localized by autoradiography.

For HPLC, the peptides are resuspended in 0.1% trifluoro-acetic acid (TFA), loaded onto a C_{18} reverse-phase column, and eluted with a gradient of acetonitrile or propan-1-ol containing 0.1% TFA. The added advantage of HPLC separation is that the peptides can then be sequenced directly to give a partial sequence of the phosphoprotein and, more especially, the sequence around the phosphorylation site (Aebersold et al., 1987).

4. Protein Phosphatases

The extent to which a protein is phosphorylated depends on the relative activities of the protein kinase and the protein phosphatase, which catalyzes the removal of the phosphate from the protein. The earlier sections of this chapter have made clear the tremendous amount of research that has focused on the phos-phorylation of proteins. By contrast, their dephosphorylation has received very little attention. In recent years, however, the importance of the phosphoprotein phosphatases in regulating the phosphorylation state of proteins has been increasingly rec-ognized (Cohen, 1989). As with the protein kinases, most of the early work has concentrated on phosphoserine and phospho-threonine-specific protein phosphatases (Cohen, 1989), but more recently, protein phosphatases specific for phosphotyrosine have been identified (Jones et al., 1989).

Four principal classes of serine- and threonine-specific pro-tein phosphatases have been purified and extensively charac-terized. These appear to account for the majority of protein phosphatase activity in eukaryotic cells (Cohen, 1989). Their classification depends on a combination of substrate speci-ficity, cation requirements, and susceptibility to inhibition by particular agents (Cohen, 1989). More recently, with the advent of molecular cloning techniques, several other phosphatases have been identified based on their sequence homologies with bio-chemically isolated and sequenced phosphatases (Cohen and Cohen, 1989).

The ability to assay the activity and types of protein phosphatases present in a particular tissue depends on the availability of a phosphorylated substrate protein. The initial classification of protein phosphatases was determined by their ability to dephosphorylate either the α or β subunit of phosphorylase kinase. Type I phosphatases preferentially dephosphorylate the β subunit, whereas the type II phosphatases prefer the α subunit. Purified subunits of phosphorylase kinase can be phosphorylated with [γ-^{32}P]-ATP using A-kinase (Section 2.1.2.1.; the unreacted ATP should be removed using gel filtration). These phosphorylated subunits can then be used to assay the ability of tissue extracts to remove [^{32}P]-phosphate from the protein in an analogous fashion to measuring the incorporation of ^{32}P]-phosphate onto a protein (Section 2.1.1.). Other phosphorylated proteins can also be used, such as the growth-associated protein, GAP-43, which is dephosphorylated by calcineurin, a type IIB phosphatase (Liu and Storm, 1989). As an alternative to measuring the amount of ^{32}P remaining in the protein, after TCA precipitation, it is easier to measure the amount of ^{32}P]-phosphate release into the TCA-soluble fraction (Brautigan and Shriner, 1988). This does, however, rely on the production of ^{32}P]-labeled protein uncontaminated with [γ-^{32}P]-ATP. This method has been used to identify seven separate phosphotyrosine-specific phosphatases in bovine brain using phosphotyrosine-casein as a substrate (Jones et al., 1989).

These methods use the phosphorylated forms of a specific protein as an indicator of phosphatase activity. One can also study the dephosphorylation of multiple, unknown phosphoproteins using SDS-PAGE in a similar manner to that used to study kinase substrates (Section 2.1.2.1.). After the kinase reaction has been completed, the kinase activity can either be halted or the specific activity of the ATP can be lowered. Kinase activity can be inhibited by removing the Mg^{2+} with addition of EDTA to a final concentration of 100 mM. This treatment does not affect the majority of phosphatase activity, since Mg^{2+} is not needed. At various times after the addition of the EDTA, the proteins are precipitated with TCA and separated by SDS-PAGE. After fluorography, the reduction in label in particular bands can be followed. Although protein phosphatases types I and IIA do not require divalent cations, types IIB and IIC require Ca^{2+}/CaM and Mg^{2+}, respectively, so the addition of EDTA will inhibit their activity. An alternative to adding EDTA is to lower the specific activity of the ATP, at least 100-fold, by the addition of excess unlabeled ATP. During the initial stages of the reaction, both kinase and phosphatase activity will be present, and the protein will be phosphorylated up to a maximum specific activity equal to that of the ATP. Upon lowering the specific activity of the ATP, both kinase and phosphatase activity will continue to be present, but the specific activity of the protein will be lowered to the new specific activity of the ATP. This has the effect of removing ^{32}P from the protein, and fluorograms will show a reduction in the intensity of the bands. By comparing the effects of the addition of EDTA and ATP, the characteristics of the phosphatases present and which substrates they act on can be revealed.

Another very useful tool for studying phosphatase activity is the tumor promotor, okadaic acid (Bialojan and Takai, 1988). This fatty-acid polyketal is a potent inhibitor of phosphatase type IIA (inhibited at 1 nM) and type I (inhibited at 10–15 nM), whereas types IIB and IIC are relatively insensitive (Bialojan and Takai, 1988). Phosphotyrosine-specific phosphatases can be inhibited by the addition of 10 μM sodium vanadate (Jones et al., 1989).

In addition to these inhibitors, nervous tissue also contains a number of endogenous proteins that act as specific protein phosphatase inhibitors (Cohen, 1989). Two of these, inhibitor I and DARPP-32 (dopamine and cAMP-regulated phosphoprotein 32; Hemmings et al., 1987), are only effective inhibitors when they have been phosphorylated by A-kinase and specifically inhibit type I phosphatases, which have a broad substrate specificity. Phosphatase type IIB has a narrow range of substrate specificities, and is particularly effective at dephosphorylating inhibitor I and DARPP-32 (Cohen, 1989). These complex interactions among kinase, phosphatase, and phosphatase inhibitors are as yet poorly understood, but show how important it is to keep in mind the roles of phosphatases when studying protein phosphorylation.

5. Physiological Function
   of Protein Phosphorylation

The ultimate goal in any study of protein phosphorylation is to answer the following general questions: Given a physiological response, is protein phosphorylation involved? Which kinase(s) is (are) involved? Which proteins are phosphorylated? In addition, an alternative approach is to start with a protein known to be involved in a cellular response, such as a receptor or an ion channel, and ask if it is phosphorylated in vivo, and if so, by what kinases and intracellular messengers. More importantly, what effect on the protein's function does phosphorylation have?

To determine whether protein phosphorylation is involved in a particular physiological response, a number of approaches can be taken. The simplest and most straightforward is to use various pharmacological agents, which either stimulate or block particular protein kinases or phosphatases, and determine their effects on the physiological response. Either these compounds need to be permeable to cell membranes so they can be used on intact cells, or the preparation has to be able to be permeabilized. Unfortunately, the list of such permeable compounds is not extensive, and does not include agents that target all the kinases and phosphatases. Phorbol esters have been shown to activate C-kinase directly (Kikkawa et al., 1984), and various Ca^{2+} ionophores and depolarizing agents will elevate intracellular Ca^{2+} levels sufficiently to activate CaM kinases (e.g., Forn and Greengard, 1978).

Most of the actions of cyclic nucleotides are believed to exert their effects via cyclic nucleotide-dependent protein kinases. Thus, physiological effects stimulated by elevation of intracellular cyclic nucleotides, by phosphodiesterase inhibitors, or by membrane-permeable cyclic nucleotide analogs are generally taken as evidence for the involvement of cyclic nucleotide-dependent protein kinases. These results should be interpreted with caution, however, since direct regulation of ion channel activity by cyclic nucleotides has been shown in photoreceptor cells (Stryer, 1986) and olfactory cells (Gold and Nakamura, 1987). Other pharmacological agents include the bioflavinoid quer-

cetin, which blocks the action of the pp60[src] tyrosine protein kinase (Graziani et al., 1983), and okadaic acid, which blocks the
action of type I and type IIA protein phosphatases (Bialojan
and Takai, 1988).

A more direct approach is to inject the kinase (or kinase
inhibitor) into a cell and then to observe changes in the cell's
physiology. This was first done in *Aplysia* bag cells, where injection of A-kinase catalytic subunit caused a broadening of the
action potential (Kaczmarek et al., 1980). Since then, direct
effects of G-kinase, C-kinase, and CaM kinase on nervous systems have been shown by intracellular injection of the kinases
(*see* Greengard, 1987).

The methods needed to identify the proteins that are
phosphorylated coincidentally with the stimulation of a
physiological response have been described in Section 2.1.2.2.
The most difficult problem, however, is to show directly that the
phosphorylation of a particular protein causes the physiological
response. A convincing example is provided by the phosphoprotein synapsin I. Synapsin I has been the subject of extensive
investigation for several years (*see* Greengard, 1987). It is found
only in the synapses of neurons and is localized primarily in
presynaptic terminals. It is phosphorylated by CaM kinase I and
II and A-kinase. Agents that increase intracellular levels of Ca^{2+}
and cAMP both stimulate the phosphorylation of synapsin I and
increase the size of postsynaptic potentials. The question that
remained from these studies was whether all or any of the
effects mediated by Ca^{2+} or cAMP were mediated by the phosphorylation of synapsin I. Direct evidence for the involvement
of synapsin I came from the squid giant synapse, a preparation
in which microelectrodes could be inserted very close to the
presynaptic terminals. Injection of dephosphosynapsin I in the
presynaptic terminals reduced the amplitude and rate of rise of
the postsynaptic potentials, whereas injection of phosphosyn-
apsin I or heat-denatured dephospho-synapsin I had no effect
(Llinas et al., 1985). These results suggest direct involvement of
synapsin I phosphorylation in regulating neurotransmitter
release, but the molecular mechanisms involved are still far
from clear.

These types of experiments are possible only on systems amenable to intracellular injections, and also depend on the ability of the injected phosphoprotein to adopt its correct cellular localization and conformation. To gain a better understanding of the physiological roles played by particular phosphoproteins, one will have to resort to molecular genetic approaches. When more phosphoproteins are identified and cloned, studies can be initiated that would involve looking for mutations in the phosphoproteins. Analysis of the effects of these mutations on the ability to phosphorylate or dephosphorylate the proteins and on neuronal function would add to our understanding of the role of phosphorylation. This type of approach, with the addition of site-directed mutagenesis, has yielded much information on the function of the β-adrenergic receptor (Lefkowitz and Caron, 1988). For example, this approach has shown which regions of the receptor are phosphorylated and that phosphorylation in the terminal 48 residues is important in desensitization of the receptor (Lefkowitz and Caron, 1988). Clearly, as more phosphoproteins are cloned, this approach will yield much exciting data.

Another approach to studying the physiological effects of protein phosphorylation is to look at identified proteins, and examine whether phosphorylation occurs and how it functionally affects the protein. This approach requires an in vitro assay for the function of the protein. Many enzymes have been shown to be phosphorylated, which affects various kinetic parameters of the protein (Krebs and Beavo, 1979). Various neurotransmitter receptors and ion channels are also phosphorylated (Levitan et al., 1983; Huganir and Greengard, 1987), but the study of the effects of phosphorylation requires a functionally active, reconstituted system. For example, the nicotinic acetylcholine receptor can be phosphorylated by A-kinase (Huganir and Greengard, 1983), C-kinase (Huganir et al., 1983), and a tyrosine kinase (Huganir et al., 1984). Using phosphorylated receptor reconstituted in lipid vesicles, both A-kinase and tyrosine kinase phosphorylation of this receptor have been shown to increase its rate of desensitization to acetylcholine (Huganir and Greengard, 1987; Hopfield et al., 1988). These types of studies must always be carried out in conjuction with in vivo studies to show that the proteins are indeed phosphorylated in vivo and that the physiological stimuli that lead to changes in cellular physiology can be explained by the changes seen in vitro.

Another way to elucidate the physiological role of protein phosphorylation is to phosphorylate the protein with ATPγS. This analog of ATP serves as a substrate for protein kinases to phosphorylate proteins, but the resulting thiophosphoproteins are very resistant to dephosphorylation by protein phosphatases (Li et al., 1988). Thus, longer-lasting phosphorylation can be achieved, which may lead to the unmasking of physiological effects. As dephosphorylation is prevented, the role of phosphatases in the system can also be studied using ATP-γ-S-labeled proteins.

Over the last few years, many advances have been made in discovering new protein substrates for kinases and new protein kinases (especially the tyrosine protein kinases), and in establishing the importance of protein phosphatases in regulating cellular activity. The major challenge now is to obtain more direct evidence for the physiological role of protein phosphorylation/dephosphorylation and to elucidate the function of the various phosphoproteins that are regulated by the various signals believed to act via protein phosphorylation.

Acknowledgments

I would like to thank James Truman, Mark Meyer, and Steve Robinow for many constructive comments on the manuscript. Any unpublished work presented here was supported by NIH grant NS25715 to James Truman.

Abbreviations

ATPγS	Adenosine 5'-[γ-thio]triphosphate
A-kinase	cAMP-dependent protein kinase
BSA	Bovine serum albumin
cAMP	Adenosine 3',5' cyclic monophosphate
cGMP	Guanosine 3',5' cyclic monosphosphate
DG	1,2-diacylglycerol

DTT   Dithiothreitol;
G-kinase   cGMP-dependent protein kinase
PAGE   Polyacrylamide gel electrophoresis
PtdSer   Phosphatidylserine
SDS   Sodium dodecyl sulfate
TCA   Trichloracetic acid

References

Aebersold R. H., Leavitt J., Saavedra R. A., Hood L. E., and Kent S. B. H. (1987) Internal amino acid sequence analysis of proteins separated by one- or two-dimensional gel electrophoresis after *in situ* protease digestion on nitrocellulose. *Proc. Natl. Acad. Sci. USA* **84,** 6970–6974.

Beavo J. A. and Mumby M. C. (1982) Cyclic AMP-dependent protein phosphorylation. *Handbook of Experimental Pharmacology* **58,** 363–392.

Beg Z. H., Stonik J. A., and Brewer H. B. (1987) Phosphorylation and modulation of the enzymatic activity of native and protease-cleaved purified hepatic 3-hydroxy-3-methylglutaryl-Coenzyme A reductase by a calcium/calmodulin-dependent protein kinase. *J. Biol. Chem.* **262,** 13,228–13,240.

Bialojan C. and Takai A. (1988) Inhibitory effects of a marine-spong toxin, okadaic acid, on protein phosphatases: Specificity and kinetics. *Biochem. J.* **256,** 283–290.

Brautigan D. L. and Shriner C. L. (1988) Methods to distinguish various types of protein phosphatase activity. *Methods Enzymol.* **159,** 339–346.

Browning M. D., Huganir R., and Greengard P. (1985) Protein phosphorylation and neuronal function. *J. Neurochem.* **45,** 11–23.

Brugge J., Cotton P., Lustig A., Yonemoto W., Lipsich L., Coussens P., Barrett J. N., Nonner D., and Keane R. W. (1987) Characterization of the altered form of the c-*src* gene product in neuronal cells. *Genes Dev.* **1,** 286,287.

Cheng N. and Sahyoun N. (1990) Neuronal tyrosine phosphorylation in growth cone glycoproteins. *J. Biol. Chem.* **265,** 2417–2420.

Cleveland D. W., Fischer S. G., Kirschner M. W., and Laemmli U. K. (1977) Peptide mapping by limited proteolysis in sodium dodecyl sulfate and analysis by gel electrophoresis. *J. Biol. Chem.* **252,** 1102–1106.

Cohen P. (1982) The role of protein phosphorylation in neuronal and hormonal control of cellular activity. *Nature* **296,** 613–620.

Cohen P. (1989) The structure and regulation of protein phosphatases. *Ann. Rev. Biochem.* **58,** 453–508.

Cohen P. and Cohen P. T. W. (1989) Protein phosphatases come of age. *J. Biol. Chem.* **264,** 21,435–21,438.

Colbran R. J., Schworer C. M., Hashimoto Y., Fong Y. L., Rich D. P., Smith M. K., and Soderling T. R. (1989) Calcium/calmodulin-dependent protein kinse II. *Biochem. J.* **258,** 313–325.

Cooper J. A. and Hunter T. (1981) Changes in protein phosphorylation in Rous sarcoma virus-transformed chicken embryo cells. *Mol. Cell. Biol.* **1**, 165–178.

Cooper J. A., Sefton B. M., and Hunter T. (1983) Detection and quantification of phosphotryosine in proteins. *Methods Enzymol.* **99**, 387–402.

Corbin J. D. and Hardman J. G. (eds.) (1983) *Hormone Action. Part F, Protein Kinases. Meth. Enzymol.* vol. 99 (Academic, New York).

DeRiemer S. A., Kaczmarek L. K., Lai Y., McGuinness T. L., and Greengard P. (1984) Calcium/calmodulin-dependent protein phosphorylation in the nervous system of *Aplysia. J. Neurochem.* **4**, 1616–1625.

Edwards S. C. and Battelle B. A. (1987) Octopamine- and cyclic AMP-stimulated phosphorylation of a protein in *Limulus* ventral and lateral eyes. *J. Neurosci.* **7**, 2811–2820.

Edwards S. C., Wishart A. C., Renninger G. H., Wiebe E. M., and Battelle B. A. (1990) Efferent innervation to *Limulus* eyes *in vivo* phosphorylates a 122 kDa protein. *Biol. Bull.* **178**, 267–278.

Elkins T., Zinn K., McAllister L., Hoffman F. M., and Goodman C. S. (1990) Genetic analysis of a *Drosophila* neural cell adhesion molecule: Interaction of fasciclin I and abelson tyrosine kinase mutations. *Cell* **60**, 565–575.

Forn J. and Greengard P. (1978) Depolarizing agents and cyclic nucleotides regulate the phosphorylation of specific neuronal proteins in rat cerebral cortex slices. *Proc. Natl. Acad. Sci. USA* **75**, 5195–5199.

Fukanaga K., Rich D. P., and Soderling T. R. (1989) Generation of the Ca^{++}-independent form of Ca^{++}/calmodulin-dependent protein kinase II in cerebellar granule cells. *J. Biol. Chem.* **264**, 21,830–21,836.

Girault J. A., Hemmings H. C., Williams K. R., Nairn A. C., and Greengard P. (1989) Phosphorylation of DARPP-32, a dopamine- and cAMP-regulated phosphoprotein, by casein kinase II. *J. Biol. Chem.* **264**, 21,748–21,759.

Gold G. H. and Nakamura T. (1987) Cyclic nucleotide-gated conductances: A new class of ion channels mediates visual and olfactory transduction. *Tends Pharmacol. Sci.* **8**, 312–316

Goy M. F., Schwartz T. L., and Kravitz E. A. (1984) Serotonin-induced protein phosphorylation in a lobster neuromuscular preparation. *J. Neurosci.* **4**, 611–626.

Graziani Y., Erikson E., and Erikson R. L. (1983) The effect of quercetin on the phosphorylation activity of the Rous sarcoma virus transforming gene product *in vitro* and *in vivo. Eur. J. Biochem.* **135**, 583–589.

Greengard P. (1976) Possible role of cyclic nucleotides and phosphorylated membrane proteins in postsynaptic actions of neurotransmitters. *Nature* **260**, 101–108.

Greengard P. (1987) Neuronal phosphoproteins, mediators of signal transduction. *Mol. Neurobiol.* **1**, 81–119.

Hames B. D. and Rickwood D. (eds) (1981) *Gel Electrophoresis of Proteins, A Practical Approach* (IRL, Oxford, UK).

Hathaway G. M. and Traugh J. A. (1983) Casein kinase II. *Methods Enzymol.* 99, 317–331.

Hauri H. P. and Bucher K. (1986) Immunobl)tting with monoclonal antibodies: Importance of the blocking solution. *Anal. Biochem.* 159, 386–389.

Haycock J. W., Browning M. D., and Greengard P. (1988) Cholinergic regulation of protein phosphorylation in bovine adrenal chromaffin cells. *Proc. Natl. Acad. Sci. USA* 85, 1677–1681.

Hemmings H. C., Walaas S. I., Ouimet C. C., and Greengard P. (1987) Dopaminergic regulation of protein phosphorylation in the striatum: DARPP-32. *Trends Neurosci.* 10, 377–383.

Hirano A. A., Greengard P., and Huganir R. L. (1988) Protein tyrosine kinase activity and its endogenous substrates in rat brain: A subcellular and regional survey. *J. Neurochem.* 50, 1447–1455.

Hoffman F. M., Fresco L. D., Hoffman-Falk H., and Shilo B. Z. (1983) Nucleotide sequences of the *Drosophila src* and *abl* homologs: conservation and variability in the *src* family oncogenes. *Cell* 35, 393–401.

Hopfield J. F., Tank D. W., Greengard P., and Huganir R. L. (1988) Functional modulation of the nicotinic acetylcholine receptor by tyrosine phosphorylation. *Nature* 336, 677–680.

Huang K. P. (1989) The mechanism of protein kinase C activation. *Trends Neurosci.* 12, 425–432.

Huganir R. L., Albert K. A., and Greengard P. (1983) Phosphorylation of the nicotine acetylcholine receptor by Ca^{2+}/phospholipid-dependent protein kinase and comparison with its phosphorylation by cAMP-dependent protein kinase. *Soc. Neurosci. Abs.* 9, 578.

Huganir R. L. and Greengard P. (1983) cAMP-dependent protein kinase phosphorylates the nicotine acetylcholine receptor. *Proc. Natl. Acad. Sci. USA* 80, 1130–1134.

Huganir R. L. and Greengard P. (1987) Regulation of receptor function by protein phosphorylation. *Trends Pharmacol.* 8, 472–477.

Huganir R. L., Miles K., and Greengard P. (1984) Phosphorylation of the nicotinic acetylcholine receptor by an endogenous tyrosine-specific protein kinase. *Proc. Natl. Acad. Sci. USA* 81, 6968–6972.

Hunter T. and Cooper J. A. (1985) Protein-tyrosine kinases. *Ann. Rev. Biochem.* 54, 897–930.

Huttner W. B. and Greengard P. (1979) Multiple phosphorylation sites in protein I and their differential regulation by cAMP and calcium. *Proc. Natl. Acad. Sci. USA* 76, 5402–5406.

Jennings K. R., Kaczmarek L. K., Hewick R. M., and Strumwasser F. (1982) Protein phosphorylation during after discharge in peptidergaic neurons of *Aplysia*. *J. Neurosci.* 2, 158–168.

Jones S. W., Erikson R. L., Ingebritsen V. M., and Ingebritsen T. S. (1989) Phosphotyrosyl-protein phosphatases. I. Separation of multiple forms from bovine brain and purification of the major form to near homogeneity. *J. Biol. Chem.* 264, 7747–7753.

Kaczmarek L. K., Jennings K. R., Strumwasser F., Nairn A., Walter U., Wilson F. D., and Greengard P. (1980) Microinjection of protein kinase enhances action potentials of bag cell neurons in culture. *Proc. Natl. Acad. Sci. USA* **77,** 7487–7492.

Kennedy M. B. (1988) Synaptic memory molecules. *Nature* **335,** 770–772.

Kennedy M. B. and Greengard P. (1981) Two calcium/calmodulin-dependent protein kinases, which are highly concentrated in brain, phosphorylate protein I at distinct sites. *Proc. Natl. Acad. Sci. USA* **78,** 1293–1297.

Kikkawa U. and Nishizuka Y. (1986) The role of protein kinase C in transmembrane signalling. *Ann. Rev. Cell Biol.* **2,** 149–178.

Kikkawa U., Kaibuchi K., Castagna M., Yumanashi J., Sano K., Tanaka Y., Miyake R., Takai Y., and Nishizuka Y. (1984) Protein phosphorylation and mechanism of action to tumor-promoting phorbol esters. *Adv. Cyclic Nucleotide Protein Phosph. Res.* **17,** 437–442.

Krebs E. G. and Beavo J. A. (1979) Phosphorylation–dephosphorylation of enzymes. *Ann. Rev. Biochem.* **48,** 923–959.

Kuo J. F. and Greengard P. (1970) Cyclic nucleotide-dependent protein kinases. VI. Isolation and partial purification of a protein kinase activated by guanosine 3',5'-monophosphate. *J. Biol. Chem.* **245,** 2493–2498.

Kuo J. F., Wyatt G. R., and Greengard P. (1971) Cyclic nucleotide-dependent protein kinases. IX. Partial purification and some properties of guanosine 3',5'-monophosphate-dependent and adenosine 3',5'-monophosphate-dependent protein kinases from various tissues and species of arthropoda. *J. Biol. Chem.* **246,** 7159–7167.

Kuo W. N. and Kuo J. F. (1976) Isolation of stimulatory modulator of guanosine 3',5'-monophosphate-dependent protein kinase from mammalian heart devoid of inhibitory modulator of adenosine 3',5'-monophosphate-dependent protein kinase. *J. Biol. Chem.* **251,** 4283–4286.

Lefkowitz R. J. and Caron M. G. (1988) Adrenergic receptors, models for the study of receptors coupled to guanine nucleotide regulatory proteins. *J. Biol. Chem.* **263,** 4993–4996.

Levitan I. B., Lemos J. R., and Novak-Hofer I. (1983) Protein phosphorylation and the regulation of ion channels. *Trends Neurosci.* **6,** 496–499.

Li H. C., Simonelli P. F., and Huan L. J. (1988) Peparation of phosphatase-resistant substrates using adenosine 5'-O-(gamma-thio) triphosphate. *Methods Enzymol.* **159,** 346–356.

Lincoln T. M. (1983) cGMP-dependent protein kinase. *Methods Enzymol.* **99,** 62–71.

Liu Y. and Storm D. R. (1989) Dephosphorylation of neuromodulin by calcineurin. *J. Biol. Chem.* **264,** 12,800–12,804.

Llinas R., McGuinness T. L., Leonard C. S., Sugimori M., and Greengard P. (1985) Intracellular injection of synapsin I or calcium/calmodulin-dependent protein kinase II alters neurotransmitter release at the squid giant synapse. *Proc. Natl. Acad. Sci. USA* **82,** 3035–3039.

Morton D. B. and Truman J. W. (1986) Substrate phosphoprotein availability regulates eclosion hormone sensitivity in an insect CNS. *Nature* **323,** 264–267.

Morton D. B. and Truman J. W. (1988) The EGPs: The eclosion hormone and cGMP-regulated phosphoproteins. I. Appearance and partial characterization in the CNS of *Manduca sexta. J. Neurosci.* **8,** 1326–1337.

Nairn A. C., Bhagat B., and Palfrey H. C. (1985a) Identification of calmodulin-dependent protein kinase III and its major M_r 100,000 substrate in mammalian tissues. *Proc. Natl. Acad. Sci. USA* **82,** 7939–7943.

Nairn A. C., Hemmings H. C., and Greengard P. (1985b) Protein kinases in the brain. *Ann. Rev. Biochem.* **54,** 931–976.

Nestler E. J. and Greengard P. (1982) Nerve impulses increase the phosphorylation state of protein I in rabbit superior cervical ganglion. *Nature* **296,** 452–454.

Nishizuka Y. (1988) The molecular heterogeneity of protein kinase C and its implications for cellular recognition. *Nature* **334,** 661–665.

Ohmstede C. A., Jensen K. F., and Sahyoun N. E. (1989) Ca^{2+}/calmodulin-dependent protein kinase enriched in cerebellar granule cells. Identification of a novel neuronal calmodulin-dependent protein kinase. *J. Biol. Chem.* **264,** 5866–5875.

Pang D. T., Wang J. K., Valtorta F., Benfanati F., and Greengard P. (1988) Protein tyrosine phosphorylation in synaptic vesicles. *Proc. Natl. Acad. Sci. USA* **85,** 762–766.

Plimmer R. H. A. (1941) Esters of phosphoric acid. IV. Phosphoryl hydroxy amino acids. *Biochem. J.* **35,** 461–469.

Rannels S. R., Beasley A., and Corbin J. D. (1983) Regulatory subunits of bovine heart and rabbit skeletal muscle cAMP-dependent protein kinase isozymes. *Methods Enzymol.* **99,** 55–62.

Reimann E. M. and Beham R. A. (1983) Catalytic subunit of cAMP-dependent protein kinase. *Methods Enzymol.* **99,** 51–55.

Rosen O. M. (1987) After insulin binds. *Science* **237,** 1452–1458.

Rotondo D., Vaughan P. F. T., and Donnellan J. F. (1987) Octopamine and cAMP stimulate protein phosphorylation in the central nervous system of *Schistocerca gregaria. Insect Biochem.* **17,** 283–290.

Schwartz J. H. and Greenberg S. M. (1987) Molecular mechanisms for memory: Second-messenger induced modifications of protein kinases in nerve cells. *Ann. Rev. Neurosci.* **10,** 459–476.

Sefton B. M., Hunter T., Beemon K., and Eckhart W. (1980) Evidence that the phosphorylation of tyrosine is essential for cellular transformation by Rous sarcoma virus. *Cell* **20,** 807–816.

Sorge L. K., Levy B. T., and Maness P. F. (1984) pp60[c-src] is developmentally regulated in the neural retina. *Cell* **36,** 249–257.

Stryer L. (1986) The cyclic GMP cacade of vision. *Ann. Rev. Neurosci.* **9,** 87–119.

Swarup G., Cohen S., and Garbers D. L. (1981) Selective dephosphorylation of proteins containing phosphotyrosine by alkaline phosphatases. *J. Biol. Chem.* **256**, 8197–8201.

Swarup G., Dasgupta J. D., and Garbers D. L. (1983) Tyrosine protein kinase activity of rat spleen and other tissues. *J. Biol. Chem.* **258**, 10,341–10,347.

Taylor S. S. (1989) cAMP-dependent protein kinase. Model for an enzyme family. *J. Biol. Chem.* **264**, 8443–8446.

Taylor S. S., Kerlavage A. R., and Zoller M. J. (1983) Affinity labeling of cAMP-dependent protein kinases. *Methods Enzymol.* **99**, 140–153.

Towbin H., Staehelin T., and Gordon J. (1979) Electrophoretic transfer of proteins from polyacrilamide gels to nitrocellulose sheets: Procedure and some applications. *Proc. Natl. Acad. Sci. USA* **76**, 4350–4354.

Ueda T. and Greengard P. (1977) Adenosine 3':5'-monophosphate-regulated phosphoprotein system of neuronal membranes. I. Solubilization, purification and some properties of an endogenous phosphoprotein. *J. Biol. Chem.* **252**, 5155–5163.

Valtorta F., Schiebler W., Jahn R., Ceccarelli B., and Greengard P. (1986) A solid-phase assay for the phosphorylation of proteins blotted on nitrocellulose filters. *Anal. Biochem.* **158**, 130–137.

Walter U. and Greengard P. (1983) Photoaffinity labeling of the regulatory subunit of cAMP-dependent protein kinase. *Methods Enzymol.* **99**, 154–162.

Whitehouse S. and Walsh D. A. (1983) Inhibitor protein of the cAMP-dependent kinase: Characteristics and purification. *Methods Enzymol.* **99**, 80–93.

Wold F. (1983) Posttranslational protein modifications: Perspective and prospectives, in *Posttranslational Covalent Modifications of Proteins* (Johnson B. C., ed.) Academic, New York, pp. 1–17.

Yang J. C., Fujitaki J. M., and Smith R. A. (1982) Separation of phosphohydroxyamino acids by high performance liquid chromatography. *Anal. Biochem.* **122**, 360–363.

Zinn K., McAllister L., and Goodman C. S. (1988) Sequence analysis and neuronal expression of fasciclin I in grasshopper and *Drosophila. Cell* **53**, 577–587.

Investigation of the Intracellular Regulators and Components of the Exocytotic Pathway

Robert D. Burgoyne

1. Introduction

Exocytotic fusion of secretory vesicles with the plasma membrane occurs in cells in either a constitutive or regulated fashion (Burgess and Kelly, 1987). Constitutive exocytosis is the mechanism by which membrane components and certain soluble proteins are released. In the regulated secretory pathway, secretory products are stored in secretory vesicles until exocytois is triggered by an intracellular signal, and is the pathway for secretion of neurotransmitters and many hormones. The regulated secretory pathways in neurons and secretory cells have a number of aspects in common (Cheek and Burgoyne, 1990), and we and many others have chosen to study the bovine adrenal chromaffin cell as a convenient model system for the investigation of regulated exocytosis (Burgoyne, 1984a; Rink and Knight, 1988; Winkler, 1988). This chapter will concentrate on work on adrenal chromaffin cells (Fig. 1) with some reference to exocytosis in other secretory cells, and will cover experimental approaches that have been used in the study of the intracellular regulators and mediators of exocytosis in chromaffin cells.

From: *Neuromethods, Vol. 20: I ntracellular Messengers*
Eds: A. Boulton, G. Baker, and C. Taylor © 1992 The Humana Press Inc.

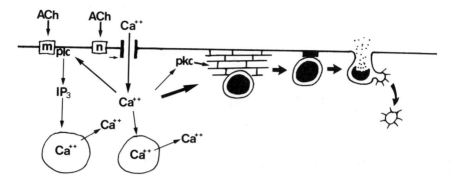

Fig. 1. General scheme showing the control of exocytosis in bovine adrenal chromaffin cells. The sources of Ca^{2+} include entry from outside and release from two separate internal stores. The cytoskeletal barrier at the cell cortex is disassembled in response to Ca^{2+} or activation of protein kinase C. The exocytotic machinery is shown as a shaded box. Following exocytosis, vesicle membrane is recovered by endocytosis through coated vesicles. Abbreviations: pIc, phosphoinositidase C; pKc, protein kinase C; ACh, acetylcholine; m, muscarinic receptor; n, nicotinic receptor.

Secretion from intact cells in reponse to agonists can be examined and manipulated in a variety of ways. Numerous pharmacological studies have examined factors involved in exocytosis, but in many cases, the drugs used were of dubious specificity, and often inhibited secretion at an early stage of stimulus–response coupling because of interference with receptor or channel function. The problem with the pharmacological approach became most evident in studies in which both secretion and the level of cytosolic free Ca^{2+} concentration ($[Ca^{2+}]_i$) were measured (e.g., Harris et al., 1986). It is clear that measurement of $[Ca^{2+}]_i$ in addition to secretion is an essential control in any experiments on intact cells. Secretion and the localization of $[Ca^{2+}]_i$ can be monitored simultaneously in single chromaffin cells (Cheek et al., 1989a) as well as cell populations. More direct information on the factors controlling exocytosis has come from the use of techniques that allow access to the intracellular sites of exocytosis and, thus, the ability to manipulate intracellular factors that may control exocytosis directly. The main approaches used with chromaffin cells have included the use of cells permeabilized by electric discharge (Baker and Knight, 1981), treatment with digitonin (Dunn and Holz, 1983), staphylococcal α-toxin (Bader et al., 1986) or streptolysin O (Sontag et al., 1988), and membrane capacitance measurement with the whole-cell, patch-clamp technique (Neher and Marty, 1982). Permeabilized cells allow control of the intracellular conditions and exocytotic secretion (Schafer et al., 1987a), and can be stimulated directly by introduction of micromolar Ca^{2+}. The patch-clamp approach has the advantage that exocytosis in single cells can be measured as well as the endocytotic recovery of vesicle membrane, which occurs concomitantly with exocytotic events.

The major physiological stimulus for exocytosis in chromaffin cells is a rise in $[Ca^{2+}]_i$. Ca^{2+}-dependent exocytosis is modulated by several intracellular factors. The mechanisms by which Ca^{2+} activates exocytosis and the way in which this is modulated still remain to be fully elucidated. In the first part of this chapter, I will discuss the nature of the factors that regulate or may be components of the Ca^{2+}-dependent exocytotic mechanism. In addition, technical approaches to the study of exocytosis and its control in chromaffin cells will be considered.

2. Intracellular Requirements for Exocytosis

In early studies using electro-permeabilized and digitonin-permeabilized bovine adrenal chromaffin cells, ATP was found to be an absolute requirement for Ca^{2+}-dependent exocytosis (Baker and Knight, 1981; Dunn and Holz, 1983). The obvious reason for a requirement for ATP is that protein phosphorylation is involved in exocytosis. Many polypeptides show increases or decreases in phosphorylation following stimulation of intact chromaffin cells (Cote et al., 1986), but the relevance of these phosphoproteins is unknown. The possibility that myosin light-chain kinase is required for exocytosis has been investigated, and the evidence suggests that this kinase does not play a key role in exocytosis (Lee et al., 1987). Protein kinase C is almost certainly involved in the control of exocytosis, and this is discussed in detail below.

One possible site for the requirement for ATP is at the level of translocation to the plasma membrane and docking. In sea urchin egg cortices, exocytosis can be initiated by Ca^{2+} alone with no requirement for ATP (Crabb and Jackson, 1985); in this system, the secretory granules are already docked at the plasma membrane. Similarly, PC12 cells permeabilized by α-toxin (Ahnert-Hilger et al., 1985) and paramecia (Vilmart-Seuwen et al., 1986) can secrete when stimulated in the absence of ATP. In these systems, too, the secretory granules are already docked at the plasma membrane. This suggests that ATP is required at an early stage in secretion, but not for the fusion event itself.

Recent experiments by Holz et al. (1989) have cast doubt on the absolute requirement of permeabilized chromaffin cells for ATP. There appears to be a small component of Ca^{2+}-dependent, ATP-independent secretion from digitonin-permeabilized chromaffin cells. This component of secretion has a similar Ca^{2+} dependence to that of ATP-dependent secretion. The secretion is very fast, occurring within 2 min (ATP-dependent secretion occurs over more than 20 min). The ATP-independent secretion is abolished by pretreatment of the cells with metabolic inhibitors, suggesting that ATP is required to prime a small subpopulation of granules that are immediately secreted in response to a rise in $[Ca^{2+}]_i$. It is interesting that a small population of granules has been shown (Burgoyne et al., 1982) to be present at the plasma membrane beyond the cytoskeletal barrier (see Section 4.). It is possible that the ATP-requiring step is the translocation of granules through this barrier to the plasma membrane.

3. The Control of Exocytosis by Intracellular Messengers

3.1. Ca^{2+}

A major signal for secretion from chromaffin cells is a rise in $[Ca^{2+}]_i$ owing either to Ca^{2+} entry (Holz et al., 1982; Kilpatrick et al., 1982) or to a lesser extent to Ca^{2+} release from intracellular stores. The nicotinic receptor-associated channel responsible for cholinergic secretion in the adrenal chromaffin cell is permeable to Ca^{2+}. The chromaffin cell possesses an L-type voltage-dependent Ca^{2+} channel, which is blocked by dihydropyridine antagonists and activated by BAY-K-8644 (Artalejo et al., 1987; Cena et al., 1989). It has been suggested that the chromaffin cell also possesses a second type of dihydropyidine-insensitive, ω-conotoxin-sensitive channel that contributes to the voltage-dependent influx of Ca^{2+} on depolarization (Rosario et al., 1989). This may explain why dihydropyridine blockers do not completely block secretion elicited by nicotinic agonists. However, Cena et al. (1989) were able to identify only a single voltage-dependent Ca^{2+} channel in chromaffin cells.

The rise in $[Ca^{2+}]_i$ after depolarization owing to nicotinic agonists or high K^+ has been monitored in chromaffin cell populations with quin-2 (Knight and Kesteven, 1983; Burgoyne, 1984b) and fura-2 (O'Sullivan and Burgoyne, 1988), and in single cells with aequorin (Cobbold et al., 1987) and fura-2 (O'Sullivan et al., 1989; Cheek et al., 1989a,b). These studies agree that the basal $[Ca^{2+}]_i$ is in the range of 50-100 nM and that the optimal dose of nicotinic agonists leads to a rise in $[Ca^{2+}]_i$ of between 200 nM and 1 μM.

The most convincing evidence for the role of $[Ca^{2+}]_i$ as the signal for catecholamine secretion comes from permeabilized cell preparations (Bader et al., 1986; Baker and Knight, 1981; Burgoyne et al., 1989b; Dunn and Holz, 1983; Knight and Baker, 1982; Sontag et al., 1988) that secrete catecholamine in response to added Ca^{2+}. The optimal $[Ca^{2+}]_i$ required to elicit secretion varies depending on the method of permeabilization. In the two most studied systems, digitonin- and electro-permeabilization, optimal secretion occurs at about 10 μM Ca^{2+}. This is considerably greater than measured values of $[Ca^{2+}]_i$ found in chromaffin cells after cholinergic stimulation.

Recent work using fura-2 to image Ca^{2+} transients in single chromaffin cells has revealed a complex Ca^{2+} signal (O'Sullivan et al., 1989; Cheek et al., 1989a,b). On stimulation with nicotine or high K^+, the rise in $[Ca^{2+}]_i$ owing to Ca^{2+} influx is initially restricted to a region beneath the plasma membrane. Following diffusion of Ca^{2+} to fill the whole of the cytoplasm, a much larger elevation of $[Ca^{2+}]_i$ occurs internally, owing to release of Ca^{2+} from internal stores. The rise in $[Ca^{2+}]_i$ detected beneath the plasma membrane was only about 40–50 nM above basal. This is surprising since, at least in the case of depolarization by high K^+, the amount of Ca^{2+} entering through channels has been calculated to be sufficient, in the absence of Ca^{2+} buffering, to raise $[Ca^{2+}]_i$ by 170 μM throughout the cell within 1 s (Artalejo et al., 1987). The apparently small rise in $[Ca^{2+}]_i$ in the subplasmalemmal region may be the result of substantial Ca^{2+} buffering and rapid extrusion of Ca^{2+} or a higher elevation of $[Ca^{2+}]_i$ very close to the plasma membrane may have been beyond the temporal (200 ms) or spatial resolution of the video-imaging system.

The Ca^{2+}-dependent release of Ca^{2+} from internal stores following nicotinic stimulation or depolarization with high K^+ could result either from $Ins(1,4,5)P_3$-mediated or Ca^{2+}-mediated release. In favor of the first mechanism, it has been demonstrated that both nicotine and high K^+ result in the generation of $Ins(1,4,5)P_3$ in bovine chromaffin cells (Eberhard and Holz, 1987; Nakaki et al., 1988) possibly because of Ca^{2+}-dependent activation of phosphoinositidase C (Eberhard and Holz, 1988). However, a process of Ca^{2+}-induced Ca^{2+} release from internal stores, which has been suggested to occur in muscle, cannot be ruled out. Similar to muscle, the adrenal chromaffin cells have caffeine-sensitive stores (Burgoyne et al., 1989a), and these stores may be functional Ca^{2+}-releasable Ca^{2+} stores, as has been shown for sympathetic neurons following depolarization with high K^+ (Lipscombe et al., 1988). Both mechanisms, Ca^{2+}-induced Ca^{2+} release and Ca^{2+}-induced $Ins(1,4,5)P_3$ formation, may be operative in chromaffin cells, but it is clear that the caffeine-sensitive store is not colocalized with the agonist-sensitive store (Burgoyne et al., 1989a).

The $Ins(1,4,5)P_3$-sensitive store in the chromaffin cell appears to be localized in the area between the plasma membrane and the nucleus, away from the secretory vesicles (Burgoyne et al., 1989a). This is the area where the endoplasmic reticulum (ER) (O'Sullivan et al., 1989) and 140-kDa Ca^{2+}-ATPase-like protein, related to the Ca^{2+}-ATPase of muscle (Burgoyne et al., 1989a) are located. A 100-kDa Ca^{2+}-ATPase-like protein, reported to be associated with the calciosome, a putative Ca^{2+}-sequestering organelle (Volpe et al., 1988), has a diffuse localization across the cell, and a similar localization was seen for the $[Ca^{2+}]_i$ rise elicited by caffeine stimulation (Burgoyne et al., 1989a). It would therefore appear that the calciosome is likely to be the caffeine-sensitive store and that the $Ins(1,4,5)P_3$-sensitive store is more likely to be associated with the ER.

Despite raising $[Ca^{2+}]_i$, muscarinic agonists elicit little or no secretion from bovine chromaffin cells (Cheek and Burgoyne, 1985; Kao and Schneider, 1985). A likely explanation is that not all chromaffin cells respond to muscarinic agonists, and in those that do respond, the rise in $[Ca^{2+}]_i$ is not close to the plasma-membrane, but is restricted to an internal region often at one pole of the cell presumably owing to release of Ca^{2+} from local-ized internal stores (Cheek et al., 1989b). Caffeine raises $[Ca^{2+}]_i$ across the entire cell body because of the release of Ca^{2+} from internal stores, but it also causes very little catecholamine secre-tion (Burgoyne et al., 1989a; Cheek et al., 1990). Also, release of Ca^{2+} from internal stores in response to thapsigargin stimulates catecholamine secretion only because it also causes a secondary influx of Ca^{2+} (Cheek and Thastrup, 1989). It appears then that the optimal Ca^{2+} signal for the triggering of secretion is a single large transient elevation of $[Ca^{2+}]_i$ with a component of the rise occurring immediately beneath the plasma membrane.

Consistent with this interpretation are findings with angio-tensin II and histamine. Whereas nicotinic stimulation results in exocytosis around the whole cell, angiotensin II, which releases Ca^{2+} from internal stores, resulting in a polarized rise in $[Ca^{2+}]_i$, leads to only a low level of catecholamine secretion (O'Sullivan and Burgoyne, 1989). The secretion is also polarized (Cheek et al., 1989b), and this may be owing to polarized Ca^{2+} entry (Cheek et al., 1991). The more potent secretagogue histamine does not open voltage-dependent channels, but nevertheless stimulates Ca^{2+} entry around the whole plasma membrane, explaining its efficacy in stimulating secretion (Cheek et al., 1991).

3.2. Cyclic Nucleotides

Cyclic AMP (cAMP) is known to be involved in the regula-tion of secretion in a number of secretory cells. In the adrenal chromaffin cell, cAMP increases biphasically on cholinergic stimulation with an initial rapid rise that falls within 1 min, and a subsequent sustained elevation of cAMP (Morita et al., 1987a). The effect of cAMP in chromaffin cells remains a subject of controversy, with different groups reporting different effects. In digitonin- and electro-permeabilized chromaffin cells, cAMP has little effect on either Ca^{2+}-dependent or Ca^{2+}-independent catecholamine release (Knight and Baker, 1982; Bittner et al., 1986). However, in intact chromaffin cells, apparently contradictory effects of cAMP have been found. Baker et al. (1986) reported that elevated cAMP following forskolin treatment inhibited nicotine- but not high K^+-induced secretion; this was shown to occur at a step distal to Ca^{2+} entry, since the rise in $[Ca^{2+}]_i$ elicited by nicotine was not affected by the elevated cAMP. Similar results were found when cAMP was elevated by cholera toxin or by using the membrane-permeant cAMP analog 8-bromo-cAMP (Cheek and Burgoyne, 1987). It was therefore proposed that cAMP was acting on an unknown intracellular messenger produced by nicotinic stimulation, which was responsible for the greater potency of nicotinic- over high K^+-induced secretion. Two other groups have reported that elevating cAMP with low doses of forskolin potentiates high K^+-induced secretion and secretion elicited by suboptimal doses of acetylcholine (Morita et al., 1987b; Marriott et al., 1988). The reports are compatible with each other, since the doses of forskolin used by Baker et al. (1986) and Cheek and Burgoyne (1987) were considerably higher. Indeed, at high doses of forskolin, the data of Morita et al. (1987b) demonstrate an inhibitory effect on acetylcholine-induced secretion. More recently, Negishi et al. (1989) have published results entirely consistent with Cheek and Burgoyne (1987), in which they found that nicotine-induced secretion was inhibited by forskolin and by dibutyryl-cAMP.

Cyclic GMP (cGMP) is generated in chromaffin cells by acetylcholine acting at a muscarinic cholinergic receptor (Schneider et al., 1979). In cultured bovine adrenal cells, the nicotinic receptor also causes an increase in cGMP (Nakaki et al., 1988). cGMP has no effect on Ca^{2+}-dependent secretion from permeabilized cells (Knight and Baker, 1982). In intact cells, elevating cGMP with nitroprusside, atrial naturietic peptide, or cell-permeable analogs, such as 8-bromo-cGMP, increases the sensitivity of chromaffin cells to nicotinic stimulation, but has no effect on secretion elicited by other agonists (O'Sullivan and Burgoyne, 1990).

4. The Cytoskeleton and Control of Exocytosis

In chromaffin cells, the secretory vesicles are anchored by a crosslinked filamentous lattice (Kondo et al., 1982), and the appears to be prevented by an electron-dense filamentous meshwork within the cell cortex (Burgoyne et al., 1988a; Burgoyne and Cheek, 1987a,b; Cheek and Burgoyne, 1990). For exocytosis to occur, the granules must be translocated to the plasma membrane. The cytoskeleton has been proposed to be involved in the mechanism of secretion in two ways: first, by acting as a contractile force to bring the granules to the plasma membrane and, second, by acting as a barrier to exocytosis.

The cytoskeleton has been proposed to act in generating an active contractile force, analogous to Ca^{2+}-activated sliding filaments in muscle. In favor of this hypothesis is the demonstration of actin, myosin, and tropomyosin in the chromaffin cell (Cheek and Burgoyne, 1990). However, the kinetics of secretion (Burgoyne et al., 1988a) can be accounted for by diffusion with no kinetic requirement for a contractile mechanism.

The actin-based cytoskeleton in the chromaffin cell is organized as a dense network of filaments under the plasma membrane. It has been proposed that this network acts as a barrier and that stimulation of the cell leads to the transient breakdown of the barrier, allowing the granules access to the site of fusion (Burgoyne et al., 1986). This hypothesis was first suggested in 1972 by Orci et al. to account for secretion from the pancreatic β-cell. Evidence for a similar mechanism has been described for a number of other secretory systems, including neutrophils, sea urchin eggs, preglomerular arterioles, macrophages, exocrine acinar cells, and synaptic nerve terminals (for review, *see* Cheek and Burgoyne, 1991).

Nicotinic, but not high K^+, stimulation of chromaffin cells leads to a transient change in the state of actin polymerization from F (filamentous) to G (globular) (Cheek and Burgoyne, 1986). The reduction in F-actin is independent of external Ca^{2+} and is inhibited by high levels of cAMP (Cheek and Burgoyne, 1987). This disassembly may be responsible for the greater efficacy of nicotine as a secretagogue compared to high K^+. This mechanism may also explain the inhibition of secretion by cAMP of nicotine, but not high K^+-induced secretion. Changes in the actin-cytoskeleton also occur in response to high K^+. It is likely that Ca^{2+} entry owing to high K^+ results in a reduction in actin crosslinking without disassembly of the filaments. Both high K^+ and nicotine evoke a reduction in the amount of actin cross-linked into the Triton-insoluble cytoskeleton (Burgoyne et al., 1989b). In the case of nicotine, this is Ca^{2+}-independent, and in the case of high K^+, it is Ca^{2+}-dependent. The reduction of actin in the Triton-insoluble cytoskeleton also occurs on elevating $[Ca^{2+}]$ in digitonin-permeabilized chromaffin cells.

The plasma membrane-associated actin-crosslinking protein fodrin also undergoes changes on stimulation. In the resting cell, fodrin is localized in the subplasmalemmal cytoskeleton. On stimulation, the fodrin rearranges into patches at the plasma membrane (Perrin and Aunis, 1985), and this redistribution may be involved in clearing the exocytotic sites at the plasma membrane. This is supported by the finding that an antibody to fodrin partially inhibits Ca^{2+}-dependent secretion from digitonin-permeabilized chromaffin cells (Perrin et al., 1987).

Additional evidence for a role for the cytoskeleton as a barrier to exocytosis comes from the effect of agents that promote disassembly of actin (cytochalasins and DNAase 1), which increase secretion in permeabilized chromaffin cells (Lelkes et al., 1986; Sontag et al., 1988). Phalloidin, which would be expected to inhibit depolymerization, inhibits secretion from digitonin-permeabilized cells. In addition, botulinum C2 toxin, which ADP-ribosylates actin and inhibits actin polymerization, has been shown to enhance secretion from PC12 cells (Matter et al., 1989).

A number of proteins could be responsible for the Ca^{2+}-dependent cytoskeletal rearrangements in chromaffin cells. Both the stable analog of GTP, GTPγS, and the phorbol ester, PMA, potentiate the Ca^{2+}-dependent cytoskeletal rearrangements in permeabilized chromaffin cells. The effect of GTPγS is blocked by neomycin, suggesting that its effects may be mediated via phosphoinositidase C (PIC) rather than through another G protein pathway. In intact cells, PMA appears to be able to initiate some rearrangement on its own, suggesting that at least part of the Ca^{2+}-dependent cytoskeletal rearrangement occurs through Ca^{2+} activation of protein kinase C (Burgoyne et al., 1989b). This could explain the requirement for ATP, for a second phase of exocytosis (Holz et al., 1989), following reorganization of the cytoskeleton and movement of secretory vesicles to the plasma membrane.

A number of cytoskeletal-associated proteins can alter the polymerization and crosslinking of the actin in a Ca^{2+}-dependent manner (for review, *see* Cheek and Burgoyne, 1990), and several of these have been found in the adrenal chromaffin cell. Gelsolin, a Ca^{2+}-dependent actin-severing protein, has been localized to the cortical cytoskeleton (Sontag et al., 1988). The actin-crosslinking protein caldesmon has also been found in the subplasmalemmal region of the chromaffin cell (Burgoyne et al., 1986). Caldesmon crosslinks actin filaments, but at high Ca^{2+} concentrations (above micromolar), it binds instead to calmodulin, releasing the actin filaments. Thus, when $[Ca^{2+}]_i$ rises on stimulation, caldesmon will cease to crosslink the actin filaments, allowing the secretory granules easier access to the plasma membrane. At resting Ca^{2+} concentrations, chromaffin granules increase the viscosity of F-actin, indicative of crosslinking of the filaments. This property is Ca^{2+}-dependent, being inhibited at micromolar Ca^{2+} (Fowler and Pollard, 1982). This Ca^{2+}-dependence may be conferred by α-actinin (Bader and Aunis, 1983).

5. Molecular Components of the Exocytotic Mechanism

Ca^{2+} initiates a number of different processes that lead to exocytosis: movement of the secretory vesicle to the plasma membrane, docking of the vesicle at the plasma membrane prior to fusion, fusion of the vesicle with the plasma membrane, and fission of the fused membranes. In some cell types, such as the sea urchin egg (Crabb and Jackson, 1985) and the PC12 cell line (Schafer et al., 1987b), the secretory vesicles are already docked at the plasma membrane, and the rise in $[Ca^{2+}]_i$ is only required to initiate fusion and fission. In chromaffin cells, most secretory vesicles are at least 300 nm from the plasma membrane, their movement restricted by a filamentous barrier (Burgoyne et al., 1982,1988a). The rise in $[Ca^{2+}]_i$ in chromaffin cells must therefore stimulate all four parts of the process to allow secretion to occur. The various aspects of the secretory process, including exocytosis, are likely to involve specific secretory vesicle and/or plasma membrane proteins that mediate membrane-membrane and membrane-cytoskeletal interactions. In addition, the basic components of the exocytotic mechanism are likely to be widespread and highly conserved given the ubiquitous nature of exocytosis. One approach that has been used in attempts to identify the molecular components of the exocytotic machinery has been to search for secretory vesicle proteins that are bona fide constituents of the secretory vesicle membrane or that become associated with the secretory vesicle following cell stimulation. Information is becoming available on the nature of such proteins from studies on secretory vesicles of a variety of mammalian secretory cells and also from nonmammalian systems, including constitutive secretory mutants of yeast and cortical granule exocytosis in the sea urchin egg. With the idea that the basic mechanisms of exocytosis are likely to be common to all cell types, attempts have been made to detect proteins common to a variety of secretory vesicle membranes with the expectation that exocytosis-specific proteins could be found. So far, no proteins have been found that can be detected on secretory vesicles from all types of secretory cells. The failure to find proteins present that are common integral membrane components of all types of secretory vesicles could be the result of one of four possible explanations. The first is that the essential exocytotic proteins have not yet been found. The second is that exocytosis in different tissues involves different secretory vesicle proteins. The third is that essential vesicle-associated proteins required for exocytosis are not integral membrane proteins of the secretory vesicle, but are loosely bound extrinsic proteins that can be lost during isolation of vesicles. The fourth is that the essential proteins are ones that only become associated with secretory vesicles following stimulation. A number of proteins, including cytoskeletal proteins, are known to be loosely associated with secretory vesicles and, in some cases, are lost on isolation of vesicles. In addition, a number of proteins can reversibly associate with secretory vesicles in a manner regulated by protein phosphorylation or by Ca^{2+}. The evidence for the involvement of various proteins in the exocytotic machinery is discussed below.

5.1. Phospholipase A_2 and Arachidonic Acid

Arachidonic acid can act as a membrane fusogen in in vitro systems, but the levels of Ca^{2+} required are unphysiologically high (Creutz, 1981). Nevertheless, arachidonic acid could have a physiological role as a fusogen, since it is produced under all conditions in which exocytosis occurs in chromaffin cells (Frye and Holz, 1984, 1985). Arachidonic acid can be generated by several pathways, including activation of phospholipase A_2, breakdown of diacylglycerol following PIC activation, or protein kinase C-mediated inhibition of reacylation of arachidonic acid. A number of inhibitors of phospholipase A_2 and arachidonic acid metabolism inhibit secretion from intact bovine chromaffin cells. However, it is difficult to determine the exact role of arachidonic acid in exocytosis, since the inhibitors of phospholipase A_2 also inhibit Ca^{2+} entry (Frye and Holz, 1983). Work on digitonin-permeabilized chromaffin cells has suggested that arachidonic acid is not involved in the stimulatory effect of GTP analogs on secretion (Morgan and Burgoyne, 1990a). In addition, the release of arachidonic acid in response to Ca^{2+} is not caused by activation of phospholipase A_2 or PIC, and is not required for Ca^{2+}-dependent exocytosis in permeabilized cells (Morgan and Burgoyne, 1990b).

5.2. Calmodulin

From immunocytochemical studies, calmodulin appears to be diffusely distributed throughout the cytoplasm of the chromaffin cell. In the presence of Ca^{2+}, calmodulin binds to purified chromaffin granule membranes (Burgoyne and Geisow, 1981; Geisow et al., 1982). However, some calmodulin will also bind to secretory granules in the absence of Ca^{2+} (Geisow and Burgoyne, 1983). The use of ^{125}I-labeled calmodulin has identified a number of Ca^{2+}-dependent calmodulin-binding proteins on the granule membrane. The two major proteins have mol wt of 53 and 65 kDa (Geisow and Burgoyne, 1983; Bader et al., 1985). The 65-kDa protein (p65) is an integral membrane component of synaptic vesicles and other secretory vesicles (Fournier and Trifaro, 1988). The function of p65 remains unknown despite its widespread distribution in secretory cells.

Addition of Ca^{2+} and calmodulin with ATP to secretory vesicles in vitro results in phosphorylation of several polypeptides (Burgoyne and Geisow, 1981). The possibilty that a calmodulin-dependent protein kinase could be involved in exocytosis is attractive given the requirement for ATP for continued exocytosis.

Calmodulin antagonists have been used extensively to try to determine the role of calmodulin in secretion. However, because of their nonspecific effects, any results should be treated with caution. The phenothiazine-derived calmodulin antagonists have been shown to inhibit secretion from chromaffin cells (Baker and Knight, 1981; Kenigsberg et al., 1982; Slepetis and Kirshner, 1982; Burgoyne et al., 1982). However, at least part of this inhibitory effect is the result of an inhibition of Ca^{2+} entry (Kenigsberg et al., 1982; Sleptis and Kirshner, 1982). In contrast, the calmodulin antagonist calmidazolium has been shown to inhibit secretion elicited by carbamylcholine and the Ca^{2+} ionophore, A23187, without affecting the rise in $[Ca^{2+}]_i$, indicating that calmodulin may have a role in secretion distal to Ca^{2+} entry (Burgoyne and Norman, 1984). Further evidence for a role for calmodulin acting at a site distal to Ca^{2+} entry comes from ultrastructural studies, which indicate that treatment of cells with trifluoperazine allows granules to accumulate at the plasma membrane, but prevents fusion (Burgoyne et al., 1982).

More convincing evidence for the role of calmodulin in secretion comes from the finding that affinity-purified calmodulin antibodies loaded into chromaffin cells by fusion with red blood cell ghosts inhibit secretion stimulated by high K^+ or acetylcholine (Kenigsberg and Trifaro, 1985). However, this does not rule out Ca^{2+} entry as one of the sites of action of calmodulin. It is also possible that the antibody–antigen complex is acting to inhibit sterically either docking of the granule to the plasma membrane or fusion itself. The antibodies do block the Ca^{2+}-dependent binding of calmodulin to granule membranes, suggesting a possible mechanism for the observed inhibition of secretion (Bader et al., 1985).

Despite extensive investigation of calmodulin, it is still not clear whether or not it is required for exocytosis. Indications that calmodulin may not be required have come from the findings that trifluoperazine had no effect on Ca^{2+}-dependent exocytosis in saponin-permeabilized chromaffin cells (Brooks and Treml, 1984), and calmodulin antagonists and anticalmodulin antibodies had no effect on exocytosis in streptolysin O-permeabilized cells (Ahnert-Hilger et al., 1989). Clearly, further work is required to determine if calmodulin is involved in exocytosis.

5.3. Secretory Vesicle Docking Protein

The initial docking of secretory vesicles onto exocytotic sites on the plasma membrane is likely to be mediated by a specific protein. One possible candidate for a docking protein in chromaffin cells is a 51-kDa plasma membrane protein that had been isolated by its ability to bind purified chromaffin cell secretory granules (Schweizer et al., 1989). Antibodies specific for this protein inhibited exocytosis in PC12 cells following their introduction by fusing red cell ghosts loaded with the antibody to the cells. In addition, exocytosis in chromaffin cells was also inhibited when the antibody was introduced via a patch pipet and exocytosis monitored using the membrane capacitance technique (Schweizer et al., 1989). The 51-kDa protein does not appear to be regulated by Ca^{2+}, and so if it is involved in secretion, it may act only at the stage of vesicle docking prior to Ca^{2+}-activated exocytosis.

5.4. Protein Kinase C

Diacylglycerol, the activator of protein kinase C (PKC), is generated via receptor-linked PIC in chromaffin cells as in other cell types (Burgess, this vol.). In the bovine adrenal chromaffin cell, the muscarinic (Fisher et al., 1981), angiotensin II (Bunn and Marley, 1989), histamine, and bradykinin (Plevin and Boarder, 1988) receptors are all linked to PIC. Cultured chromaffin cells also have a Ca^{2+}-activatable PIC, so that stimulation by high K^+ or nicotine, in the presence of external Ca^{2+}, results in the generation of Ins(1,4,5)P_3 and diacylglycerol; this does not occur in the absence of external Ca^{2+} in the case of high K^+ (Eberhard and Holz, 1987,1988). PIC can also be activated in digitonin-permeabilized chromaffin cells by raising the $[Ca^{2+}]_i$ (Eberhard and Holz, 1987). It would therefore appear that PIC is activated in all cases where Ca^{2+}-dependent secretion occurs.

In many cell types, activation of PKC has a marked stimulatory effect on secretion (Rink et al., 1983). In rat adrenal gland, activation of PKC potentiates Ca^{2+} influx and secretion owing to nicotine or high K^+ (Wakade et al., 1986). Activation of PKC by phorbol esters in intact bovine adrenal chromaffin cells has been reported to elicit little (Brocklehurst et al., 1985) or no secretion (Knight and Baker, 1983; Burgoyne and Norman, 1984).

In permeabilized chromaffin cells, phorbol esters increase the affinity of the exocytotic machinery for Ca^{2+} (Knight and Baker, 1983; Pocotte et al., 1985). Protein kinase C is also activated directly by Ca^{2+} in permeabilized cells (TerBush et al., 1988) and Ca^{2+} stimulates phosphorylation of the same proteins that are phosphorylated in response to phorbol esters (Lee and Holz, 1986). The question of whether PKC activation is an absolute requirement for exocytosis has been addressed by the use of a number of different inhibitors. Inhibition of PKC by sphingosine or by downregulation of PKC by long-term treatment with phorbol esters partially inhibited Ca^{2+}-dependent secretion from digitonin-permeabilized cells (Burgoyne et al., 1988b). In another study, PKC inhibitors were shown to inhibit secretion from electro-permeabilized chromaffin cells partially (Knight et al., 1989). Some secretion remains even after treatment of cells with high concentrations of PKC inhibitors, and so these findings suggests that PKC is involved in Ca^{2+}-dependent secretion in a modulatory role, but is not essential for exocytosis.

5.5. Guanine Nucleotide-Binding Proteins

The possible involvement of guanine nucleotide-binding proteins (G proteins) in exocytosis has been investigated using nonhydrolyzable analogs of GTP, such as GTPγS and GMP-PNP, which bind to GTP-binding proteins and constitutively activate them. GDPβS, an analog of GDP, has also been used to lock G proteins in an inactive state. The use of these compounds in permeabilized secretory cells, initially by Gomperts and coworkers, has indicated that G proteins may be involved in the control of exocytosis at stages other than transmembrane signal transduction. In permeabilized neutrophils, GTP analogs are able to elicit Ca^{2+}-independent secretion that does not appear to be associated with activation of protein kinase C (Barrowman et al., 1986). A similar effect has been shown in several other cell types (Burgoyne, 1987). The experiments have led to the idea that there is a G protein, termed G_e, directly involved in exocytosis (Gomperts, 1986).

In adrenal chromaffin cells, GTPγS and GMP-PNP have been shown to cause Ca^{2+}-independent secretion in digitonin-permeabilized chromaffin cells (Bittner et al., 1986). Burgoyne et al. (1989b) showed that GMP-PNP elicited Ca^{2+}-independent secretion, but that GTPγS simply increased the Ca^{2+}-sensitivity of secretion in a manner similar to activation of PKC with PMA. More recently, it has been shown that a range of GTP analogs are able to elicit Ca^{2+}-independent exocytosis in chromaffin cells and that these effects are not the result of activation of known intracellular messenger pathways, including generation of arachidonic acid (Morgan and Burgoyne, 1990a). In contrast with these reports are the findings of Knight and Baker (1985), who demonstrated that GTPγS inhibited Ca^{2+}-dependent secretion from bovine chromaffin cells, whereas it potentiated Ca^{2+}-dependent secretion from chicken adrenal chromaffin cells. The reason for the differences in these findings is unclear. Although it is clear that activation of G proteins in chromaffin cells exerts a regulatory effect on exocytosis, it is unlikely that such proteins are essential for Ca^{2+}-dependent exocytosis because GDPβS has little effect on Ca^{2+}-stimulated exocytosis in chromaffin cells (Morgan and Burgoyne, 1990a).

The finding that GTP effects in secretory cells are apparently independent of any known intracellular messenger has led to an interest in the "*ras*-like" class of small guanine nucleotide-binding proteins. Ha-*ras* microinjected into mast cells resulted in exocytotic degranulation (Bar-Sagi and Gomperts, 1988). The most convincing evidence for a role in exocytosis for GTP-binding proteins comes from work on temperature-sensitive yeast secretory mutants. The SEC4 mutant has a block in secretion at a late stage prior to exocytotic fusion, leading to a buildup of secretory vesicles in the cytoplasm. The SEC4 gene product is a 23.5-kDa protein with 32% homology with *ras* (Salminen and Novick, 1987), and the SEC4 protein can bind GTP on nitrocellulose blots (Goud et al., 1988). Wild-type SEC4 is found on the cytoplasmic faces of both the plasma membrane and the secretory vesicle, and it appears to be essential for constitutive exocytosis. Small GTP-binding proteins have been found in many tissues and cell types. In chromaffin cells, these proteins are found on the cytoplasmic face of the secretory vesicles and appear to consist of a family of at least 4–5 proteins with molecular masses of 20–24 kDa (Burgoyne and Morgan, 1989). It seems possible that these proteins may serve analogous roles to the yeast GTP-binding proteins in regulating vesicular traffic and exocytosis.

5.6. Phosphorylation, Dephosphorylation, and Exocytosis

As described above, ATP is essential for the priming and, in some cases, maintenance of exocytosis, suggesting that protein phosphorylation may be important. In the presence of the ATP analog ATPγS, protein kinases thiophosphorylate proteins that then become resistant to dephosphorylation by phosphatases (*see* chapters by Moreton and Potter, this vol). ATPγS inhibits exocytosis in saponin-permeabilized chromaffin cells, and the interpretation of these data was that phosphorylation may be needed to prime exocytotic sites, but that exocytotic fusion required dephosphorylation of a protein (Brooks et al., 1984). The idea that dephosphorylation is involved in exocytosis has also been suggested by work on other systems.

In the mast cell, ATP is not essential for Ca^{2+}-dependent secretion, but it increases the Ca^{2+}-sensitivity of the permeabilized cell. When ATP is present with GTPγS, the lag before onset of secretion elicited by Ca^{2+} is markedly increased. This has been interpreted as evidence that a phosphatase activated by a G protein is involved at a late step in exocytosis (Tatham and Gomperts, 1989). In *Paramecium*, dephosphorylation of a 65-kDa protein occurs prior to exocytosis (Zieseniss and Plattner, 1985), and this dephosphorylation appears to be caused by calcineurin, a Ca^{2+}-calmodulin-dependent protein phosphatase. Microinjection of anticalcineurin antibodies blocks exocytosis, although microinjection of the active Ca^{2+}–calmodulin–calcineurin complex elicits exocytosis in *Paramecium* (Momayezi et al., 1987). As with mast cells, ATP in *Paramecium* inhibits exocytosis in a ghost preparation, but is required for secretion in the intact organism (Vilmart-Seuwen et al., 1986). Thus, for two secretory systems, ATP appears to be involved in priming exocytosis, but later steps appear to involve an ATP-inhibitable event, possibly dephosphorylation. The existence of the 63–65-kDa *Paramecium* protein (parafusin), which undergoes dephosphorylation, in yeast and in mammalian tissues, such as rat brain, thyroid, liver, and adrenal gland (Satir et al., 1989), suggests that it may be a highly conserved protein and intimately involved with exocytosis, though as yet, no data are available on its function in other secretory cells.

5.7. Involvement of Calpactin in Exocytosis

The adrenal medulla contains many proteins that bind reversibly with secretory vesicle membranes and phospholipid bilayers in the presence of micromolar concentrations of Ca^{2+} (Burgoyne, 1990; Creutz et al., 1983,1987; Geisow and Burgoyne, 1982). A major family of these proteins is the annexins. The family consists of at least eight different proteins known as annexins I–VIII, including lipocortin (annexin I), calpactin (annexin II), p70 (annexin VI), and synexin (annexin VII) (for review, *see* Burgoyne and Geisow, 1989). This family of proteins has been isolated from a wide range of mammalian tissues, and chromaffin cells contain several of these proteins. Each member of the family contains a conserved sequence of 70 amino acids repeated four times (except for p70, which contains eight repeating units). In addition to binding Ca^{2+}, lipocortin and calpactin are substrates for a number of kinases, including PKC, cAMP-dependent protein kinase, calmodulin-dependent protein kinase, and tyrosine kinases with phosphorylation occurring in the variable amino terminal domain of the proteins. The function of two of the annexin family, synexin and calpactin, has been studied intensively in the chromaffin cell, and there is now evidence that calpactin may be required for exocytosis possibly as the Ca^{2+} receptor that mediates membrane fusion.

Synexin was first isolated from adrenal medulla by Creutz et al. (1978). This 47-kDa protein lowers the $[Ca^{2+}]$ required to aggregate chromaffin cell granules (Creutz et al., 1978). However, it seems unlikely that synexin-mediated aggregation and fusion could occur in vitro, since the $[Ca^{2+}]$ required is unphysiologically high. Pollard has developed a hypothesis, the "hydrophobic bridge," which envisages the insertion of synexin into the plasma membrane and granule membrane to promote fusion (Pollard et al., 1988). However, these effects require high $[Ca^{2+}]$. Optimal secretion is achieved at around 10 μM in permeabilized chromaffin cells. It therefore seems unlikely that synexin acting alone could have a role in Ca^{2+}-dependent secretion.

Calpactin is a more likely candidate for the Ca^{2+} receptor in exocytosis. Calpactin is bound to the inner surface of the plasma membrane in chromaffin cells (Burgoyne and Cheek, 1987a) and other cell types, and can exist as a monomer (p36) or a heterotetramer of calpactin and another protein of 11-kDa ($p36_2p11_2$). The calpactin monomer and tetramer aggregate membranes at physiologically relevant $[Ca^{2+}]$. The tetramer is the most potent protein in aggregating granules with a threshold for $[Ca^{2+}]$ of 0.7 μM and half-maximal aggregation occurring at 1.8 μM (Drust and Creutz, 1988). Calpactin is the only annexin to aggregate granules at a $[Ca^{2+}]$, which is physiologically realistic (Burgoyne, 1988). Following nicotinic stimulation, calpactin becomes phosphorylated on alkali-sensitive sites (possibly serine residues) (Creutz et al., 1987).

Following digitonin-permeabilization, calpactin (Ali et al., 1989) and other cytosolic proteins (Sarafian et al., 1987) leak out of the chromaffin cells. This is accompanied by a loss of secretory response to elevated Ca^{2+}. At least part of this response can be reconstituted by addition of the calpactin tetramer or the p36 subunit. This reconstitution can be inhibited by an affinity-purified antibody to p36 (Ali et al., 1989). Other annexins were not active in this assay, and it appears that the activity of calpactin is dependent on its unique N-terminal domain. Controlled proteolytic cleavage of the N-terminus generates the core, p33, which can still bind to lipid bilayers, but is inactive in reconstituting exocytosis (Ali and Burgoyne, 1990). The N-terminal region is likely to be an important regulatory domain of calpactin required for full biological activity. In addition to the reconstitution experiments, it was found that secretion from permeabilized chromaffin cells was inhibited by a synthetic peptide (CAMK-GAGTDEGALIEI-LASR) based on the conserved repeating unit of the annexin family (Ali et al., 1989). The peptide also inhibits the reconstituting effect of exogenous p36. This evidence suggests that calpactin is involved in the mechanism of Ca^{2+}-dependent exocytosis in permeabilized chromaffin cells. Since calpactin can aggregate secretory vesicles and cause membrane fusion in vitro (Drust and Creutz, 1988), it is possible that calpactin itself can result in membrane fusion in intact cells following its binding to secretory granules in the presence of elevated $[Ca^{2+}]$.

Further evidence suggesting that calpactin plays a role in the late stages of the secretory process has come from ultrastructural studies. Nakata et al. (1990) found, using immunogold labeling, that calpactin was concentrated at the sites of exocytosis in stimulated chromaffin cells. In addition, in rapidly frozen and deep-etched material, it was seen that aggregation of chromaffin granules or liposomes by calpactin was the result of the formation of 6–10 nm long crosslinking filaments consisting of calpactin. Essentially identical filaments were found to link secretory granules to the plasma membrane at the sites of exocytosis in stimulated cells, suggesting that calpactin was responsible for a late stage of interaction between granules and plasma membrane. Calpactin is therefore a major candidate for a molecular component of the exocytotic machinery, but its mode of action in intact cells remains to be determined (Fig. 2).

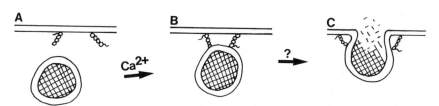

Fig. 2. Proposed involvement of calpactin in exocytosis. In resting cells (A) calpactin, consisting of four conserved domains and an N-terminal tail, is localized to the inner surface of the plasma membrane. Following a rise in $[Ca^{2+}]_i$, secretory vesicles become crosslinked to the plasma membrane via calpactin (B), and this leads on to exocytotic fusion and fission of membranes to allow release of vesicle contents (C).

6. Technical Aspects
in the Study of Exocytosis in Chromaffin Cells

From the preceding review, it should be clear that major progress in the understanding of the intracellular factors involved in the control and mechanisms of exocytosis has come from determination of intracellular messenger levels in conjunction with secretion, and even more, from the use of permeabilized cells that have allowed manipulation of both small molecules and proteins within them. The potentially powerful technique of patch-clamp capacitance measurement has been little exploited probably because of the technical difficulty of the technique. Instead, many laboratories have used methods for the bulk permeabilization of chromaffin cells or other cell types. In the case of chromaffin cells, most information is available for the relatively straightforward technique of digitonin-permeabilization of cultured cells. In the following sections, the methods used in my laboratory for the study of intact and digitonin-permeabilized chromaffin cells will be detailed. As well as the methods for the determination of secretion, the techniques used for the determination of $[Ca^{2+}]_i$ in cell populations and changes in the state of the cytoskeleton are also described.

6.1. Isolation and Culture of Chromaffin Cells

Fresh bovine adrenal glands are transported to the laboratory from a local abbattoir within 1 h of slaughter. Chromaffin cells are dissociated from isolated medullae by enzymatic digestion in Ca^{2+}- and Mg^{2+}-free Krebs-Ringer buffer containing 145 mM NaCl, 5 mM KCl, 1.2 mM NaH$_2$PO$_4$, 10 mM glucose, and 20 mM HEPES, pH 7.4 (Buffer A) as follows.

1. Inject gland three times with Buffer A.
2. Inject gland with 5 mL of 0.1% protease (Sigma P5417) in Buffer A.
3. Incubate gland at 37°C for 15 min.
4. Repeat Steps 2 and 3.
5. Dissect out medulla and chop into fine pieces with scissors.
6. Put medulla into a tissue culture flask (75 cm^2) with 10 mL per gland of 0.1% collagenase (Sigma C0130) in Buffer A, gas neck of flask with 5% O_2/95% CO_2, and then shake at 120 rpm at 37°C on a shaking water bath.
7. After 30 min, filter dissociated cells through muslin.
8. Centrifuge filtrate at 1000 rpm in a bench centrifuge for 10 min.
9. Resuspend pellets by flicking tube in Buffer A.
10. Centrifuge at 400 rpm for 10 min.
11. Resuspend pellets in Buffer A as in 9, but then underlay with 5 mL of 4% bovine serum albumin in Buffer A. Repeat Step 10.
12. Resuspend pellets in 10 mL Buffer A, filter through single muslin, and count cells.
13. Repeat Step 10, and resuspend pellets in culture medium and plate as required.

(N.B. all buffers must be well gassed with 5% O_2/ 95% CO_2 just before use.)

For culture, chromaffin cells can be purified by differential plating in culture medium (Dulbecco's Modified Eagle's Medium with 25 mM HEPES, 10% fetal calf serum, 8 μM fluorodeoxyuridine, 50 μg/mL gentamycin, 10 μM cytosine arabinoside, 2.5 μg/mL fungizone, 25 U/mL penicillin, 25 μg/mL streptomycin). The cell suspension is incubated at 37°C for 90 min in a tissue culture flask to allow attachment of nonchromaffin cells. Medium containing unattached cells is removed and used to plate out cells for secretion experiments in 24-well trays at a density of 0.7–1 × 10⁶ cells/well and used after 2–7 d in culture. Cell density is determined by counting cells that exclude Trypan blue (0.1%) in a hemocytometer under phase-contrast optics. Live cells appear phase-bright and clearly distinguishable from the stained dead cells. Despite the presence of mitotic inhibitors, extensive growth of nonchromaffin cells can still occur in cultures plated directly on plastic. However, when cultures are grown on plastic pretreated with poly-L-lysine (10 µg/mL in DMEM for 30 min) contaminating, dividing nonchromaffin cells die within 2–3 d, and even after 6 d, the cultures contain at least 90–95% chromaffin cells. Chromaffin cells can be identified by vital staining with 0.3 mg/mL neutral red, which is taken up into secretory granules (Livett, 1984).

6.2. Determination of Catecholamine Secretion

For the use of freshly isolated cells for secretion experiments, cells after isolation (Step 11) are washed twice by centrifugation in buffer A containing 1.3 mM MgCl$_2$, 3 mM CaCl$_2$, and 0.5% bovine serum albumin (buffer B), and preincubated for 1 h in buffer B. The cells are washed twice with buffer B and stimulated by addition of cells to microcentrifuge tubes containing secretagogues in buffer B. Stimulation with 55 mM KCl is carried out by adding iso-osmotic buffer in which KCl replaced an equal concentration of NaCl. After a fixed time, the reaction is stopped by addition of an equal volume of ice-cold buffer A containing 1.3 mM MgCl$_2$, 1 mM EGTA, and 0.5% bovine serum albumin, and the cells centrifuged for 2 min at 16,000g in a microcentrifuge. Aliqouts of supernatant are taken for catecholamine assay. For experiments using cultures, the culture medium is removed, and cells are washed with buffer B and stimulated by addition of buffer containing secretagogues. After a fixed time, catecholamine release is determined by removal of buffer from wells, which is centrifuged for 2 min at 16,000g in a microcentrifuge before duplicate aliquots are assayed for total catecholamine using a fluorimetric assay (Von Euler and Floding, 1955). Total catecholamine remaining within the cells is determined after release of catecholamines with 1% Triton X-100. Since the absolute levels of catecholamine released vary among cell preparations, the secretion data are calculated and expressed as percentage of total cellular catecholamine released.

6.3. Digitonin-Permeabilization of Cell Cultures

Cultures are washed twice with buffer A containing 1.3 mM MgCl$_2$ and permeabilized by addition of 20 μM digitonin (Calbiochem, Nottingham, UK) in 139 mM K glutamate, 2 mM ATP, 2 mM MgCl$_2$, 5 mM EGTA, and 20 mM PIPES with appropriate amounts of CaCl$_2$ to give the calculated free Ca^{2+} concentration at pH 6.5. Standard Ca^{2+} buffers for secretory experiments had free Ca^{2+} concentrations of 0 (no added CaCl$_2$), 0.1 μM (0.43 mM CaCl$_2$), 0.3 μM (0.64 mM CaCl$_2$), 1.0 μM (1.66 mM CaCl$_2$), 3.0 μM (3 mM CaCl$_2$), 10 μM (4.16 mM CaCl$_2$), and 30 μM (4.7 mM CaCl$_2$). Permeabilization buffers can be prepared in bulk and stored for prolonged periods at –20°C before use. Where appropriate, other agents, such as GTP analogs or phorbol esters, can be present throughout the incubation in permeabilization buffer. In some cases, cells can be prepermeabilized for 6 min before stimulation with Ca^{2+} in the presence or absence of other agents to ensure that added compounds do not affect permeabilization. After a fixed time, aliquots of buffer are removed, centrifuged at 16,000g for 2 min, and assayed for released catecholamines.

6.4. Assay of Cytoskeletal Actin

In the assay of cytoskeletal actin, both cultured and freshly isolated cells can be used. In the case of cultured cells (with or without permeabilization), the medium is removed after stimulation of the cells and the cells extracted in Triton extraction buffer (1% Triton X-100, 50 mM KCl, 5 mM EGTA, 2 mM MgCl$_2$, 2 mM phenylmethylsulfonyl fluoride, 10 mM Tris-HCl, pH 7.0) for 5 min on ice. The Triton-insoluble cytoskeletal residues are solubilized in 150–225 μL/well of SDS-solubilization buffer (1.25% Na dodecyl sulfate, 2 mM EDTA, 10% sucrose, 10% glycerol, 1% 2-mercaptoethanol, 125 mM Tris-HCl, pH 6.8) and boiled immediately for 2 min. Cytoskeletal actin and catecholamine secretion can be assayed in parallel using the same cultures. With freshly isolated cells, aliquots of cells (9 × 10^5 cells) are added to secretagogues in microcentrifuge tubes. After incubation at room temperature for 0–90 s, an equal vol of ice-cold buffer A containing 20 mM EGTA is added. The cells are kept on ice, pelleted by centrifugation at 4°C in a microcentrifuge at 16,000g for 10 s, and resuspended in Triton extraction buffer. After incubation for 5 min on ice, the Triton-insoluble cytoskeletons are pelleted by centrifugation at 4°C at 16,000g for 1 min, and solubilized and boiled in SDS-solubilization buffer. Samples are analyzed, immediately or after freezing and storage for 24 h, for actin content by SDS-polyacrylamide gel electrophoresis on mini slab gels. In all experiments, samples from each experimental condition should be run on the same slab gel, and duplicate slab gels run together. The amount of cytoskeletal actin can be quantitated as peak area after densitometry of Coomassie blue-stained gels.

6.5. Assay of Cytosolic Ca²⁺ Concentration in Cell Populations

Cells can be used immediately after isolation as for secretion experiments or used after culture in 60 mm diameter plastic Petri dishes at 3×10^6/dish. Cells used for measurement of $[Ca^{2+}]_i$ are first preplated for 2 h to remove contaminating nonchromaffin cells. For $[Ca^{2+}]_i$ measurements using the fluorescent dye fura-2, cells from 3-d cultures are gently scraped from the surface of the plastic Petri dish and washed twice by centrifugation at 100g for 10 min, after which over 90% of the cells are viable as assessed by Trypan Blue exclusion. Cultured cells in suspension or freshly isolated cells are incubated with 2 μM fura-2-acetoxymethylester for 30 min at room temperature in buffer B. The cells are washed and left for 45 min before use. Fluorescence measurements on cell suspensions (10^6/mL) are carried out in buffer B and calibrated following the experiment by lysis with digitonin (200 μM final) to give F_{max} followed by addition of EGTA (40 mM final) from a stock solution of 400 mM EGTA brought to pH 9 by addition of NaOH. The concentration of free Ca²⁺ can be calculated from the equation

$$[Ca^{2+}]_i = K_d(F - F_{min})/(F_{max} - F)$$

where K_d is 135 nM at 20°C. Further details of Ca²⁺ indicators and of the techniques used for single-cell imaging are included in the chapters by Poenie and Moreton.

7. Conclusions and Future Prospects

From the kinds of experimental approaches outlined above, a great deal has been learned about signaling within secretory cells, such as chromaffin cells (Fig. 1) and the intracellular factors that regulate exocytosis. Many factors have been identified that are required for, or that modulate, secretion, and the Ca^{2+}-binding protein calpactin has emerged as a candidate for a Ca^{2+} receptor in exocytosis (Fig. 2). The most rapid progress over the past few years has come from the application of methods that allow access to exocytotic sites through membrane permeabilization. These approaches have been applied to secretory cells by many laboratories, but have yet to be applied to any great extent to the study of neurotransmitter release from neurons. One technique that may yield important results in the future for the study of neurotransmitter release from synaptosomes is permeabilization by a freeze/thaw protocol (Nichols et al., 1989, 1990). This preparation has been used to introduce protein kinases or their inhibitors into synaptosomes where they regulate protein phosphorylation and neurotransmitter release.

Ca^{2+} is the central controlling signal for exocytosis in chromaffin cells and neurons. It seems likely that in these cells other intracellular pathways (involving cyclic nucleotides, protein kinase C, G proteins, and so on) will turn out to have only modulatory roles in controlling the Ca^{2+}-sensitivity of the exocytotic pathway. The cytoskeleton plays an important role in regulating exocytosis by binding secretory vesicles, or by forming a cortical barrier to prevent exocytosis until the correct signal is received for disassembly of the cytoskeleton and exocytosis. In chromaffin cells, the cortical cytoskeleton is controlled by a number of factors, including Ca^{2+}, and Ca^{2+} is also required for exocytotic fusion. In neurons, the control of the binding of synaptic vesicles within the cytoskeleton (Hirokawa et al., 1989) involves the vesicle-specific phosphoprotein synapsin I (De Camilli and Greengard, 1986). It has been suggested that phosphorylation of synapsin I results in the release of synaptic vesicles to allow their movement to the active zone ready for exocytosis following Ca^{2+} influx. The major challenge in the study of exocytosis is the determination of the molecular mechanism of the exocytotic fusion process. The exocytotic machinery must consist of, at least, a Ca^{2+}-sensing element and an element that acts as a membrane fusogen. These elements could reside in the same molecule. The difficulty lies in distinguishing those components that are essential for exocytosis from those that have purely regulatory roles or are involved in earlier steps in the secretory process, such as rearrangement of the cytoskeleton, secretory vesicle movement, or vesicle docking. The use of permeabilized cells in which components essential for exocytosis are lost as a result of leakage may allow reconstitution of the secretory machinery by introduction of exogenous proteins. This approach has shown that the Ca^{2+}-binding protein calpactin is required for Ca^{2+}-dependent exocytosis in chromaffin cells. Calpactin is capable of fusing membranes in vitro, though whether it functions as a Ca^{2+}-sensor or fusogen in the intact cell still remains to be determined, nor is it known if it is required for exocytosis in any other cell types. Further investigation of the actions of botulinum and tetanus toxins may provide further clues about the components of the exocytotic machinery. These toxins inhibit exocytosis in chromaffin cells (Adam-Vizi et al., 1988; Bittner and Holz, 1988; Knight et al., 1985; Penner et al., 1986) and in neurons. It is also possible that they may act close to the site of exocytosis, but this is not certain. However, there is as yet no information about the nature of their intracellular target.

A complete understanding of the components and regulators of the exocytotic machinery will require the development of techniques for the manipulation of intracellular conditions that will allow investigation of exocytosis in response to natural agonists in intact cells. Further progress would also be helped by the development of an in vitro fusion system in which purified components (plasma membrane, secretory vesicle, and so on) could be reconstituted to allow precise investigation of the essential components of exocytosis. Development of such a system has often been attempted and has proven to be difficult. Molecular biological approaches to manipulate gene expression may also become important in this field in the near future.

Acknowledgment

R. D. B. wishes to thank the MRC for continued support of this work.

Abbreviations

ATPγS	Adenosine 5'-[γ-thio]triphosphate
$[Ca^{2+}]_i$	Intracellular free Ca^{2+} concentration
cAMP	Adenosine 3',5' cyclic monophosphate
cGMP	Guanosine 3',5' cyclic monophosphate
GDP	Guanosine 5'-diphosphate
GDPβS	Guanosine 5'-[β-thio]diphosphate
GMP-PNP	Guanosine 5'-[βγ-imido]triphosphate
GTP	Guanosine 5'-triphosphate
GTPγS	Guanosine 5'-[γ-thio]triphosphate
Ins(1,4,5)P_3	Inositol 1,4,5-trisphosphate
PIC	Phosphoinositidase C (= phospholipase C)
PKC	Protein kinase C;
PMA	Phorbol 12-myristate 13-acetate
SDS	Sodium dodecyl sulfate

References

Adam-Vizi V., Rosener S., Aktories K., and Knight D. E. (1988) Botulinum toxin-induced ADP-ribosylation and inhibition of exocytosis are unrelated events. *FEBS Lett.* **238**, 277–280.

Ahnert-Hilger G., Bhakdi S., and Gratzl M. (1985) Minimal requirement for exocytosis: A study using PC12 cells permeabilized with staphylococcal alpha-toxin. *J. Biol. Chem.* **260**, 12,730–12,734.

Ahnert-Hilger G., Bader M.-F., Bhakdi S., and Gratzl M. (1989) Introduction of macromolecules into bovine medullary chromaffin cells and rat pheo- chromocytoma cells (PC12) by permeabilization with streptolysin O-inhibitory effect of tetanus toxin on catecholamine secretion. *J. Neuro-chem.* **52**, 1751–1758.

Ali S. M. and Burgoyne R. D. (1990) These stimulatory effect of calpactin (annexin II) on calcium-dependent exocytosis in chromaffin cells: Requirement for both the N-terminal and core domains of p36 and ATP. *Cell Signal.* **2**, 265–276.

Ali S. M., Geisow M. J., and Burgoyne R. D. (1989) A role for calpactin in calcium-dependent exocytosis in adrenal chromaffin cells. *Nature* **340,** 313–315.

Artalejo C. R., Garcia A. G., and Aunis D. (1987) Chromaffin cell calcium channel kinetics measured isotopically through fast calcium, strontium and barium fluxes. *J. Biol. Chem.* **262,** 915–926.

Bader M.-F. and Aunis D. (1983) The 97-KD-actinin-like protein in chromaffin granule membranes from adrenal medulla: Evidence for localization on the cytoplasmic surface and for binding to actin filament. *Neurosci.* **8,** 165–181.

Bader M.-F., Hikita T., and Trifaro J. M. (1985) Calcium-dependent calmodulin binding to chromaffin granule membranes: Presence of a 65-kilodalton calmodulin-binding protein. *J. Neurochem.* **44,** 526–539.

Bader M.-F., Thierse D., Aunis D., Ahnert-Hilger G., and Gratzl M. (1986) Characterization of hormone and protein release from α-toxin-permeabilized chromaffin cells in primary culture. *J. Biol. Chem.* **261,** 5777–5783.

Baker E. M., Cheek T. R., and Burgoyne R. D. (1986) Cyclic AMP inhibits secretion from bovine adrenal chromaffin cells evoked by carbamylcholine but not high K⁺. *Biochim. Biophys. Acta* **846,** 167–173.

Baker P. F. and Knight D. E. (1981) Calcium control of exocytosis and endocytosis in bovine adrenal medullary cells. *Phil. Trans. R. Soc. Lond. [Biol.]* **296,** 83–103.

Barrowman M. M., Cockcroft S., and Gomperts B. D. (1986) Two roles for guanine nucleotides in the stimulus-secretion sequence of neutrophils. *Nature* **319,** 504–507.

Bar-Sagi D. and Gomperts B. D. (1988) Stimulation of exocytotic degranulation by microinjection of the *ras* oncogene protein into rat mast cells. *Oncogene* **3,** 463–469.

Bittner M. A. and Holz R. W. (1988) Effects of tetanus toxin on catecholamine release from intact and digitonin-permeabilized chromaffin cells. *J. Neurochem.* **51,** 451–456.

Bittner M. A., Holz R. W., and Neubig R. R. (1986) Guanine nucleotide effects on catecholamine secretion from digitonin-permeabilized adrenal chromaffin cells. *J. Biol. Chem.* **261,** 10,182–10,188.

Brocklehurst K. W., Morita K., and Pollard H. B. (1985) Characterization of protein kinase C and its role in catecholamine secretion from bovine adrenal medullary cells. *Biochem. J.* **228,** 35–42.

Brooks J. C. and Treml S. (1984) Effect of trifluoperazine and calmodulin on catecholamine secretion by saponin-skinned cultured chromaffin cells. *Life Sci.* **34,** 669–674.

Brooks J. C., Treml S., and Brooks M. (1984) Thiophosphorylation prevents catecholamine secretion by chemically skinned chromaffin cells. *Life Sci.* **35,** 569–574.

Bunn S. J. and Marley P. D. (1989) Effects of angiotensin II on cultured, bovine adrenal medullary cells. *Neuropeptides* **13,** 121–132.

Burgess T. L. and Kelly R. B. (1987) Constitutive and regulated secretion of proteins. *Ann. Rev. Cell Biol.* **3**, 143–193.

Burgoyne R. D. (1984a) Mechanisms of secretion from adrenal chromaffin cells. *Biochim. Biophys. Acta* **779**, 201–216.

Burgoyne R. D. (1984b) The relationship between secretion and intracellular free calcium in bovine adrenal chromaffin cells. *Biosci. Rep.* **4**, 605–611.

Burgoyne R. D. (1987) Control of exocytosis. *Nature* **328**, 112,113.

Burgoyne R. D. (1988) Calpactin in exocytosis? *Nature* **331**, 20.

Burgoyne R. D. (1990) Secretory vesicle-associated proteins and their role in exocytosis. *Ann. Rev. Physiol.* **52**, 647–659.

Burgoyne R. D. and Cheek T. R. (1987a) Reorganisation of peripheral actin filaments as a prelude to exocytosis. *Biosci. Rep.* **7**, 281–288.

Burgoyne R. D. and Cheek T. R. (1987b). Role of fodrin in secretion. *Nature* **326**, 448.

Burgoyne R. D. and Geisow M.-J. (1981) Specific binding of ^{125}I-calmodulin to and protein phosphorylation in adrenal chromaffin granule membranes. *FEBS Lett.* **131**, 127–131.

Burgoyne R. D. and Geisow M. J. (1989) The annexin family of calcium-binding proteins. *Cell Calcium* **10**, 1–10.

Burgoyne R. D. and Morgan A. (1989) Low molecular weight GTP-binding proteins of adrenal chromaffin cells are present on the secretory granule. *FEBS Lett.* **245**, 122–126.

Burgoyne R. D. and Norman K. M. (1984) Effect of calmidazolium and phorbol ester on catecholamine secretion from adrenal chromaffin cells. *Biochim. Biophys. Acta* **805**, 37–43.

Burgoyne R. D., Cheek T. R., and Norman K. M. (1986) Identification of a secretory granule-binding protein as caldesmon. *Nature* **319**, 68–70.

Burgoyne R. D., Cheek T. R., O'Sullivan A. J., and Richards R. C. (1988a) Control of the cytoskeleton during secretion, in *Molecular Mechanisms in Secretion* (Thorn N. A., Treiman M., and Petersen O. H., eds.), Munksgaard, Copenhagen, pp. 612–627.

Burgoyne R. D., Morgan A., and O'Sullivan A. J. (1988b) A major role for protein kinase C in calcium activated exocytosis in permeabilized adrenal chromaffin cells. *FEBS Lett.* **238**, 151–155.

Burgoyne R. D., Geisow M. J., and Barron J. (1982) Dissection of stages in exocytosis in adrenal chromaffin cells with the use of trifluoperazine. *Proc. R. Soc. Lond. [Biol.]* **216**, 111–115.

Burgoyne R. D., Cheek T. R., Morgan A., O'Sullivan A. J., Moreton R. B., Berridge M. J., Mata A., Colyer J., Lee A. G., and East J. M. (1989a) Distribution of two distinct Ca^{2+}-ATPase-like proteins and their relationship to the agonist-sensitive calcium store in bovine adrenal chromaffin cells. *Nature* **342**, 72–74.

Burgoyne R. D., Morgan A., and O'Sullivan A. J. (1989b) The control of cytoskeletal actin and exocytosis in intact and permeabilized chromaffin cells: Role of calcium and protein kinase C. *Cell. Signal.* **1**, 323–334.

Cena V., Stutzin A., and Rojas E. (1989) Effects of calcium and Bay K-8644 on calcium currents in adrenal medullary chromaffin cells. *J. Memb. Biol.* **112**, 255–265.

Cheek T. R. and R. D. Burgoyne. (1985) Effect of activation of muscarinic receptors on intracellular free calcium and secretion in bovine adrenal chromaffin cells. *Biochim. Biophys. Acta* **846**, 167–173.

Cheek T. R. and Burgoyne R. D. (1986) Nicotine evoked disassembly of cortical actin filaments in bovine adrenal chromaffin cells. *FEBS Lett.* **207**, 110–113.

Cheek T. R. and Burgoyne R. D. (1987) Cyclic AMP inhibits both nicotine stimulated actin disassembly and catecholamine secretion from bovine adrenal chromaffin cells. *J. Biol. Chem.* **262**, 11,663–11,666.

Cheek T. R. and Burgoyne R. D. (1991) The cytoskeleton in secretion and neurotransmitter release, in *The Neuronal Cytoskeleton* (Burgoyne R. D., ed.), Liss, New York, pp. 309–325.

Cheek T. R. and Thastrup O. (1989) Internal Ca^{2+} mobilization and secretion in bovine adrenal chromaffin cells. *Cell Calcium* **10**, 213–221.

Cheek T. R., Jackson T. R., O'Sullivan A. J., Moreton R. B., Berridge M. J., and Burgoyne R. D. (1989a) Simultaneous measurement of cytosolic calcium and secretion in single bovine adrenal chromaffin cells by fluorescent imaging of fura-2 in co-cultured cells. *J. Cell Biol.* **109**, 1219–1227.

Cheek T. R., O'Sullivan A. J., Moreton R. B., Berridge M. J., and Burgoyne R. D. (1989b) Spatial localization of the stimulus-induced rise in cytosolic Ca^{2+} in bovine adrenal chromaffin cells: Distinct nicotinic and muscarinic patterns. *FEBS Lett.* **247**, 429–434.

Cheek T. R., O'Sullivan A. J., Moreton R. B., Berridge M. J., and Burgoyne R. D. (1990) The caffeine-sensitive store in bovine adrenal chromaffin cells: An examination of its role in triggereing secretion and Ca^{2+} homeostasis. *FEBS Lett.* **266**, 91–95.

Cheek T. R., O'Sullivan,A. J., Moreton R. B., Berridge M. J., and Burgoyne R. D. (1991) Submitted for publication.

Cobbold P. H., Cheek T. R., Cuthbertson K. S. R., and Burgoyne R. D. (1987) Calcium transients in single adrenal chromaffin cells detected with aequorin. *FEBS Lett.* **211**, 44–48.

Cote A., Doucet J.-P., and Trifaro J.-M. (1986) Phosphorylation and dephosphorylation of chromaffin cell proteins in response to stimulation. *Neurosci.* **19**, 629–645.

Crabb J. H. and Jackson R. C. (1985) In vitro reconstitution of exocytosis from plasma membrane and isolated secretory vesicles. *J. Cell Biol.* **101**, 2263–2273.

Creutz C. E. (1981) *Cis*-unsaturated fatty acids induce the fusion of chromaffin granule aggregated by synexin. *J. Cell Biol.* **91**, 247–256.

Creutz C. E., Pazoles C. J., and Pollard H. B. (1978) Identification and purification of an adrenal medullary protein (synexin) that causes calcium-dependent aggregation of isolated chromaffin granules. *J. Biol. Chem.* **253**, 2858–2866.

Creutz C. E., Dowling L. G., Sando J. J., Villar-Palasi C., Whipple J. H., and Zaks W. J. (1983) Characterisation of the chromobindins. Soluble proteins that bind to the chromaffin granule membrane in the presence of Ca^{2+}. *J. Biol. Chem.* **258,** 14,664–14,674.

Creutz C. E., Zaks W. J., Hamman H. C., Crane S., Martin W. H., Gould K. L., Oddie K. M., and Parsons S. J. (1987) Identification of chromaffin granule-binding proteins. Relationship of the chromobindins to calelectrin, synhibin and the tyrosine kinase substrates p35 and p36. *J. Biol. Chem.* **262,** 1860–1868.

De Camilli P. and Greengard P. (1986) Synapsin I, a synaptic vesicle-associated neuronal phosphoprotein. *Biochem. Pharmacol.* **35,** 4349–4357.

Drust D. S. and Creutz C. E. (1988) Aggregation of chromaffin granule by calpactin at micromolar levels of calcium. *Nature* **331,** 88–91.

Dunn L. A. and Holz R. W. (1983) Catecholamine secretion from digitonin-treated adrenal medullary chromaffin cells. *J. Biol. Chem.* **258,** 4989–4993.

Eberhard D. A. and Holz R. W. (1987) Cholinergic stimulation of inositol phosphate formation in bovine adrenal chromaffin cells: Distinct nicotinic and muscarinic mechanisms. *J. Neurochem.* **49,** 1634–1643.

Eberhard D. A. and Holz R. W. (1988) Intracellular Ca^{2+} activates phospholipase C. *Trends Neurosci.* **11,** 517–520.

Fisher S. K., Holz R. W., and Agranoff B. W. (1980) Muscarinic receptors in chromaffin cell cultures mediate enhanced phospholipid labelling but not catecholamine secretion. *J. Neurochem.* **37,** 491–497.

Fournier S. and Trifaro J.-M. (1988) A similar calmodulin-binding protein expressed in chromaffin, synaptic, and neurohypophyseal secretory vesicles. *J. Neurochem.* **50,** 27–37.

Fowler V. M. and Pollard H. B. (1982) Chromaffin granule membrane-F-actin interactions are calcium sensitive. *Nature* **295,** 336–339.

Frye R. A. and Holz R. W. (1983) Phospholipase A_2 inhibitors block catecholamine secretion and calcium uptake in cultures bovine adrenal medullary cells. *Mol. Pharmacol.* **23,** 547–550.

Frye R. A. and Holz R. W. (1984) The relationship between arachidonic acid release and catecholamine secretion from cultured bovine adrenal chromaffin cells. *J. Neurochem.* **43,** 146–150.

Frye R. A. and Holz R. W. (1985) Arachidonic acid release and catecholamine secretion from digitonin-permeabilized chromaffin cells. Effects of micromolar calcium, phorbol ester, and protein alkylating agents. *J. Neurochem.* **44,** 265–273.

Geisow M. J. and Burgoyne R. D. (1982) Calcium-dependent binding of cytosolic proteins by chromaffin granules from adrenal medulla. *J. Neurochem.* **38,** 1735–1741.

Geisow M. J. and Burgoyne R. D. (1983) Recruitment of cytosolic proteins to a secretory granule membrane depends on Ca^{2+}-calmodulin. *Nature* **301,** 432–435.

Geisow M. J., Burgoyne R. D., and Harris A. (1982) Interaction of calmodulin with adrenal chromaffin granule membranes. *FEBS Lett.* **143,** 69–72.

Gomperts B. D. (1986) Calcium shares the limelight in stimulus-secretion coupling. *Trends Biochem. Sci.* **11**, 290–292.

Goud B., Salminen A., Walworth N. C., and Novick P. (1988) A GTP-binding protein required for secretion rapidly associates with secretory vesicles and the plasma membrane in yeast. *Cell* **53**, 753–768.

Harris B., Cheek T. R., and Burgoyne R. D. (1986) Effects of metalloendoprotease inhibitors on secretion and intracellular free calcium in bovine adrenal chromaffin cells. *Biochim. Biophys. Acta* **889**, 1–5.

Hirokawa N., Sobue K., Kanda K., Harada A., and Yorifuji H. (1989) The cytoskeletal architecture of the presynaptic terminal and the molecular structure of synapsin I. *J. Cell. Biol.* **108**, 111–126.

Holz R. W. Senter R. A., and Frye R. A. (1982) Relationship between Ca^{2+} uptake and catecholamine secretion in primary dissociated cultures of adrenal medulla. *J. Neurochem.* **39**, 635–645.

Holz R. W., Bittner M. A., Peppers S. C., Senter R. A., and Eberhard D. A. (1989) MgATP-independent and MgATP-dependent exocytosis. Evidence that MgATP primes adrenal chromaffin cells to undergo exocytosis. *J. Biol. Chem.* **264**, 5412–5419.

Kao L. S. and Schneider A. S. (1985) Mucarinic receptors on bovine chromaffin cells mediate a rise in cytosolic calcium that is independent of extracellular calcium. *J. Biol. Chem.* **260**, 2019–2022.

Kenigsberg R. L. and Trifaro J. M. (1985) Microinjection of calmodulin antibodies into cultured chromaffin cells blocks catecholamine release in response to stimulation. *Neurosci.* **14**, 335–347.

Kenigsberg R. L., Cote A., and Trifaro J. M. (1982) Trifluoperazine, a calmodulin inhibitor, blocks secretion in cultured chromaffin cells at a step distal from calcium entry. *Neurosci.* **7**, 2277–2281.

Kilpatrick D. L., Slepetis R. J., Corcoran J. J., and Kirshner N. (1982) Calcium uptake and catecholamine secretion by cultured bovine adrenal medulla cells. *J. Neurochem.* **38**, 427–435.

Knight D. E. and Baker P. F. (1982) Calcium dependence of catecholamine release from bovine adrenal medullary cells after exposure to intense electric fields. *J. Memb. Biol.* **68**, 107–140.

Knight D. E. and Baker P. F. (1983) The phorbol ester TPA increases the af-finity of exocytosis for calcium in leaky adrenal medullary cells. *FEBS Lett.* **160**, 98–100.

Knight D. E. and Baker P. F. (1985) Guanine nucleotides and Ca-dependent exocytosis. Studies on two adrenal cell preparations. *FEBS Lett.* **189**, 345–349.

Knight D. E. and Kesteven N. T. (1983) Evoked transient intracellular free Ca^{2+} changes and secretion in isolated bovine adrenal medullary cell. *Proc. R. Soc. Lond. [Biol.]* **218**, 177–199.

Knight D. E., Sugden D., and Baker P. F. (1989) Evidence implicating protein kinase C in exocytosis from electropermeabilized bovine chromaffin cells. *J. Memb. Biol.* **104**, 21–34.

Knight D. E., Tonge D. A., and Baker P. F. (1985) Inhibition of exocytosis in bovine adrenal medullary cells by botulinum toxin type D. *Nature* 317, 719–721.

Kondo H., Wolosewick J. J., and Pappas G. D. (1982) The microtrabecular lattice of the adrenal medulla revealed by polyethylene glycol embedding and stereo electron microscopy. *J. Neurosci.* 2, 57–65.

Lee S. A. and Holz R. W. (1986) Protein phosphorylation and secretion in digitonin-permeabilized adrenal chromaffin cells. Effects of micromolar Ca^{2+}, phorbol esters and diacylglycerol. *J. Biol. Chem.* 261, 17,089–17,098.

Lee S. A., Holz R. W., and Hathaway D. R. (1987) Effects of purified myosin light chain kinase on myosin light chain phosphorylation and catecholamine secretion in digitonin-permeabilised chromaffin cells. *Biosci. Rep.* 7, 323–332.

Lelkes P. I., Friedman J. E., Rosenheck K., and Oplatka A. (1986) Destabilization of actin filaments as a requirement for the secretion of catecholamines from permeabilized chromaffin cells. *FEBS Lett.* 208, 357–363.

Lipscombe D., Madison D. V., Poenie M., Reuter H., Tien R. W., and Tsien R. Y. (1988) Imaging of cytosolic Ca^{2+} transients arising from Ca^{2+} stores and Ca^{2+} channels in sympathetic neurons. *Neuron* 1, 355–365.

Livett B. G. (1984) Adrenal medullary cells in vitro. *Physiol. Rev.* 64, 1103–1161.

Marriott D., Adams M., and Boarder M. R. (1988) Effect of forskolin and prostaglandin E_1 on stimulus secretion coupling in cultured bovine adrenal chromaffin cells. *J. Neurochem.* 50, 616–623.

Matter K., Dreyev F., and Aktories F. (1989) Actin involvement in exocytosis from PC12 cells: Studies on the influence of botulinum toxin C2 on stimulated noradrenaline release. *J. Neurochem.* 52, 370–376.

Momayezi M., Lumpert C. J., Kersen H., Gras U., Plattner H., Krinks M. H., and Klee C. B. (1987) Exocytosis induction in *Paramecium tetraurelia* cells by exogenous phosphoprotein phosphatase in vivo and in vitro: Possible involvement of calcineurin in exocytotic membrane fusion. *J. Cell Biol.* 105, 181–189.

Morgan A. and Burgoyne R. D. (1990a) Stimulation of calcium-independent catecholamine secretion from digitonin-permeabilized bovine adrenal chromaffin cells by guanine nucleotide analogues. Relationship to arachidonate release. *Biochem. J.* 269, 521–526.

Morgan A. and Burgoyne R. D. (1990b) Relationship between arachidonic acid release and calcium-dependent exocytosis in digitonin-permeabilized bovine adrenal chromaffin cells. *Biochem. J.* 271, 571–574.

Morita K., Dohi T., Kitayama S., Koyama Y., and Tsijimoto A. (1987a) Enhancement of stimulation-evoked catecholamine release from cultured bovine adrenal chromaffin cells by forskolin. *J. Neurochem.* 48, 243–247.

Morita K., Dohi T., Kitayama S., Koyama Y., and Tsijimoto A. (1987b) Stimulation-evoked Ca^{2+} fluxes in cultures bovine adrenal chromaffin cells are enhanced by forskolin. *J. Neurochem.* 48, 248–252.

Nakata T., Sobue K., and Hirokawa N. (1990) Conformational change and localization of calpactin I complex involved in exocytosis as revealed by quick-freeze, deep-etch electron microscopy and immunocytochemistry. *J. Cell Biol.* **110**, 13–25.

Nakaki T., Sasakawa N., Yamamoto S., and Kato R. (1988) Functional shift from muscarinic to nicotinic cholinergic receptors involved in inositol trisphosphate and cyclic GMP accumulation during the primary culture of adrenal chromaffin cells. *Biochem. J.* **251**, 397–403.

Negishi M., Ito S., and Hayaishi O. (1989) Prostaglandin E receptors in bovine adrenal medulla are coupled to adenylate cyclase via G_i and to phosphoinositide metabolism in a pertussis toxin-insensitive manner. *J. Biol. Chem.* **264**, 3916–3923.

Neher E. and Marty A. (1982) Discrete changes of cell membrane capacitance observed under conditions of enhanced secretion in bovine adrenal chromaffin cells. *Proc. Natl. Acad. Sci. USA* **79**, 6712–6716.

Nichols R. A., Wu W. C.-S., Haycock J. W., and Greengard P. (1989) Introduction of impermeant molecules into synaptosomes using freeze/thaw permeabilisation. *J. Neurochem.* **52**, 521–529.

Nichols R. A., Sihra T. S., Czernik A. J., Nairn A., and Greengard P. (1990) Calcium/calmodulin-dependent protein kinase II increases glutamate and noradrenaline release from synaptosomes. *Nature* **343**, 647–651.

Orci L., Gabbay K. H., and Malaisse W. J. (1972) Pancreatic B-cell web: Its possible role in insulin secretion. *Science* **175**, 1128–1130.

O'Sullivan A. J. and Burgoyne R. D. (1988) Thhe role of cytoplasmic pH in the inhibitory action of high osmolarity on secretion from adrenal chromaffin cells. *Biochim. Biophys. Acta* **969**, 211–216.

O'Sullivan A. J. and Burgoyne R. D. (1989) A comparison of bradykinin, angiotensin II and muscarinic stimulation of cultured bovine adrenal chromaffin cells. *Biosci. Rep.* **9**, 243–252.

O'Sullivan A. J. and Burgoyne R. D. (1990) Cyclic GMP modulates nicotine-stimulated secretion in cultured bovine adrenal chromaffin cells: Effects of 8-bromo-cGMP, atrial natriuretic peptide and nitroprusside (nitric oxide). *J. Neurochem.* **54**, 1805–1808.

O'Sullivan A. J., Cheek T. R., Moreton R. B., Berridge M. J., and Burgoyne R. D. (1989) Localization and heterogeneity of agonist-induced changes in cytosolic calcium concentration in single bovine adrenal chromaffin cells from video-imaging of fura-2. *EMBO J.* **8**, 401–411.

Penner R., Neher E., and Dreyer F. (1986) Intracellularly injected tetanus toxin inhibits exocytosis in bovine adrenal chromaffin cells. *Nature* **324**, 76–78.

Perrin D. and Aunis D. (1985) Reorganization of α-fodrin induced by stimulation in secretory cells. *Nature* **314**, 589–592.

Perrin D., Langley O. K., and Aunis D. (1987) Anti α-fodrin inhibits secretion from permeabilized chromaffin cells. *Nature* **326**, 498–501.

Plevin R. and Boarder M. J. (1988) Stimulation of formation of inositol phosphates in primary cultures of bovine adrenal chromaffin cells by angiotensin II, histamine, bradykinin, and carbachol. *J. Neurochem.* **51**, 634–641.

Pocotte S. L., Frye R. A., Senter R. A., Terbuh D. R., Lee S. A., and Holz R. W. (1985) Effects of phorbol ester on catecholamine secretion and protein phoshphorylation in adrenal medullary cell cultures. *Proc. Natl. Acad. Sci. USA* **82**, 930–934.

Pollard H. B., Burns A. L., and Rojas E. (1988) A molecular basis for synexin-driven calcium-dependent membrane fusion. *J. Exp. Biol.* **139**, 267–286.

Rink T. J. and Knight D. E. (1988) Stimulus-secretion coupling: A perspective highlighting the contributions of Peter Baker. *J. Exp. Biol.* **139**, 1–30.

Rink T. J., Sanchez A., and Hallam T. J. (1983) Diacylglycerol and phorbol ester stimulate secretion without raising cytoplasmic free calcium in human platelets. *Nature* **305**, 317–319.

Rosario L. M., Soria B., Feuerstein G., and Pollard H. B. (1989) Voltage-sensitive calcium flux into bovine chromaffin cells occurs through dihydropyridine-sensitive and dihydropyridine- and Ω-conotoxin-insensitive pathways. *Neurosci.* **29**, 735–747.

Salminen A. and Novick P. J. (1987) A *ras*-like protein is required for a post-golgi event in yeast secretion. *Cell* **49**, 527–538.

Sarafian T., Aunis D., and Bader M.-F. (1987) Loss of proteins from digitonin-permeabilised adrenal chromaffin cells essential for exocytosis. *J. Biol. Chem.* **262**, 16,671–16,676.

Satir B. H., Hamasaki T., Reichman M., and Murtaugh T. J. (1989) Species distribution of a phosphoprotein (parafusin) involved in exocytosis. *Proc. Natl. Acad. Sci. USA* **86**, 930–932.

Schafer T., Karli U. O., Gratwohl E. K.-M., Schweizer F. E., and Burger M. M. (1987a) Digitonin-permeabilized cells are exocytosis competent. *J. Neurochem.* **49**, 1696–1707.

Schafer T., Karli U. O., Schweizer F. E., and Burger M. M. (1987b) Docking of chromaffin granules—a necessary step in exocytosis. *Biosci. Rep.* **7**, 269–279.

Schneider A. S., Cline H. T., and Lemaire S. (1979) Rapid rise in cyclic GMP accompanies catecholamine secretion in suspensions of isolated adrenal chromaffin cells. *Life Sci.* **24**, 1389–1394.

Schweizer F. E., Schafer T., Tapparelli C., Grob M., Karli U. O., Heumann R., Thoenen H., Bookman R. J., and Burger M. M. (1989) Inhibition of exocytosis by intracellularly applied antibodies against a chromaffin granule-binding protein. *Nature* **339**, 709–712.

Slepetis R. and Kirshner N. (1982) Inhibition of $^{45}Ca^{2+}$ uptake and catecholamine secretion by phenothiazines and pimozide in adrenal medulla cell cultures. *Cell Calcium* **3**, 183–190.

Sontag J. M., Aunis D., and Bader M.-F. (1988) Peripheral actin filaments control calcium-mediated catecholamine release from streptolysin O-permeabilised chromaffin cells. *Eur. J. Cell Biol.* **46**, 316–326.

Tatham P. E. and Gomperts B. D. (1989) ATP-inhibits onset of exocytosis in permeabilised mast cells. *Biosci. Rep.* **9**, 99–109.

Terbush D. R., Bittner M. A., and Holz R. W. (1988) Ca^{2+} influx causes rapid translocation of protein kinase C to membranes. Studies of the effects of secretagogues in adrenal chromaffin cells. *J. Biol. Chem.* **263,** 18,873–18,879.

Vilmart-Seuwen J., Kersken H., Sturzl R., and Plattner, H. (1986) ATP keeps exocytosis sites in a primed state but is not required for membrane fusion: An analysis with *Paramecium* cells in vivo and in vitro. *J. Cell Biol.* **103,** 1279–1288.

Volpe P., Krause K. H., Hashimoto S., Zorzato F., Pozzan T., Meldolesi J., and Lew D. P. (1988) "Calciosome," a cytoplasmic organelle: The inositol 1,4,5-trisphosphate-sensitive Ca^{2+} store of nonmuscle cells? *Proc. Natl. Acad. Sci. USA* **85,** 1091–1095.

Von Euler U. S. and Floding I. (1955) A fluorimetric micromethod for differential estimation of adrenaline and noradrenaline. *Acta Physiol. Scand.* **(Suppl.)118,** 45–56.

Wakade A. R., Malhotra R. K., and Wakade T. D. (1986) Phorbol ester facilitates ^{45}Ca^{2+} accumulation and catecholamine secretion by nicotine and excess K^{+} but not by muscarine in rat adrenal medulla. *Nature* **321,** 698–700.

Winkler H. (1988) Occurrence and mechanism of exocytosis in adrenal medulla and sympathetic nerve, in *Handbook of Experimental Pharmacology* (Trendelenburg U. and Weiner N., eds.), Springer-Verlag, Berlin, pp. 43–118.

Zieseniss E. and Plattner H. (1985) Synchronous exocytosis in *Paramecium* cells involves very rapid (<1 sec), reversible dephosphorylation of a 65kDa phosphoprotein in exocytosis-competent strains. *J. Cell Biol.* **101,** 2028–2035.

Intracellular Messengers
in Vertebrate Photoreceptors

Electrophysiological Techniques

Hugh R. Matthews and Trevor D. Lamb

1. Introduction

Rod and cone photoreceptors transduce the absorption of a photon of light into an electrical response. Within the last decade, our understanding of photoreceptors has increased enormously, and this growth in understanding has resulted in large part from the application of new electrophysiological techniques to these cells. The biochemistry and physiology of phototransduction have been extensively reviewed elsewhere (Stryer, 1986, 1988; Pugh and Cobbs, 1986; Pugh, 1987; Liebman et al., 1987; Hurley, 1987; Fain and Matthews, 1990; McNaughton, 1990; Pugh and Lamb, 1990); in this chapter, we concern ourselves with the electrophysiological techniques themselves.

The outer segment of the vertebrate rod is specialized for phototransduction, and comprises a stack of membranous disks with which the photopigment rhodopsin and other transduction proteins are associated. The inner segment contains the conventional metabolic machinery of the cell (Fig. 1A). In darkness, a steady dark current flows in across the outer-segment membrane, and out through various voltage-gated conductances in

From: *Neuromethods, Vol. 20: Intracellular Messengers*
Eds: A. Boulton, G. Baker, and C. Taylor ©1992 The Humana Press Inc.

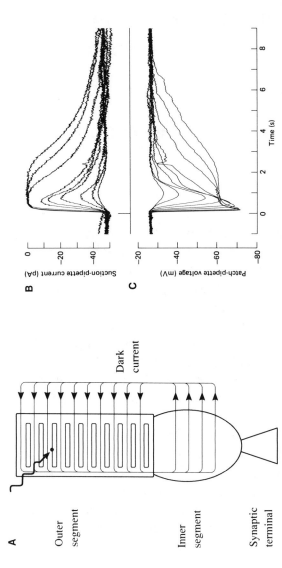

Fig. 1. A. Schematic diagram of a rod photoreceptor showing the principal elements of the cell. The outer segment contains a stack of about 1000 membranous disks, in which the photosensitive pigment rhodopsin is embedded; several other important proteins, notably the G protein and the phosphodiesterase (PDE), are also associated with the disk membrane, whereas a number of other proteins involved in transduction are soluble. The inner segment contains the nucleus, mitochondria, and other organelles of metabolism. In darkness, a circulating current flows as shown. **B**. Recordings of the photocurrent. The circulating dark current is reduced by illumination, and this reduction is graded with the intensity of illumination. C. Simultaneous measurements of intracellular voltage from the same cell. Responses in B and C were recorded using the dual-electrode technique illustrated in Fig. 5, from a salamander rod stimulated with brief flashes for which the intensity ranged from 10 to 3500 photoisomerizations. Reproduced with permission from Lamb et al. (1986).

the inner-segment membrane. Stimulation by light results in a graded reduction in this circulating current (Fig. 1B), whereby the degree of suppression increases with increasing flash intensity until, for the brightest flashes, the dark current is completely suppressed. The voltage-gated conductances of the inner segment (reviewed in Owen, 1987) shape this photocurrent response to yield the photovoltage response (Fig. 1C), which is delivered to the synaptic terminal for transmission to higher-order cells.

Phototransduction is initiated by the absorption of a photon of light by rhodopsin, located in the disk membrane. The resultant decrease in conductance, however, takes place in the outer-segment plasma membrane, which in rods is topologically separate from the disk membrane. This observation has long been taken as evidence for the involvement of an intracellular messenger of excitation, which is now known to be cGMP (Fesenko et al., 1985). Not only excitation, but also light adaptation, in photoreceptors has been shown to involve a diffusible cytoplasmic messenger (Donner and Hemilä, 1978; Bastian and Fain, 1979), and recent evidence has shown this messenger to be Ca^{2+} (Torre et al., 1986; Matthews et al., 1988; Nakatani and Yau, 1988c).

This chapter is concerned with the electrophysiological techniques and experimental approaches used to determine the identities of these intracellular messengers, and to investigate the means by which they operate. In the next section, we present an overview of the techniques used for electrical recording from photoreceptors, and then in Section 3., we provide details of specific techniques, based on the use of suction-pipets and patch-pipets. Subsequently, we describe the application of these methods to the measurement of photoresponses during the incorporation of substances into the cytoplasm (Section 4.), and during external perfusion of the light-sensitive outer segment with test solutions (Section 5.).

We would emphasize that our coverage is restricted to electrophysiological studies. It would be quite beyond the scope of this chapter to attempt to deal with biochemical, molecular biological, and optical methods of studying the internal messengers involved in phototransduction.

2. Overview of Electrical Recordings from Photoreceptors

2.1. Early Approaches

Many different electrophysiological techniques have been used to record the responses of vertebrate photoreceptors to light. The earliest studies, carried out using microelectrodes to record the intracellular voltage, yielded a wealth of information about the nature and properties of the voltage responses of these cells to illumination (reviewed in Tomita, 1970; Kaneko, 1979). However, the shaping of the voltage responses by various conductances in the inner segment complicates interpretation of such recordings. A more direct approach is to record the photocurrent itself. Since the current–voltage relation of the outer-segment membrane is very shallow over the physiological voltage range, the photocurrent may be viewed as the output of the transduction mechanism, uncontaminated by any modifications introduced by the inner segment.

The first recordings of photocurrent were obtained by measuring the extracellular voltage gradients in the intact retina resulting from the circulating current (Hagins et al., 1970). That approach has, however, been supplanted by the suction-pipet technique (Baylor et al., 1979), which allows the photocurrent responses of individual rods or cones to be recorded. In the last decade, the most useful electrophysiological techniques for the study of intracellular messengers in photoreceptors have proven to be the suction-pipet and patch-pipet techniques, which are described in detail below.

2.2. Preparation of Isolated Photoreceptors

Much of the work on transduction in vertebrate photoreceptors has been carried out on the rods and cones of amphibia, primarily because the large size and relative robustness of their photoreceptors make recording from individual cells easier than is the case with mammals. The most widely used species have been the toad *(Bufo marinus)* and the larval tiger salamander *(Ambystoma tigrinum)*; the latter has the advantage that the isolation of individual photoreceptors is relatively straightforward (Werblin, 1978). More recently, it has become possible to study the smaller and more fragile rods and cones of mammals (Baylor et al., 1984,1987; Kraft, 1988; Tamura et al., 1989; Matthews, 1990,1991; Schnapf et al., 1990; Ratto et al., 1991).

The most common means of obtaining isolated photoreceptors is through mechanical dissociation of the retina (e.g., Werblin, 1978; Baylor et al., 1979; Attwell et al., 1982; Hodgkin et al., 1985; Lamb et al., 1986), although enzymatic dissociations have also been used (Bader et al., 1979; MacLeish et al., 1984). In our laboratory, the isolation of single photoreceptors is carried out as follows. The dark-adapted animal is killed by a method appropriate to the species (e.g., decapitation and pithing for amphibia), and both eyes are removed under dim red light. All subsequent stages of the dissection are carried out under infrared illumination, and the preparation is viewed through an infrared image converter coupled to the dissecting microscope. The eyeballs are hemisected with a razor blade, and each piece of eyecup is subdivided into several pieces. The pieces of eyecup are transferred to a Ringer-filled Petri dish, which has been prepared with a layer of cured Sylgard silicone rubber (about 2 mm deep) in its base. Under Ringer, the retina is removed from the piece of eyecup using fine forceps, and is chopped delicately and repeatedly with a small piece of razor blade. This procedure yields a combination of small fragments of retina together with a number of isolated rods and cones; in the larval salamander, the yield of individual photoreceptors is far higher than in other species. The preparation is then transferred to the recording chamber by pipet.

2.3. Microscopy and Optical Stimulation

The techniques for microscopy and for micromanipulation of the cells are conventional, and have been described previously (Baylor et al., 1979; Lamb et al., 1986). As illustrated schematically in Fig. 2A, the recording chamber is mounted on the stage of an inverted compound microscope, and the preparation is viewed by an infrared-sensitive television system. To minimize stimulation by visible light, the microscope and mi-

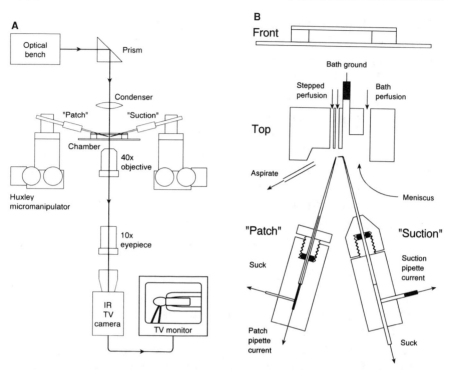

Fig. 2. **A**. Schematic diagram of the experimental setup. The recording chamber is mounted on the stage of an inverted compound microscope and is viewed from below by means of an infrared-sensitive television camera. Illumination from an optical bench is focused onto the preparation via the microscope condenser. Suction pipets and patch pipets enter the open front of the chamber, and are positioned using micromanipulators placed on each side of the microscope. The microscope, manipulators, and head-stage amplifiers are enclosed within a light-tight Faraday cage. **B**. Detail of the recording chamber and pipet holders. The chamber is made from two silanized glass slides or coverslips, separated vertically by about 3 mm. Perfusate enters the rear of the chamber (*see also* Fig. 13) and is removed from the front by a needle connected to a suction line; in this way, it is possible to maintain a stable and roughly vertical meniscus. The pipets are each bent near their tips to facilitate close approach to each other and to the bottom of the chamber.

cromanipulators are mounted inside a light-tight Faraday cage, within a room lit only by dim red light, and the cells are viewed with infrared illumination ($\lambda > 850$ nm). Illumination (both infrared and visible) is delivered via the microscope condenser from an optical bench (Baylor and Hodgkin, 1973; Baylor et al.,

1979) external to the Faraday cage. The wavelength, intensity, duration, and geometry of the optical stimuli may be controlled by narrow-band interference filters, by neutral density filters, by digitally-timed shutters, and by appropriate field stops, respectively.

2.4. Recording Chamber and Cell Manipulation

To provide a high-quality optical path for light stimulation, in addition to the optical path for viewing, the recording chamber used in electrophysiological experiments on photoreceptors is typically constructed with glass slides (or coverslips) both at the top and the bottom (*see* Fig. 2B). In this design, the front of the chamber is left open, for insertion of the recording pipets, and the Ringer solution is held in place by surface tension. Perfusing solutions typically enter from the rear of the chamber, and excess fluid is sucked away from the free surface at the front of the chamber. It is usual to silanize the glass from which the chamber is constructed (*see below*) to help prevent adhesion of cells. In certain experiments, however, where it is desirable that the photoreceptors be stuck down, yet undamaged, it is possible to insert into the bottom of the chamber a small piece of glass coverslip coated with Concanavalin A (e.g., Crawford et al., 1989).

After the preparation of chopped retina has been pipeted into the chamber, allowed to settle for several minutes, and then washed with clean perfusate, the recording pipet(s) may be introduced and a suitable cell located. By viewing the television monitor (external to the Faraday cage), the material at the base of the chamber may be scanned in order to locate a suitable undamaged isolated photoreceptor. Since all such manipulations are performed in darkness, it is very helpful to have available a readout of the X-Y position of the microscope stage, and this may be accomplished with a pair of linear position transducers connected to digital panel meters. In this way, it is straightforward to return later to cells that have previously been found, or to move to the position of perfusion pipes and so on. Once a suitable cell has been located, the recording techniques described in Section 3. may be employed.

To minimize electrical transients in experiments involving perfusion of the chamber, it is helpful to incorporate a bath-clamp (or virtual ground) circuit, which reduces the impedance from the bathing solution to ground (Baylor et al., 1984; Hodgkin et al., 1984). It is also helpful if the investigator is grounded.

2.5. Data Storage and Reduction

The methods used for the subsequent storage and analysis of the electrical responses recorded by these techniques are fairly standard. As an example, we describe here the approach used in our laboratory. The suction-pipet current is recorded using a conventional current-to-voltage converter (Baylor et al., 1979), whereas the patch-pipet current is recorded using a commercial patch-clamp amplifier. The suction-pipet and patch-pipet signals are low-pass filtered to prevent aliasing, and are then digitized and stored on the hard disk of a minicomputer (Lamb, 1983) for subsequent analysis. Since photoreceptor responses to light are slow, especially in the rods and cones of amphibia, a recording bandwidth of only a few tens to hundreds of Hertz is required. Additionally, though, the signals are recorded at much greater bandwidth using a digital tape recorder based on a commercial digital audio processor and videocassette recorder (Lamb, 1985).

3. Recording Techniques

3.1. Suction-Pipet Technique

3.1.1. Outline of the Technique

The suction-pipet technique, developed by Baylor et al. (1979), allows the recording of the photocurrent responses from individual rods and cones. In its original form, the outer segment of a single rod protruding from a small piece of retina is drawn into a tight-fitting glass pipet, as illustrated for a toad rod in Fig. 3A. Using the suction-pipet technique, stable recordings of the photocurrent of individual photoreceptors can be made for periods of several hours. It is for this reason that the technique has become the principal means for studying the electrical responses of rods and cones to light. An interesting advantage

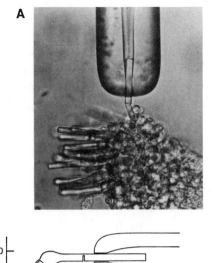

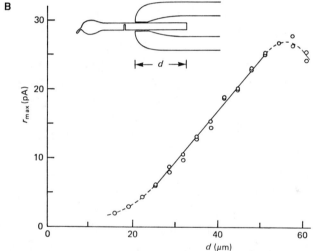

Fig. 3. **A.** Photomicrograph of the suction-pipet method. A toad rod, still connected to a small piece of retina, has its outer segment drawn partway into a snugly fitting glass suction pipet, which is used to record the circulating current. **B.** Dependence of circulating dark current on the length d of outer segment drawn into the suction pipet. The linear dependence is consistent with a uniform density of circulating dark current along the length of the outer segment. A and B reproduced with permission from Baylor et al. (1979).

of a technique such as this, which allows a high optical resolution, is that the spatial distribution of the light stimulus may be controlled very accurately; this has permitted examination of the extent of spread of the internal messengers of excitation and adaptation within the outer segment (Lamb et al., 1981).

In a variant of the original technique, the inner segment (rather than the outer segment) of an isolated cell is drawn into the suction-pipet, leaving the outer segment exposed to the bath solution. This approach allows rapid exchange of the solution bathing the outer segment, as will be described in Section 5.

3.1.2. Fabrication of Suction Pipets

Suction pipets are fabricated by a multistage process illustrated in Fig. 4. First, a pipet is pulled from medium-wall borosilicate glass using a conventional microelectrode puller (Fig. 4A). Subsequent stages of fabrication are performed using a compound microscope equipped with a measuring graticule. Second, the pipet is scored very finely with a diamond knife and is broken off square (Fig. 4B) at an external diameter of approx 50 µm for large-lumen pipets. Third, the pipet is fire-polished under visual control using an electrically heated platinum filament, until the desired internal diameter is achieved (Fig. 4C). The appropriate diameter depends on the species and on the type of photoreceptor to be recorded from, varying from <2 µm for mammalian rods or cones (Baylor et al., 1984; Tamura et al., 1989) to about 12 µm for salamander rods (Lamb et al., 1986; Cobbs and Pugh, 1987; Hodgkin et al., 1987). The smaller-lumen suction-pipets used to record from mammalian rods can alternatively be pulled in much the same way as patch-pipets are made, without the need for separate scribing and breaking.

To prevent the cell from sticking to the interior of the pipet, it is necessary to coat the glass surface with silane. This is achieved by heating the pipets to 200°C and exposing them to tri-*n*-butylchlorosilane vapor.

Finally, it may be desirable to bend each pipet near its tip for some experiments, for example, to allow the approach of a patch-pipet (*see* Section 3.3.) or to facilitate rapid solution changes (*see* Section 5.1.). This can be done by locally heating the pipet at the desired point and then applying a bending moment using a suitably designed jig (Hodgkin et al., 1984; H. R. Matthews, 1986).

3.1.3. Entry of the Cell into the Suction-Pipet

Once the preparation has been introduced into the chamber and a suitable cell has been found (*see* Section 2.4.) and

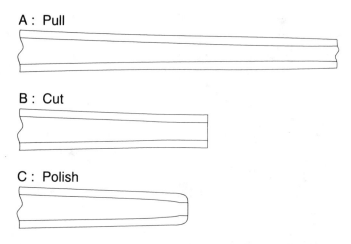

Fig. 4. Steps in the fabrication of a suction pipet. **A.** A conventional intra-cellular-type micropipet is pulled, and **B.** is cut off square at an appropriate diameter. This is achieved (under microscopic visualization) by scoring the pipet lightly with a diamond knife and then bending its tip. Next, using a heated Pt wire, the pipet is melted back to provide a lumen of the required diameter, **C.** For this process, it is advisable to cover the microscope objective with a coverslip, and to coat the heating element with molten glass (by dipping it once into powdered glass) to protect the objective from vaporized metal. Finally (not shown), the pipet is silanized and, if necessary, bent near its tip.

examined under high magnification, the cell may be drawn into the suction-pipet by application of gentle suction to the pipet holder. The suction is applied by hand, using a micrometer connected to an oil-filled precision syringe (Baylor et al., 1979); it is helpful to monitor the level of suction using a sensitive pressure transducer coupled to the suction line. During entry of the cell into the pipet, the electrical resistance of the pipet tip is usually monitored by applying square-wave voltage pulses. The pipet resistance typically increases by 3–5x when the outer segment is drawn in, thus constraining the majority of the dark current from the portion of the outer-segment contained within the pipet to flow through the recording apparatus. Once the cell has been drawn in to the desired position (as in Fig. 3A or Fig. 5), the

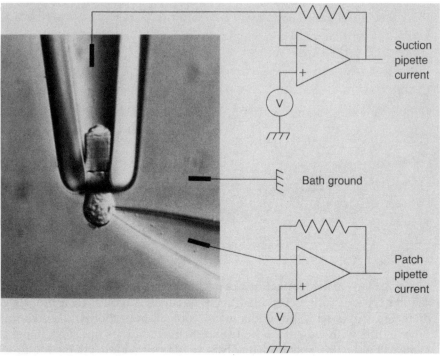

Fig. 5. The simultaneous suction-pipet and patch-pipet technique. A salamander rod is shown with its outer segment drawn into a suction pipet and with a patch pipet sealed onto the protruding inner segment. The schematic diagram shows the two pipets connected to current-to-voltage converters (i.e., both pipets are voltage-clamped). For the responses illustrated in Fig. 1, the patch-clamp amplifier was switched to current-clamp mode after rupture of the membrane patch, so that it recorded the intracellular voltage, while the suction-pipet recorded the circulating current. Modified from Lamb et al. (1986).

pressure is restored to zero, and recording may then commence. Light-induced changes in circulating current can be recorded by stimulating the cell with visible light, rather than the infrared light with which it is viewed: The family of flash responses illustrated in Fig. 1B was obtained using the suction-pipet method.

If the ciliary neck of the photoreceptor is abruptly severed, the current rapidly relaxes to the same level as recorded during the saturated light response, showing that bright light can completely suppress the circulating current (Baylor et al., 1979). Figure 3B shows that the magnitude of the response to saturating light recorded by the suction-pipet technique varies linearly with the length of outer segment drawn into the pipet, consistent with an approximately uniform density of current along the outer segment (Baylor et al., 1979). With the cell in the optimal position, the suction-pipet collects 60–80% of the total circulating current (Baylor et al., 1979; Lamb et al., 1986).

3.2. Patch-Clamp Technique

The patch-clamp technique, developed by Hamill et al. (1981), has been widely applied to the study of intracellular messengers in photoreceptors. In this technique, a fire-polished glass pipet is "sealed" onto the membrane of the cell, producing a seal resistance of many gigohms. The methods for patch-clamp recording have been reviewed extensively elsewhere (*see* references in Sakmann and Neher, 1983), so here we shall concern ourselves only with details specific to the application of the technique to photoreceptors.

Gigohm seals can be established on salamander photoreceptors without the need for enzymatic treatment of the cell membrane. Seals have been obtained on either the outer or inner segment of rods by numerous groups (Bodoia and Detwiler, 1985; Fesenko et al., 1985; Gray and Attwell, 1985; Cobbs and Pugh, 1985; Matthews et al., 1985; Haynes et al., 1986; Lamb et al., 1986; McNaughton et al., 1986; G. Matthews, 1986, 1987; Zimmerman and Baylor, 1986; Matthews and Watanabe, 1987, 1988; Lagnado et al., 1988; Cervetto et al., 1989; Stern et al., 1986; Menini, 1990). In the case of cones, however, seal formation has mainly been restricted to the inner segment (Cobbs et al., 1985; Matthews et al., 1990); with the highly invaginated outer-segment membrane, it has only been possible to obtain seals at the very tip of the outer segment (Haynes and Yau, 1985, 1990).

All four of the recording configurations described by Hamill et al. (1981) have been used successfully on photoreceptors. Excised "inside-out" patches of outer-segment membrane have provided a conclusive demonstration that the "light-sensitive" conductance is activated by cGMP (Fesenko et al., 1985; discussed in Section 3.6.), whereas "cell-attached" patches have been used to study the properties of this conductance *in situ*. The "whole-cell" configuration has proven valuable for the study both of transduction and of adaptation, often in conjunction with simultaneous suction-pipet recording, as described in the next section. Finally, the excised "outside-out" configuration is often established after a period of intracellular dialysis during whole-cell recording, but in this case, the cell left behind is usually of more interest than the patch, since the substances introduced from the patch-pipet are then trapped within the cytoplasm. We now discuss in detail the simultaneous suction-pipet and whole-cell patch-pipet technique, but we leave for later consideration the excised-patch and cell-attached patch techniques (Section 3.6.).

3.3. Simultaneous Suction-Pipet and Whole-Cell Technique

3.3.1. Outline of the Simultaneous Recording Technique

A powerful approach to the study of transduction in the intact photoreceptor is the simultaneous use of the suction-pipet and whole-cell patch-pipet techniques (*see*, e.g., Lamb et al., 1986; Cobbs and Pugh, 1987). This method is illustrated in Fig. 5, which shows a salamander rod with its outer segment drawn into a suction-pipet and with a patch-pipet sealed onto the protruding inner segment. Once a gigohm seal has been established, the patch of membrane at the pipet tip can be ruptured by gentle suction to establish the whole-cell recording configuration. In a variant of this technique, it is possible to draw the inner segment of the cell into the suction-pipet and to seal the patch-pipet onto the exposed outer segment (*see* Section 5.1.). A major advantage of the dual-electrode technique is that the suction-pipet can be used to provide a continuous record of the response of the cell (to illumination or to solution changes) before, during, and after the introduction into the cytoplasm of a substance from the patch-pipet (*see* next section).

Once the whole-cell configuration has been established, it is possible to record with the patch-pipet either under voltage-clamp or under current-clamp. Since the current–voltage relation of the outer segment is very shallow in the physiological voltage range (Bader et al., 1979; Bodoia and Detwiler, 1985; Baylor and Nunn, 1986), there is little to choose between the two modes when recording the relatively slow responses to dim flashes. The voltage-clamp mode has the advantage of ensuring that capacitive currents and changes in voltage-sensitive conductances are eliminated, whereas the current-clamp mode has the advantage of permitting the simultaneous measurement of intracellular voltage and circulating current via the patch and suction-pipets, respectively (*see* Fig. 1B,C). When studying the interaction between the internal messenger (cGMP) and the ion channels, however, it becomes important to use the voltage-clamp mode, since otherwise the onset phase of the response to intense flashes is obscured by the cell's capacitive time constant of ca. 10–20 ms in rods (Cobbs and Pugh, 1987). Similarly, in recordings from cones, the voltage-clamp mode is greatly to be preferred, because the slope conductance of the outer segment is significant (Haynes and Yau, 1985), and because the cone's capacitive time constant is even greater than that for rods (Lamb et al., 1989).

3.3.2. Entry of Substances into the Cytoplasm

A major application of the simultaneous recording technique is the use of the patch-pipet to dialyze the photoreceptor cytoplasm (*see* Section 4.). The substance to be introduced can normally be included in the pseudo-intracellular solution filling the patch-pipet; once the membrane patch has been ruptured, the contents of the pipet and the cytoplasm begin to exchange by aqueous diffusion (Fenwick et al., 1982).

It is possible to demonstrate the entry into the cytoplasm of substances contained in the patch-pipet by several methods. Most directly, it is possible to include a fluorescent probe, such as carboxyfluorescein or fluorescently-labeled cGMP, and to visualize its movement by standard fluorescence microscopy techniques (Fig. 6, from Pugh et al., 1990; *see also* Baylor, 1988). Alternatively, the entry of the pipet contents into the cell can be demonstrated by using a substance that has a well-defined effect on the transduction mechanism and by following the onset of this effect as the substance diffuses into the cytoplasm. For

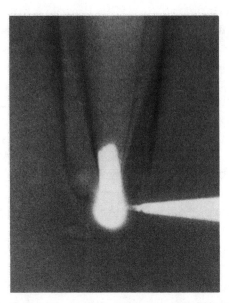

Fig. 6. Entry of a fluorescent marker from the whole-cell patch pipet. This epi-fluorescence image was taken 2 min after the beginning of entry of 5(6)-carboxyfluorescein from the patch pipet, which contained 5 mM of the dye. It shows intense fluorescence from the inner segment, together with less intense fluorescence from the outer segment; much of this quantitative difference arises from the considerably smaller water space of the outer segment owing to the presence of the dense stack of disks. The outline of a cone still attached to the rod inner segment can just be made out. Reproduced with permission from Pugh et al. (1990).

example, the introduction of cGMP from a patch-pipet results in a massive increase in circulating current as the nucleotide reaches the outer segment (Matthews et al., 1985; Cobbs and Pugh, 1985; Cobbs et al., 1985; Sather and Detwiler, 1987; Cameron and Pugh, 1990; see also Section 4.3.). Similarly, the entry of Na^+ results in a progressive decline in the dark current (Lamb et al., 1986; Schmidt et al., 1990), probably as a result of the elevation of intracellular $[Ca^{2+}]$ induced by the lowered Na^+ gradient. These lines of evidence suggest that small ions and molecules included in the patch-pipet solution reach a steady concentration in the photoreceptor cytoplasm within a few minutes. Larger molecules move in more slowly, in line with their lower diffusion coefficients.

During whole-cell recording, not only do substances present in the patch-pipet enter the cytoplasm, but substances present in the cytoplasm similarly diffuse out into the patch-pipet. Such gradual "washout" of macromolecules is widely believed to lead to the progressive running down of normal function in many cell types (*see*, e.g., Marty and Neher, 1983). This problem can be avoided by restricting the whole-cell recording to a brief period and then gently pulling the pipet away from the cell membrane, which reseals. In this way, a substance in the patch-pipet may be trapped within the cytoplasm of the cell (Torre et al., 1986; McNaughton et al., 1986; Lamb and Matthews, 1988a,b). The required sequence of manipulations is precisely the same as that used to produce an outside-out membrane patch, but as we noted above, in this case it is the cell rather than the patch that is of interest.

3.4. Whole-Cell Recording from Isolated Outer Segments

An alternative approach is to use the whole-cell patch technique to record from an isolated rod outer segment rather than from an intact rod (Sather and Detwiler, 1987; Lagnado et al., 1988). With this approach, a suction-pipet is sometimes used simply to hold the outer segment steady during manipulations with the patch-pipet and during superfusion. In the whole-cell configuration, the patch-pipet acts as a substitute for the absent inner segment, and provides a diffusional sink for the Na^+, which enters through the outer-segment conductance. It must also provide a source of nucleotides, such as ATP, GTP, and cGMP, which are needed to maintain the function of the transduction mechanism (Sather and Detwiler, 1987).

This approach offers two significant advantages stemming from the absence of the inner segment. First, for studying the biochemistry of transduction, there is no longer the possibility that enzymes in the inner segment will modify substances introduced via the patch-pipet. Second, in the absence of the voltage-gated conductances of the inner segment, the remaining ionic pathways comprise only the light-sensitive conductance (Baylor and Lamb, 1982; Baylor and Nunn, 1986) and the Na^+–Ca^{2+} exchanger (Lagnado et al., 1988; Cervetto et al., 1989); this greatly facilitates the study of Na^+–Ca^{2+} exchange in rods.

3.5. "Truncated Cell" Technique

The truncated cell technique, developed by Yau and Nakatani (1985b; Nakatani and Yau, 1988b), allows more rapid internal dialysis of the entire rod outer segment. In this technique, illustrated in Fig. 7, the rod outer segment is drawn part way into a suction-pipet. The inner segment and basal portion of the outer segment are then sheared off with a glass probe, leaving the "truncated" outer segment in the suction-pipet with its interior accessible to the bath solution, which is then changed from Ringer to a pseudo-intracellular solution containing cGMP. By switching the bathing solution, a succession of different substances can be introduced into the outer segment, while the suction-pipet is used to record the response to light. Conceptually, the truncated cell technique is therefore rather similar to whole-cell patch-clamp recording, in that it allows dialysis of the rod cytoplasm. Its main advantage over whole-cell recording is the ability to switch rapidly between different dialysis solutions; its main drawback is the inability to control the transmembrane voltage, because of the high-leak conductance between the pipet and the membrane of the outer segment.

3.6. Excised Patch and Cell-Attached Patch Recordings

The techniques of excised-patch and cell-attached recording are standard, and here we will simply summarize their recent application to photoreceptors.

3.6.1. cGMP-Dependence
of the Outer-Segment Conductance

The excised-patch technique has been used extensively to determine the nature and action of the messenger of excitation in rods and cones. The identity of the internal messenger of excitation had long been a matter of controversy (reviewed by Pugh and Cobbs, 1986; Pugh, 1987; Pugh and Lamb, 1990). Two competing candidates had been put forward: Ca^{2+} and cGMP. The

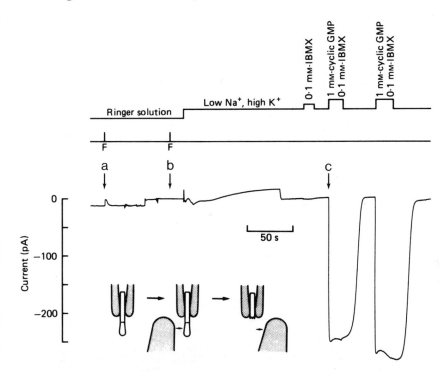

Fig. 7. Truncated cell technique. The inset shows a rod drawn into the suction pipet (left), the approach of the glass probe (middle), and the "truncated" outer segment after the remainder of the cell has been sheared off (right). The experimental trace shows (a) the photoresponse of the original cell, (b) the absence of a response from the truncated outer segment bathed in Ringer solution, and (c) the development of a circulating current upon switching the bathing solution to one containing cGMP and the phosphodiesterase inhibitor IBMX; other experiments (not shown) demonstrated that this current was light-suppressible. Reproduced with permission from Nakatani and Yau, 1988b).

controversy was finally resolved by Fesenko et al. (1985), who showed that cGMP could directly influence the outer-segment membrane conductance. They superfused excised inside-out patches of rod outer-segment membrane with test solutions, as shown in Fig. 8. When the cytoplasmic face of the patch was exposed to cGMP, the conductance of the patch increased, whereas exposure to other cyclic nucleotides or to altered $[Ca^{2+}]$ had no effect. Furthermore, this cGMP-activated conductance

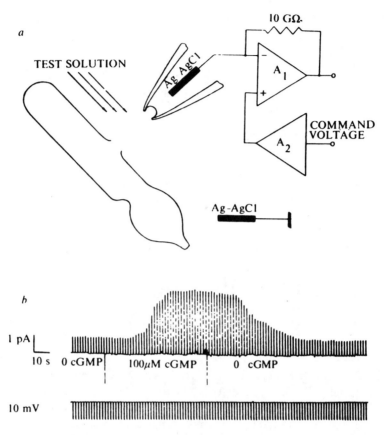

Fig. 8. The excised-patch technique with a rod photoreceptor, and the cGMP-induced currents that are recorded. **A.** Method: After forming a seal on the outer-segment membrane, the patch pipet is abruptly pulled away to form an excised "inside-out" patch in which the cytoplasmic surface faces the bathing solution. **B.** The current elicited by voltage pulses of 10 mV before, during, and after perfusion with 100 μM cGMP. Reproduced with permission from Fesenko et al. (1985).

exhibited electrical properties similar to those of the light-dependent conductance in the intact rod (Bader et al., 1979; Baylor and Nunn, 1986).

These experiments have subsequently been confirmed and extended by many other groups (reviewed by Yau and Baylor, 1989). The electrical and ionic properties of the light-sensitive conductance recorded in the cell-attached configuration have been shown to be virtually identical to those of the cGMP-activated conductance recorded in the excised inside-out configuration (G. Matthews, 1986, 1987; Matthews and Watanabe, 1987, 1988), indicating that the conductance is the same in the two cases. All of these observations (reviewed in Yau and Baylor, 1989) provided powerful support for the hypothesis that cGMP functions as the intracellular messenger of excitation in rod photoreceptors, and that the light response results from a light-induced decrease in cGMP concentration; accordingly, cGMP is sometimes referred to as a "negative transmitter." Similar experiments on salamander cones indicate that, in these cells, the messenger of excitation is again cGMP (Haynes and Yau, 1985).

3.6.2. Properties of the Outer-Segment Channel

The electrical, ionic, and molecular properties of the outer-segment conductance have been studied intensively using the excised inside-out patch technique. It has been shown that cGMP interacts cooperatively with the conductance, exhibiting a Hill coefficient of 2–3 (references in Yau and Baylor, 1989), and a half-activating concentration of 10–50 μM cGMP. Furthermore, the requirement of the channel for cGMP appears to be highly specific (for example, cAMP is at least 30× less effective), though a number of cGMP analogs are even more potent than cGMP itself (*see,* e.g., Zimmerman et al., 1985).

The ionic selectivity of the channel and the blocking effect of divalent cations have been studied by Nunn (1987), Zimmerman and Baylor (1988), Menini (1990), and Colamartino et al. (1991). These studies support previous results from intact rods (*see* Section 5.2.) that the channel is fairly unselective for Na^+ and K^+, but that it preferentially carries (and is blocked by) divalent cations. In normal Ringer solution, the channel is blocked for about 98% of the time by divalent ions, but by removing all divalent cations from the bathing media, it has been possible to investigate the activity at the single-channel level (Haynes et al., 1986; Zimmerman and Baylor, 1986). Additionally, the kinetics of activation of the conductance has been examined by flash photolysis of "caged" cGMP (Karpen et al., 1988; *see* chapter by Parker, this vol.), yielding results consistent with three diffusion-controlled cGMP binding steps. These topics have recently been reviewed by Yau and Baylor (1989), and will not be covered here.

4. Incorporation of Agents into the Cytoplasm

4.1. Overview of Methods

The methods used in electrophysiological experiments to incorporate agents into the cytoplasm of cells may be divided into several categories. Of the various methods, three categories have been successfully applied to photoreceptors, and two of these have already been described in this chapter. First, the whole-cell patch-pipet technique (either alone or in combination with suction-pipet recording) is suitable for incorporating most soluble agents, although difficulties are experienced with large proteins; the technique has been described in Sections 3.2.–3.4. Second, the truncated cell technique, described in Section 3.5., is also suitable for most soluble agents, but suffers from the limitation that the intracellular voltage must be held near zero because of the low seal resistance. The third category involves the use of membrane-permeant agents superfused externally; this approach is described subsequently in Section 5.3., after we introduce the techniques used for the rapid switching of external solutions. In the following sections, we describe the application of these techniques to the incorporation of a variety of different agents into the photoreceptor cytoplasm.

A class of method used in biochemical and optical experiments for the incorporation of substances into photoreceptors is the permeabilization of the cell membrane; this may be accomplished by freeze-thawing (*see*, e.g., Uhl et al., 1990) or with high electric fields (*see*, e.g., Gray-Keller et al., 1990). However, since this approach has not yet been applied in electrophysiological experiments, it will not be described here. A more conventional technique, the pressure injection of substances from an intracellular micropipet, has not proven successful in experiments on isolated vertebrate photoreceptors (probably because of damage to the cell membrane) and will therefore not be described here either.

4.2. Ca²⁺ Chelators

The incorporation of Ca^{2+} chelators into rods and cones has been used to investigate the role of light-induced changes of cytoplasmic [Ca^{2+}] in modulating the light response. The introduction of a Ca^{2+} chelator would be expected to slow changes in free [Ca^{2+}] and, thereby, to retard the actions of Ca^{2+}. There are two ways in which Ca^{2+} chelators have been introduced into photoreceptors.

The first approach is to use the whole-cell patch-clamp technique to introduce chelator from the patch-pipet by aqueous diffusion (Matthews et al., 1985; Lamb et al., 1986; Torre et al., 1986; Lamb and Matthews, 1988a). This approach is illustrated in Fig. 9, which shows the progressive changes in the light response that take place following the start of whole-cell recording, when the patch-pipet contains 10 mM of the Ca^{2+} chelator BAPTA (Tsien, 1980). By withdrawing the patch-pipet after a period of whole-cell recording, the chelator can be trapped in the cytoplasm (as described in Section 3.3.), yielding a cell with an elevated Ca^{2+}-buffering capacity (Torre et al., 1986; Lamb and Matthews, 1988a). The fact that the Ca^{2+}-buffering capacity of the cytoplasm has been substantially elevated may be demonstrated by exposing the outer segment to a low-Ca^{2+} solution (*see* Section 5.2.).

The second approach is to introduce the Ca^{2+} chelator in the form of the membrane-permeant acetoxymethyl ester (Tsien, 1981; *see* chapter by Poenie). The ester diffuses freely across the cell membrane and is hydrolyzed within the cell by intracellular esterases to yield the free chelator. This technique is widely used as a means of introducing fluorescent indicators into cells (*see* Tsien, 1989 for review). When the method is applied to photoreceptors, it is normally necessary to enrich the incubation medium with ascorbate and pyruvate to minimize the toxic effects of the formaldehyde released during ester hydrolysis (Garcìa-Sancho, 1984; Korenbrot and Brown, 1984). This approach has been used to incorporate the chelator Quin-2 into amphibian rods (Korenbrot and Miller, 1986) and BAPTA into mammalian rods (Matthews, 1990,1991), and it has also been used to introduce the fluorescent indicator Fura–2 into rods in the intact retina, as a means of measuring the cytoplasmic [Ca^{2+}] (Ratto et al., 1988).

The introduction of Ca^{2+} chelators into the rod cytoplasm provided evidence that excluded Ca^{2+} as the messenger of excitation (Matthews et al., 1985; Lamb et al., 1986), because the

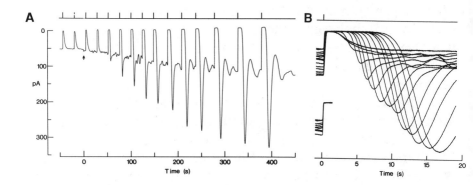

Fig. 9. Photoresponses recorded with the suction pipet during incorporation of the Ca^{2+} buffer BAPTA into the rod from a whole-cell patch pipet. A.
Continuous record; arrow indicates time of patch rupture. B. Superimposed
response to bright flashes, from A. Inset shows the rising phase of the first
nine responses. The patch pipet contained 10 mM BAPTA in a pseudointra-
cellular solution; within the pipet, the free Ca^{2+} concentration was ca. $10^{-9}M$.
Reproduced with permission from Matthews et al., 1985.

chelators had minimal effect on the rising phase of the light
response (see Fig. 9B). Instead, the experiments revealed a role
for $[Ca^{2+}]_i$ in light adaptation, since the onset of adaptation was
profoundly slowed in the presence of trapped chelator (Torre et
al., 1986). These results therefore suggested that Ca^{2+} might act
as the messenger of adaptation in photoreceptors, an idea that
has been greatly strengthened by evidence obtained from other
methods of opposing changes in $[Ca^{2+}]_i$ (see Section 5.3.).

4.3. cGMP

One aim of introducing cGMP into the intact photoreceptor
is to manipulate the cGMP-activated conductance in situ. In early
experiments, cGMP (or an analog) was introduced into rods by
iontophoresis from an intracellular micropipet (Miller and Nicol,
1979; MacLeish et al., 1984). More recently, the preferred methods of introduction have been the whole-cell patch-clamp technique (Cobbs and Pugh, 1985; Cobbs et al., 1985; Matthews et
al., 1985; Zimmerman et al., 1985; Cameron and Pugh, 1990) and
the truncated cell technique (Yau and Nakatani, 1985b; Nakatani
and Yau, 1988b). Figure 10 shows that the effect of incorporat-

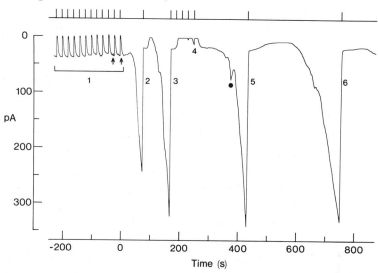

Fig. 10. Incorporation of cGMP into a rod photoreceptor. Responses to saturating flashes were recorded using a suction pipet, whereas cGMP was introduced from a patch pipet containing 2 mM of the nucleotide in a pseudo-intracellular solution. The initial 12 responses (marked 1) are controls. At the first arrow, the patch pipet was sealed onto the inner-segment membrane, and at the second arrow, the underlying patch of membrane was ruptured. At this point, the sequence of flashes was interrupted, and the circulating current quickly built up as the cGMP diffused to the outer segment. Flashes of the original intensity were again given at intervals and completely suppressed the large circulating current. For the responses marked 2 and 3, the current did not fall instantaneously to zero, because the large influx of Ca^{2+} accompanying the elevated circulating current led to an elevated Na$^+$–Ca^{2+} exchange transient. Reproduced with permission from Matthews et al., 1985.

ing cGMP into a rod from a patch-pipet is to increase the circulating current greatly. Note, however, that this increased current is still suppressible by light, further supporting the identity of the light-dependent and cGMP-activated conductances. As would be expected, light is much less effective at suppressing the current increase induced by the hydrolysis-resistant cGMP analog 8-bromo-cGMP (MacLeish et al., 1984; Zimmerman et al., 1985; Cameron and Pugh, 1990).

The truncated cell technique offers the advantage over whole-cell patch-clamp recording that it readily allows the presentation to the cytoplasmic surface of a number of different concentrations of cGMP during a single experiment. Therefore, this technique can be used to investigate the concentration-dependence of activation of the conductance *in situ* (Yau and Nakatani, 1985b; Nakatani and Yau, 1988b). However, since both of these techniques introduce the cyclic nucleotide at a single point, it is not possible to achieve a uniform concentration along the outer segment (Cameron and Pugh, 1990), unless the resting dark rate of hydrolysis by the phosphodiesterase is reduced using a phosphodiesterase inhibitor (Nakatani and Yau, 1988b) or unless a weakly hydrolyzable analog, such as 8-bromo-cGMP, is used (MacLeish et al., 1984; Zimmerman et al., 1985; Cameron and Pugh, 1990). Indeed, the existence of spatial nonuniformity in the current density with point injection of cGMP has been used to provide a measure of the longitudinal diffusion of cGMP in the cytoplasm (Cameron and Pugh, 1990).

4.4. Nucleotide Analogs
and Other Low-Molecular-Weight Drugs

The introduction of nucleotide analogs into photoreceptors is often used as a means of interfering with the normal operation of the transduction cascade controlling the concentration of cGMP. An example is provided by the use of hydrolysis-resistant analogs of GTP to probe the inactivation of the GTP-binding protein transducin following illumination. Figure 11 shows that the incorporation of GTPγS during a brief period of whole-cell recording results in an enormous prolongation of the response to a subsequent intense flash recorded by the suction-pipet (Lamb and Matthews, 1986, 1988b; Kondo and Miller, 1989). Similar effects are seen in recordings from isolated outer segments upon incorporating GTPγS from a patch-pipet (Sather and Detwiler, 1987). These results indicate that the inactivation of transducin can become rate-limiting for the recovery of the flash response in vivo and suggest that the speed of GTP hydrolysis by the GTPase activity of transducin is faster in vivo than has hitherto been measured in vitro (discussed in Lamb and Matthews, 1988b; *see also* Vuong and Chabre, 1990; Uhl et al., 1990).

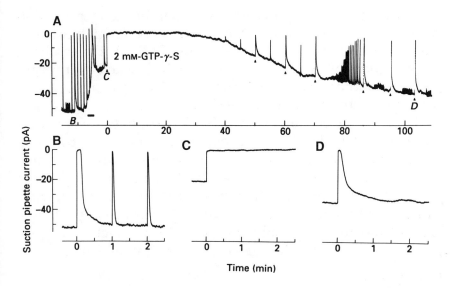

Fig. 11. Incorporation of GTPγS into a rod photoreceptor. (A) Continuous recording of suction-pipet current over a period of nearly 2 h. During the time indicated by the bar, a patch pipet containing 2 m*M* GTPγS was in the whole-cell configuration on the inner segment; at the end of the bar, the patch pipet was gently withdrawn, so that the cell membrane sealed over and the GTPγS was trapped in the cytoplasm. Arrowheads mark the timing of intense flashes delivering 4×10^5 photoisomerizations; responses at three different times are replotted below in B, C, and D on a faster time base. In control conditions (B), the response to the intense flash began recovering within 15 s and was fully recovered within 1 min. After trapping GTPγS in the cytoplasm (C), the same flash elicited a response that did not begin recovering for 30 min. After 100 min (D), the response to the intense flash was similar to the original response. Reproduced with permission from Lamb and Matthews (1988b).

More extensive biochemical manipulations can be carried out by applying the whole-cell technique to isolated outer segments or by using the truncated cell technique. These approaches have the advantage, when introducing metabolically labile nucleotides, that the enzymatically active inner segment is absent, thereby removing a possible means of breakdown. Using these techniques, the requirements of the transduction mechanism for ATP and GTP have been investigated (Yau and Nakatani, 1985b; Nakatani and Yau, 1988b; Sather and Detwiler, 1987).

Other blockers of specific stages within the transduction mechanism can be introduced using these techniques. For example, the endogenous formation of cGMP via the synthetic enzyme guanylyl cyclase can effectively be blocked by incorporating imidodiphosphate into the rod outer segment (Detwiler et al., 1989); this agent in fact acts by inhibiting the pyrophosphatase, and the elevated levels of pyrophosphate then prevent the synthesis of cGMP from proceeding in the forward direction. As new biochemical agents that interfere with the transduction cascade become available, their incorporation into photoreceptors will allow the gap between biochemistry in vitro and the transduction process in vivo to be progressively bridged. This is likely to be especially important for unraveling the processes responsible for the recovery of the light response.

4.5. Proteins

Proteins can most readily be incorporated into photoreceptors by using the whole-cell patch-pipet technique. There are, however, two problems associated with the use of this technique. The first problem is that proteins tend to contaminate the glass of the patch-pipet tip, making the establishment of a gigohm seal impossible. This problem can be circumvented by changing the solution within the patch-pipet after the seal has been formed by infusing the solution from a fine tube inserted near its tip (Cull-Candy et al., 1980; Soejima and Noma, 1984; McNaughton et al., 1986; Lapointe and Szabo, 1987; Neher and Eckert, 1988; Tang et al., 1990); a suitable micro-pump has been described by Lamb and Matthews (1985). The second problem is that the diffusion coefficient of a large protein is so low that a very long period of whole-cell recording is necessary in order to obtain a useful concentration of protein in the cytoplasm.

These techniques could be used to introduce proteins that might be expected to interact with specific components of the transduction mechanism, for example, antibodies to any of the proteins involved in the transduction mechanism (reviewed in Hurley, 1987; see, e.g., Hamm and Bownds, 1984) or those proteins themselves. A drawback of this approach is that many of these antibodies and native proteins are too large to introduce in sufficient quantity to have a significant effect. To minimize this difficulty, proteins that are of low molecular weight, yet effective at low concentration, would be the best candidates for incorporation. One obvious protein meeting these requirements would be the regulatory γ-subunit of the phosphodiesterase (PDE).

The most successful application of the incorporation of a protein into photoreceptors so far has been the introduction of the luminescent Ca^{2+}-indicator aequorin into salamander rods by the whole-cell technique (McNaughton et al., 1986). For this ca. 20-kDa protein to diffuse into the cytoplasm in sufficient quantity, the patch-pipet needs to be in the whole-cell configuration for some 10 min. As illustrated in Fig. 12, from McNaughton et al. (1986), the method has been used to measure the elevated levels of $[Ca^{2+}]_i$ resulting from manipulations (such as superfusion with phosphodiesterase inhibitors) that lead to increased Ca^{2+} influx. In the resting state in normal Ringer solution, the light emission from the aequorin is below the limit of detection, and from calibrations, it can be shown that the resting free cytoplasmic Ca^{2+} concentration must be no more than 0.6 μM.

If the cell is exposed to the phosphodiesterase inhibitor 3-isobutyl–1-methylxanthine (IBMX), the hydrolysis of cGMP is reduced and the cGMP concentration increases, causing the opening of ion channels and a consequent increase in the influx of Ca^{2+}. This is shown in Fig. 12A by the ramping of circulating current, and in Fig. 12B by the increased light emission from aequorin. In the traces labeled 2 in that figure, the increased light emission from the protein directly stimulated the photoreceptor, causing closure of the light-sensitive pathway; in the traces labeled 1, a bright flash was presented and the normal light response was accompanied by a rapid reduction in $[Ca^{2+}]_i$. Panel C shows the calculated $[Ca^{2+}]_i$, which at the time of the flash was about 8 μM, a factor of more than tenfold above its normal resting level.

Hence, these experiments have demonstrated directly that $[Ca^{2+}]_i$ declines during illumination, although it has not so far proven possible to resolve the normal dark resting level or to measure the concentration to which $[Ca^{2+}]_i$ falls during the nor-

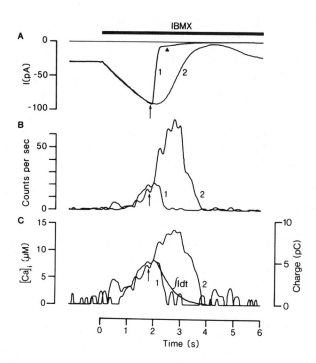

Fig. 12. Photocurrent and light emission from a rod containing the Ca^{2+}-sensitive photoprotein aequorin, upon exposure to the phosphodiesterase inhibitor IBMX. In response 1, a bright flash was delivered at the arrow; in response 2, no external light was applied. A. Photocurrent recorded with a suction pipet. B. Light emission recorded by a photomultiplier. C. Ca^{2+}_i calculated from trace B. Reproduced with permission from McNaughton et al. (1986).

mal light response. Furthermore, the experiments have provided a means of directly examining the rate of operation of the Na^+–Ca^{2+} exchanger as a function of $[Ca^{2+}]_i$ (Cervetto et al., 1987).

5. External Perfusion of the Outer Segment

5.1. Techniques for Rapid Solution Changes

Changing the solution bathing the rod outer segment provides a powerful, albeit indirect, means of perturbing the transduction mechanism. For example, it is possible to change the concentrations of permeant ions rapidly or to deliver a membrane-permeant blocker of a specific stage in transduction. These and other applications are described in Sections 5.2. and 5.3. Solution changes can also be applied to the inside-out patch or truncated outer-segment preparations to study the effect of a given substance when applied directly on the cytoplasmic side; these approaches have been described in Sections 3.5. and 3.6.

Much has been learned from the relatively slow solution changes that result following alteration of the bulk solution entering the recording chamber (*see*, e.g., Yau et al., 1981; Hodgkin et al., 1984). However, a more powerful approach is to carry out solution changes that are rapid in comparison with the photo-transduction process and, thereby, resolve the kinetics of the response of the transduction mechanism to this perturbation.

Many techniques exist for rapidly and repeatably changing the solution bathing a single cell or a membrane patch. Such a change may be achieved by the rapid onset of flow from an orifice aimed at the cell being recorded from (e.g., Fenwick et al., 1982), by changing the composition of a single flowing stream of solution (e.g., Matthews and Watanabe, 1988), or by translating the cell across the interface between two flowing streams of solution. The last approach, which derives from the "pan pipe" technique of Yellen (1982), has been widely used for effecting rapid changes in the solution bathing the photoreceptor outer segment during suction-pipet recording and will be described in detail here.

This technique for rapid solution changes was developed by Hodgkin et al. (1985) and is illustrated in Fig. 13. The inner segment of a single rod (or cone) is drawn into the suction-pipet, leaving the outer segment exposed to the bath solution. The suction-pipet is positioned so as to place the cell near the interface between two flowing streams of solution emerging from tubes built into the rear of the recording chamber (*see also* Fig. 2B). Solution changes are carried out by rapidly "stepping" this interface across the cell by translating the entire chamber relative to the suction-pipet; for reproducibility, this translation is triggered electrically. The step may be driven by compressed air (Hodgkin et al., 1985), by a solenoid (Lamb and Matthews, 1988a), or by a stepping motor (Hodgkin et al., 1987; Matthews, 1988). In Fig. 13,

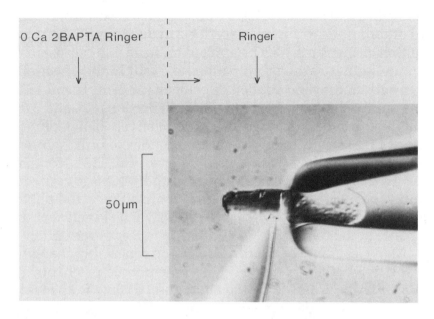

0 Ca 2BAPTA Ringer   Ringer

50 μm

Fig. 13. Technique for rapidly changing the perfusate bathing the outer segment of a rod. The inner segment of the rod is drawn into a suction-pipet in order to record the circulating current, leaving the outer segment exposed to the bathing solution. The cell is positioned near the interface between two flowing streams of perfusate. To change solutions, the interface between these two streams is rapidly stepped across the cell. A "whole-cell" patch pipet is sealed onto the outer segment of the cell to dialyze the cytoplasm.

a "whole-cell" patch-pipet was also sealed onto the exposed outer segment for intracellular dialysis.

Using this technique, solution changes may be carried out that are rapid in comparison with the kinetics of the photoresponse. The time to change solutions represents only the time taken to translate the interface between the flowing streams across the cell (not the duration of the entire "step"). The time for completion of the solution change can be assessed by stepping an "empty" suction-pipet between solutions of very different conductance (Hodgkin et al., 1985). More routinely, it can be estimated from the rise time of the junction currents that flow across the "seal" between the cell and the suction-pipet, in response to the change in junction potential between the solu-

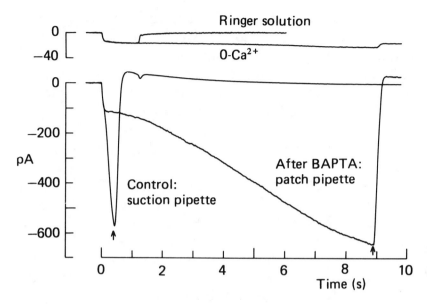

Fig. 14. Rapid reduction of $[Ca^{2+}]_o$ from 1 mM to $10^{-9}M$, under control conditions and after incorporation of BAPTA into the cytoplasm from a patch pipet; the latter trace was recorded with the patch pipet, while it was still in the voltage-clamped "whole-cell" configuration. A bright flash was delivered in each case at the time indicated by the arrow. Reproduced with permission from Lamb and Matthews (1988a).

tions in the pipet and bath (*see* top trace in Fig. 14). As assessed from these junction currents, the solution change is normally complete within 30–100 ms.

5.2. Alteration of External Ions

Approaches in which the ionic composition of the external solution is altered can be divided into two main classes. The first is the investigation of the ionic selectivity of the outer-segment conductance in the intact rod or cone (Hodgkin et al., 1984, 1985; Yau and Nakatani, 1984a ; Menini et al., 1988; Nakatani and Yau, 1988a). These experiments demonstrated that the outer-segment conductance is relatively unselective between monovalent cations and preferentially selective for Ca^{2+} (reviewed in Yau and Baylor, 1989). These results have been complemented by the more recent experiments on the selectivity of the outer-segment conductance in excised patches, which were discussed in Section 3.6. The second approach is to use external changes in ionic composition to investigate and manipulate the control of $[Ca^{2+}]_i$ by the photoreceptor. This approach has led to an understanding of the significance of Ca^{2+} in the vertebrate photoreceptor and will be discussed in detail in the remainder of this section.

The $[Ca^{2+}]_i$ in vertebrate photoreceptors is believed to be controlled by the balance between Ca^{2+} influx through the outer-segment conductance (Yau et al., 1981; Capovilla et al., 1983; Hodgkin et al., 1984, 1985; Yau and Nakatani, 1984a) and Ca^{2+} efflux through a Na^+–Ca^{2+} exchange mechanism (Yau and Nakatani, 1984b; Hodgkin et al., 1985). If this dynamic balance is perturbed by modifying Ca^{2+} influx or efflux, then $[Ca^{2+}]_i$ will change accordingly. Note that this is precisely what takes place during the light response, when the influx through the outer-segment conductance decreases and the efflux through the exchanger continues, resulting in a light-induced decrease in $[Ca^{2+}]_i$ (Yau and Nakatani, 1985a; McNaughton et al., 1986; Ratto et al., 1988).

The Na^+–Ca^{2+} exchanger has been extensively studied by manipulating the external ionic composition. The electrogenic nature of its operation (reviewed by Lagnado and McNaughton, 1990) allows the rate of exchange to be recorded directly by a suction or whole-cell patch-pipet following the complete suppression of the outer-segment conductance by bright light. External ionic substitutions have demonstrated the specific requirement for external Na^+ and have allowed the driving gradient for the exchanger to be manipulated (Yau and Nakatani, 1984b, 1985a; Hodgkin et al., 1985, 1987; Hodgkin and Nunn, 1987; Lagnado et al., 1988; Nakatani and Yau, 1988a). External solution changes can be used to "load" the photoreceptor with Ca^{2+}, the subsequent extrusion of which can be studied. As was described in Section 3.4., the isolated outer segment is an especially suitable preparation for these experiments, since the outer-segment membrane contains only the cGMP-gated conductance and the exchanger. Most recently, such approaches have indicated the involvement of the outward K^+ gradient as well as the inward Na^+ gradient in driving the transport of Ca^{2+} by the exchanger (Cervetto et al., 1989).

It has long been known that reduction of the external Ca^{2+} concentration $[Ca^{2+}]_o$ results in a large increase in the circulating current (Hagins et al., 1970; Yau et al., 1981; Hodgkin et al., 1984). This is illustrated in Fig. 14, which shows that the increase in current following a rapid reduction in $[Ca^{2+}]_o$ consists of two components (Lamb and Matthews, 1988a). The first component is as rapid as the solution change itself, and is believed to result from the direct interaction of external Ca^{2+} with the outer-segment conductance (Lamb and Matthews, 1988a; Nakatani and Yau, 1988a). The second component can be greatly slowed by the introduction of the Ca^{2+} chelator BAPTA from a patch-pipet, indicating that this component is brought about by the reduction in $[Ca^{2+}]_i$; as in the normal light response, this drop in $[Ca^{2+}]_i$ results from continued Ca^{2+} extrusion by the exchanger following the cessation of Ca^{2+} influx through the outer-segment conductance. It is believed that this reduction in $[Ca^{2+}]_i$ relieves inhibition on guanylyl cyclase (Koch and Stryer, 1988; Lolley and Racz, 1982; Pepe et al., 1986), the synthetic enzyme that produces cGMP, thereby greatly elevating cGMP concentration when $[Ca^{2+}]_i$ is artificially reduced in darkness.

If the extrusion of Ca^{2+} by the Na^+–Ca^{2+} exchanger is prevented by substituting Li^+ for external Na^+ and thereby removing the driving gradient, then the circulating current declines approximately exponentially in darkness (Hodgkin et al., 1985; Hodgkin and Nunn, 1988), as illustrated in Fig. 15A. This decline is believed to result from the rapid inhibition of guanylyl cyclase by the elevation of $[Ca^{2+}]_i$ consequent upon the inhibition of the exchanger. The rate constant of this decline therefore represents the velocity of the phosphodiesterase, which hydrolyzes cGMP. As shown in Fig. 15, the light-induced increase in this rate constant can be used as a measure of the acceleration of phosphodiesterase velocity during the light response.

It is possible to "clamp" $[Ca^{2+}]_i$ by combining these two methods, i.e., by greatly reducing $[Ca^{2+}]_o$, thereby minimizing Ca^{2+} influx, and simultaneously removing external Na^+ and substituting for it an ion, such as Li^+ or guanidinium, which cannot support Ca^{2+} extrusion by the exchanger, thereby minimizing Ca^{2+} efflux. When the solution bathing the outer segment is rapidly changed to such a "low-Ca^{2+}/0-Na^+" solution, $[Ca^{2+}]_i$ remains

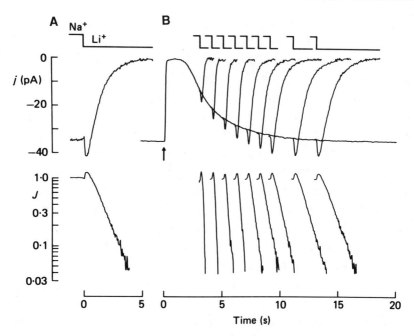

Fig. 15. Rapid replacement of external Na⁺ by Li⁺. Suction-pipet current is plotted on linear (above) and logarithmic (below) scales, in response to switching to Li⁺ either in darkness (A) or at different times after a flash delivering 6000 photoisomerizations (B). Reproduced with permission from Hodgkin and Nunn (1988).

near the value it had before the solution change for some 10–15 s (Matthews et al., 1988; Nakatani and Yau, 1988c; Fain et al., 1989; Matthews et al., 1990). When this approach is used to prevent light-induced changes in $[Ca^{2+}]_i$, the responses both to steady light (Fig. 16a) and to dim flashes (Fig. 16c,d) continue to rise for longer than in control, and are therefore larger and reach peak later. It can be shown that, under these conditions, where $[Ca^{2+}]_i$ cannot change, all of the manifestations of light adaptation are abolished, and that both rods and cones simply saturate compressively. These results are qualitatively consistent with the slowing of adaptation following the incorporation of Ca^{2+} chelator from a patch-pipet (Torre et al., 1986; see Section 4.2.) and indicate that cytoplasmic Ca^{2+} acts as the messenger of light adaptation in vertebrate photoreceptors (for review, see Fain and Matthews, 1990).

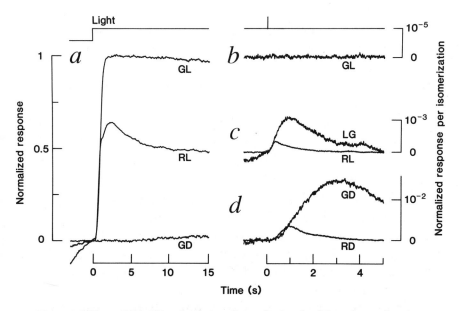

Fig. 16. Effect of rapidly changing the solution bathing the rod outer segment from normal Ringer to low-Ca^{2+}/0-Na^+ solution on the responses to steps (*a*) and flashes (*b–d*) of light. Traces are identified as D, in darkness or L, in steady light (8.1 photons $\mu m^{-1} s^{-1}$); and R, in Ringer or G, in low-Ca^{2+}/0-Na^+, solution containing 3 μM free Ca^{2+} and 110 mM guanidinium. A bright flash fails to elicit a measurable response when the steady light is presented during exposure to low-Ca^{2+}/0-Na^+ solution (panel *b*, trace GL). When low-Ca^{2+}/0-Na^+ solution is presented after equilibration to the background light (*c*), the response to a dim flash (trace LG) rises further than in control (trace RL) along a common rising phase, resulting in an increase in sensitivity. A similar change in the dim flash response is seen when low-Ca^{2+}/0-Na^+ solution is presented in darkness (*d*, traces GD and RD). Reproduced with permission from Matthews et al. (1988).

5.3. Superfusion with Drugs

Superfusion of the photoreceptor can also be used to deliver compounds with some defined action on transduction. Such drugs can be subdivided into those with an external and those with an internal action. Externally acting drugs can directly affect only the outer-segment conductance or the Na^+–Ca^{2+} exchanger, whereas internally acting drugs may affect any of the components of the transduction mechanism. There may also be substantial differences among different drugs in the time-course of their action. Internally acting drugs may either be rapidly membrane permeant or, alternatively, may be slowly internalized by some other mechanism. Rapid superfusion techniques are only necessary when applying a fast-acting drug. Examples of the application of each of these classes of drugs are given below.

The outer-segment conductance is blocked by the organic Ca^{2+} antagonist L-*cis*-diltiazem. This compound is effective when applied to the cytoplasmic face of inside-out patches and somewhat less so when applied externally to the intact rod (Stern et al., 1986). The block by L-*cis*-diltiazem is voltage-sensitive, taking place nearly halfway across the membrane field from the cytoplasmic face (McLatchie and Matthews, 1990). The Na^+–Ca^{2+} exchange is partially blocked by externally applied quinidine (Lagnado and McNaughton, 1987), though little is known about the site of its action.

The best example of a drug that is rapidly membrane-permeant and exerts a well-defined effect on the transduction mechanism is the phosphodiesterase inhibitor IBMX. External application of IBMX rapidly inhibits the phosphodiesterase, leading to an elevated circulating current (Cervetto and McNaughton, 1986) as the consequence of an increased cGMP concentration. The initial rate of this increase reflects the velocity of guanylyl cyclase (Hodgkin and Nunn, 1988), allowing the acceleration of the cyclase during illumination to be investigated. The increased current in the presence of IBMX can also be used as a convenient means of loading the photoreceptor with Ca^{2+} (*see* Section 4.5.).

Pertussis toxin provides an example of an internally acting agent that is believed to be slowly internalized by receptor-mediated endocytosis. External application of pertussis toxin results in depolarization, and a diminution and slowing of light responses (Falk and Shiells, 1987). These results are broadly consistent with an inactivation of the GTP-binding protein transducin by pertussis toxin. However, it is probably more appropriate to introduce such slowly incorporated molecules directly into the cytoplasm by one of the techniques described in Section 4.

6. Summary

Over the last decade, a number of new electrophysiological techniques have been developed that have been crucial to the advancement of our knowledge of the role of intracellular messengers in phototransduction. In this chapter, we have attempted to describe the most important aspects of these techniques and their applications to the study of phototransduction.

Abbreviations

$[Ca^{2+}]_o$	Extracellular free Ca^{2+} concentration
$[Ca^{2+}]_i$	Intracellular free Ca^{2+} concentration
cGMP	Guanosine 3',5' cyclic monophosphate
GTPγS	Guanosine 5'-[γ-thio]triphosphate
IBMX	3-Isobutyl-1-methylxanthine

References

Attwell D., Werblin F. S., and Wilson M. (1982) The properties of single cones isolated from the tiger salamander retina. *J. Physiol.* **328,** 259–283.

Bader C. R., MacLeish P. R., and Schwartz E. A. (1979) A voltage-clamp study of the light response in solitary rods of the tiger salamander. *J. Physiol.* **296,** 1–26.

Bastian B. L. and Fain G. L. (1979) Light adaptation in toad rods: requirement for an internal messenger which is not calcium. *J. Physiol.* **297,** 493–520.

Baylor D. A. (1988) The light regulated ionic channel of retinal rod cells. *Proceedings of the Retina Research Foundation Symposium* **1,** 31–40.

Baylor D. A. and Hodgkin A. L. (1973) Detection and resolution of visual stimuli by turtle photoreceptors. *J. Physiol.* **234,** 163–198.

Baylor D. A. and Lamb T. D. (1982) Local effects of bleaching in retinal rods of the toad. *J. Physiol.* **328,** 49–71.

Baylor D. A. and Nunn B. J. (1986) Electrical properties of the light-sensitive conductance of salamander rods. *J. Physiol.* **371,** 115–145.

Baylor D. A., Lamb T. D., and Yau K.-W. (1979) The membrane current of single rod outer segments. *J. Physiol.* **288,** 589–611.

Baylor D. A., Nunn B. J., and Schnapf J. L. (1984) The photocurrent, noise and spectral sensitivity of rods of the monkey *Macaca fascicularis. J. Physiol.* **357,** 575–607.

Baylor D. A., Nunn B. J., and Schnapf J. L. (1987) Spectral sensitivity of cones of the monkey *Macaca fascicularis*. *J. Physiol.* **390,** 145–160.

Bodoia R. D. and Detwiler P. B. (1985) Patch-clamp recordings of the light-sensitive dark noise in retinal rod from the lizard and frog. *J. Physiol.* **367,** 183–216.

Cameron D. A. and Pugh E. N., Jr. (1990) The magnitude, time course and spatial distribution of current induced in salamander rods by cyclic guanine nucleotides. *J. Physiol.* **430,** 419–439.

Capovilla M., Caretta A., Cervetto L., and Torre V. (1983) Ionic movements through light-sensitive channels of toad rods. *J. Physiol.* **343,** 295–310.

Cervetto L. and McNaugton P. A. (1986) The effects of phosphodiesterase inhibitors and lanthanum ions on the light-sensitive current of toad retinal rods. *J. Physiol.* **370,** 91–109.

Cervetto L., Lagnado L., and McNaughton P. A. (1987) Activation of the Na:Ca exchange in salamander rods by intracellular Ca. *J. Physiol.* **382,** 135P.

Cervetto L., Lagnado L., Perry R. J., Robinson D. W., and McNaughton P. A. (1989) Extrusion of calcium from rod outer segments is driven by both sodium and potassium gradients. *Nature* **337,** 740-743.

Cobbs W. H. and Pugh E. N., Jr. (1985) Cyclic GMP can increase rod outer segment light-sensitive current 10-fold without delay of excitation. *Nature* **313,** 585–587.

Cobbs W. H. and Pugh E. N., Jr. (1987) Kinetics and components of the flash photocurrent of isolated retinal rods of the larval salamander *Ambystoma tigrinum*. *J. Physiol.* **394,** 529–572.

Cobbs W. H., Barkdoll A. E. III, and Pugh E. N., Jr. (1985) Cyclic GMP increases photocurrent and light sensitivity of retinal cones. *Nature* **305,** 50–53.

Colamortino G., Menini A., and Torre V. (1991) Blockage and permeation of divalent cations through the cyclic GMP-activated channel from retinal rods. *J. Physiol.* (in press).

Crawford A. C., Evans M. G., and Fettiplace R. (1989) Activation and adaptation of transducer currents in turtle hair cells. *J. Physiol.* **419,** 405–434.

Cull-Candy S. G., Miledi R., and Parker I. (1980) Single glutamate-activated channels recorded from locust muscle fibres with perfused patch-clamp electrodes. *J. Physiol.* **321,** 195–210.

Detwiler P. B., Rispoli G., Sather W. A., and Ertel E. A. (1989) Intracellular biochemical manipulation of phototransduction in detached rod outer segments. *Biophys. J.* **55,** 581a.

Donner K. O. and Hemilä S. (1978) Excitation and adaptation in the vertebrate rod photoreceptor. *Med. Biol.* **56,** 52–63.

Fain G. L. and Matthews H. R. (1990) Calcium and the mechanism of light adaptation in vertebrate photoreceptors. *Trends Neurosci.* **13,** 378–384.

Fain G. L., Lamb T. D., Matthews H. R., and Murphy R. L. W. (1989) Cytoplasmic calcium as the messenger for light adaptation in salamander rods. *J. Physiol.* **416,** 215–243.

Falk G. and Shiells R. A. (1987) Effect of G-protein inactivation of salamander rods. *J. Physiol.* **390**, 161P.

Fenwick E. M., Marty A., and Neher E. (1982) A patch-clamp study of bovine chromaffin cells and their sensitivity to acetylcholine. *J. Physiol.* **331**, 577–597.

Fesenko E. E., Kolesnikov S. S., and Lyubarsky A. L. (1985) Induction by cyclic GMP of cationic conductance in plasma membrane of retinal rod outer segment. *Nature* **313**, 310–313.

Garcìa-Sancho J. (1984) Inhibition of glycolysis in the human erythrocyte by formaldehyde and Ca-chelator esters. *J. Physiol.* **357**, 60P.

Gray P. and Attwell D. (1985) Kinetics of light-sensitive changes in vertebrate photoreceptors. *Proc. R. Soc. [B]* **223**, 379–388.

Gray-Keller M. P., Biernbaum M. S., and Bownds M. D. (1990) Transducin activation in electropermeabilized frog rod outer segments is highly amplified and a portion equivalent to phosphodiesterase remains membrane bound. *J. Biol. Chem.* **265**, 15,323–15,332.

Hagins W. A., Penn R. D., and Yoshikami S. (1970) Dark current and photocurrent in retinal rods. *Biophys. J.* **10**, 380–412.

Hamill O. P., Marty A., Neher E., Sakmann B., and Sigworth F. (1981) Improved patch-clamp techniques for high-resolution current recording from cells and cell-free membrane patches. *Pflügers Arch.* **391**, 85–100.

Hamm H. E. and Bownds M. D. (1984) A monoclonal antibody to guanine nucleotide binding protein inhibits the light-activated cyclic GMP pathway in frog rod outer segments. *J. Gen. Physiol.* **84**, 265–280.

Haynes L. W. and Yau K.-W. (1985) Cyclic GMP-sensitive conductance in outer segment membrane of catfish cones. *Nature* **317**, 61–64.

Haynes L. W. and Yau K.-W. (1990) Single-channel measurements from the cyclic GMP-activated conductance of catfish retinal cones. *J. Physiol.* **429**, 451–481.

Haynes L. W., Kay A. R., and Yau K.-W. (1986) Single cyclic GMP-activated channel activity in excised patches of rod outer segment membrane. *Nature* **321**, 66–70.

Hodgkin A. L. and Nunn B. J. (1987) The effect of ions on sodium-calcium exchange in salamander rods. *J. Physiol.* **391**, 371–398.

Hodgkin A. L. and Nunn B. J. (1988) Control of light-sensitive current in salamander rods. *J. Physiol.* **403**, 439–471.

Hodgkin A. L., McNaughton P. A., and Nunn B. J. (1985) The ionic selectivity and calcium dependence of the light-sensitive pathway in toad rods. *J. Physiol.* **358**, 447–468.

Hodgkin A. L., McNaughton P. A., and Nunn B. J. (1987) Measurement of sodium-calcium exchange in salamander rods. *J. Physiol.* **391**, 347–370.

Hodgkin A. L., McNaughton P. A., Nunn B. J., and Yau K.-W. (1984) Effect of ions on retinal rods from *Bufo marinus. J. Physiol.* **350**, 649–680.

Hurley J. B. (1987) Molecular properties of the cGMP cascade of vertebrate photoreceptors. *Ann. Rev. Physiol.* **49**, 793–812.

512   *Matthews and Lamb*

Kaneko A. (1979) Physiology of the retina. *Ann. Rev. Neurosci.* **2**, 169–191.
Karpen J. W., Zimmerman A. L., Stryer L., and Baylor D. A. (1988) Gating kinetics of the cyclic GMP-activated channel of retinal rods: flash photolysis and voltage clamp studies. *Proc. Natl. Acad. Sci. USA* **85**, 1287–1291.
Koch K.-W. and Stryer L. (1988) Highly cooperative feedback control of retinal rod guanylate cyclase by calcium ions. *Nature* **334**, 64–66.
Kondo H. and Miller W. H. (1989) Rod light adaptation may be mediated by acceleration of the phosphodiesterase-guanylate cyclase cycle. *Proc. Natl. Acad. Sci. USA* **85**, 1322–1326.
Korenbrot J. I. and Brown J. E. (1984) The use of Quin 2 as an intracellular Ca indicator in single cells and small cell assemblies. *J. Gen. Physiol.* **84**, 7a.
Korenbrot J. I. and Miller D. L. (1986) Calcium ions act as a modulator of intracellular information flow in retinal rod phototransduction. *Neurosci. Res. Suppl.* **4**, S11–S34.
Kraft T. W. (1988) Photocurrents of cone photoreceptors of the golden-mantled ground squirrel. *J. Physiol.* **404**, 199–213.
Lagnado L. and McNaughton P. A. (1987) Inhibition of Na:Ca exchange in isolated salamander rods by quinindine. *J. Physiol.* **390**, 163P.
Lagnado L. and McNaughton P. A. (1990) Electrogenic properties of the Na:Ca exchange. *J. Membr. Biol.* **113**, 177–191.
Lagnado L., Cervetto L., and McNaughton P. A. (1988) Ion transport by the Na–Ca exchange in isolated rod outer segments. *Proc. Natl. Acad. Sci. USA* **85**, 4548–4552.
Lamb T. D. (1983) "ANADISK": A disk recorder for analogue signals. *J. Physiol.* **343**, 16–17P.
Lamb T. D. (1985) An inexpensive digital tape recorder suitable for neurophysiological signals. *J. Neurosci. Methods* **15**, 1–13.
Lamb T. D. and Matthews H. R. (1985) A precision microperfusion apparatus. *J. Physiol.* **369**, 4P.
Lamb T. D. and Matthews H. R. (1986) Incorporation of hydrolysis-resistant analogues of GTP into rod photoreceptors isolated from the tiger salamander. *J. Physiol.* **381**, 58P.
Lamb T. D. and Matthews H. R. (1988a) External and internal actions in the response of salamander rods to altered external calcium concentration. *J. Physiol.* **403**, 473–494.
Lamb T. D. and Matthews H. R. (1988b) Incorporation of analogues of GTP and GDP into rod photoreceptors isolated from the tiger salamander. *J. Physiol.* **407**, 463–488.
Lamb T. D., Matthews H. R., and Murphy R. L. W. (1989) Cell capacitance limits the responses of cone photoreceptors isolated from the tiger salamander retina. *J. Physiol.* **414**, 51P.
Lamb T. D., Matthews H. R., and Torre V. (1986) Incorporation of calcium buffers into salamander retinal rods: a rejection of the calcium hypothesis of phototransduction. *J. Physiol.* **372**, 315–349.

Lamb T. D., McNaughton P. A., and Yau K.-W. (1981) Spatial spread of activation and background desensitization in toad rod outer segments. *J. Physiol.* **319**, 463–496.

Lapointe J.-Y. and Szabo G. (1987) A novel holder allowing internal perfusion of patch-clamp pipettes. *Pflügers Arch.* **410**, 212–216.

Liebman P. A., Parker K. R., and Dratz E. A. (1987) The molecular mechanism of visual excitation and its relation to the structure and composition of the rod outer segment. *Ann. Rev. Physiol.* **49**, 765–791.

Lolley R. N. and Racz E. (1982) Calcium modulation of cyclic GMP synthesis in rat visual cells. *Vision Res.* **22**, 1481-1486.

MacLeish P. R., Schwartz E. A., and Tachibana M. (1984) Control of the generator current in solitary rods of the *Ambystoma tigrinum* retina. *J. Physiol.* **348**, 645–664.

Marty A. and Neher E. (1983) Tight-seal whole-cell recording, in *Single Channel Recording* (Sakmann B. and Neher E., eds.), Plenum, New York, pp. 107–122.

Matthews G. (1986) Comparison of the light-sensitive and cyclic GMP-sensitive conductances of the rod photoreceptor: noise characteristics. *J. Neurosci.* **6**, 2521–2526.

Matthews G. (1987) Single channel recordings demonstrate that cGMP opens the light-sensitive ion channel of the rod photoreceptor. *Proc. Natl. Acad. Sci. USA* **84**, 299–302.

Matthews G. and Watanabe S.-I. (1987) Properties of ion channels closed by light and opened by guanosine 3',5'-cyclic monophosphate in toad retinal rods. *J. Physiol.* **389**, 691–716.

Matthews G. and Watanabe S.-I. (1988) Activation of single ion channels from toad retinal rod inner segments by cyclic GMP: concentration dependence. *J. Physiol.* **403**, 389–405.

Matthews H. R. (1986) A simple microforge for bending the tips of patch-pipettes. *J. Physiol.* **381**, 8P.

Matthews H. R. (1988) A linear stepping motor microscope stage translator for fast solution changes. *J. Physiol.* **396**, 14P.

Matthews H. R. (1990) Evidence implicating cytoplasmic calcium concentration as the messenger for light adaptation in rod photoreceptors isolated from the guinea-pig retina. *J. Physiol.* **425**, 48P.

Mathews, H. R. (1991) Incorporation of chelator into guinea-pig rod shows that calcium mediates mammalian photoreceptor light adaptation. *J. Physiol.* **436**, 93–105.

Matthews H. R., Torre V., and Lamb T. D. (1985) Effects on the photoresponse of calcium buffers and cyclic GMP incorporated into the cytoplasm of retinal rods. *Nature* **313**, 582–585.

Matthews H. R., Fain G. L., Murphy R. L. W., and Lamb T. D. (1990) Light adaptation in cones of the salamander: a role for cytoplasmic calcium concentration. *J. Physiol.* **420**, 447–469.

Matthews H. R., Murphy R. L. W., Fain G. L., and Lamb T. D. (1988) Photoreceptor light adaptation is mediated by cytoplasmic calcium concentration. *Nature* **334**, 67–69.

McLatchie L. M. and Matthews H. R. (1990) Voltage dependent block by l-*cis*-diltiazem of the cyclic-GMP activated conductance of rods isolated from the tiger salamander retina. *J. Physiol.* **430**, 129P.

McNaughton P. A. (1990) Light response of vertebrate photoreceptors. *Physiol. Rev.* **70**, 847–883.

McNaughton P. A., Cervetto L., and Nunn B. J. (1986) Measurement of the intracellular free calcium concentration in salamander rods. *Nature* **322**, 261–263.

Menini A. (1990) Currents carried by monovalent cations through cyclic-GMP-activated channels in excised patches from salamander rods. *J. Physiol.* **424**, 167–185.

Menini A., Rispoli G., and Torre V. (1988) The ionic selectivity of the light-sensitive current in isolated rods of the tiger salamander. *J. Physiol.* **402**, 279–300.

Miller W. H. and Nicol G. D. (1979) Evidence that cGMP regulates membrane potential in rod photoreceptors. *Nature* **280**, 64–66.

Nakatani K. and Yau K.-W. (1988a) Calcium and magnesium fluxes across the plasma membrane of the toad rod outer segment. *J. Physiol.* **395**, 695–729.

Nakatani K. and Yau K.-W. (1988b) Guanosine 3',5'-cyclic monophosphate-activated conductance studied in a truncated rod outer segment of the toad. *J. Physiol.* **395**, 731–753.

Nakatani K. and Yau K.-W. (1988c) Calcium and light adaptation in retinal rods and cones. *Nature* **334**, 69–71.

Neher E. and Eckert R. (1988) *Calcium and Ion Channel Modulation* (Grinnell A. D., Armstrong D., and Jackson M. D., eds.), Plenum, New York.

Nunn B. J. (1987) Ionic permeability ratios of the cyclic GMP-activated conductance in the outer-segment membrane of salamander rods. *J. Physiol.* **394**, 17P.

Owen W. G. (1987) Ionic conductances in rod photoreceptors. *Ann. Rev. Physiol.* **49**, 743–764.

Pepe I. M., Panfoli I., and Cugnoli C. (1986) Guanylate-cyclase in rod outer segments of the toad retina. *FEBS Lett.* **203**, 73–76.

Pugh E. N., Jr. (1987) The nature and identity of the internal excitational transmitter of vertebrate phototransduction. *Ann. Rev. Physiol.* **49**, 715–741.

Pugh E. N., Jr. and Cobbs W. H. (1986) Visual transduction in vertebrate rods and cones: A tale of two transmitters, calcium and cyclic GMP. *Vision Res.* **26**, 1613–1643.

Pugh E. N., Jr. and Lamb T. D. (1990) Cyclic GMP and calcium: the internal messengers for excitation and adaptation in vertebrate photoreceptors. *Vision Res.* **30**, 1923–1948.

Pugh E. N., Jr., Cobbs W. H., and Tanner J. D. (1990) Phototransduction in vertebrate rods: the electrophysiological approach to the cGMP theory, in *National Research Council Committee on Vision Frontiers of Visual Science, Proceedings of the 1988 Symposium*. National Academy of Sciences Press, Washington.

Ratto G. M., Payne R., Owen W. G., and Tsien R. Y. (1988) The concentration of cytosolic free Ca^{2+} in vertebrate rod outer segments measured with fura2. *J. Neurosci.* **8**, 3240–3246.

Ratto G. M., Robinson D. W., Yan B., and McNaughton P. A. (1991) Development of the light response in neonatal mammalian rods. *Nature* **351**, 654–657.

Sakmann B. and Neher E. (1983) *Single-Channel Recording* (Plenum, New York).

Sather W. A. and Detwiler P. B. (1987) Intracellular biochemical manipulation of phototransduction in detached rod outer segments. *Proc. Natl. Acad. Sci. USA* **84**, 9290–9294.

Schmidt K.-F. Nöll G. N., and Baumann C. (1990) Effect of intracellularly applied sodium ions on the dark voltage of isolated retinal rods. *Visual Neurosci.* **4**, 331–336.

Schnapf J. L., Nunn B. J., Meister M., and Baylor D. A. (1990) Visual transduction in cones of the monkey *Macaca fascicularis. J. Physiol.* **427**, 681–713.

Soejima M. and Noma A. (1984) Mode of regulation of the ACh-sensitive K-channel by the muscarinic receptor in rabbit atrial cells. *Pflügers Arch.* **400**, 424–431.

Stern J. H., Kaupp U. B., and MacLeish P. R. (1986) Control of the light-regulated current in rod photoreceptors by cyclic GMP, calcium, and l-*cis*-diltiazem. *Proc. Natl. Acad. Sci. USA* **83**, 1163–1167.

Stryer L. (1986) The cyclic nucleotide cascade of vision. *Ann. Rev. Neurosci.* **9**, 87–119.

Stryer L. (1988) Molecular basis of visual excitation. *Cold Spring Harbor Symp. Quant. Biol.* **53**, 283–294.

Tamura T., Nakatani K., and Yau K.-W. (1989) Light adaptation in cat retinal rods. *Science* **245**, 755–758.

Tang J. M., Wang J., Quandt F. N., and Eisenberg R. S. (1990) Perfusing pipettes. *Pflügers Arch.* **416**, 347–350.

Tomita T. (1970) Electrical activity of vertebrate photoreceptors. *Q. Rev. Biophys.* **3**, 179–222.

Torre V., Matthews H. R., and Lamb T. D. (1986) Role of calcium in regulating the cyclic GMP cascade of phototransduction in retinal rods. *Proc. Natl. Acad. Sci. USA* **83**, 7109–7113.

Tsien R. Y. (1980) New calcium indicators and buffers with high selectivity against magnesium and protons: design, synthesis and properties of prototype structures. *Biochemistry* **19**, 2396–2404.

Tsien R. Y. (1981) A non-disruptive technique for loading calcium buffers and indicators into cells. *Nature* **290,** 527, 528.

Tsien R. Y. (1989) Fluorescent probes of cell signalling. *Ann. Rev. Neurosci.* **12,** 227–253.

Uhl R., Wagner R., and Ryba N. (1990) Watching G proteins at work. *Trends Neurosci.* **13,** 64–70.

Vuong T. M. and Chabre M. (1990) Subsecond deactivation of transducin by endogenous GTP hydrolysis. *Nature* **346,** 71–74.

Werblin F. S. (1978) Transmission along and between rods in the tiger salamander retina. *J. Physiol.* **280,** 449–470.

Yau K.-W. and Baylor D. A. (1989) Cyclic GMP-activated conductance of retinal photoreceptor cells. *Ann. Rev. Neurosci.* **12,** 289–327.

Yau K.-W. and Nakatani K. (1984a) Cation selectivity of light-sensitive current in retinal rods. *Nature* **309,** 352–354.

Yau K.-W. and Nakatani K. (1984b) Electrogenic Na–Ca exchange in retinal rod outer segment. *Nature* **311,** 661–663.

Yau K.-W. and Nakatani K. (1985a) Light-induced reduction of cytoplasmic free calcium in retinal rod outer segment. *Nature* **313,** 579–582.

Yau K.-W. and Nakatani K. (1985b) Light-suppressible, cyclic GMP-sensitive conductance in the plasma membrane of a truncated rod outer segment. *Nature* **317,** 252–255.

Yau K.-W., McNaughton P. A., and Hodgkin A. L. (1981) Effects of ions on the light-sensitive current of retinal rods. *Nature* **292,** 502–505.

Yellen G. (1982) Single Ca^{2+}-activated nonselective cation channels in neuroblastoma. *Nature* **296,** 357–359.

Zimmerman A. L. and Baylor D. A. (1986) Cyclic GMP-sensitive conductance of retinal rods consists of aqueous pores. *Nature* **321,** 70–72.

Zimmerman A. L. and Baylor D. A. (1988) Ionic permeation in the cGMP-activated channel of retinal rods. *Biophys. J.* **53,** 472a.

Zimmerman A. L., Yamanaka G., Eckstein F., Baylor D. A., and Stryer L. (1985) Interaction of hydrolysis-resistant analogs of cyclic GMP with the phosphodiesterase and light-sensitive channel of retinal rod outer segments. *Proc. Natl. Acad. Sci. USA* **82,** 8813–8817.

Intracellular Messengers in Invertebrate Photoreceptors Studied in Mutant Flies

Baruch Minke and Zvi Selinger

1. Introduction: The Inositol Lipid Cascade of Vision

Invertebrate phototransduction is a central subject in neuroscience that includes several key topics with implications to a wide variety of cells and tissues. The phototransduction cascade begins with isomerization of the retinoid chromophore by photons that activate the photopigment. The activated photopigment facilitates the exchange of bound guanosine 5'-diphosphate (GDP) with cytoplasmic guanosine 5'-triphosphate (GTP) via a guanine nucleotide-binding protein (G protein). The activated G protein, probably the α subunit, activates, in turn, a membrane-bound phosphoinositidase C (PIC) enzyme, which catalyzes the hydrolysis of phosphatidylinositol 4,5 bisphosphate (PtdInsP_2) into two intracellular messengers: hydrophilic inositol 1,4,5-trisphosphate (InsP_3) and hydrophobic diacylglycerol (DG) (Fein et al., 1984; Brown et al., 1984; Payne, 1986; Payne et al., 1988; Tsuda, 1987; Paulsen et al., 1987; Selinger and Minke, 1988; Montell, 1989). The role of DG in photoreceptors is not clear, but InsP_3 releases Ca^{2+} from intracellular compartments mainly from the submicrovillar cisternae (SMC), which is part of the smooth endoplasmic reticulum (Brown and Rubin, 1984; Payne et al., 1986b; Baumann and Walz, 1989). Ca^{2+} functions as a messenger for light adaptation (Lisman and Brown, 1972; Brown and Blinks, 1974) and possibly for excitation (Stieve and Bruns, 1980; Bolsover and Brown, 1985; Stieve, 1986; Payne et al., 1986a, 1988).

From: *Neuromethods, Vol. 20: Intracellular Messengers*
Eds: A. Boulton, G. Baker, and C. Taylor © 1992 The Humana Press Inc.

The nature of the light-sensitive ionic channels and their activator is still unknown. Both Ca^{2+} and guanosine 3',5' cyclic monophosphate (cGMP) (Johnson et al., 1986; Saibil, 1984) have been suggested as putative internal transmitters that open the channels.

For the photoreceptor to operate as an efficient photon counter, each step in the cascade needs to have an inactivation mechanism. Putative inactivating mechanisms have been found for each of the above stages: The photopigment in its active meta-rhodopsin (M) state is possibly inactivated through a combination of phosphorylation by rhodopsin kinase (Paulsen and Bentrop, 1984; Lisman, 1985; Minke, 1986) and binding of photo-receptor-specific arrestin, as in vertebrates (Yamada et al., 1990; Smith et al., 1990). The G protein is inactivated by a GTP-hydro-lyzing activity, an intrinsic property of its α subunit (Cassel and Selinger, 1976; Calhoon et al., 1980). PIC activity is controlled by the level of available $PtdInsP_2$ or Ca^{2+}, and inactivated by inactivation of the G protein (Devary et al., 1987). $InsP_3$ level is regulated by a very efficient $InsP_3$ 5-phosphatase, which hydrolyzes $InsP_3$ to $Ins(1,4)P_2$ (Devary et al., 1987), or possibly by phosphorylation of $InsP_3$ by a kinase to form $Ins(1,3,4,5)P_4$ (Berridge and Irvine, 1989). The intracellular Ca^{2+} concentration ($[Ca^{2+}]_i$) is maintained by controlling its release from the SMC at the level of the $InsP_3$-sensitive Ca^{2+} channel via a negative feedback of Ca^{2+} itself (Payne et al., 1988, 1990).$[Ca^{2+}]$ is also controlled by intracellular buffering mechanisms and by a powerful Na^+– Ca^{2+} exchanger (Minke and Tsacopoulos, 1986; O'Day and Gray-Keller, 1989; Ziegler and Walz, 1989), which uses the Na^+ gradient to extrude Ca^{2+} from the photoreceptors. The photoreceptors operate over a very wide range of light intensities: They are sensitive to single photons on the one hand and to bright light on the other. This is achieved by an automatic gain control known as light and dark adaptation (Lisman and Brown, 1972), which is obtained by complex mechanisms of feed-forward and feed-back regulatory loops (Lisman and Brown, 1975; Bolsover and Brown, 1985; Payne and Fein, 1986; Grzywacz et al., 1988; Suss et al., 1989). These regulatory loops operate by controlling the rate of occurrence and mainly the magnitude of the unitary responses to single photons (quantum bumps), which sum to produce the receptor potential (Dodge et al., 1968; Wong, 1978; for review, *see* Stieve, 1986). The principal (but not the only) messenger for light adaptation is Ca^{2+} (Lisman and Brown, 1972, 1975; Stieve, 1986). Figure 1 summarizes the primary sequence of the phototransduction cascade.

The requirement to translate light into electrical signals in a reliable, fast, and efficient manner over a wide range of light intensities dictates a concentrated and highly organized enzymatic machinery containing large quantities of key proteins within the photoreceptors. Similar enzymatic machineries are found in several hormonal systems in a variety of cells, but the high concentrations of the signaling proteins in photoreceptors together with their ability to be light-activated makes the photoreceptor an ideal choice for studying intracellular messenger systems. For example, in turkey erythrocyte, a favorite cell in which to study the β-adrenergic system (Cassel and Selinger, 1977), there are $\sim 10^3$ β-adrenergic receptors/cell, compared to 10^8 receptors (rhodopsin molecules)/photoreceptor, and the latter can be turned on and off simply by using different wavelengths of light (*see below*).

A common strategy to unravel complex pathways similar to the phototransduction cascade has been to dissect the system by single-point mutations, each affecting a discrete step of the enzyme cascade. Genetic analysis has been applied to the visual system of *Drosophila*, yielding over 100 photoreceptor potential mutants. These mutants fall into more than 15 mutual complementation groups*, thus setting a minimal number of gene products that are involved in the phototransduction process (Pak, 1979). Figure 1 demonstrates several mutations that affect some of the key proteins in the phototransduction cascade. We have initiated combined electrophysiological, pharmacological, and

---

*A complementation test consists of testing the phenotype of a fly heterozygous for two independently isolated mutations. If the fly displays the mutant phenotype, the mutations are said not to "complement." A group of noncomplementing mutations defines a complementation group. All of the mutations within a given complementation group are regarded as affecting the same gene (Pak, 1979).

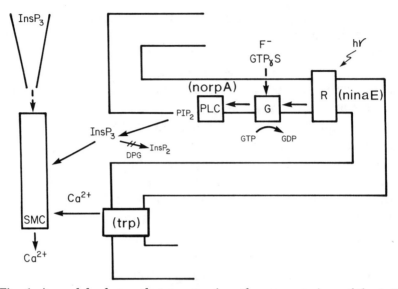

Fig. 1. A model scheme that summarizes the current view of the initial steps in the phototransduction cascade in the microvilli of invertebrates. Mutations affecting various stages in the cascade are indicated in brackets. (The scheme is a modification of a similar scheme of Payne, 1986; reproduced from Suss et al., 1989, with permission from The Rockefeller University Press and the authors.)

biochemical studies of phototransduction in *Drosophila* and *Drosophila* mutants. The relatively small diameter of *Drosophila* photoreceptors (e.g., ~3 µm) makes it a difficult preparation for several electrophysiological methods. We therefore took advantage of the similar characteristics of the photoreceptors of other Diptera and have also used the larger flies *Musca*, *Lucilia*, and *Calliphora* in our experiments.

2. Anatomy of the Retina

The fly eye is composed of many repeat units referred to as ommatidia. There are about 700 ommatidia in *Drosophila* eye, ~3000 in *Musca*, and ~5000 in *Calliphora* (Hardie, 1985). Each ommatidium contains a dioptric apparatus (Fig. 2) composed of transparent chitinous cuticle (C) and an extracellular fluid-filled cavity, the pseudocone (PC). The floor of the cavity is formed by four Semper cells (SZ) and the walls by two primary pigment

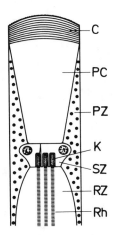

Fig. 2. Distal portion of a *Musca* ommatidium showing the optical apparatus. C, cornea; PC, pseudocone; PZ, primary pigment cell; K, rhabdomere
cap; SZ, Semper cell; RZ, retinula cell; Rh, rhabdomere. (Reproduced from
Kirschfeld, 1967, with permission from Springer Verlag and the author.)

cells (PZ), which together circle the pseudocone. Below this rigid
structure of the optical apparatus lie the photoreceptor cells (RZ).
There are eight photoreceptor cells in each ommatidium (Fig. 3)
arranged in a precise asymmetrical trapezoid manner, which
allow the identification of individual cells. The large peripheral
(R1–6) photoreceptors are all similar and extend the entire length
of the retina (Fig. 3b). The central cells (R7/8) are smaller (Fig.
3b). The distal two-thirds of the ommatidium center is formed
by R7, and the proximal third by R8. Each photoreceptor has a
light-sensitive membrane containing the photopigment, which
forms tightly packed microvilli, arranged in a structure called a
rhabdomere (Fig. 3a, stripes; Fig. 3b, black area). This highly ordered structure (in the fly) forms a waveguide that is widely
exploited for optical methods (*see below*). The central rhabdomeres
in each ommatidium (R7/8) are part of a different class of photoreceptors. The rhabdomeres of these photoreceptors are located
one above the other (Fig. 3b) (R7 is the upper one) forming one
continuous waveguide (for review, *see* Hardie, 1985).

The axons of receptors R1–6 cross a connective tissue (1–2
μm thick), the basement membrane, and synapse on the second-

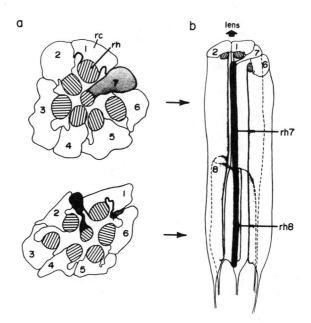

Fig. 3. Schematic diagram showing the positions of eight retinula cells (numbered as shown) and their rhabdomeres within an ommatidium—a, presents crosssectional views at the levels indicated by the arrows. In b, retinula cells R3–5 are omitted for clarity; rc, retinula cell; rh, rhabdomere. (Reproduced from Schinz et al., 1982, with permission from the Rockefeller University Press and the authors.)

order lamina (La) neuron via a sign-inverting histamine-mediated synapse (Hardie, 1989). The basement membrane, which delineates the proximal boundary of the retina, limits the permeability to large molecules possibly forming a "blood–brain barrier" (Shaw, 1978).

The central cell axons bypass the lamina and synapse in the medulla (Me; *see* Fig. 4). Each ommatidium is surrounded along its length by six secondary pigment (glia) cells that contain numerous pigment granules (0.1 µm in diameter) consisting mainly of ommochrome and pteridine, and having an optical screening function. The secondary pigment cells also have a major role in osmotic and ionic buffering of the extracellular space around the photoreceptors (Coles and Tsacopoulos, 1981). These cells are very fragile, collapsing under the mildest osmotic shock (Shaw, 1978; for a detailed review, *see* Hardie, 1985).

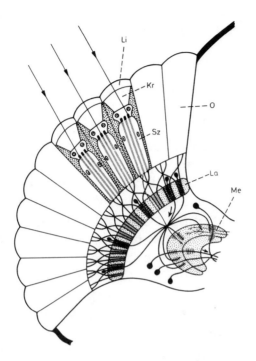

Fig. 4. Schematic drawing of a horizontal section through a fly's compound eye with parts of the optic lobe demonstrating the neural superposition principle. Li, corneal lens; Kr, pseudocone; Sz, photoreceptor cell; O, ommatidium; La, lamina; Me, medulla. (Reproduced from Kirschfeld, 1971, with permission from Springer Verlag and the author.)

A widely used strategy for studying phototransduction is to eliminate the yellowish-brown screening pigment from the pigment granules, which are located in both the pigment cells and the photoreceptors, by one or two mutations (Lindsley and Grell, 1968). The electrophysiological and structural properties of the white- and red-eyed flies are similar. White-eyed mutants are available for all the fly species mentioned above. The advantage of using white-eyed flies is the elimination of a red filter with variable absorption properties in front of the eye (the pigment granules migrate in the photoreceptor cells toward the rhabdomere on illumination). Some key electrical as well as autofluorescence responses can be measured only in white-eyed flies. On the other hand, pigment migration can serve as an experimental tool, and some optical measurements are easier to perform on red-eyed flies (for reviews, see Franceschini, 1975, 1983; Stavenga and Schwemer, 1984).

3. The Molecular Stages of the Phototransduction Cascade

3.1. The Photopigment

The photopigment (a protein analog of the hormone receptor) is an integral membrane protein in which absorption of light quanta results in conformational changes leading to activation of the second stage in the cascade. In vertebrate rods and cones, photoisomerization of the chromophore 11-cis retinal leads to dissociation of the retinal chromophore from the protein opsin (Wald, 1936). Regeneration of rhodopsin is achieved after an extended period by a series of chemical reactions, some of which are carried out outside the photoreceptor cells (Dowling, 1960). In invertebrate photopigments, photoisomerization of the 11-cis isomer (R) results in the production of a dark-stable intermediate, metarhodopsin (all-trans M) (for review, see Hamdorf, 1979). Isomerization of the chromophore from all-trans (M) to the original 11-cis isomer can be done only by light (Schwemer, 1984): Light absorption by M regenerates R. Some M molecules can be converted to an unstable form (Mu), which is degraded (see below). The chromophore of the Dipteran photopigment is 3-hydroxyretinal (Vogt, 1983; for review, see Kirschfeld, 1986).[*]

The absorption maxima of Musca R1–6 receptors is ~490 nm and ~580 nm for R and M, respectively. The absorption maxima of the central photoreceptors of Musca can be of several types. There are three types of R7 photoreceptors: R7r, which is similar to R1–6 in all properties; R7p, which is a UV receptor with 335-nm and 460-nm peak absorption of R and M, respectively; and R7y, with 430-nm and 505-nm peak absorption of R and M, respectively. The lower R8 can be of three similar types: R8r, with similar absorption maxima to R1–6; R8p, with 460-nm R (unknown M); and R8y, with 520-nm R and unknown M (for review, see Hardie, 1985). All photoreceptors, except for R7p

---

[*]A 3-hydroxy-retinal-derived visual pigment should be called "xanthopsin" (Vogt, 1983). Nevertheless, we have used the more familiar term rhodopsin.

and R8p, have additional prominent UV absorption maximum at ~350 nm owing to a sensitizing pigment, which is 3-hydroxy-retinol (for review, *see* Kirschfeld, 1986).

3.1.1. Methods to Study the Photopigment

The response to light can be measured from the intact fly using its behavior or its electrical response. The electrical response can be recorded from the whole eye extracellularly by the electroretinogram (ERG) or from a single cell intracellularly. The light response can also be recorded from in vitro preparations using sliced eye by intracellular or extracellular measurements during superfusion of the eye. Recently, current measurements from isolated ommatidia in suction pipets have also been used. The precise and sophisticated optical properties of the fly eye have been used for photometric measurements. The autofluorescence of the photopigment has been used for in vivo microspectro-fluorometry. Morphological methods, such as freeze-fracture electron microscopy, have also been used to study the photo-pigment. All the above methods were exploited to study photopigment-deficient mutants.

3.1.1.1. ELECTROPHYSIOLOGY: THE ELECTRORETINOGRAM (ERG)

The ERG measures the extracellular current flow in the eye following or during illumination. It is composed of several components (Fig. 5, *see* Pak, 1975). The main component is a corneal negative potential arising from summation of (1) the extracellularly recorded receptor potentials and (2) a slow response of the pigment cells. The latter is a depolarization initiated by accumulation of K^+ in the extracellular space during the receptor potential in the photoreceptors. This response of the pigment cells is substantial only during bright light (compare 5A to 5C), and it is abolished in the sliced superfused eye (Fig. 5B; Minke, 1982). Additional components in the ERG are the "on" and "off" transients (arrows, Fig. 5C) appearing shortly after the onset and offset of the receptor potential (Pak, 1975). The polarity of the "on" transient is reversed relative to the receptor potential because of the sign inverting synapse between the R1–6 axons and the large monopolar neurons (Fig. 4). The ERG of the fly is very large relative to similar signals in other eyes (it can reach 30 mV in the fly compared to several μV in vertebrates). It is therefore a conve-

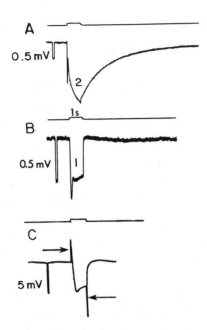

Fig. 5. The effect of perfusion on the ERG waveform. The fig. shows ERG waveforms of a *Drosophila* sliced eye to 550-nm green light at the beginning of the perfusion (upper trace A) and after 15 min of perfusion (middle trace B). The bottom trace (C) shows the ERG response in intact fly to dim (3.0 log attenuation) light showing the "on" and "off" transients. (Reproduced from Minke, 1982, with permission from The Rockefeller University Press.)

nient signal for many studies, especially for screening for visual mutants. The major disadvantage of the ERG is its heterogeneous cellular origin, which distorts the waveforms of the typical response of the various cells. Accordingly, the ERG cannot be used to analyze the waveform of the photoreceptor potential or the lamina cells; and it does not reflect unitary responses to single photons and their summation (photoreceptor noise).

Numerous studies using intracellular recordings in the fly have been published (*see*, for example, reviews by Pak, 1979; Hamdorf, 1979; Hillman et al., 1983; Hardie, 1985), and such recordings have been the main method to study the various cell types in fly retina (*see* Hardie, 1985). The main advantage of the intracellular techniques is that, when the indifferent electrode is located close to the recording pipet, it records the undistorted waveform of the receptor potential of single cells. In addition, the quantal responses to single photons (quantum bumps) or their summation (noisy depolarization) is also recorded. These signals are of major importance in studying phototransduction (*see below*). The drawback is the difficulty in maintaining long, stable recording from the small *Drosophila* photoreceptors especially during perfusion of the retina. Voltage-clamp, current measurements, and efficient injection of chemicals into the cells are not feasible with this technique, but various methods of current measurements using suction pipets overcome these difficulties.

3.1.1.2. Visualization of Photoreceptors by Optical Methods

The precise and sophisticated optics of the fly compund eye has been revealed by numerous studies (for reviews, *see* Kirschfeld, 1967; Franceschini, 1975; Stavenga, 1979; Kunze, 1979; Land, 1981). From the vast literature on optical studies, we will briefly describe only the most useful methods for genetic studies.

The fly belongs to the category of compound eyes with optics forming apposition images (Kunze, 1979) and to the subcategory of open rhabdomere, whose rhabdomeres in each ommatidium are separate and not fused (Fig. 3). With an eye slice preparation of the fly, it is possible to visualize individual rhabdomeres behind each facet under physiological illumination. Using a point source, it can then be seen that, in general, only one rhabdomere in an ommatidium is illuminated and, further, that in neighboring ommatidia different rhabdomeres are illuminated (de Vries and Kuiper, 1958). A precise mapping of the receptive fields of the individual rhabdomeres (Kirschfeld, 1967) and anatomical tracing of the retinula cell axons (Trujillo-Cenoz and Melamed, 1966; Braitenberg, 1967) indicated that one point (or parallel rays) simultaneously illuminates seven rhabdomeres, each in a different ommatidium, in an accurately repeatable sequence. Furthermore, the axons of these seven photoreceptors converge on the same second-order cells in the lamina cartridges (Fig. 4). The most important optical requirement of the above neural superposition is that the interommatidial angle should equal the angle between the visual axes of adjacent rhabdomeres (Fig. 4; *see* Hardie, 1985 for a review). Measurements

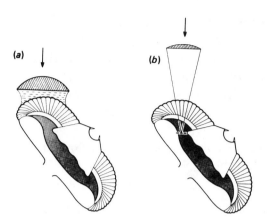

Fig. 6. Two methods used in combination with epifluorescence microscopy for viewing the receptor cells in the living insect. (a) The cornea is
optically neutralized with a drop of water, and the microscope (obj. 25 ×/
0.65, water-immersion) is focused on to the distal rhabdomere endings. (b)
Magnified and superimposed virtual images of rhabdomere patterns as seen
when focusing a (low-power) microscope at the level of the center of curvature of the eye ("deep pseudopupil"). (After Franceschini and Kirschfeld,
1971; Franceschini, 1975. Reproduced with permission from Springer Verlag
and the authors.)

by Pick (1977) in *Musca* show that this is not strictly true; nevertheless, the above superposition principle still holds.

There is no need to slice the fly eye in order to visualize the
receptor cells in the living eye. Two optical methods can be used
to visualize the rhabdomeres (Fig. 6): (a) The cornea can be optically neutralized with a drop of water while the microscope
objective (obj 25×/0.65; water-immersion) is focused on the
distal rhabdomere endings (Fig. 6a); (b) magnified and superimposed virtual images of the rhabdomere patterns are observed
when focusing a low-power microscope at the level of the center
of curvature of the eye (Fig. 6b). The image is called the deep
pseudopupil (DPP) (Franceschini and Kirschfeld, 1971). A single
facet acts as a magnifying glass and gives magnified virtual
images of its seven distal rhabdomere endings along its optical
axis. The same holds for neighboring facets. Since the optical
axes of all these facets cross one another at the center of curvature of the eye, there appears at this point a superposition of vir-

tual images of (identical) rhabdomere patterns. In the eye of higher Diptera, the DPP consists of seven large spots observable when focusing a low-power microscope deep into the eye (Fig. 7) (Franceschini and Kirschfeld, 1971; Franceschini, 1975).

3.1.1.3. PHOTOMETRIC MEASUREMENTS OF ABSORPTION SPECTRA

Photometric methods are used to determine directly or indirectly the absorption spectra of various visual-pigment states and the kinetics of the transitions among them. Photometric measurements are also used to determine schemes of the visual-pigment systems with their various pigment states and transitions (for reviews, *see* Hamdorf, 1979; Hillman et al., 1983).

These methods include:

1. Absorption spectrum in a detergent extract.
2. Absorption spectrum of the intact membrane using microspectrophotometry (MSP).
3. Dichroic absorption—measuring absorption of polarized light.

3.1.2. The Signals Used to Characterize the Photopigment

THE EARLY RECEPTOR POTENTIAL: Absorption changes in the photopigment molecules after illumination are usually accompanied by a redistribution of charges in the pigment molecules (Cone, 1967). This charge displacement induces a spread of current that can be recorded extracellularly (Brown and Murakami, 1964) or intracellularly (Brown et al., 1967). The recorded potential is called the fast photovoltage or early receptor potential (ERP). The ERP is used to measure the kinetics of pigment transitions and pigment-population changes *in situ*. When light is absorbed by R and M, the intracellular ERP is negative and positive, respectively. The polarities are reversed when the ERP is recorded extracellularly (for review, *see* Hillman et al., 1983).

THE META RHODOPSIN POTENTIAL: In the fly, Pak and Lidington (1974) found a fast potential that arose from light absorption by M. Like the ERP, this M potential has been used to characterize a visual-pigment system of the fly. Intracellular recordings from photoreceptors and second-order neurons of the fly have indi-

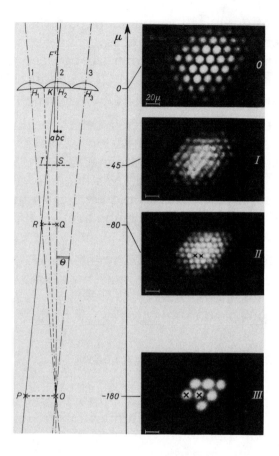

Fig. 7. Formation of the deep pseudopupil (level III) and of two other image superpositions (levels I and II) in the depth of the eye of a living and intact *Drosophila*. Drawing on the left: schematic longitudinal section of three ommatidia. Virtual images of rhabdomere ending *a* are given by lens 2 on axis F'P. These points define three depths at which superposition of virtual images should occur as a consequence of the formation of similar images (at the same points T,R,P) by neighboring lenses. Photos on the right: (objective LEITZ UM 20, sin μ = 0.22). Level 0—corneal pseudopupil observed when the microscope is focused on the cornea. Level I— first superposition of virtual images of (nonhomologous) rhabdomere endings (microscope focused 45 μm under the corneal level). Level II—second superposition of virtual images of (nonhomologous) rhabdomere endings (focus: 80 μm under the corneal level). Level III—third superposition of virtual images of (homologous) rhabdomere endings (focus: 180 μm under the corneal level). (Reproduced from Franceschini and Kirschfeld, 1971, with permission of Springer Verlag and the authors.)

cated that the M potential is predominantly a response of the second-order neurons. The M potential in the second-order neurons (M_2 phase) is induced by the positive ERP (M_1 phase), which arises from selective absorption by M in the photoreceptors (Fig. 8; Stephenson and Pak, 1980; Minke and Kirschfeld, 1980). The M_1 phase has all the characteristics of an ERP. The M_2 phase is very sensitive to the experimental conditions and is quickly abolished by anoxia, CO_2 (Fig. 8D), or by cooling the eye. Nevertheless, the M_2 phase is a very useful signal to characterize the photopigment, since it is easy to record and has a relatively large amplitude (>5 mV; Fig. 8E) in ERG recordings. The M potential reflects selectively the activation of the metarhodopsin in receptors R1–6.

THE LATE RECEPTOR POTENTIAL: The late receptor potential (LRP) is the physiological response to light of the photoreceptor cells. It arises from the light-induced opening of the ionic channels in the plasma membrane of the photoreceptors. The LRP is the last step of the phototransduction cascade and, therefore, reflects only indirectly the activation of the various stages of the cascade. The relationship between the amplitude of the LRP during prolonged light and the intensity of the stimulus is strongly affected, in a complex time-dependent manner, by the history and intensity of illumination (Millecchia and Mauro, 1969). The major advantage of using the LRP to study the photopigment is the large amplification existing between photon absorption and the conductance change underlying the LRP. This amplification enables the detection of the voltage response to the absorption of a single photon by a single R molecule. Accordingly, in *Drosophila* mutants with a 10^6-fold reduction in photopigment level, the remaining 10^2 molecules are still capable of giving close to a saturating response (Stephenson et al., 1983; Johnson and Pak, 1986). No other signal is as sensitive to light as the LRP.

THE PROLONGED DEPOLARIZING AFTER POTENTIAL: (PDA), like the LRP, arises from light-induced opening of ionic channels in the plasma membrane. However, in contrast to the LRP, which quickly declines to baseline after the cessation of the light stimulus, the PDA is a depolarization that continues long after light offset (Fig. 9A) (*see* Hamdorf, 1979; Hillman et al., 1983 for

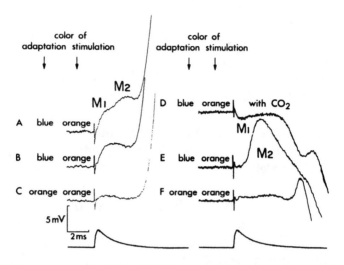

Fig. 8. The components of the initial responses to a constant orange light, after various adaptations as indicated, recorded intracellularly in a single cell (left column) and extracellularly from the ERG (right column) in a single white-eyed fly, *Chrysomya*. (A) The M_1, M_2 phases were illustrated when the indifferent electrode was placed on the thorax. (B) The same as in trace A, but with indifferent electrode in the photoreceptor layer. The M_2 phase disappeared. (C) The same as in trace B, but after orange adaptation. (D) Initial ERG under low level of CO_2, which abolished the M_2 phase. (E) The same as in trace D, but 5 min after CO_2 application had been interrupted. The response recovered within 1 min. (F) The same as in trace E, but after saturating orange adaptation (reproduced from Minke and Kirschfeld, 1980, with permission of The Rockefeller University Press and the authors).

reviews). The PDA has been a major tool to screen for visual mutants of *Drosophila* (Pak, 1979).

The PDA is observed only when a considerable amount of pigment (>20%) is converted from rhodopsin (R) to its dark-stable photoproduct acid metarhodopsin (M). Figure 9 shows that the larger the amount of R to M conversion, the longer the PDA. The duration of the PDA depends in a supralinear manner on the *amount* of light (i.e., the duration of the stimulus multiplied by the light intensity) (Fig. 9B). The PDA can be depressed at any time by pigment conversion from M to R. The degree of

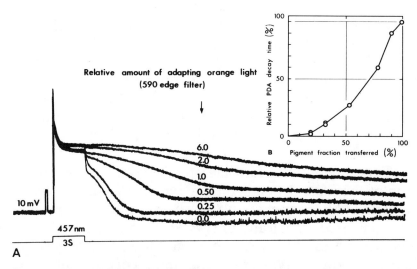

Fig. 9. The dependence of the prolonged depolarizing afterpotential (PDA) on the amount of pigment conversion from rhodopsin (R) to metarhodopsin (M) by a constant blue light. The PDA was recorded intracellularly from white-eyed *Calliphora*. A: Superimposed traces. Before each trace, the cell received orange lights (590 edge filter) with variable amounts of photons as indicated above the corresponding trace. B: The dependence of the relative decay time of the various PDAs to 1/e of the steady-state amplitude on the relative amount of pigment conversion by the orange adaptation. (Reproduced from Minke, 1986, with permission from Springer Verlag).

PDA depression depends on the amount of M to R conversion (Hochstein et al., 1973). After the depression of maximal PDA, additional PDA can be induced immediately by R to M conversion. The situation is more complex when the PDA-depressing light is given following the decline of the PDA (Hochstein et al., 1973). When R to M conversion is maximal, a maximal PDA is induced. Additional strong light stimuli, which do not change the distribution of the pigment between R and M, do not affect the duration of the PDA, but only induce light-coincident receptor potentials (LRP), which are superimposed on the PDA (Hamdorf, 1979). The duration of a maximal PDA varies greatly among different species. Even in closely related species of the fly, such as *Calliphora* and *Drosophila*, the duration of maximal PDA can be very different: a few minutes in *Calliphora*, (Hamdorf,

1979) as compared to a few hours in *Drosophila* (for review, *see* Minke, 1986). The duration of a maximal PDA in a specific photoreceptor is tightly linked to the total amount of its visual pigment. When the amount of the visual pigment is reduced by vitamin A deprivation (Stark and Zitzmann, 1976; Hamdorf, 1979) or by mutation (Larrivee et al., 1981), the duration of the maximal PDA is reduced in correlation with the amount of pigment present until the PDA (but not the LRP) is completely abolished.

Experiments with barnacle lateral eye indicate that the duration of the PDA depends on the *concentration* of the pigment converted (Almagor et al., 1986). When a localized strong laser light converted a small amount (~4%) of pigment from R to M in a restricted area of the cell, a low-amplitude, but long-duration PDA was induced. When the same amount of pigment was converted from R to M by a light that was diffused all over the cell, no detectable PDA was observed. A much larger amount of pigment (i.e., 30%) had to be converted by a diffused light in order to induce a detectable PDA. These observations, together with the effects of low pigment concentration and the supralinear dependence of the PDA on pigment conversion, suggest that PDA is induced when *neighboring* pigment molecules are converted from R to M. Possibly, the inactivation of the photopigment is drastically slowed under such conditions.

AUTOFLUORESCENCE: Autofluorescence of the fly photopigment was best studied using microspectrofluorometry in vivo (MSF). The first reports came from Franceschini (1977; *Musca*) and Stark et al. (1977; *Drosophila*), who independently discovered a clearly fluorescing deep pseudopupil in the eyes of flies. This fluorescence was only present in flies raised on a vitamin A-rich diet; it was absent or reduced in vitamin A-deprived flies (Stark et al., 1979). This fluorescence arose from the photopigment of R1–6 receptors. Subsequent work revealed that, in the fly, metarhodopsin and the sensitizing pigment fluoresce, whereas rhodopsin fluoresces negligibly. Hence, by measuring fluorescence one can monitor changes in metarhodopsin content induced by the excitation beam (for reviews, *see* Stavenga and Schwemer, 1984; Franceschini, 1983).

Apart from metarhodopsin, another fly visual pigment state (M') fluorescing in the red exists (Franceschini et al., 1981). The excitation spectrum of M', peaking at 570 nm, is very similar to the absorbance spectrum of M. The emission spectrum peaks at 660 nm. The shape of the emission spectrum is essentially identical to that of M, but the emission of M' is ≥4× that of M (Franceschini, 1983; Stavenga and Schwemer, 1984).

The autofluorescence of individual photoreceptors in an optically neutralized cornea in vivo is a very useful method to identify the different types of photoreceptors in the living fly (Franceschini et al., 1981). The blue excitation simultaneously reveals four classes of cells: red R1–6, green R7s, black R7s, and red R7s. The red-fluorescing R7s appear to be only in the male eye (*Musca*), as revealed by fluorescence mappings of the retinal mosaics (Franceschini et al., 1981; Hardie et al., 1981). All these fluorescence emissions arise from different carotenoid-based pigments contained in the rhabdomeres: They are all absent or much reduced in flies (*Musca* and *Drosophila* white, *Calliphora* chalky) whose larvae have been raised on a β-carotene- and vitamin A-deficient medium (Franceschini, 1977; Franceschini et al., 1981; Stark et al., 1977, 1979).

3.1.3. Morphology and Biochemistry

The photopigment molecules in a complex with other proteins can be recognized as particles in freeze-fracture preparations of the photoreceptor membrane (mainly at the protoplasmic face; Fig. 10). Detailed procedures of freeze-fracture electron microscopy (EM) are found in many studies (*see*, for example, Schinz et al., 1982). Just as in vertebrate rods, the microvillar membranes are packed with particles approx 8 nm in diameter (Fig. 10). A number of studies have shown that the density of these particles is related to the visual pigment concentration, which may be manipulated by diet (Boschek and Hamdorf, 1976; Schwemer, 1979; Harris et al., 1977) or mutations (Larrivee et al., 1981; Schinz et al., 1982).

The concentration of fly photopigment can be reduced to <1% by raising the flies on a synthetic medium lacking retinoids (Goldsmith et al., 1964; Stark and Zitzmann, 1976; Stephenson et

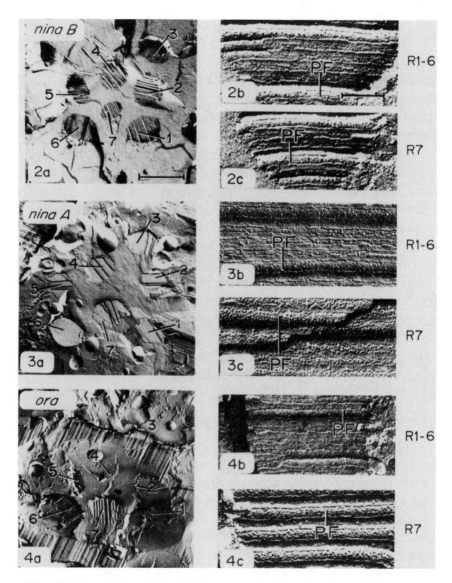

Fig. 10. Freeze-fracture replicas of ommatidia crossfractured at a distal level displaying the seven rhabdomeres (labeled 1–7), obtained from $ninaB^{P315}$ (2a), $ninaA^{P228}$ (3a), and ora^{JK84} (4a); and enlarged views of portions of rhabdomeres of the peripheral retinula cells R1–6 (2b, $ninaB$; 3b, $ninaA$; 4b, ora) and of the central retinula cell R7 (2c, $ninaB$; 3c, $ninaA$, 4c, ora). Bar: 2 μm for 2a, 3a, 4a; 0.2 μm for 2 b and c, 3b and c, and 4b and c. PF, protoplasmic face. (Reproduced form Schinz et al., 1982, with permission from The Rockefeller University Press and the authors.)

al., 1983). Stocks of vitamin A-deprived *Drosophila* can be maintained for many generations without loss of viability. The photopigment content of the deprived *Drosophila* can be restored by retinoid supplementation either to the medium in which the flies grow (Stark and Zitzmann, 1986; Minke and Kirschfeld, 1979; Stephenson et al., 1983; Isono et al., 1988) or by putting drops of carotenoids (dissolved in ethanol) on the eye and keeping the flies in light for ~2 d (Giovannucci and Stephenson, 1986, 1987). An additional method to reduce the photopigment content is to raise the flies under continuous green (>550 nm) light for several days (for details, *see* Fig. 10, Schwemer, 1979, 1984). The visual pigment concentration was varied in a detailed study (Schwemer and Henning, 1984) by a range of procedures, each of which affected the particle density accordingly. However, Schwemer and Henning (1984) estimate that there is a residual population (ca. 35%) of particles that must correspond to some other proteins. Particle counts in flies with a full visual pigment content vary between 3000–6000/μm^2 (*Calliphora* 4500/μm^2 [Schwemer and Henning, 1984], 6000/μm^2 [Chi and Carlson, 1979]; *Musca* 3800–6000/μm^2 [Boschek and Hamdorf, 1976]; *Drosophila* 4200/μm^2 [Harris et al., 1977], or ~3000/μm^2 [Schinz et al., 1982]) (Good particles per micron^2).

Vertebrate rhodopsins undergo phosphorylation of their C-terminal, *Ser* and *Thr* residues by a specific rhodopsin kinase. This phosphorylation is a prerequisite to subsequent inhibition of the catalytic interaction of rhodopsin and the G protein (for review, *see* Stryer, 1986).

Paulsen and Bentrop (1984) (*see also* Bentrop and Paulsen, 1986) showed that *Calliphora* rhodopsin also underwent light-dependent phosphorylation. Since it is possible to convert 100% of the photopigment to the rhodopsin state, the phosphorylation was shown to be confined to the metarhodopsin state, whereas subsequent dephosphorylation required regeneration to rhodopsin and the presence of Ca^{2+}. However, neither the kinase nor the phosphatase was characterized in these studies.

3.1.4. Mutations Affecting the Photopigment

One of the most powerful and fruitful tools to study the structure and function of the photopigment has been to apply genetic and molecular genetic techniques. Identification of the genes and gene products involved in photopigment synthesis and/or function by isolating photopigment-deficient mutants has provided invaluable information about the photopigment system of the fly. The characterization of these photopigment-deficient mutants required the use of almost all the experimental methods described above.

One of the prerequisites for the isolation of useful mutants is that the mutations affecting the process under investigation be nonlethal. The methods for mutagenesis and isolation of the *Drosophila* mutants have been described in detail by Pak (1979). Fortunately, a considerable number of key proteins necessary for phototransduction turned out to be photoreceptor-specific proteins; mutations affecting them were not therefore lethal.

All photopigment-deficient mutants that have been isolated on the second and third chromosomes of *Drosophila** were characterized by the PDA, which was the main tool to screen for these mutants. This required the use of white-eyed flies only. As mentioned in Section 3.1.2., a reduction in the photopigment content is reflected in an unusually short or abnormal PDA. Once the mutants were isolated, the reduced photopigment content was verified by in vivo and in vitro spectrophotometry, by using the M potential and by measuring the density of pigment-containing particles visualized in freeze fracture (Pak, 1979). In principle, fluorometric techniques could also be used.

The photopigment-deficient mutants can be classified into three main groups:

1. Mutants with defects in opsin structure or in its insertion into the membrane;
2. Mutants with defects in the chromophore cycle; and
3. Mutants with defects in proteins that indirectly affect the content of the photopigment via the structure of the rhabdomere.

Drosophila melanogaster has four pairs of chromosomes. Of these four, the two large autosomes, chromosomes II and III, are thought to carry as much as 80% of the genetic information.

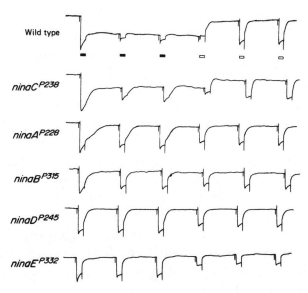

Fig. 11. ERG records from wild-type and from one representative allele of each of the five *nina* complementation groups. Records are arranged in order of departure from wild-type. Stimulus protocol was three unattenuated blue stimuli (filled bars under wild-type record) followed by three unattenuated orange stimuli (open bars) at 30-s intervals. Stimulus duration was 4 s, and calibration pulse shown before each response was 5 mV. (Reproduced from Stephenson et al., 1983, with permission from the authors.)

All these mutants are called *n*either *i*nactivation *n*or *a*fterpotential *(nina)* (Pak, 1979). The ERG phenotypes of the main complementation groups that have been studied in detail are given in Fig. 11.

3.1.4.1. MUTATIONS AFFECTING OPSIN

Different classes of photopigments in the fly differ in the opsin, but not in the chromophore. Accordingly, mutations that affect the opsin are expected to be specific to a specific class of photoreceptors, whereas mutations in the chromophore cycle are expected to affect all classes of photoreceptors.

The third chromosome *ninaE* mutant was originally classified as an "opsin mutant" because of three main observations all made in the lab of Pak:

1. The elimination of the PDA was confined to R1–6 and not the UV photoreceptor R7.

2. Freeze-fracture studies showed selective reduction of particles in R1–6, but not in the central photoreceptors (Fig. 10). This fact was most pronounced in the ora[JK84] isolated by Koeing and Merriam (1977), which was found to be an allele of the *ninaE* locus by Stephenson et al. (1983). In this mutant, R1–6 cells show only vestigial rhabdomeres with extremely low content of rhodopsin particles, whereas R7 and R8 are normal (Fig. 10).

3. The *ninaE* locus showed gene dosage effect. Namely, the amount of opsin produced depends on the number of copies of the structural gene for the opsin present in the fly (O'Brien and MacIntyre, 1978). It was found that heterozygotes (*ninaE*/+) to the *ninaE* mutation have approx one-half of R1–6 rhodopsin as wild-type flies. The amount of R1–6 rhodopsin in some homozygote alleles was <1% of normal (Stephenson et al., 1983).

The *ninaE* gene was cloned and sequenced independently by O'Tousa et al. (1985) and Zuker et al. (1985) on the basis of its homology to bovine opsin cDNA. The amino acid sequence deduced from nucleotide sequence is 373 residues long, and the polypeptide chain contains seven hydrophobic segments that appear to correspond to the seven transmembrane segments characteristics of other rhodopsins. Three regions of *Drosophila* rhodopsin are highly conserved with corresponding domains of bovine rhodopsin, suggesting an important role for these polypeptide regions.

On the basis of homology to the *ninaE* gene, three other *Drosophila* genes encoding for an additional three opsins have been isolated and characterized (for review, *see* Montell, 1989). The genes encoding Rh3 and Rh4 opsins were found to be expressed in nonoverlapping sets of the central R7 photoreceptor cells (Zuker et al., 1987; Montell et al., 1987; Fryxell and Meyerowitz, 1987). The remaining opsin Rh2 was found to be expressed in the ocelli of *Drosophila* (Pollock and Benzer, 1988; Feiler et al., 1988). The opsins of R8 receptors have not been cloned yet. One of the most striking genetic manipulations of the genes encoding the photopigments was the use of *Drosophila* mutants transformed with a chimeric gene that expresses a foreign visual pigment (Feiler et al., 1988). The pigment, which is normally located in R1–6 cells, was replaced by the pigment of the ocelli. The misexpressed opsin in R1–6 cells resulted in a functional photopigment, as revealed by the PDA and behavior with spectral characteristics of the ocellar photopigment (Feiler et al., 1988).

The second chromosome *ninaA* mutant also fulfills the above criteria of being defective in the opsin and not in the chromophore. The drastic (99%) reduction in the photopigment level in some alleles is confined (in all alleles) to R1–6 only, as revealed by reduction in rhodopsin particles in freeze fracture in R1–6 cells only (Fig. 10) and by the lack of, or short PDA in R1–6 (Fig. 11) and normal PDA in R7 receptors (Larrivee et al., 1981; Stephenson et al., 1983). Unlike the *ninaE*, the *ninaA* gene did not show the gene-dosage effect with respect to rhodopsin concentration. In addition, the reduction of rhodopsin levels in R1–6 photoreceptors of *ninaA* mutant flies was shown not to be the result of reduced expression of the R1–6 opsin gene, but seems to be the result of a defect in some aspect of posttranslational processing of the *ninaE* opsin.

The *ninaA* gene was cloned and sequenced by different techniques in two laboratories (Shieh et al., 1989; Schneuwly et al., 1989). The *ninaA* gene encodes a 227 amino acid protein that has 40% amino acid sequence identity with the vertebrate cyclosporin A-binding protein cyclophilin. Cyclophilin has been shown to be a prolyl-*cis-trans* isomerase that catalyzes, in vitro, rate-limiting steps in the folding of a number of proteins. It has therefore been suggested that the isomerase activity of the *ninaA* protein is necessary for the correct folding and the stability of rhodopsin in R1–6 receptors of *Drosophila*.

3.1.4.2. MUTATIONS AFFECTING THE CHROMOPHORE CYCLE

A considerable fraction of the M molecules is degraded (at a rate of $k = 0.3.h^{-1}$), in spite of the thermal stability of fly metarhodopsin. The fate of the chromophore can be summarized by the scheme proposed for the fly by Schwemer (1988) (Fig. 12). Upon degradation of the M_u state, all-*trans* 3-hydroxy retinal (AT 3-OH-AL) and the sensitizing pigment 3-hydroxy retinol (3-OH-OL) are released from the opsin. Both are transported to a store distal to the retina, where light enters the eye.

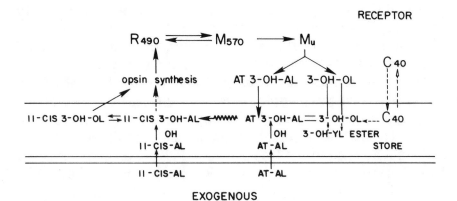

Fig. 12. Flow diagram of 3-hydroxy retinoids in the eye of the blowfly (for explanation, *see* text). AL: aldehyde; OL; alcohol; C_{40}: C_{40} carotenoids; R490: R1–6 rhodopsin; M570: R1–6 metarhodopsin; M_u: M-state, which is degraded YL retinil ester; AT all-*trans*. The wavy arrow and R490 = M570 are phototransitions. (Reproduced from Schwemer, 1988, with permission from the author.)

The tissue in which the retinoids are stored has not yet been identified, but preliminary data suggest that the retinoids are stored in the primary pigment cells. The visual pigment chromophore is stored as aldehyde, but part of the alcohol seems to be esterified while in storage (3-OH-YL ESTER). Illumination with short-wavelength light (wavy arrow, in Fig. 12) leads to a highly stereospecific isomerization of the all-*trans* aldehyde to the 11-*cis* form (11-*CIS* 3-OH-AL), thus supporting the hypothesis that a retinochrome-like pigment (Hara and Hara, 1972) is involved in the isomerization process. Subsequently, the 11-*cis* aldehyde is reduced to the alcohol (11-*cis* 3-OH-OL), which is then transported to the photoreceptor cell. Hydrolysis of retinyl esters (3-OH-YL ESTER) seems also to take place in the store. Further experiments demonstrated that exogenous 11-*cis* and all-*trans* retinal are hydroxylated, and are in part reduced to the corresponding alcohols. Although they have not yet been identified, it is assumed that the transport of 3-hydroxy retinoids between photoreceptor and the store is achieved by specific retinoid-binding proteins. Soluble proteins that may serve this function have been isolated from honeybee (Schwemer et al., 1984) and squid retinae (Ozaki et al., 1987). The chromophore cycle involves a relatively large number of processes and proteins, and therefore is expected to yield a number of mutants defective in the chromophore cycle.

The *ninaB* mutant showed a reduced number of rhodopsin particles in freeze-fractured EM in all photoreceptors (Fig. 10) and absence of PDA in R7 (*ninaD* was not examined for these parameters), thus suggesting a reduced photopigment in all photoreceptors. Both *ninaB* and *ninaD* mutants are characterized by low visual pigment concentration as determined by MSP and the M potential. A defect in retinoid metabolism was suggested by retinoid supplementation experiments to the food or directly to the eye (Stephenson et al., 1983).

By applying retinol, all-*trans* retinal, or 11-*cis* retinal directly to the eye, it has been shown that *ninaD* can be rescued from phenotype at least as fast as vitamin A-deprived wild-type (Giovannucci and Stephenson, 1986). More recent experiments with *ninaB* gave similar results. Using an HPLC procedure, it had been shown by the above authors that peaks corresponding to isomers of 3-hydroxy retinal and 3-hydroxy retinol were greatly attenuated in whole-head extracts of *ninaD* grown on cornmeal medium. When *ninaD* was raised on β-carotene-enriched medium, the carotenoid profiles of heads, abdomens, and larvae did not differ significantly from similarly treated vitamin A-deprived wild-type. This is consistent with Stephenson's hypothesis that the *defect* in this mutant *lies outside the head*. When *ninaB* was raised on carotene-enriched medium, however, the heads showed additional peaks that eluted with the same k' values as lutein, zeaxanthin, and β-carotene standards. Since *ninaB* is not rescuable by *carotene* (Stephenson et al., 1983), this may be indicative of a *metabolic block in the head*.

3.1.4.3. A MUTATION AFFECTING THE STRUCTURE OF THE RHABDOMERE

The *ninaC* is a second chromosome mutant of *Drosophila*, which was isolated on the basis of abnormal PDA (Stephenson et al., 1983; Fig. 11). The reduced rhodopsin level in *ninaC* (to ~30% of normal) appears to be mainly owing to a reduction in the diameter of the rhabdomeres (Matsumoto et al., 1987). The microvilli in the rhabdomere contain a cytoskeletal infrastructure revealed in EM studies as an electron-dense region. In *ninaC* the microvilli are shorter in all photoreceptors than normal, and the cytoskeletal electron-dense regions seen in wild-type flies are markedly reduced (Matsumoto et al., 1987). The *ninaC* gene is expressed as two alternatively spliced RNAs encoding two proteins of 1135 and 1501 amino acids (Montell and Rubin, 1988). Comparison of the deduced amino acid sequences with previously sequenced genes and proteins revealed a novel protein with two regions of homology with known proteins. The region near the N-terminus of both proteins (266 amino acids) is similar to the catalytic portion of protein kinases. This part is jointed to a 725-amino acid region with homology to the entire globular head region of myosin I called myosin heavy chain. The physiological role of the *ninaC*, a novel myosin kinase protein, is not known.

3.2. The Guanine Nucleotide-Binding Protein (G Protein)

The role of the G protein in phototransduction and its properties can be studied directly using biochemical methods in cell-free membrane preparations of the fly eye (Blumenfeld et al., 1985). To demonstrate that activation of a G protein is necessary for the physiological response, the biochemical measurements need to be supplemented by pharmacological studies of intact photoreceptors in vivo or in vitro. In pharmacological experiments, the involvement of the G protein in the phototransduction cascade can be implicated indirectly via chemical excitation of the photoreceptor cells in the dark by all chemical agents *(see below)* known to interact specifically with the G protein. No *Drosophila* mutant with a deficient G protein has so far been isolated.

Numerous studies of G proteins in many cells and tissues (including photoreceptors) have indicated that AIF_4^- and the hydrolysis-resistant analogs of GTP, guanosine 5'-[γ-thio] triphosphate (GTPγS) and guanosine 5'-[βγ-imido] triphosphate (GMP-PNP), can activate the G protein directly (Cassel and Selinger, 1977; Bigay et al., 1985).

To demonstrate the involvement of a G protein in fly photo-transduction, photoreceptors of *Musca* and *Lucilia* were super-fused, or the intact retina was injected with solutions containing either F⁻ or one of the GTP analogs, GMP-PNP and GTPγS. These agents have been reported to generate discrete voltage fluctuations (bumps) when injected into *Limulus* photoreceptors (Fein and Corson, 1979, 1981; Bolsover and Brown, 1982). All three agents produced noisy depolarizations in *Musca* and *Lucilia* photoreceptors when applied to the retina. Although individual bumps were too small to be resolved, the power spectra of voltage noise induced by these agents resembled closely those of light-induced noise. F⁻-induced noise was reduced by adaptation following strong illumination. Detailed analysis indicated that the F⁻-induced bumps were 3–5x smaller than light bumps and were less strongly affected by adaptation (Minke and Stephenson, 1985). F⁻, GMP-PNP, and GTPγS appear to generate noise in *Musca* and *Lucilia* photoreceptors through a common effect on the pathway of phototransduction via activation of the G protein.

In *Limulus* ventral photoreceptors, inhibition of the G protein was studied in addition to its excitation (Bolsover and Brown, 1982; Fein, 1986; Kirkwood et al., 1989). Injection of guanosine 5'-[β-thio] diphosphate (GDPβS), an analog of GDP (Eckstein et al., 1979), into the photoreceptors inhibits the light response at a stage prior to $InsP_3$ action (Fein, 1986) and reduces the frequency of the light-induced bumps (Kirkwood et al., 1989). Similar experiments were not performed in the fly.

The participation of a light-activated G protein in photo-transduction can best be demonstrated by biochemical studies in cell-free membrane preparations of fly eyes. In membrane preparations of fly eye, assay of light-dependent GTP hydrolysis (GTPase) activity serves as a functional tool to study the interaction between the photopigment and the G protein. It is necessary to use the reaction conditions developed to assay the catecholamine-stimulated GTPase activity of the adenylyl cyclase system (Cassel and Selinger, 1976), since the rhabdomere membrane preparations are not as homogenous and free of contaminations as the outer-segment membranes of the vertebrate rods.

In assaying the hydrolysis of [γ^{32}P]-GTP, the nonspecific nucleoside triphosphatases is greatly diminished by employing low concentrations of labeled GTP (0.1–0.5 μm), by the presence of large excess of unlabeled ATP (0.2–0.5 mM), and by the addition of a nucleoside triphosphate regeneration system (creatine phosphate and creatine kinase) that prevents the transfer of labeled phosphate from GTP to ADP (Cassel and Selinger, 1976). Further selectivity in light-dependent GTP hydrolysis can be achieved in *Musca* and *Calliphora* eye membrane preparation by conducting the GTPase assay at 0–4°C (water ice bath). Low temperature preferentially decreases the nonspecific GTP hydrolysis, whereas the metarhodopsin-stimulated hydrolysis of GTP is decreased to a much lesser extent (Blumenfeld et al., 1985).

The number of G proteins in a photoreceptor cell can be estimated by monitoring the time course of [^{35}S]-GTPγS binding to eye membrane preparations preilluminated with either blue or red light. The incremental binding of GTPγS after blue illumination (maximal pigment in the M state) is relatively fast and saturates after 5 min, whereas GTPγS binding to membranes preilluminated with red light is slow and progresses linearly over 10 min. This background value is subtracted from the total binding after blue illumination to give the binding to the G protein resulting from interaction with metarhodopsin.

Using *Musca* eye-membrane preparations, Devary et al. (1987) have found that binding assays of GTPγS over a range of concentrations from $10^{-8}M$ to $10^{-7}M$ all saturate at 10 pmol of GTPγS bound/mg membrane protein, indicating that this value is the total number of sites in the membrane. From this value, they have calculated that a peripheral *Musca* photoreceptor cell contains 1.6×10^{6} G protein molecules. The ratio of rhodopsin to G protein molecules per photoreceptor cell (*Musca*) is ~40:1 (Blumenfeld, 1988).

The affinity label that was developed to identify G$_s$ (Pfeuffer, 1977) was used to identify the fly photoreceptor G protein. In these experiments, [α^{32}P]-azidoanilido GTP (10 nM) was incubated for 20 min in the cold (4°C) with eye membrane preparations that were preilluminated with either red or blue lights. Following removal of free nucleotide, the bound nucleotide was covalently attached to the proteins by illumination with mercury lamp, and the labeled proteins were visualized by SDS polyacrylamide gel electrophoresis (PAGE) and autoradiography (Devary et al., 1987). The combination of low concentration of the affinity-labeled GTP analog and the large excess of unlabeled ATP results in very specific labeling. In the *Musca* rhabdomeral membrane preparation, preillumination with blue light and affinity labeling with [α^{32}P]-azidoanilido GTP revealed a labeled band corresponding to a protein of 41 kDa. In dark adapted membranes or membranes illuminated with red light, an additional 39-kDa protein band was labeled, apparently representing a GTP-binding protein that did not interact with meta rhodopsin. Paulsen and Bentrop (1986) found that a 41 kDa protein of *Calliphora* fly rhabdomeric membranes was ADP-ribosylated by cholera toxin following illumination with red light.

3.3. Phosphoinositidase C

Pharmacological studies of PIC in the fly involve the demonstration that InsP_3 can mimic light excitation and that inhibition of InsP_3 hydrolysis by the InsP_3 5-phosphatase inhibitor (2,3 bisphosphoglyceric acid—DPG) enhanced both the light response and excitation by InsP_3. The same methods that were used to study the G protein were also used to study InsP_3 action (Devary et al., 1987; Suss et al., 1989). The facilitation of the light-induced noise by InsP_3 or by combination of InsP_3 + DPG, and the relatively large amplitudes of the InsP_3-induced bumps make it much easier to demonstrate the involvement of PIC in excitation of intact photoreceptors relative to excitation by chemical activation of the G protein. Figure 13 demonstrates InsP_3 + DPG-induced excitation and adaptation in *Lucilia* photoreceptors.

Assay of PIC activity is performed in two stages. In the first stage, phosphoinositides are labeled with [^{3}H]-inositol using dissected preparations that contain intact photoreceptors. Long incubation (4 h) with [^{3}H]-inositol is needed to achieve enough incorporation of [^{3}H]-inositol, and the labeling is critically dependent on efficient oxygenation, which is achieved by incubation under 100% O_2 and gentle shaking. At the end of the incubation period, membranes are prepared and stored under liquid nitro-

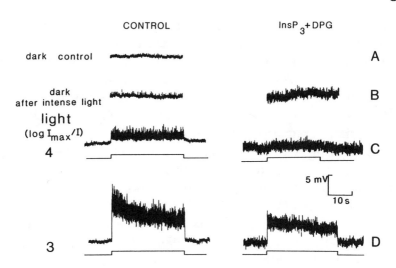

Fig. 13. Ins(1,4,5) P_3 + DPG excite normal *Lucilia* photoreceptors in the dark. (Row A) A trace showing the noise level in the dark-adapted cell. (Row B) The noise level 5 min after intense (5-s) maximal-intensity white light before, (left) and after (right) injections of InsP_3 (1 mM) + DPG (50 mM) in Ringer's solution combined with the intense white light, which is required to introduce hydrophilic molecules into the cell. (Rows C, D) The response to dim lights before (left) and after application of InsP_3 + DPG (right). (Reproduced from Suss et al., 1989 with permission from The Rockefeller University Press and the authors.)

gen (Devary et al., 1987). [³H]-inositol-labeled membranes are then stimulated, and hydrolysis of phosphoinositides is monitored by measurement of production of water-soluble inositol phosphates after separation by chromatography on Dowex columns (Downes and Michell, 1981; *see* chapter by Godfrey in this vol.).

To demonstrate light- and GTP-dependent breakdown of phosphoinositides, it is essential to bring the free Ca²⁺ concentration to 50 nM or less with EGTA buffer and to include in the reaction mixture ATP and a nucleoside triphosphate-regeneration system. The presence of ATP enables the replenishment of the hydrolyzed polyphosphoinositides by conversion of phosphatidylinositol to PtdIns$P$ and PtdInsP_2 by endogenous phosphoinositide kinases. Under this condition, it is difficult to show any effect of added GTP. On the other hand, using GTPγS and GDPβS as an activator and an inhibitor of the G protein, respectively, and DPG as inhibitor of inositol trisphosphate 5-phosphatase, it was demonstrated that the light-dependent phosphoinositidase C activity is controlled by a G protein, that the major product formed is $Ins(1,4,5)P_3$, and that this product is rapidly hydrolyzed by a specific phosphomonoesterase (Devary et al., 1987).

Key evidence that light-activated PIC is a necessary step in visual excitation in the fly was provided by the isolation of a putative PIC gene of *Drosophila*—the *no receptor potential* A (*norpA*) mutant (Bloomquist et al., 1988). The X-linked *norpA* mutant (Pak et al., 1969; Hotta and Benzer, 1970) had long been a strong candidate for transduction-defective mutant because of its drastically reduced receptor potential. The reduction in receptor potential amplitude in *norpA* mutants is variable in different alleles. In the strongest alleles, the light response is totally abolished, whereas the photopigment is normal (Pak et al., 1976). The G protein (Heichel, Minke, and Selinger, unpublished) and the ionic channel activity as reflected in the shape and size of the quantum bumps (Pak, 1979) seem also to be normal. A correlation between the *norpA* mutant and PIC was suggested by Inoue et al. (1985), who reported that PIC activity, which is abundant in normal *Drosophila* eyes, is drastically reduced in *norpA* mutants. Moreover, in several *norpA* alleles tested, the degree of reduction in PIC activity was found to be correlated with behavioral defects as well as with reduced ERG amplitude (Inoue et al., 1985). However, to make the above findings more definitive, a demonstration of light-activated PIC was necessary. The correlation between light-activated PIC and the *norpA* mutation was provided by electrophysiological (a correlation to excitation; Deland and Pak, 1973) and biochemical studies of the *norpA* allele *norpA*[H52] (Selinger and Minke, 1988). The *norpA*[H52] was found to be reversibly temperature-sensitive. At permissive temperature (18°C), the ERG of this allele is normal. However, raising the temperature above 35°C abolishes the ERG instantaneously. If the eye is not exposed to the elevated temperature too long (<3 min), the ERG recovers by lowering the temperature once again (Deland and Pak, 1973). Light-dependent activation of PIC in the *norpA*[H52] and wild-type *Drosophila* indicated that, although PIC activity was similar in wild-type and the mutant at the permissive temperature, it was abolished in the mutant, but not in wild-type fly at the restrictive temperature. The normal light-dependent GTPase and DG kinase activities in norpA and deficient PIC in norpA strongly suggest that light-activated PIC is deficient in the norpA mutant (Selinger and Minke, 1988).

The most conclusive evidence that the norpA gene encodes PIC came from the cloning and sequencing of the norpA locus of Drosophila (Bloomquist et al., 1988). The putative norpA protein is composed of 1095 amino acid residues and has extensive sequence similarity to a PIC amino acid sequence from bovine brain (Stahl et al., 1988). The norpA mutant thus provides critical evidence that the phosphoinositide signaling system is an essential component of the phototransduction machinery in the fly by showing that no phototransduction takes place in the absence of functional PIC.

4. Light-Induced Reduction in Excitation Efficiency Owing to the *trp* and *nss* Mutations

The *t*ransient *r*eceptor *p*otential (*trp*) mutant of Drosophila (Cosens and Manning, 1969; Minke et al., 1975; Lo and Pak, 1981; Minke, 1982; Montell et al., 1985; Montell and Rubin, 1989; Wong et al., 1989) and the *n*o *s*teady-*s*tate (*nss*) mutant of the sheep blowfly Lucilia (Howard, 1984; Barash et al., 1988) can be useful for dissecting invertebrate phototransduction. In the third chromosome *trp* mutant of Drosophila and the *nss* mutant of Lucilia, the receptor potential appears almost normal in response to a flash, but quickly decays to baseline during prolonged illumination (Fig. 14).

The effects of the *nss* mutation on the light response of Lucilia are very similar to the effects of the *trp* mutation on the photoreceptor potential of Drosophila (Barash et al., 1988). Noise analysis and voltage measurements indicate that the decay of the receptor potential is caused by a severe reduction in the rate of occurrence of the quantum bumps. The bumps are only slightly

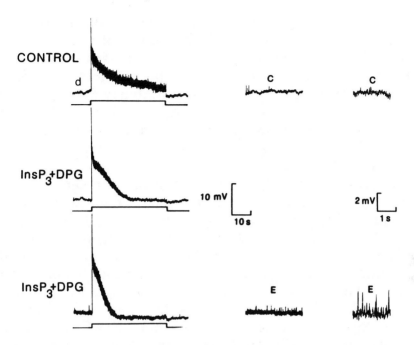

Fig. 14. InsP$_3$ + DPG facilitate the light response of the *nss* mutant, accelerate its decline to baseline, and induce noise in the dark. (Left column) Intracellular recordings from a dark-adapted (3-min) photoreceptor of the *nss* mutant before injection (CONTROL) showing a response to an orange light and to the same orange light after InsP$_3$ (1 mM) and DPG (50 mM) were injected, combined with 20 s of maximal-intensity white light (second trace). The third trace shows the response of the same cell to the same stimulus after additional application of InsP$_3$ + DPG combined with maximal-intensity white illuminations. The middle column shows recordings of noise in the dark 1 min after the cessation of maximal-intensity white light in the control (C) and 1 min after the 10 s of white light combined with InsP$_3$ + DPG were applied (E). The right column shows enlarged segments of the traces in the middle column. (Reproduced from Suss et al., 1989, with permission from The Rockefeller University Press and the authors.)

modified in shape and amplitude during the decline of the response to light of medium intensity. The *trp* and *nss* mutations affect the triggering mechanism of the bump by making the absorbed photons ineffective, possibly as a result of the abnormally low [Ca^{2+}]$_i$ found in the photoreceptors of these mutants (Suss et al., 1989).

The *nss* mutant was the main mutant used in pharmacological studies (Suss et al., 1989), because *Lucilia* is more accessible to electrophysiological studies because of the much larger diameter of its photoreceptors. Experiments on the *nss* and *trp* mutants using various combinations of dim background lights and prolonged, more intense test light revealed that the rate at which the receptor potential decays, in response to continuous illumination with test light, is very sensitive to the presence of dim background light. In both mutants, background light considerably accelerates the decay of the receptor potential. It was therefore possible to use chemical excitation equivalent to a dim background light to localize the phototransduction step that is modified by the *nss* mutation. Presumably, agents that mimic the effect of dim background light on the *nss* mutant act on the transduction pathway at a site prior to that which is modified by the mutation. The experiments depicted in Fig. 14 show that introduction of $InsP_3$ + DPG into the *nss Lucilia* photoreceptor cells excites the photoreceptors in the dark (trace E vs C) and accelerates the rate of decline of the receptor potential in response to subsequent continuous illumination in a manner similar to that of background light. It therefore appears that the *nss* gene product operates at a late stage of the phototransduction pathway, subsequent to production of $InsP_3$, consistent with $[Ca^{2+}]_i$ deficiency underlying the mutation phenotype.

The *trp* gene was identified by rescuing the *trp* phenotype by introduction of a 6.5-kb genomic sequence by germ-line transformation (Montell et al., 1985). The *trp* gene encodes a novel 1275 amino acid protein with eight putative transmembrane domains (Montell and Rubin, 1989; Wong et al., 1989). Analysis of several mutant alleles indicates that the *trp* phenotype arises from the complete absence of protein rather than production of a defective gene product (for review, *see* Montell, 1989).

Application of La^{3+}, a known inhibitor of Ca^{2+}-binding proteins to the extracellular space of the blowfly *Calliphora*, converts the wild-type response to prolonged illumination into a *trp* or *nss*-like response (Hochstrate, 1989). Application of La^{3+} to *Drosophila Musca* and *Lucilia* flies gives similar results (Suss, 1990). In addition, shot noise analysis indicated that a combination of light and La^{3+} caused a large reduction in the rate of occurrence of the quantum bumps, which sum to produce the photoreceptor potential (Suss, 1990). This effect is very similar to the effects of the *trp* and *nss* mutations of *Drosophila* and *Lucilia* flies on the quantum bump rate. La^{3+} applied to the *nss* mutant had little effect on the bump rate, thus suggesting specificity in its action. It is therefore possible that both La^{3+} and the *trp* or *nss* mutations inhibit Ca^{2+} transport into fly photoreceptors, a process apparently essential for light excitation *(see* Fig. 1).The mechanism of receptor-stimulated Ca^{2+} entry in other cells is an important question (Berridge and Irvine, 1989), and the *trp* or *nss* mutations may provide invaluable insight.

5. Concluding Remarks

Invertebrate photoreceptors use the ubiquitous phosphoinositide signaling system (Berridge, 1987) to mediate phototransduction. It is therefore likely that the information gained from studies of phototransduction in the fly photoreceptors will be relevant to many cells and to a variety of biological phenomena. The photoreceptors of the fly have a number of advantages over other cellular systems, since the functional requirements of the photoreceptor cells resulted in a concentrated and highly organized enzymatic machinery that could be studied as multidisciplinary approaches. Only in fly photoreceptors can one apply electrophysiology, photometry, biochemistry, morphology, genetics, and molecular genetic tools to the same preparation. The present chapter illustrates the variety of methods that have been used to study the intracellular messenger system of fly photoreceptors with emphasis on the genetic approach.

The use of visual mutants of *Drosophila* in physiological, biochemical, and molecular genetic studies has already resulted in cloning and sequencing of a number of key proteins: PIC, cyclophilin, two types of arrestin (S antigen), protein kinase, a novel myosin kinase (*ninaC*), and a putative novel Ca^{2+} transporter protein (*trp*). Further applications of these methods will surely reveal not only the molecular details of invertebrate phototransduction, but also more generally the properties of the phosphoinositide signaling pathway.

Abbreviations

$[Ca^{2+}]_i$	Intracellular free Ca^{2+} concentration
cGMP	Guanosine 3',5' cyclic monophosphate
DPG	2,3-bisphosphoglycerate
GDP	Guanosine 5'-diphosphate
GTP	Guanosine 5'-triphosphate
$InsP_2$	Inositol bisphosphate (isomer specified in text)
$InsP_3$	Inositol trisphosphate (isomer specified in text)
$InsP_4$	Inositol tetrakisphosphate (isomer specified in text)
PIC	Phosphoinositidase C (= phospholipase C)
PKC	Protein kinase C (= C kinase)
$PtdInsP_2$	Phosphatidylinositol bisphosphate (isomer specified in text)
SMC	Submicrovillar cisternae

References

Almagor E., Hillman P., and Minke B. (1986) Spatial properties of the prolonged depolarizing afterpotential in barnacle photoreceptors. I. The induction process. *J. Gen. Physiol.* **87,** 391–405.

Barash S., Suss E., Stavenga D. G., Rubinstein C. T., Selinger Z., and Minke B. (1988) Light reduces the excitation efficiency in the *nss* mutant of the sheep blowfly *Lucilia. J. Gen. Physiol.* **92,** 307–330.

Baumann O. and Walz B. (1989) Calcium and inositol polyphosphate-sensitivity of the calcium-sequestering endoplasmic reticulum in the photoreceptor cells of the honeybee drone. *J. Comp. Physiol. A.* **165,** 627–636.

Bentrop J. and Paulsen R. (1986) Light-modulated ADP ribosilation protein phosphorylation and protein binding in isolated fly receptor membranes. *Eur. J. Biochem.* **161,** 61–67.

Berridge M. J. (1987) Inositol triphosphate and diacylglycerol: Two interacting second messengers. *Ann. Rev. Biochem.* **56,** 159–193.

Berridge M. J. and Irvine R. F. (1989) Inositol phosphates and cell signalling. *Nature* **341,** 197–205.

Bigay J., Deterre P., Pfister C., and Chabre M. (1985) Fluoroaluminates activate transducin-GDP by mimicking the phosphate of GTP in its binding site. *FEBS Lett.* **191,** 181–185.

Bloomquist B. T., Shortridge R. D., Schneuwly S., Pedrew M., Montell C., Steller H., Rubin G., and Pak W. L. (1988) Isolation of a putative phospholipase C gene of *Drosophila norpA,* and its role in phototransduction. *Cell* **54,** 723–733.

Blumenfeld A (1988) The role of guanine nucleotide binding protein in the mechanism of phototransduction. Ph.D. thesis. The Hebrew University of Jerusalem.

Blumenfeld A., Erusalimsky J., Heichal O., Selinger Z., and Minke B. (1985) Light-activated guanosinetriphosphatase in *Musca* eye membranes resembles the prolonged depolarizing afterpotential in photoreceptor cells. *Proc. Natl. Acad. Sci. USA* **82**, 7116–7120.

Bolsover S. R. and Brown J. E. (1982) Injection of guanosine and adenosine nucleotides into *Limulus* ventral photoreceptor cells. *J. Physiol.* **322**, 325–342.

Bolsover S. R. and Brown J. E. (1985) Calcium ion, an intracellular messenger of light adaptation also participates in excitation of *Limulus* photoreceptors. *J. Physiol.* **364**, 381–393.

Boschek C. B. and Hamdorf K. (1976) Rhodopsin particles in the photoreceptor membrane of an insect. *Z. Naturforsch.* **31c**, 763.

Braitenberg V. (1967) Patterns of projection in the visual system of the fly. I. Retina-lamina projections. *Exp. Brain Res.* **16**, 184–209.

Brown J. E. and Blinks J. R. (1974) Changes in intracellular free calcium concentration during illumination of invertebrate photoreceptors: Detection with aequorin. *J. Gen. Physiol.* **64**, 643–665.

Brown J. E. and Rubin J. L. (1984) A direct demonstration that inositoltrisphosphate induces an increase in intracellular calcium in *Limulus* photoreceptors. *Biochem. Biophys. Res. Commun.* **125**, 1137–1142.

Brown J. E., Murray J. R., and Smith T. G. (1967) Photoelectric potential from photoreceptor cells in ventral eye of *Limulus*. *Science* **158**, 665, 666.

Brown J. E., Rubin L. J., Ghalayini A. J., Tarver A. P., Irvine R. F., Berridge M. J., and Anderson R. E. (1984) Myoinositol polyphosphate may be a messenger for visual excitation in *Limulus* photoreceptors. *Nature* **311**, 160–163.

Brown K. T. and Murakami M. (1964) A new receptor potential of the monkey retina with no detectable latency. *Nature (Lond.)* **201**, 626–628.

Calhoon R., Tsuda M., and Ebrey T.G. (1980) A light-activated GTPase from octopus photoreceptors. *Biochem. Biophys. Res. Commun.* **94**, 1452–1457.

Cassel D. and Selinger Z. (1976) Catecholamine-stimulated GTPase activity in turkey erythrocyte membranes. *Biochim. Biophys. Acta* **452**, 538–551.

Cassel D. and Selinger Z. (1977) Mechanism of adenylate cyclase activation by cholera toxin: Inhibition of GTP hydrolysis at the regulatory site. *Proc. Natl. Acad. Sci. USA* **74**, 3307–3311.

Chi C. and Carlson S. D. (1979) Ordered membrane particles in rhabdomeric microvilli of the housefly *(Musca domestica L.) J. Morphol.* **161**, 309–322.

Coles J. A. and Tsacopoulos M. (1981) Ionic and possible metabolic interactions between sensory neurones and glial cells in the retina of the honeybee drone. *J. Exp. Biol.* **95**, 75–92.

Cone R. A. (1967) Early receptor potential: Photoreversible charge displacement in rhodopsin. *Science* **155**, 1128–1131.

Cosens D. J. and Manning A. (1969) Abnormal electroretinogram from a *Drosophila* mutant. *Nature* **224**, 285–287.

Deland M. C. and Pak W. L. (1973) Reversible temperature sensitive photo-transduction mutant of *Drosophila melangoaster*. *Nature New Biol.* **244**, 184–186.

Devary O., Heichal O., Blumenfeld A., Cassel D. Suss E., Barash S., Rubinstein C. T., Minke B., and Selinger Z. (1987) Coupling of photoexcited rhodopsin to inositol phospholipid hydrolysis in fly photoreceptors. *Proc. Natl. Acad. Sci. USA* **84**, 6939–6943.

de Vries H. L. and Kuiper J. W. (1958) Optics of the insect eye. Conference on photoreception. *Ann. N.Y. Acad. Sci.* **74**, 196–203.

Dodge F. A., Jr., Knight B. W., and Toyoda J. (1968) Voltage noise in *Limulus* visual cells. *Science* **160**, 88–90.

Dowling J. E. (1960) Chemistry of visual adaptation in the rat. *Nature* **188**, 114–116.

Downes C. P. and Michell B. H. (1981) The poly-phosphoinositide phosphodiesterase of erythrocyte membrane. *Biochem. J.* **198**, 133–140.

Eckstein F., Cassel D., Levkovitz H., Lowe M., and Selinger Z. (1979) Guanosine 5-O-(2-thiodiphosphate): An inhibitor of adenylate cyclase stimulation by guanine nucleotides and fluoride ions. *J. Biol. Chem.* **254**, 9829–9834.

Feiler R., Harris W. A., Kirschfeld K., Wehrhahn C., and Zuker C. S. (1988) Targeted misexpression of a *Drosophila* opsin gene leads to altered visual function. *Nature* **333**, 737–741.

Fein A. (1986) Blockade of visual excitation and adaptation in *Limulus* photoreceptors by GDP-β-S. *Science* **232**, 1543–1545.

Fein A. and Corson D. W. (1979) Both photons and fluoride ions excite *Limulus* ventral photoreceptors. *Science* **204**, 77–79.

Fein A. and Corson D. W. (1981) Excitation of *Limulus* photoreceptors by vanadate and by a hydrolysis-resistant analogue of guanosine triphosphate. *Science* **212**, 555–557.

Fein A., Payne R., Corson D. W., Berridge M. J., and Irvine R. F. (1984) Photoreceptor excitation and adaptation by inositol 1,4 ,5-trisphosphate. *Nature* **311**, 157–160.

Franceschini N. (1975) Sampling of the visual environment by the compound eye of the fly: Fundamentals and applications, in *Photoreceptor Optics* (Snyder A. W. and Menzel R., eds.), Springer-Verlag, Berlin, pp. 98–125.

Franceschini N. (1977) *In-vivo* fluorescence of the rhabdomeres in an insect eye. *Proc. Int. Union Physiol. Sc. XIII.* XXVIIth Congr. Paris 237.

Franceschini N. (1983) *In-vivo* microspectrofluorimetry of visual pigments, in *The Biology of Photoreception* (Cosens D. J. and Vince-Price D., eds.), *Soc. Exp. Biol.* Cambridge University Press, Cambridge, UK, pp. 53–85.

Franceschini N. and Kirschfeld K. (1971) Les phénoménes de pseudopupille dans l'oeil composé de *Drosophila*. *Kybernetik* **9**, 159–182.

Franceschini N., Kirschfeld K., and Minke B. (1981) Fluorescence of photo-receptor cells observed *in vivo*. *Science* **213**, 1264–1267.

Fryxell K. J. and Meyerowitz E. M. (1987) An opsin gene that is expressed only in the R7 photoreceptor of *Drosophila*. *EMBO J.* **6**, 443–451.

Giovannucci D. R. and Stephenson R. S. (1986) Effect of ninaD^{246} mutation on vitamin A metabolism in *Drosophila* eye. *Invest. Ophthalmol. Visual Sci.* **27(Suppl.)**, 191.

Giovannucci D. R. and Stephenson R. S. (1987) Analysis of HPLC of *Drosophila* mutations affecting vitamin A metabolism. *Invest. Ophthalmol. Visual Sci.* **28(Suppl.)**, 253.

Goldsmith T. H., Barker R. J., and Cohen C. F. (1964) Sensitivity of visual receptors of carotenoid depleted flies: A vitamin A deficiency in an invertebrate. *Science* **146**, 65–67.

Grzywacz N. M., Hillman P.and Knight B. W. (1988) The quantal source of area superlinearity of flash responses in *Limulus* photoreceptors. *J. Gen. Physiol.* **91**, 659–684.

Hamdorf K. (1979) The physiology of invertebrate visual pigment, in *Handbook of Sensory Physiology*, vol. VII/6A (Autrum H., ed.), Springer-Verlag, Berlin, pp. 145–224.

Hara T. and Hara R. (1972) Cephalopod retinochrome, in *Handbook of Sensory Physiology* vol. VII/1 (Dartnall H. J. A., ed.), Springer-Verlag, Berlin, pp. 720–746.

Hardie R. C. (1985) Functional organization of the fly retina, in *Progress in Sensory Physiology* (Autrum H., Ottoson D., Perl E. R., Schmidt R. F., Shimazu H., and Willis W. D., eds.), Springer-Verlag, New York, pp. 1–79.

Hardie R. C. (1989) Neurotransmitters in compound eyes, in *Facets of Vision* (Stavenga D. G. and Hardie R. C., eds.), Springer-Verlag, Berlin, pp. 235–256.

Hardie R. C., Franceschini N., Ribi W., and Kirschfeld K. (1981) Distribution and properties of sex-specific photoreceptors in the fly *Musca domestica*. *J. Comp. Physiol.* **145**, 139–152.

Harris W. A., Ready D. F., Lipson E. D., Hudspeth A. J., and Stark W. S. (1977) Vitamin A deprivation and *Drosophila* photopigments. *Nature (Lond.)* **266**, 648–650.

Hillman P., Hochstein S., and Minke B. (1983) Transduction in invertebrate photoreceptors: The role of pigment bistability. *Physiol. Rev.* **63**, 668–772.

Hochstein S., Minke B., and Hillman P. (1973) Antagonistic components of the late receptor potential in the barnacle photoreceptor arising from different stages of the pigment process. *J. Gen. Physiol.* **62**, 105–128.

Hochstrate P. (1989) Lanthanum mimics the *trp* photoreceptor mutant of *Drosophila* in the blowfly *Calliphora*. *J. Comp. Physiol. A.* **166**, 179–188.

Hotta Y. and Benzer S. (1970) Genetic dissection of the *Drosophila* nervous system by means of mosaics. *Proc. Natl. Acad. Sci. USA* **67**, 1156–1163.

Howard J. (1984) Calcium enables photoreceptor pigment migration in a mutant fly. *J. Exp. Biol.* **113**, 471–475.

Inoue H., Yoshioka T., and Hotta Y. (1985) A genetic study of inositol trisphosphate involvement in phototransduction using *Drosophila* mutants. *Biochem. Biophys. Res. Commun.* **132**, 513–519.

Isono K., Tanimura T., Oda Y., and Tsukahara Y. (1988) Dependency on light and vitamin A derivatives of the biogenesis of 3-hydroxyretinal and visual pigment in the compound eyes of *Drosophila melangoaster*. *J. Gen. Physiol.* **92**, 587–600.

Johnson E. C. and Pak W. L. (1986) Electrophysiological study of *Drosophila* rhodopsin mutants. *J. Gen. Physiol.* **88**, 651–673.

Johnson E. C., Robinson P. R., and Lisman J. E. (1986) Cyclic GMP is involved in the excitation of invertebrate photoreceptors. *Nature* **324**, 468–470.

Kirkwood A., Weiner D., and Lisman J. E. (1989) An estimate of the number of G regulatory proteins activated per excited rhodopsin in living *Limulus* ventral photoreceptors. *Proc. Natl. Acad. Sci. USA* **86**, 3872–3876.

Kirschfeld K. (1967) Die Projektion der optischen Umwelt auf das Raster der Rhabdomere im Komplexauge von *Musca*. *Exp. Brain Res.* **3**, 248–270.

Kirschfeld K. (1971) Aufnahme und Verarbeitung optischer Dafen im Komplexauge von Insekten. *Naturwissenschaften* **58**, 201–209.

Kirschfeld K. (1986) Activation of visual pigment: Chromophore structure and function, in *The Molecular Mechanism of Photoreception* (Stieve, H., ed.), Springer-Verlag, Berlin, pp. 31–49.

Koeing J. and Merriam J. R. (1977) Autosomal ERG mutants. *Drosophila Information Service* **52**, 50, 51.

Kunze P. (1979) Apposition and superposition eyes, in *Handbook of Sensory Physiology*, vol. VII/6A (Autrum H., ed.), Springer-Verlag, Berlin, pp. 441–502.

Land M. F. (1981) Optics and vision in invertebrates, in *Handbook of Sensory Physiology* vol. VII/6B (Autrum H., ed.), Springer-Verlag, Berlin, pp. 471–592.

Larrivee D. C., Conrad S. K., Stephenson R. S., and Pak W. L. (1981) Mutation that selectively affects rhodopsin concentration in the peripheral photoreceptors of *Drosophila melangoaster*. *J. Gen. Physiol.* **78**, 521–545.

Lindsley D. L. and Grell E. H. (1968) *Genetic Variations of Drosophila Melangoaster*. Carnegie Institution of Washington, Washington, DC.

Lisman J. E. (1985) The role of metarhodopsin in the generation of spontaneous quantum bumps in ultraviolet receptors of *Limulus* median eye. Evidence for reverse reactions into an active state. *J. Gen. Physiol.* **85**, 171–187.

Lisman J. E. and Brown J. E. (1972) The effect of intracellular iontophoretic injection of calcium and sodium ions on the light response of *Limulus* ventral photoreceptors. *J. Gen. Physiol.* **59**, 701–719.

Lisman J. E. and Brown J. E. (1975) Effects of intracellular injection of calcium buffers on light adaptation in *Limulus* ventral photoreceptors. *J. Gen. Physiol.* **66**, 489–506.

Lo M.-V. C. and Pak W. L. (1981) Light-induced pigment granule migration in the retinular cells of *Drosophila melanogster:* Comparison of wild type with ERG-defective mutant. *J. Gen. Physiol.* **77**, 155–175.

Matsumoto H., Isono K., Pye Q., and Pak W. L. (1987) Gene encoding cytoskeletal proteins in *Drosophila* rhabdomeres. *Proc. Natl. Acad. Sci. USA* **84**, 985–989.

Millecchia R. and Mauro A. (1969) The ventral photoreceptor cells of *Limulus* II. The basic photoresponse. *J. Gen. Physiol.* **54**, 310–330.

Minke B. (1982) Light-induced reduction in excitation efficiency in the *trp* mutant of *Drosophila. J. Gen. Physiol.* **79**, 361–385.

Minke B. (1986) Photopigment-dependent adaptation in invertebrates—Implications for vertebrates, in *The Molecular Mechanism of Photoreception* (Stieve, H., ed.), Springer-Verlag, Berlin, pp. 241–265.

Minke B. and Kirschfeld K. (1979) The contribution of a sensitizing pigment to the photosensitivity spectra of fly rhodopsin and metarhodopsin. *J. Gen. Physiol.* **73**, 517–540.

Minke B. and Kirschfeld K. (1980) Fast electrical potentials arising from activation of metarhodopsin in the fly. *J. Gen. Physiol.* **75**, 381–402.

Minke B. and Stephenson R. S. (1985) The characteristics of chemically induced noise in *Musca* photoreceptors. *J. Comp. Physiol.* **156**, 339–356.

Minke B. and Tsacopoulos M. (1986) Light induced sodium dependent accumulation of calcium and potassium in the extracellular space of bee retina. *Vision Res.* **26**, 679–690.

Minke B., Wu C.-F., and Pak W. L. (1975) Induction of photoreceptor voltage noise in the dark in a *Drosophila* mutant. *Nature* **258**, 84–87.

Montell C. (1989) Molecular genetics of *Drosophila* vision. *Bioessays* **11**, 43–48.

Montell C. and Rubin G. M. (1988) The *Drosophila ninaC* locus encodes two photoreceptor cell specific proteins with domains homologous to protein kinases and the myosin heavy chain head. *Cell* **52**, 757–772.

Montell C. and Rubin G. M. (1989) Molecular characterization of *Drosophila trp* locus: A putative integral membrane protein required for phototransduction. *Neuron* **2**, 1313–1323.

Montell C., Jones K., Hafen E., and Rubin G. (1985) Rescue of the *Drosophila* phototransduction mutation *trp* by germline transformation. *Science* **230**, 1040–1043.

Montell C., Jones K., Zuker C. S., and Rubin G. (1987) A second opsin gene expressed in the ultraviolet-sensitive R7 photoreceptor cells of *Drosophila melangoaster. J. Neurosci.* **7**, 1558–1566.

O'Brien S. J. and MacIntyre R. J. (1978) Genetics and biochemistry of enzymes and specific proteins of *Drosophila*, in *The Genetics and Biology of Drosophila*, vol. 2a (Ashburner M. and Wright T. F. R., eds.), Academic, New York, pp. 396–551.

O'Day M. and Gray-Keller M. P. (1989) Evidence for electrogenic Na^+/Ca^{2+} exchange in *Limulus* ventral photoreceptors. *J. Gen. Physiol.* **93**, 473–492.

O'Tousa J. E., Baehr W., Martin R. L., Hirsh J., Pak W. L., and Applebury M. L. (1985) The *Drosophila ninaE* gene encodes an opsin. *Cell* **40**, 839–850.

Ozaki K., Terakita A., Hara R., and Hara T. (1987) Isolation and characterization of a retinal-binding protein from the squid retina. *Vision Res.* **27**, 1057–1070.

Pak W. L. (1975) Mutations affecting the vision of *Drosophila melangoaster*, in *Handbook of Genetics*, vol. 3 (King R. C., ed.), Plenum, New York, pp. 703–733.

Pak W. L. (1979) Study of photoreceptor function using *Drosophila* mutants, in *Neurogenetics: Genetic Approaches to the Nervous System* (Breakefield X., ed.), Elsevier, North-Holland, New York, pp. 67–99.

Pak W. L. and Lidington K. J. (1974) Fast electrical potential from a long-lived long-wavelength photoproduct of fly visual pigment. *J. Gen. Physiol.* **63**, 740–756.

Pak W. L., Grossfield J., and White N.V. (1969) Nonphototactic mutants in a study of vision in *Drosophila*. *Nature* **222**, 351–354.

Pak W. L., Ostroy S. E., Deland M., and Wu C.-F. (1976) Photoreceptor mutant of *Drosophila:* Is protein involved in intermediate steps of photo-transduction? *Science* **194**, 956–959.

Paulsen R. and Bentrop J. (1984) Reversible phosphorylation of opsin induced by irradiation of blowfly retinae. *J. Comp. Physiol. A.* **155**, 39–45.

Paulsen R., Bentrop J., Baurenschmitt H. T., Bockerg D., and Peters K. (1987) Phototransduction in invertebrate: Component of cascade mechanism in fly photoreceptors. *Photobiochem. Photobiophys.* **(Suppl.)** 261–272.

Payne R. (1986) Phototransduction by microvillar photoreceptors of invertebrates: Mediation of visual cascade by inositol trisphosphate. *Photobiochem. Photobiophys.* **13**, 373–397.

Payne R. and Fein A. (1986) The initial response of *Limulus* ventral photoreceptors to bright flashes. Released calcium as a synergist to excitation. *J. Gen. Physiol.* **87**, 243–269.

Payne R., Corson D. W., and Fein A. (1986a) Pressure injection of calcium both excites and adapts *Limulus* ventral photoreceptors. *J. Gen. Physiol.* **88**, 107–126.

Payne R., Corson D. W., Fein A., and Berridge M. J. (1986b) Excitation and adaptation of *Limulus* ventral photoreceptors by inositiol 1,4,5-trisphosphate result from a rise in intracellular calcium. *J. Gen. Physiol.* **88**, 127–142.

Payne R., Flores T. M., and Fein A. (1990) Feedback inhibition by calcium limits the release of calcium by inositol trisphosphate in *Limulus* ventral photoreceptors. *Neuron* **4**, 547–555.

Payne R., Waltz B., Levy S., and Fein A. (1988) The localization of calcium release by inositol trisphosphate in *Limulus* photoreceptors and its control by negative feedback. *Phil. Trans. Roy. Soc. Lond. B. Biol. Sci.* **320**, 359–379.

Pfeuffer T. (1977) GTP-binding protein in membranes and the control of adenylate cyclase activity. *J. Biol. Chem.* **252,** 7224–7234.

Pick B. (1977). Specific misalignments of rhabdomere visual axes in the neural superposition eye of dipteran flies. *Biol. Cybern.* **26,** 215–224.

Pollock J. A. and Benzer S. (1988) Transcript localization of four opsin genes in the three visual organs of *Drosophila*; Rh2 is ocellus specific. *Nature* **333,** 779–782.

Saibil H. R. (1984) A light-stimulated increase of cyclic GMP in squid photoreceptors. *FEBS Lett.* **168,** 213–216.

Schinz R. H., Lo M.-V. C., Larrivee D. C., and Pak W. L. (1982) Freeze-fracture study of the *Drosophila* photoreceptor membrane: Mutations affecting membrane particle density. *J. Cell Biol.* **93,** 961–969.

Schneuwly S., Shortridge R. D., Larrivee D. C., Ono T., Ozaki M., and Pak W. L. (1989) *Drosophila ninaA* gene encodes an eye-specific cyclophilin (cyclosporine A binding protein). *Proc. Natl. Acad. Sci. USA* **86,** 5390–5394.

Schwemer J. (1979) Molekulare Grundlagen der Photorezeption bei der Schmeissfliege *Calliphora erythrocephala* Meig. Habilitationsschrift, Ruhr University, Bochum.

Schwemer J. (1984) Renewal of visual pigment in photoreceptors of the blowfly. *J. Comp. Physiol. A.* **154,** 535–547.

Schwemer J. (1988) Cycle of 3-hydroxy retinoids in an insect eye, in *Molecular Physiology of Retinal Proteins* (Hara T., ed.), Yamada Science Foundation, Japan, pp. 299–304.

Schwemer J. and Henning U. (1984) Morphological correlates of visual pigment turnover in photoreceptors of the fly, *Calliphora erythrocephala. Cell Tissue Res.* **236,** 293–303.

Schwemer J., Pepe I. M., Paulsen R., and Cugnoli C. (1984) Light-induced *trans-cis* isomerization of retinal by a protein from the honeybee retina. *J. Comp. Physiol. A.* **154,** 549–554.

Selinger Z. and Minke B. (1988) Inositol lipid cascade of vision studied in mutant flies. *Cold Spring Harbor Symposia on Quant. Biol.* **LIII,** 333–341.

Shaw S. R. (1978) The extracellular space and blood-eye barrier in an insect retina: An ultrastructural study. *Cell Tissue Res.* **188,** 35–61.

Shieh B.-H., Stamnes M. A., Seavello S., Harris G. L. and Zuker C. S. (1989) The *ninaA* gene required for visual transduction in *Drosophila* encodes a homologue of cyclosporin A-binding protein. *Nature* **338,** 67–70.

Smith D. P., Shieh B.-H., and Zuker C. S. (1990) Isolation and structure of an arrestin gene from *Drosophila. Proc. Natl. Acad. Sci. USA* **87,** 1003–1007.

Stahl M. L., Ferenz C. R. Kelleher K. L. Kriz R. W., and Knopf J. L. (1988) Sequence similarity of phospholipase C with the non-catalytic region of *src. Nature* **332,** 269–272.

Stark W. S. and Zitzmann W. G. (1976) Isolation of adaptation mechanisms and photopigment spectra by vitamin A deprivation in *Drosophila. J. Comp. Physiol.* **105,** 15–27.

Stark W. S., Ivanyshyn A. M., and Greenberg R. M. (1977) Sensitivity and photopigments of R1–6, a two-peaked photoreceptor in *Drosophila*, *Calliphora* and *Musca*. *J. Comp. Physiol.* 121, 289–305.

Stark W. S., Stavenga D. G., and Kruizinga B. (1979) Fly photoreceptor fluorescence is related to UV sensitivity. *Nature* 280, 581–583.

Stavenga D. G. (1979) Pseudopupils of compound eyes, in *Handbook of Sensory Physiology*, vol.VII/6A (Autrum H., ed.), Springer-Verlag, Berlin, pp. 357–439.

Stavenga D. G. and Schwemer J. (1984) Visual pigments of invertebrates, in *Photoreception and Vision in Invertebrates* (Ali M. A., ed.), Plenum, New York, pp. 11–61.

Stephenson R. S. and Pak W. L. (1980) Heterogenic components of a fast electrical potential in *Drosophila* compound eye and their relation to visual pigment photoconversion. *J. Gen. Physiol.* 75, 353–379.

Stephenson R. S., O'Tousa J., Scavarda N. J., Randall L. L., and Pak W. L. (1983) *Drosophila* mutants with reduced rhodopsin content, in *The Biology of Photoreception* (Cosens D. and Vince-Price D., eds.), Cambridge University Press, Cambridge, UK, pp. 477–501.

Stieve H. (1986) Bumps, the elementary excitatory responses of invertebrates, in *The Molecular Mechanism of Photoreception* (Stieve H., ed.), Springer-Verlag, Berlin, pp. 199–230.

Stieve H. and Bruns M. (1980) Dependence of bump rate and bump size in *Limulus* ventral nerve photoreceptor on light adaptation and calcium concentration. *Biophys. Struc. Mechanism* 6, 271–285.

Stryer L. (1986) Cyclic GMP cascade of vision. *Ann. Rev. Neurosci.* 9, 87–119.

Suss E. (1990) Electrophysiological studies of second messengers action in fly phototransduction. Ph.D. thesis. The Hebrew University of Jerusalem.

Suss E., Barash S., Stavenga D. G., Stieve H., Selinger Z., and Minke B. (1989) Chemical excitation and inactivation in photoreceptors of the fly mutants *trp* and *nss*. *J. Gen. Physiol.* 94, 465–491.

Trujillo-Cenoz O. and Melamed J. (1966) Electron microscope observations on the peripheral and intermediate retinas of Dipterans, in *The Functional Organization of the Compound Eye* (Symp. Wenner-Gren Center) (Bernhard C. G., ed.), Pergamon, Oxford, pp. 339–361.

Tsuda M. (1987) Photoreception and phototransduction in invertebrate photoreceptors. *Photochem. Photobiol.* 45, 915–931.

Vogt K. (1983) Is the fly visual pigment a rhodopsin? *Z. Naturforsch.* 38C, 329–333.

Wald G. (1936) Carotenoids and the visual cycle. *J. Gen. Physiol.* 19, 351–371.

Wong F. (1978) Nature of light-induced conductance changes in ventral photoreceptors of *Limulus*. *Nature* 276, 76–79.

Wong F., Schaefer E. L., Roop B. C., LaMendola J. N., Johnson-Seaton D., and Shao D. (1989) Proper function of the *Drosophila trp* gene product during pupal development is important for normal visual transduction in adult. *Neuron* 3, 81–94.

Yamada Y., Takeuchi Y., Komori N., Kobayashi H., Sakai Y., Hotta Y., and Matsumoto H. (1990) A 49-Kilodalton phosphoprotein in the *Drosophila* photoreceptors is an arrestin homolog. *Science* 248, 483–486.

Ziegler A. and Walz B. (1989) Analysis of extracellular calcium and volume changes in the compound eye of the honeybee drone, *Apis mellifera. J. Comp. Physiol. A.* 165, 697–709.

Zuker C. S., Cowman A. F., and Rubin G. M. (1985) Isolation and structure of a rhodopsin gene from *D. melangoaster. Cell* 40, 851–858.

Zuker C. S., Montell C., Jones K. R., Laverty T., and Rubin G. M. (1987) A rhodopsin gene expressed in photoreceptor cell R7 of the *Drosophila* eye: Homologies with other signal transducing molecules. *J. Neurosci.* 7, 1550–1557.

Index